De Gruyter Expositions in Mathematics 57

Simple Lie Algebras over Fields of Positive Characteristic

III. Completion of the Classification

by

Helmut Strade

De Gruyter

Mathematics Subject Classification 2010: 17-02, 17B50, 17B20, 17B05.

ISBN 978-3-11-026298-8
e-ISBN 978-3-11-026301-5
ISSN 0938-6572

Library of Congress Cataloging-in-Publication Data

A CIP catalog record for this book has been applied for at the Library of Congress.

Bibliographic information published by the Deutsche Nationalbibliothek

The Deutsche Nationalbibliothek lists this publication in the Deutsche Nationalbibliografie; detailed bibliographic data are available in the Internet at http://dnb.dnb.de.

Typesetting: PTP-Berlin Protago-TEX-Production GmbH, www.ptp-berlin.eu
Printing and binding: Hubert & Co. GmbH & Co. KG, Göttingen
∞ Printed on acid-free paper

Printed in Germany

www.degruyter.com

Diesen Band widme ich meinen großartigen Enkelkindern

Jakob Leander
Bennett Samuel
Frida Louisa
Leo Joshua
Lotta Magdalena

Contents

Introduction

In the present third (and final) volume of *Simple Lie Algebras over Fields of Positive Characteristic* we conclude the proof of the Classification Theorem announced in the Introduction of Volume 1. There we also outlined the proof. Accordingly, in a first important step the Classification Theorem was proved in Volume 2 for simple Lie algebras of absolute toral rank 2.

It is a well-known fact that LIE's theorem on the trigonalizability of solvable Lie algebras fails to hold in positive characteristic. This failure is one of the major obstructions in the classification theory over fields of positive characteristic. Historically every advance made concerning this problem allowed progress in the classification of simple Lie algebras ([Wil 77, Theorem 2.1], [Str 89/2, Theorem 3.1], [Pre 94, Theorem 1]). Applying the determination of the simple Lie algebras of absolute toral rank 2, one can obtain further substantial information on the trigonalizability of subalgebras and the $[p]$-nilpotency of elements (Theorem 17.3.1), and this result then allows us to list the possibilities of 2-sections in simple Lie algebras with respect to a torus of maximal toral rank (Chapter 17.4).

In Chapter 18 we consider the case when a torus T of maximal dimension exists for which $C_L(T)$ is not trigonalizable. The list of 2-sections is considerably reduced. This result then enables the construction of a subalgebra of codimension 5, which allows the application of the Weak Recognition Theorem. As a result, all algebras with this property are Melikian.

In Chapter 19 we consider the case when a torus T of maximal dimension exists for which all 1-sections $L(\alpha, T)$ are solvable. Again the list of 2-sections is considerably reduced. This result then enables us to show that every solvable section $L(\alpha_1, \dots, \alpha_t, T)$ is abelian. If there is a root space which is at least 2-dimensional, then one constructs a subalgebra which allows the application of the Weak Recognition Theorem. As a result, all algebras with this property are of Special type $S(m; \underline{n}; \Phi(\tau))^{(1)}$. If all root spaces are 1-dimensional, then one derives the explicit multiplication table with respect to a suitable basis, and obtains that the algebras under consideration are the algebras $L(G, 0, f)$ of R.E. BLOCK. These algebras are of Hamiltonian type.

In the general case (Chapter 20) one has to make the right choice of the torus T of maximal toral rank. This is done by toral switching (Chapter 1.5). Every 1-section with respect to this torus has a subalgebra uniquely determined by internal properties. The particular choice of the torus implies that the space $Q(L,T)$ spanned by these subalgebras again is a subalgebra. If $Q(L,T)$ coincides with L, then one can determine the 2-sections completely and apply SELIGMAN's characterization of classical algebras (Theorem 4.1.2). Otherwise one defines a filtration by use of $Q(L,T)$ and considers the associated graded algebra. It turns out that the socle of this graded algebra is of the form $S \otimes \mathcal{O}(m;\underline{1})$, where S is graded and simple. Applying KAC' Recognition Theorem, one concludes that this graded algebra is a restricted Lie algebra of the Cartan type. Applying some deformation techniques then gives that the original algebra is also of Cartan type.

As a general setting L denotes a finite dimensional simple Lie algebra over an algebraically closed field F of characteristic $p > 3$. The semisimple p-envelope of L is mostly considered a subalgebra of Der L and is denoted by $L_{[p]}$. Moreover, T stands for a torus of maximal dimension in $L_{[p]}$. Put $H := C_L(T)$ as well as $\tilde{H} := C_{L_{[p]}}(T)$. Although H and $\tilde{H}$ depend on the choice of T, it will always be clear from the context which torus T is meant, and therefore this dependency will not be indicated.

Chapter 16

Miscellaneous

In this chapter we treat several different topics. First we collect some supplements to earlier chapters on tori in Cartan type Lie algebras. Secondly, we describe the 2-dimensional tori in $H(2;\underline{1};\Phi(1))_{[p]}$. As in the algebras $H(2;(1,2))^{(1)}{}_{[p]}$ (see §10.2) and $H(2;\underline{1};\Phi(\tau))^{(1)}{}_{[p]}$ (see §10.3) it is easier to normalize such a torus by an automorphism of $W(2;\underline{1})$ and then determine the multiplication in the respective isomorphic copy of the algebra rather than to work in the original algebra itself. In a third section we describe the central extensions of $H(2;\underline{1};\Phi(\tau))^{(1)}$, and some irreducible representations of an algebra $\mathfrak{g}$ where $\mathfrak{g}/\operatorname{rad}\mathfrak{g}$ is isomorphic to $H(2;\underline{1};\Phi(\tau))^{(1)}{}_{[p]}$. Under conditions which are given by the needs of the Classification Proof we apply these results to split off the radical and derive some information on p-characters. Finally, we work in Melikian algebras. First we show that all Melikian algebras possess a symmetric invariant bilinear form. As a consequence, the central extensions of the smallest Melikian algebra $\mathcal{M}(1,1)$ split. Finally, the Classification Proof will need a strong result on the p-character of representations of $\mathcal{M}(1,1)$ of dimension ≤ 125. One can deduce this from a remarkable paper on representations of graded Lie algebras by A. A. Premet and S. Skryabin [P-Sk 99]. However, here we will only use elementary methods to prove the required statement on p-characters.

16.1 Tori in some simple Lie algebras

In this section we observe that without changing the proofs at all one can include an additional statement into Theorems 7.5.5, 7.5.8, and 7.5.13.

Theorem 16.1.1 (cf. Theorem 7.5.5).

(1) *For every maximal torus T of $CS(m;\underline{1})$ there is $\mu \in \operatorname{Aut}\mathcal{O}(m;\underline{1})$ such that $\Phi_\mu \in \operatorname{Aut} CS(m;\underline{1})$ and $\Phi_\mu(T)$ is one of*

$$\Big(\sum_{i=1}^{s} F(1+x_i)\partial_i\Big) \oplus \Big(\sum_{i=s+1}^{m} F x_i\partial_i\Big), \quad 0 \leq s \leq m.$$

In addition, if $t_1, \dots, t_s$ are toral elements in T such that $T = (\oplus_{i=1}^s Ft_i) \oplus \big(T \cap W(m;\underline{1})_{(0)}\big)$, then μ can be chosen such that

$$\Phi_\mu(t_i) = (1 + x_i)\partial_i - (1 - \operatorname{div}(t_i))x_m\partial_m$$

for $1 \le i \le s$.

(2) *For every maximal torus T of $S(m;\underline{1})^{(1)}$ there is $\mu \in \operatorname{Aut} \mathcal{O}(m;\underline{1})$ such that $\Phi_\mu \in \operatorname{Aut} S(m;\underline{1})^{(1)}$ and $\Phi_\mu(T)$ is one of*

$$\Big(\sum_{i=1}^{s} F((1 + x_i)\partial_i - x_m\partial_m)\Big) \oplus \Big(\sum_{i=s+1}^{m-1} F(x_i\partial_i - x_m\partial_m)\Big), \quad 0 \le s \le m-1.$$

In addition, if $t_1, \dots, t_s$ are toral elements in T such that $T = (\oplus_{i=1}^s Ft_i) \oplus \big(T \cap W(m;\underline{1})_{(0)}\big)$, then μ can be chosen such that

$$\Phi_\mu(t_i) = (1 + x_i)\partial_i - x_m\partial_m$$

for $1 \le i \le s$.

Theorem 16.1.2 (cf. Theorem 7.5.8). *For every maximal torus T of $H(2r;\underline{1})^{(2)}$ there is $\mu \in \operatorname{Aut} \mathcal{O}(2r;\underline{1})$ such that $\Phi_\mu \in \operatorname{Aut} H(2r;\underline{1})$ and $\Phi_\mu(T)$ is one of*

$$\Big(\sum_{i=1}^{s} FD_H((1 + x_i)x_{i+r})\Big) \oplus \Big(\sum_{i=s+1}^{r} FD_H(x_ix_{i+r})\Big), \quad 0 \le s \le r.$$

In addition, if $t_1, \dots, t_s$ are toral elements in T such that $T = (\oplus_{i=1}^s Ft_i) \oplus \big(T \cap W(2r;\underline{1})_{(0)}\big)$, then μ can be chosen such that

$$\Phi_\mu(t_i) = D_H((1 + x_i)x_{i+r})$$

for $1 \le i \le s$.

Theorem 16.1.3 (cf. Theorem 7.5.13). *For every torus $T \subset K(2r + 1;\underline{1})_{(-1)}$ there is $\mu \in \operatorname{Aut} \mathcal{O}(2r + 1;\underline{1})$ such that $\Phi_\mu \in \operatorname{Aut} K(2r + 1;\underline{1})$ and $\Phi_\mu(T)$ is contained in*

$$\Big(\sum_{i=1}^{s} FD_K((1+x_i)x_{i+r})\Big) + \Big(\sum_{i=s+1}^{r} FD_K(x_ix_{i+r})\Big) + FD_K(x_{2r+1} + \sum_{i=1}^{s} x_ix_{i+r}),$$

where $s = \dim T/T \cap W(2r + 1;\underline{1})_{(0)} \le r$.
In addition, if $t_1, \dots, t_s$ are toral elements in T such that $T = (\oplus_{i=1}^s Ft_i) \oplus \big(T \cap W(2r + 1;\underline{1})_{(0)}\big)$, then μ can be chosen such that

$$\Phi_\mu(t_i) = D_K((1 + x_i)x_{i+r})$$

for $1 \le i \le s$.

The proof of Theorem 10.1.4 on the absolute toral rank of Hamiltonian algebras in two generators also works for Lie algebras of Special type.

Theorem 16.1.4. *The following holds.*

(1) $TR(S(m;\underline{n})^{(1)}) = (\sum n_i) - 1.$

(2) $TR(S(m;\underline{n};\Phi(\tau))^{(1)}) = \sum n_i.$

(3) $TR(S(m;\underline{n})^{(1)}) = \sum n_i.$

Proof. We adopt the proof of Theorem 10.1.4.

Let us prove part (1). Theorem 7.5.5 covers the case $\underline{n} = \underline{1}$. So we may assume $n_m > 1$. Put (with the notations of §10.1)

$$\omega := f_{(n_1),1}(x_1)\cdots f_{(n_{m-1}),1}(x_{m-1}) f_{(n_m-1),1}(x_m) df_{(n_1),-1}(x_1)\wedge\cdots\wedge df_{(n_{m-1}),-1}(x_{m-1})\wedge df_{(n_m-1),-1}(x_m).$$

Find by definition $g \in \mathcal{O}(m;(n_1,\dots,n_{m-1},n_m-1))_{(1)}$ so that $\omega = (-1)^m(1+g)\omega_S$. Choose an element $\tilde{g} \in \mathcal{O}(m;\underline{n})_{(2)}$ such that $\partial_m(\tilde{g}) = g$, and define an isomorphism of $\mathcal{O}(m;\underline{n})$ by

$$\varphi(x_i) := x_i \ \forall i < m, \quad \varphi(x_m) := x_m + \tilde{g}.$$

Then we have $\varphi(\omega_S) = (-1)^m\omega$. We conclude $S(m;\underline{n};\omega) \cong S(m;\underline{n})$. Proceed now as in the proof of Theorem 10.1.4(1).

To prove (2), we put, as in the proof of Theorem 10.1.4(2),

$$\omega' := \Big(\prod_{i=1}^{m} f_{(n_i),1}(x_i)\Big) df_{(n_1),-1}(x_1)\wedge\cdots\wedge df_{(n_m),-1}(x_m).$$

By definition there is $g \in \mathcal{O}(m;\underline{n})_{(1)}$ such that $\omega' = (-1)^m(1+g)\omega_S$. Then $\omega' \in \Omega(m;\underline{n})$.

By part (1) of this theorem the algebra $S(m;\underline{n};\omega')^{(\infty)}$ is not isomorphic to $S(m;\underline{n})^{(1)}$. Suppose it is isomorphic to $S(m;\underline{n};\Phi(l))$. Due to Theorem 6.3.6(2) there is $\varphi \in \mathrm{Aut}_c\,\mathcal{O}(m;\underline{n})$ such that $\varphi(\omega') \in F^*\exp(x_l^{(p)})\omega_S$. On the other hand one has $\varphi(\omega') = (-1)^m(1+\varphi(g))J(\varphi)\omega_S$, where $(-1)^m(1+\varphi(g))J(\varphi) \in \mathcal{O}(m;\underline{n})$. Thus this is impossible. Then $S(m;\underline{n};\omega')^{(\infty)}$ is isomorphic to $S(m;\underline{n};\Phi(\tau))^{(1)}$ by Theorem 6.8.3. Consequently,

$$\sum_{i=1}^{m} n_i \le TR(S(m;\underline{n};\Phi(\tau))^{(1)}) \le TR(W(m;\underline{n})) = \sum_{i=1}^{m} n_i.$$

This is claim (2).

To prove assertion (3) one may again use the same arguments as in the proof of Theorem 10.1.4(3). □

It is difficult to describe the absolute toral rank of a Hamiltonian algebra, because it strongly depends on the isomorphism type of the algebra. Some comments on this problem can be found in §20.8. A particular case, however, will become important later in Chapter 19. Namely, let G denote an elementary p-group of order p^m and $(\alpha_1,\dots,\alpha_m)$ an $\mathbb{F}_p$-basis, $f : G\times G\to F$ a non-degenerate skew-symmetric $\mathbb{F}_p$-bilinear form, and $L=\oplus_{\alpha\in G\setminus\{0\}} Fu_\alpha$ a (p^m-1)-dimensional vector space. Define a product

$$[u_\alpha,u_\beta] := f(\alpha,\beta)u_{\alpha+\beta} \text{ if } \alpha+\beta\neq 0, \qquad [u_\alpha,u_{-\alpha}] := 0.$$

This definition is given in [Blo 58] as the algebra $L(G,0,f)$. We are going to embed $L(G,0,f)$ into $W(m;\underline{1})$. Put $z_k := 1+x_k$, $z^\alpha := \prod_{i=1}^m z_i^{r_i}$ for $\alpha=\sum r_i\alpha_i$, and

$$\Theta : L(G,0,f)\to W(m;\underline{1}),\ \Theta(u_\alpha) := \sum_{k=1}^m z^\alpha f(\alpha,\alpha_k)z_k\partial_k \in W(m;\underline{1}). \quad (16.1.1)$$

One computes for $\alpha=\sum r_i\alpha_i$ and $\beta=\sum s_i\alpha_i$ the product

$$\begin{aligned}
[\Theta(u_\alpha),\Theta(u_\beta)] &= \Big[\sum_{i=1}^m z^\alpha f(\alpha,\alpha_i)z_i\partial_i, \sum_{j=1}^m z^\beta f(\beta,\alpha_j)z_j\partial_j\Big]\\
&= \sum_{i,j=1}^m z^\alpha f(\alpha,\alpha_i)s_i z^\beta f(\beta,\alpha_j)z_j\partial_j\\
&\quad - \sum_{i,j=1}^m z^\beta f(\beta,\alpha_j)r_j z^\alpha f(\alpha,\alpha_i)z_i\partial_i\\
&= \sum_{k=1}^m\sum_{i=1}^m z^{\alpha+\beta}\big(f(\alpha,\alpha_i)s_i f(\beta,\alpha_k) - f(\beta,\alpha_i)r_i f(\alpha,\alpha_k)\big)z_k\partial_k\\
&= \sum_{k=1}^m z^{\alpha+\beta}\big(f(\alpha,\beta)f(\beta,\alpha_k) - f(\beta,\alpha)f(\alpha,\alpha_k)\big)z_k\partial_k\\
&= \sum_{k=1}^m f(\alpha,\beta)z^{\alpha+\beta}f(\alpha+\beta,\alpha_k)z_k\partial_k = \Theta([u_\alpha,u_\beta]).
\end{aligned}$$

This shows that Θ is an injective homomorphism of Lie algebras.

Theorem 16.1.5. *Let $(\alpha_1,\dots,\alpha_m)$ be an $\mathbb{F}_p$-basis of G. Define toral derivations of $L(G,0,f)$ by*

$$d_i(u_\alpha) := r_i u_\alpha \textit{ for } \alpha=\sum r_i\alpha_i, \quad i=1,\dots,m.$$

Put $T := \sum_{i=1}^{m} Fd_i$. *Then* T *is a torus in the minimal* p*-envelope* $L(G,0,f)_{[p]}$ *of maximal dimension. One has*

$$L(G,0,f)_{[p]} = L(G,0,f) \oplus T, \quad C_{L(G,0,f)}(T) = \{0\}.$$

In particular, $TR(L(G,0,f)) = m$ *holds.*

Proof. Since f is biadditive and skew-symmetric, one easily computes $(\operatorname{ad} u_\alpha)^p(u_\beta) = f(\alpha,\beta)^p u_\beta$. If one considers the minimal p-envelope of $L(G,0,f)$ in its derivation algebra, then the above computation gives $u_\alpha{}^{[p]} = \sum_{k=1}^{m} f(\alpha,\alpha_k)^p d_k$. Let T' denote the torus generated by $\{u_\alpha{}^{[p]^j} \mid \alpha \in G,\ j > 0\}$. The above shows that $T' \subset T$. If $T' \neq T$, then $\dim T' =: q < m$. Let $(t_1,\dots,t_q)$ be a basis of T' consisting of toral elements. Then $(\alpha_i(t_j))_{\substack{1\le i\le m\\ 1\le j\le q}}$ is a singular matrix with entries in $\mathbb{F}_p$, and therefore there is $\gamma \in G$ such that $\gamma \neq 0$ and $\gamma(T') = 0$. But this means

$$0 = \gamma(u_\alpha{}^{[p]})u_\gamma = [u_\alpha{}^{[p]}, u_\gamma] = f(\alpha,\gamma)^p u_\gamma$$

for all $\alpha \in G$, which contradicts the assumption that f is non-degenerate. Hence $T' = T$. Consequently, $TR(L(G,0,f)) = TR(W(m;\underline{1})) = m$ and T is a torus of maximal dimension in $L(G,0,f)_{[p]}$.

Clearly, $C_{L(G,0,f)}(T) = \{0\}$ and $L(G,0,f) \cap T = \{0\}$ hold. Finally, we observe that $L(G,0,f) + T$ is closed under the p-mapping. Hence it coincides with the minimal p-envelope. □

16.2 Maximal tori in $H(2;\underline{1};\Phi(1))_{[p]}$

In §10.4 we determined the orbits of semisimple elements in $H(2;\underline{1};\Phi(1))$. The question how these fit into a 2-dimensional torus has been left open. We address that problem here by using a method which goes back to [B-W 88] and [P-S 04]. Namely, we normalize the torus in $W(2;\underline{1})$ and then determine the multiplication in an isomorphic copy of $H(2;\underline{1};\Phi(1))$.

In the following, let $\mathfrak{t}$ denote a 2-dimensional torus of $H(2;\underline{1};\Phi(1))_{[p]} = H(2;\underline{1};\Phi(1)) + \mathfrak{t}$ (Theorem 10.4.6(1)). Choose according to Theorem 7.5.1 an automorphism $\sigma \in \operatorname{Aut}\mathcal{O}(2;\underline{1})$ such that

$$\tilde{\mathfrak{t}} := \Phi_\sigma(\mathfrak{t}) := \sigma \circ \mathfrak{t} \circ \sigma^{-1} = F(\delta_1 + x_1)\partial_1 \oplus F(\delta_2 + x_2)\partial_2, \quad \delta_i \in \{0,1\} \tag{16.2.1}$$

and, in addition, for a chosen toral element $t_1 \in \mathfrak{t} \setminus W(2;\underline{1})_{(0)}$,

$$\Phi_\sigma(t_1) = (1 + x_1)\partial_1 \quad \text{(or if preferable)} \quad \Phi_\sigma(t_1) = (1 + x_2)\partial_2 \tag{16.2.2}$$

hold. Put $z_i := \delta_i + x_i$. With this choice we set

$$\omega := \sigma(\omega_{H,1}), \quad \mathfrak{g} := \Phi_\sigma\big(H(2;\underline{1};\Phi(1))\big) = H(2;\underline{1};\omega) \tag{16.2.3}$$

(see Volume 1, p. 337). Then $\omega = 2q dx_1 \wedge d_2$, where

$$q = \frac{1}{2}(\exp \sigma(x_1)^{(p)})\big(\partial_1(\sigma(x_1))\partial_2(\sigma(x_2)) - \partial_1(\sigma(x_2))\partial_2(\sigma(x_1))\big).$$

By definition, $\exp \sigma(x_1)^{(p)}$ is invertible in $\mathcal{O}((2))$. Moreover, since σ is an automorphism of $\mathcal{O}(2;\underline{1})$, $\partial_1(\sigma(x_1))\partial_2(\sigma(x_2)) - \partial_1(\sigma(x_2))\partial_2(\sigma(x_1))$ is invertible in $\mathcal{O}(2;\underline{1})$. Note that

$$\partial_i(q) = \frac{1}{2}(\exp \sigma(x_1)^{(p)})\partial_i\Big(\sigma(x_1)^{(p)} + \partial_1(\sigma(x_1))\partial_2(\sigma(x_2)) - \partial_1(\sigma(x_2))\partial_2(\sigma(x_1))\Big).$$

As $\partial_i(\sigma(x_1)^{(p)}) = \sigma(x_1)^{(p-1)}\partial_i(\sigma(x_1)) \in \mathcal{O}(2;\underline{1})$, one has

$$q_i := q^{-1}\partial_i(q) \in \mathcal{O}(2;\underline{1}).$$

As in Volume 1, pp. 336–340 we set $\omega = \omega_{1,2}dx_1 \wedge dx_2 + \omega_{2,1}dx_2 \wedge dx_1$ with $\omega_{1,2} = -\omega_{2,1} = q$ and put

$$(g_{i,j}) = (\omega_{i,j})^{-1} = \begin{pmatrix} 0 & -q^{-1} \\ q^{-1} & 0 \end{pmatrix}.$$

Due to Theorem 6.5.6 one has

$$H((2;\omega)) = \{g_{1,2}\partial_1(f)\partial_2 + g_{2,1}\partial_2(f)\partial_1 \mid f \in \mathcal{O}((2))\}.$$

We have to determine $\mathfrak{g} = H(2;\underline{1};\omega) = H((2;\omega)) \cap W(2;\underline{1})$. In order to do so, we set according to Volume 1, p. 338, Equation (6.5.4)

$$\mathcal{D} : \mathcal{O}((2)) \to H((2;\omega)),\ \mathcal{D}(f) := D_{H,\omega}(qf) = q^{-1}\big(\partial_1(qf)\partial_2 - \partial_2(qf)\partial_1\big).$$

Note that

$$q^{-1}\partial_i(qf) = q^{-1}\partial_i(q)f + \partial_i(f) = q_i f + \partial_i(f) \in \mathcal{O}(2;\underline{1}) \text{ if } f \in \mathcal{O}(2;\underline{1}).$$

Therefore $\mathcal{D}(\mathcal{O}(2;\underline{1})) \subset \mathfrak{g}$. On the other hand, suppose $f \in (\ker \mathcal{D}) \cap \mathcal{O}(2;\underline{1})$. Then $\partial_1(qf) = \partial_2(qf) = 0$, and hence $qf \in F$, and $f \in Fq^{-1}$. As $q \notin \mathcal{O}(2;\underline{1})$, this gives $f = 0$. Computing dimensions one obtains

$$\mathfrak{g} = \mathcal{D}(\mathcal{O}(2;\underline{1})).$$

Since $\mathfrak{t}(\omega_{H,1}) \subset F\omega_{H,1}$, we have $z_i\partial_i(\omega) \in F\omega$ for $i = 1,2$. This gives $z_i\partial_i(q)+q \in Fq$, hence $z_i\partial_i(q) = \gamma_i q$ for some $\gamma_i \in F$. Consequently,

$$z_i\partial_i(q_j) = z_i\partial_i(q^{-1}\partial_j(q)) = -\delta_{i,j}q_j, \quad i,j = 1,2.$$

Since $q_j \in \mathcal{O}(2;\underline{1})$, this gives $q_j = r_j z_j^{p-1}$ for some $r_j \in F$. As a result,

$$\mathcal{D}(f) = \big(r_1 z_1^{p-1} f + \partial_1(f)\big)\partial_2 - \big(r_2 z_2^{p-1} f + \partial_2(f)\big)\partial_1. \tag{16.2.4}$$

Suppose $(r_1, r_2) \in \mathbb{F}_p\delta_1 \times \mathbb{F}_p\delta_2$. Choose $0 \le i, j \le p-1$ satisfying

$$i = 0 \text{ if } \delta_1 = 0,\ i \equiv -r_1\ (p) \text{ if } \delta_1 = 1; \quad j = 0 \text{ if } \delta_2 = 0,\ j \equiv -r_2\ (p) \text{ if } \delta_2 = 1.$$

Then $z_1^i z_2^j$ is a well-defined nonzero element of $\mathcal{O}(2;\underline{1})$. If $\delta_1 = 0$, then $r_1 = i = 0$ by assumption, whence $r_1 z_1^{p-1} z_1^i z_2^j + \partial_1(z_1^i z_2^j) = 0$. If $\delta_1 = 1$, then $r_1 z_1^{p-1} z_1^i z_2^j + \partial_1(z_1^i z_2^j) = r_1 z_1^{i-1} z_2^j - r_1 z_1^{i-1} z_2^j = 0$ as well. The same argument holds for the index 2, hence $\mathcal{D}(z_1^i z_2^j) = 0$ holds. We mentioned, however, that $(\ker\ \mathcal{D}) \cap \mathcal{O}(2;\underline{1}) = \{0\}$. Let us summarize

$$r_1, r_2 \in F, \quad \delta_1, \delta_2 \in \{0, 1\}, \quad (r_1, r_2) \notin \mathbb{F}_p\delta_1 \times \mathbb{F}_p\delta_2. \tag{16.2.5}$$

Our first result on maximal tori and multiplication coefficients is the following theorem.

Theorem 16.2.1.

(1) *Let $\mathfrak{t} \subset H(2;\underline{1};\Phi(1))_{[p]}$ be a 2-dimensional torus. There are $\sigma \in \operatorname{Aut}\mathcal{O}(2;\underline{1})$, $\delta_1, \delta_2 \in \{0, 1\}$ and $r_1, r_2 \in F$ such that Equation* (16.2.5) *and (with $z_i := \delta_i + x_i$)*

$$\Phi_\sigma(\mathfrak{t}) = F z_1\partial_1 \oplus F z_2\partial_2,$$
$$\Phi_\sigma\big(H(2;\underline{1};\Phi(1))\big) = \Big\{\big(r_1 z_1^{p-1} f + \partial_1(f)\big)\partial_2 - \big(r_2 z_2^{p-1} f + \partial_2(f)\big)\partial_1 \mid f \in \mathcal{O}(2;\underline{1})\Big\}$$

hold.

(2) *Suppose $\delta_1, \delta_2 \in \{0, 1\}$ and $r_1, r_2 \in F$ satisfy Equation (16.2.5). There are a 2-dimensional torus $\mathfrak{t} \subset H(2;\underline{1};\Phi(1))_{[p]}$ and $\sigma \in \operatorname{Aut}\mathcal{O}(2;\underline{1})$ such that (with $z_i := \delta_i + x_i$)*

$$\Phi_\sigma(\mathfrak{t}) = F z_1\partial_1 \oplus F z_2\partial_2,$$
$$\Phi_\sigma\big(H(2;\underline{1};\Phi(1))\big) = \Big\{\big(r_1 z_1^{p-1} f + \partial_1(f)\big)\partial_2 - \big(r_2 z_2^{p-1} f + \partial_2(f)\big)\partial_1 \mid f \in \mathcal{O}(2;\underline{1})\Big\}$$

hold.

Proof. The first part has already been proved by the preceding deliberations. So let $\delta_1, \delta_2 \in \{0, 1\}$ and $r_1, r_2 \in F$ satisfy Equation (16.2.5). If $\delta_1 = 0$, then set $h_1 := \exp(-r_1 x_1^{(p)})$. We have $\partial_1(h_1) = h_1(-r_1 x_1^{(p-1)}) = h_1 r_1 z_1^{p-1}$. If $\delta_1 = 1$, then set $h_1 := \exp\big(\sum_{l=1}^{p} r_1(-1)^{l-1}(l-1)!x_1^{(l)}\big)$ and obtain

$$\begin{aligned}\partial_1(h_1) &= h_1\Big(\sum_{l=1}^{p} r_1(-1)^{l-1}(l-1)!x_1^{(l-1)}\Big) = h_1 r_1\Big(\sum_{l=0}^{p-1}(-1)^l x_1^l\Big)\\ &= h_1 r_1 z_1^{p-1}.\end{aligned}$$

Proceed similarly for z_2 and put $q := h_1 h_2 \in \mathcal{O}((2))^*$. Then

$$q_i := q^{-1}\partial_i(q) = r_i z_i^{p-1}$$

for $i = 1, 2$. Set $\omega := 2q dx_1 \wedge dx_2$ and recall that $q \in \mathcal{O}((2))^*$, $q^{-1}dq = q_1 dx_1 + q_2 dx_2 \in \Omega^1(2; \underline{1})$. Due to Definition 6.4.1 (Volume 1, pp. 314 and 302) ω is a Hamiltonian form subordinate to $\mathcal{O}(2; \underline{1})$. It is of the second type as $r_1 \neq 0$ or $r_2 \neq 0$ and therefore $h_1 h_2 \notin \mathcal{O}(2; \underline{1})$. Due to Theorem 6.4.11(3) there is $\sigma \in \operatorname{Aut}_c \mathcal{O}(2; \underline{1})$ such that $\omega = \sigma(\omega_{H,l})$ for some $l = 1, 2$, while Theorem 6.4.11(2) shows that we may take $l = 1$. Note that $\operatorname{Aut}_c \mathcal{O}(2; \underline{1}) = \operatorname{Aut} \mathcal{O}(2; \underline{1})$ by definition.

Define $\mathcal{D}(f)$ by Equation (16.2.4). Suppose $f \in (\ker \mathcal{D}) \cap \mathcal{O}(2; \underline{1})$. Then

$$\partial_i(f) = -r_i z_i^{p-1} f, \quad i = 1, 2.$$

Write $f = \sum_{r,s=0}^{p-1} a_{r,s} z_1^r z_2^s$. If $\delta_1 = 0$, the former equation implies $a_{r,s} = 0$ for all $r \neq 0$, and this implies that only $r_1 f = 0$ is possible. Then either $f = 0$ or $r_1 = 0 \in \mathbb{F}_p \delta_1$. If $\delta_1 = 1$, then we have $z_1 \partial_1(f) = -r_1 f$. This gives either $f = 0$ or $r_1 \in \mathbb{F}_p = \mathbb{F}_p \delta_1$. Due to Equation (16.2.4) we conclude that only $f = 0$ is possible, whence $(\ker \mathcal{D}) \cap \mathcal{O}(2; \underline{1}) = \{0\}$.

Next we compute

$$\begin{aligned}
\mathcal{D}(f)(\omega) &= 2\big((r_1 z_1^{p-1} f + \partial_1(f))\partial_2(q) - (r_2 z_2^{p-1} f + \partial_2(f))\partial_1(q)\big) dx_1 \wedge dx_2 \\
&\quad + 2q\big(r_1 z_1^{p-1}\partial_2(f) + \partial_2\partial_1(f) - r_2 z_2^{p-1}\partial_1(f) - \partial_1\partial_2(f)\big) dx_1 \wedge dx_2 \\
&= 2\big((r_1 z_1^{p-1} f + \partial_1(f)) q r_2 z_2^{p-1} - (r_2 z_2^{p-1} f + \partial_2(f)) q r_1 z_1^{p-1}\big) dx_1 \wedge dx_2 \\
&\quad + 2q\big(r_1 z_1^{p-1}\partial_2(f) + \partial_2\partial_1(f) - r_2 z_2^{p-1}\partial_1(f) - \partial_1\partial_2(f)\big) dx_1 \wedge dx_2 \\
&= 0.
\end{aligned}$$

Then $\mathcal{D}(f) \in H(2; \underline{1}; \omega) = \Phi_\sigma\big(H(2; \underline{1}; \Phi(1))\big)$ for all $f \in \mathcal{O}(2; \underline{1})$. A dimension argument yields

$$\Phi_\sigma\big(H(2; \underline{1}; \Phi(1))\big) = \mathcal{D}(\mathcal{O}(2; \underline{1})).$$

Finally we observe that

$$\begin{aligned}
z_i \partial_i(\omega) &= z_i \partial_i(2q dx_1 \wedge dx_2) = z_i \partial_i(2q) dx_1 \wedge dx_2 + 2q dx_1 \wedge dx_2 \\
&= 2(z_i q r_i z_i^{p-1} + q) dx_1 \wedge dx_2 = (r_i \delta_i + 1)\omega.
\end{aligned}$$

Then $z_i \partial_i \in \operatorname{Der} \Phi_\sigma\big(H(2; \underline{1}; \Phi(1))\big)$. As a consequence,

$$\mathfrak{t} := \Phi_\sigma^{-1}\big(F z_1 \partial_1 \oplus F z_2 \partial_2\big) \subset \operatorname{Der} H(2; \underline{1}; \Phi(1)) = H(2; \underline{1}; \Phi(1))_{[p]}. \qquad \square$$

Next we are going to describe the root space decomposition with respect to any 2-dimensional torus $\mathfrak{t}$ and do this by turning it into a torus

$$\tilde{\mathfrak{t}} := \Phi_\sigma(\mathfrak{t}) = F z_1 \partial_1 \oplus F z_2 \partial_2$$

as in Theorem 16.2.1(1). Define roots $(\epsilon_i)_{i=1,2}$ dual to $(z_j\partial_j)_{j=1,2}$. Let $\mathcal{D}$ be as in Equation (16.2.4) and set $\mathfrak{g} := \Phi_\sigma\big(H(2;\underline{1};\Phi(1))\big) = \mathcal{D}(\mathcal{O}(2;\underline{1}))$. Theorem 10.4.6 states that every root space in $H(2;\underline{1};\Phi(1))$, hence in $\mathfrak{g}$, is 1-dimensional. Then it is easy to see that

$$\mathfrak{g}_{(a-1)\epsilon_1+(b-1)\epsilon_2} = F\mathcal{D}(z_1^a z_2^b), \quad \tilde{\mathfrak{t}}\cap\mathfrak{g} = F\mathcal{D}(z_1z_2).$$

Proposition 16.2.2.

(1) *Suppose* $\tilde{\mathfrak{t}}\cap\mathfrak{g} = Fz_1\partial_1$. *Then* $\delta_1 = 1$, $r_1 = -1$, *and* $\mathfrak{g}(\epsilon_2)$ *is abelian. Moreover, if* $\delta_2 = 1$, *then every* $\mathfrak{g}_{j\epsilon_2}$ $(j \in \mathbb{F}_p)$ *is spanned by an element which acts non-nilpotently on* $\mathfrak{g}$. *In any case, one has*

$$\mathfrak{g}(\epsilon_2) \subset W(2;\underline{1})_{(0)} + \tilde{\mathfrak{t}}\cap(\ker\epsilon_2).$$

(2) *Suppose* $\tilde{\mathfrak{t}}\cap\mathfrak{g} \neq Fz_1\partial_1$. *Then* $\mathfrak{g}(\epsilon_2) \cong W(1;\underline{1})$ *under an isomorphism* Ψ *which maps* $\tilde{\mathfrak{t}}\cap\mathfrak{g}$ *onto* $F(\delta_2 + x)\partial$. *One has*

$$\Psi^{-1}(W(1;\underline{1})_{(0)}) = \mathfrak{g}(\epsilon_2)\cap\big(W(2;\underline{1})_{(0)} + \tilde{\mathfrak{t}}\cap(\ker\epsilon_2)\big).$$

Proof. Recall that

$$\mathfrak{g}(\epsilon_2) = \sum_{j=0}^{p-1} F\mathcal{D}(z_1z_2^j) = \sum_{j=0}^{p-1} F\big((r_1\delta_1+1)z_2^j\partial_2 - (r_2z_1z_2^{p-1+j} + jz_1z_2^{j-1})\partial_1\big),$$

$$\tilde{\mathfrak{t}}\cap\mathfrak{g} = F\mathcal{D}(z_1z_2) = F\big((r_1\delta_1+1)z_2\partial_2 - (r_2\delta_2+1)z_1\partial_1\big).$$

(a) If $\tilde{\mathfrak{t}}\cap\mathfrak{g} = Fz_1\partial_1$, then $r_1\delta_1 + 1 = 0$. This is possible only if $\delta_1 = 1$ and $r_1 = -1$. Consequently,

$$\mathfrak{g}(\epsilon_2) = Fr_2z_1z_2^{p-1}\partial_1 + \sum_{j=1}^{p-1} F(r_2\delta_2 + j)z_1z_2^{j-1}\partial_1$$

is abelian. As $r_1 \in \mathbb{F}_p\delta_1$ in the present case, Equation (16.2.5) implies $r_2 \notin \mathbb{F}_p\delta_2$. Consequently, $r_2 \neq 0$, $r_2 \notin \mathbb{F}_p$ if $\delta_2 = 1$, and hence $r_2\delta_2 + j \neq 0$ for all $j \neq 0$. This gives $\mathfrak{g}_{(j-1)\epsilon_2} = Fz_1z_2^{j-1}\partial_1$ for all $j \in \{1,\dots,p\}$. If $\delta_2 = 1$, then $\big(\operatorname{ad} z_1z_2^{j-1}\partial_1\big)^p = \operatorname{ad} z_1\partial_1$. Thus every $\mathcal{D}(z_1z_2^j)$ acts non-nilpotently. Finally,

$$\begin{aligned}\mathfrak{g}(\epsilon_2) &= Fz_1z_2^{p-1}\partial_1 + \sum_{j=1}^{p-1} Fz_1z_2^{j-1}\partial_1\\ &\equiv F\delta_2^{p-1}z_1\partial_1 + \sum_{j=1}^{p-1} F\delta_2^{j-1}z_1\partial_1 \pmod{W(2;\underline{1})_{(0)}}\\ &\subset Fz_1\partial_1 = \tilde{\mathfrak{t}}\cap(\ker\epsilon_2).\end{aligned}$$

(b) If $\tilde{\mathfrak{t}} \cap \mathfrak{g} \neq F z_1 \partial_1$, then $r_1\delta_1 + 1 \neq 0$. Set

$$e_{-1} := \frac{1}{r_1\delta_1 + 1}\mathcal{D}(z_1) = \partial_2 - \frac{r_2}{r_1\delta_1 + 1} z_1 z_2^{p-1}\partial_1,$$
$$e_j := \frac{1}{r_1\delta_1 + 1}\mathcal{D}(z_1 z_2^{j+1}) = z_2^{j+1}\partial_2 - \frac{r_2\delta_2 + j + 1}{r_1\delta_1 + 1} z_1 z_2^j \partial_1,\ 0 \le j \le p-2.$$

For $i, j \geq 0$ one computes

$$\begin{aligned}[e_i, e_j] &= (j-i) z_2^{i+j+1}\partial_2 - \frac{r_2\delta_2 + j + 1}{r_1\delta_1 + 1} j z_1 z_2^{i+j}\partial_1 + \frac{r_2\delta_2 + i + 1}{r_1\delta_1 + 1} i z_1 z_2^{i+j}\partial_1 \\ &= \begin{cases} (j-i)e_{i+j} & \text{if } i+j \le p-2, \\ 0 & \text{if } i+j > p-2,\ \delta_2 = 0, \\ (j-i)e_{i+j-p}, & \text{if } i+j > p-2,\ \delta_2 = 1, \end{cases}\end{aligned}$$

$$\begin{aligned}[e_{-1}, e_j] &= (j+1) z_2^j\partial_2 - \frac{r_2\delta_2 + j + 1}{r_1\delta_1 + 1} j z_1 z_2^{j-1}\partial_1 + \frac{r_2}{r_1\delta_1 + 1}(-1) z_1 z_2^{p-1+j}\partial_1 \\ &= \begin{cases} \partial_2 - \frac{r_2}{r_1\delta_1 + 1} z_1 z_2^{p-1}\partial_1 = e_{-1} & \text{if } j = 0, \\ (j+1)e_{j-1} & \text{if } j > 0. \end{cases}\end{aligned}$$

The mapping $e_{j-1} \mapsto (\delta_2 + x)^j\partial$ establishes an isomorphism $\Psi : \mathfrak{g}(\epsilon_2) \cong W(1;\underline{1})$ which maps $\tilde{\mathfrak{t}} \cap \mathfrak{g} = F e_0$ onto $F(\delta_2 + x)\partial$. If $\delta_2 = 0$, then $\Psi^{-1}(W(1;\underline{1})_{(0)}) = \sum_{j \geq 0} F e_j$ while $F(e_0 + \frac{1}{r_1\delta_1+1} z_1\partial_1) + \sum_{j>0} F e_j \subset W(2;\underline{1})_{(0)}$. If $\delta_2 = 1$, then $\Psi^{-1}(W(1;\underline{1})_{(0)}) = \sum_{j\geq 0} F(e_j - e_{-1})$ while $(e_j - e_{-1}) + \frac{j+1}{r_1\delta_1+1} z_1\partial_1 \in W(2;\underline{1})_{(0)}$. This proves, for both cases, $\Psi^{-1}(W(1;\underline{1})_{(0)}) \subset W(2;\underline{1})_{(0)} + F z_1\partial_1 = W(2;\underline{1})_{(0)} + \tilde{\mathfrak{t}} \cap (\ker \epsilon_2)$. As $e_{-1} \notin W(2;\underline{1})_{(0)} + F z_1\partial_1$, the final claim follows. □

Let us turn these results to the original algebra $H(2;\underline{1};\Phi(1))$ and any 2-dimensional torus $\mathfrak{t}$.

Theorem 16.2.3. *Let $\mathfrak{t} \subset H(2;\underline{1};\Phi(1))_{[p]}$ be a 2-dimensional torus and assume that $\mathfrak{t} \cap H(2;\underline{1};\Phi(1))$ is not a torus. The following holds.*

(1) *For any non-zero root $\alpha \in \Gamma(\mathfrak{t}, H(2;\underline{1};\Phi(1)))$ one has*

$$H(2;\underline{1};\Phi(1)))(\alpha) \cong W(1;\underline{1})$$

under an isomorphism which maps

$$\big(H(2;\underline{1};\Phi(1))(\alpha)\big) \cap \big(W(2;\underline{1})_{(0)} + \mathfrak{t} \cap (\ker\alpha)\big)$$

onto $W(1;\underline{1})_{(0)}$.

(2) *There exists an improper root.*

Proof. Let α be a non-zero root and choose a toral element t_α such that $\mathfrak{t}\cap(\ker\alpha) = Ft_\alpha$. Since $\mathfrak{t}\cap H(2;\underline{1};\Phi(1))$ is 1-dimensional (Theorem 10.4.6(1)) and is not a torus by assumption, then $t_\alpha \notin \mathfrak{t}\cap H(2;\underline{1};\Phi(1))$.

Suppose $t_\alpha \in W(2;\underline{1})_{(0)}$. Since $H(2;\underline{1};\Phi(1))_{(0)}$ is closed under $[p]$-th powers (Lemma 7.1.1(3)) and $\mathfrak{t}\cap H(2;\underline{1};\Phi(1))$ is not a torus by assumption, one has $\mathfrak{t}\cap H(2;\underline{1};\Phi(1)) \not\subset H(2;\underline{1};\Phi(1))_{(0)}$. Then $\mathfrak{t}\not\subset W(2;\underline{1})_{(0)}$. So there is a toral element $t_0 \notin W(2;\underline{1})_{(0)}$ in $\mathfrak{t}$. Find according to Equations (16.2.1) and (16.2.2) an automorphism σ such that $\Phi_\sigma(t_0) = (1+x_2)\partial_2$ and $\tilde{\mathfrak{t}} = F(\delta_1+x_1)\partial_1 \oplus F(1+x_2)\partial_2$. Then $\Phi_\sigma(t_\alpha) = r(\delta_1+x_1)\partial_1 + s(1+x_2)\partial_2$ with $r,s\in F$. As $\Phi_\sigma(t_\alpha)\in W(2;\underline{1})_{(0)}$, one has $s=0$ and $\delta_1=0$. Proposition 16.2.2(2) applies to $\Phi_\sigma\big(H(2;\underline{1};\Phi(1))(\alpha)\big) = \mathfrak{g}(\epsilon_2)$ and shows that $H(2;\underline{1};\Phi(1))(\alpha)\cong W(1;\underline{1})$ is improper Witt.

If $t_\alpha \notin W(2;\underline{1})_{(0)}$, then we find according to Equations (16.2.1) and (16.2.2) an automorphism σ such that $\Phi_\sigma(t_\alpha) = (1+x_1)\partial_1$ and $\tilde{\mathfrak{t}} = F(1+x_1)\partial_1 \oplus F(\delta_2+x_2)\partial_2$. Proposition 16.2.2(2) applies to $\Phi_\sigma\big(H(2;\underline{1};\Phi(1))(\alpha)\big) = \mathfrak{g}(\epsilon_2)$ and shows that $H(2;\underline{1};\Phi(1))(\alpha)\cong W(1;\underline{1})$ is Witt. Since $\Gamma(\mathfrak{t}, H(2;\underline{1};\Phi(1)))$ is a 2-dimensional $\mathbb{F}_p$-vector space (Theorem 10.4.6(2)), there is a root β which satisfies $\beta((\delta_2+x_2)\partial_2)=0$. Arguing as in the first case gives that β is improper Witt.

Consequently, in either case Proposition 16.2.2(2) applies, α is Witt, and there is an improper root. To prove the final statement, we observe that in case (2) of Proposition 16.2.2 one has that $\big(H(2;\underline{1};\Phi(1))(\alpha)\big)\cap\big(W(2;\underline{1})_{(0)}+\mathfrak{t}\cap(\ker\alpha)\big) = \Phi_\sigma^{-1}\big(\mathfrak{g}(\alpha)\cap \big(W(2;\underline{1})_{(0)}+\mathfrak{t}\cap(\ker\epsilon_2)\big)\big)$ is the inverse image of $W(1;\underline{1})_{(0)}$. □

Theorem 16.2.4. *Let $\mathfrak{t}\subset H(2;\underline{1};\Phi(1))_{[p]}$ be a 2-dimensional torus and assume that $\mathfrak{t}\cap H(2;\underline{1};\Phi(1))$ is a torus satisfying $\mathfrak{t}\not\subset W(2;\underline{1})_{(0)}$. The following holds.*

(1) *There is a root $\alpha\in\Gamma(\mathfrak{t}, H(2;\underline{1};\Phi(1)))\setminus\{0\}$ satisfying $\alpha(\mathfrak{t}\cap H(2;\underline{1};\Phi(1)))=0$. For every such root $H(2;\underline{1};\Phi(1))(\alpha)$ is abelian. Moreover, $H(2;\underline{1};\Phi(1))(\alpha)\subset W(2;\underline{1})_{(0)}+\mathfrak{t}\cap(\ker\alpha)$.*
(2) *There is a root $\alpha\in\Gamma(\mathfrak{t}, H(2;\underline{1};\Phi(1)))$ satisfying $\alpha(\mathfrak{t}\cap H(2;\underline{1};\Phi(1)))\neq 0$. Every such root satisfies $H(2;\underline{1};\Phi(1))(\alpha)\cong W(1;\underline{1})$. Moreover, $\big(H(2;\underline{1};\Phi(1))(\alpha)\big)\cap\big(W(2;\underline{1})_{(0)}+\mathfrak{t}\cap(\ker\alpha)\big)$ is the inverse image of $W(1;\underline{1})_{(0)}$.*
(3) *There exists an improper root.*

Proof. (1) Let $t_1\in\mathfrak{t}$ be a toral element which spans $\mathfrak{t}\cap H(2;\underline{1};\Phi(1))$. Since $\Gamma(\mathfrak{t}, H(2;\underline{1};\Phi(1)))$ is a 2-dimensional $\mathbb{F}_p$-vector space (Theorem 10.4.6(2)), there is a nonzero root α which satisfies $\alpha(t_1)=0$.

Suppose $t_1\in W(2;\underline{1})_{(0)}$. By assumption there is a toral element $t_0\in\mathfrak{t}\setminus W(2;\underline{1})_{(0)}$. Find according to Equations (16.2.1) and (16.2.2) an automorphism σ such that $\Phi_\sigma(t_0) = (1+x_2)\partial_2$ and $\tilde{\mathfrak{t}} = F(\delta_1+x_1)\partial_1\oplus F(1+x_2)\partial_2$. Then $\Phi_\sigma(t_1) = r(\delta_1+x_1)\partial_1 + s(1+x_2)\partial_2$ with $r,s\in F$. As $\Phi_\sigma(t_1)\in W(2;\underline{1})_{(0)}$, one has $s=0$ and $\delta_1=0$. Proposition 16.2.2(1) shows that this case is impossible.

Therefore $t_1 \notin W(2;\underline{1})_{(0)}$. Find according to Equations (16.2.1) and (16.2.2) an automorphism σ such that $\Phi_\sigma(t_1) = (1+x_1)\partial_1$ and $\tilde{\mathfrak{t}} = F(1+x_1)\partial_1 \oplus F(\delta_2 + x_2)\partial_2$. Proposition 16.2.2(1) yields a proof of claim (1).

(2), (3) Since $\Gamma(\mathfrak{t}, H(2;\underline{1};\Phi(1)))$ is a 2-dimensional $\mathbb{F}_p$-vector space (Theorem 10.4.6(2)), there is a root $\alpha \in \Gamma(\mathfrak{t}, H(2;\underline{1};\Phi(1)))$ satisfying $\alpha(\mathfrak{t} \cap H(2;\underline{1};\Phi(1))) \neq 0$. Let α be any such root. Choose a toral element t_α such that $\mathfrak{t} \cap (\ker \alpha) = Ft_\alpha$.

We now adopt, using the assumption $\mathfrak{t} \not\subset W(2;\underline{1})_{(0)}$, the proof of Theorem 16.2.3 to obtain a proof of claims (2) and (3). □

Theorem 16.2.5. *Let $\mathfrak{t} \subset H(2;\underline{1};\Phi(1))_{[p]}$ be a 2-dimensional torus and assume $\mathfrak{t} \subset W(2;\underline{1})_{(0)}$. The following holds.*

(1) *There are 2 independent roots α_1, α_2 which are proper Witt. For these roots $H(2;\underline{1};\Phi(1))(\alpha_i) \cap H(2;\underline{1};\Phi(1))_{(0)}$ is the unique maximal subalgebra of codimension 1.*

(2) *After an adjustment let $-\alpha_i$ be the roots sticking out of $H(2;\underline{1};\Phi(1))_{(0)}(\alpha_i)$ $(i = 1, 2)$. The root $\alpha_2 - \alpha_1$ is classical. Moreover,*

$$H(2;\underline{1};\Phi(1))(\alpha_2 - \alpha_1) \subset H(2;\underline{1};\Phi(1))_{(0)}$$

and

$$\operatorname{rad} H(2;\underline{1};\Phi(1))(\alpha_2 - \alpha_1) \subset H(2;\underline{1};\Phi(1))_{(1)} + \mathfrak{t}.$$

(3) *All other 1-sections are solvable and $H(2;\underline{1};\Phi(1))(\mu) \subset H(2;\underline{1};\Phi(1))_{(1)} + \mathfrak{t}$ holds for these roots μ.*

Proof. Choose according to Equations (16.2.1) and (16.2.2) an automorphism σ such that $\tilde{\mathfrak{t}} = Fx_1\partial_1 \oplus Fx_2\partial_2$. Let t_α be a toral element such that $\mathfrak{t} \cap (\ker \alpha) = Ft_\alpha$. Write $t_\alpha = rx_1\partial_1 + sx_2\partial_2$, where $r, s \in \mathbb{F}_p$.

If $s = 0$, then we have to determine $\mathfrak{g}(\epsilon_2)$. Part (1) of Proposition 16.2.2 cannot occur. Part (2) shows that $H(2;\underline{1};\Phi(1))(\alpha)$ is Witt. In addition, since $\delta_2 = 0$, ϵ_2 is proper Witt. It follows that α is proper and that $H(2;\underline{1};\Phi(1))(\alpha) \cap W(2;\underline{1})_{(0)} = H(2;\underline{1};\Phi(1))_{(0)}(\alpha)$ is the unique maximal subalgebra of codimension 1. If $r = 0$, then we obtain the same results by symmetry.

Let α_1, α_2 denote the respective roots on $\mathfrak{t}$ and $-\alpha_1$, $-\alpha_2$ be the respective roots sticking out of $H(2;\underline{1};\Phi(1))_{(0)}$.

Suppose $r = s \neq 0$. Then it can only be that $\alpha \in \mathbb{F}_p^*(\alpha_2 - \alpha_1)$. Observe that $t_\alpha = r(x_1\partial_1 + x_2\partial_2)$ acts invertibly on $W(2;\underline{1})/W(2;\underline{1})_{(0)}$ and centralizes the space $W(2;\underline{1})_{(0)}/W(2;\underline{1})_{(1)}$. Hence $H(2;\underline{1};\Phi(1))(\alpha) \subset H(2;\underline{1};\Phi(1))_{(0)}$ and

$$H(2;\underline{1};\Phi(1))(\alpha)/\operatorname{rad} H(2;\underline{1};\Phi(1))(\alpha) \cong H(2;\underline{1};\Phi(1))_{(0)}/H(2;\underline{1};\Phi(1))_{(1)} \cong \mathfrak{sl}(2).$$

This is case (2) of the theorem.

Suppose $r, s, r-s \neq 0$. In this case it is not hard to compute that $C_{W(2;\underline{1})}(t_\alpha) \subset W(2;\underline{1})_{(1)} + \mathfrak{t}$. This is case (3) of the theorem. □

If $\delta_1 = 0$, $\delta_2 = 1$, $r_1 = 1$, $r_2 \notin \mathbb{F}_p$, then Equation (16.2.5) is satisfied and $\tilde{\mathfrak{t}} \cap \mathfrak{g} = F((1+x_2)\partial_2 - (r_2+1)x_1\partial_1)$. In this case $\tilde{\mathfrak{t}} \cap \mathfrak{g}$ is not a torus.

If $\delta_1 = 0$, $\delta_2 = 1$, $r_1 = 1$, $r_2 = -1$, then Equation (16.2.5) is satisfied and $\tilde{\mathfrak{t}} \cap \mathfrak{g} = F(1+x_2)\partial_2$. In this case $\tilde{\mathfrak{t}} \not\subset W(2;\underline{1})_{(0)}$ and $\tilde{\mathfrak{t}} \cap \mathfrak{g}$ is a torus.

If $\delta_1 = \delta_2 = 0$, $r_1 = r_2 = 1$, then Equation (16.2.5) is satisfied and $\tilde{\mathfrak{t}} \subset W(2;\underline{1})_{(0)}$. Therefore the cases described in the previous three theorems really do occur.

16.3 Representations of $H(2;\underline{1};\Phi(\tau))^{(1)}$

Let us repeat some deliberations from Chapter 10.3. One denotes $x^{(\tau(\underline{1}))} = x_1^{(p-1)}x_2^{(p-1)}$, and considers the Hamiltonian form $\omega = (1 + x^{(\tau(\underline{1}))})dx_1 \wedge dx_2$ as well as

$$H(2;\underline{1};\Phi(\tau)) = H(2;\underline{1};\omega), \quad \omega = (1 + x^{(\tau(\underline{1}))})dx_1 \wedge dx_2.$$

Theorem 10.3.1 states that the minimal p-envelope of $H(2;\underline{1};\Phi(\tau))^{(1)}$ is the full derivation algebra of $H(2;\underline{1};\Phi(\tau))^{(1)}$, and in the proof of that theorem (Volume 2, p. 41) it is mentioned that

$$\operatorname{Der} H(2;\underline{1};\Phi(\tau))^{(1)} = H(2;\underline{1};\Phi(\tau)).$$

The algebras in question are described (see Volume 2, p. 40, Equation (10.3.1) ff.) as

$$H(2;\underline{1};\Phi(\tau)) = \{D_{H,\omega}(f) \mid f \in \mathcal{O}(2;\underline{1}) + Fx_1^{(p)} + Fx_2^{(p)}\},$$
$$H(2;\underline{1};\Phi(\tau))^{(1)} = \{D_{H,\omega}(f) \mid f \in \mathcal{O}(2;\underline{1})\},$$

where

$$D_{H,\omega}(f) = 2(1 - x^{(\tau(\underline{1}))})(\partial_1(f)\partial_2 - \partial_2(f)\partial_1).$$

One easily concludes that

$$D_{H,\omega}(x_1^{(a)}x_2^{(b)}) = 2(x_1^{(a-1)}x_2^{(b)}\partial_2 - x_1^{(a)}x_2^{(b-1)}\partial_1) \text{ if } a+b>1$$

(we set $x_i^{(r)} = 0$ if $r < 0$) and

$$D_{H,\omega}(x_1) = 2(1 - x^{(\tau(\underline{1}))})\partial_2, \quad D_{H,\omega}(x_2) = -2(1 - x^{(\tau(\underline{1}))})\partial_1.$$

Volume 1, pp. 17/18 and Equation (1.1.1) immediately imply

$$\begin{aligned} D_{H,\omega}(x_1)^{[p]} &= 2(\partial_2 - x^{(\tau(\underline{1}))}\partial_2)^{[p]} = -2(\operatorname{ad}\partial_2)^{p-1}(x^{(\tau(\underline{1}))}\partial_2) \\ &= -2x_1^{(p-1)}\partial_2 = -D_{H,\omega}(x_1^{(p)}), \\ D_{H,\omega}(x_2)^{[p]} &= 2(\partial_1 - x^{(\tau(\underline{1}))}\partial_1)^{[p]} = -2(\operatorname{ad}\partial_1)^{p-1}(x^{(\tau(\underline{1}))}\partial_1) \\ &= -2x_2^{(p-1)}\partial_1 = D_{H,\omega}(x_2^{(p)}). \end{aligned}$$

Recall that $H(2;\underline{1};\Phi(\tau))^{(1)}{}_{(0)} = H(2;\underline{1};\Phi(\tau))^{(1)} \cap W(2;\underline{1})_{(0)}$. We will need some information about this subalgebra. An analogue is well known for the restricted Hamiltonian algebra of dimension p^2-2.

Proposition 16.3.1.

(1) $\big(H(2;\underline{1};\Phi(\tau))^{(1)}{}_{(0)}\big)^{(1)} = \operatorname{span}\{D_{H,\omega}(x_1^{(a)}x_2^{(b)}) \mid 0 \le a,b \le p-1,\ 2 \le a+b \le 2p-3\}$;

(2) $\big(H(2;\underline{1};\Phi(\tau))^{(1)}{}_{(0)}\big)^{(1)}$ *is generated as an algebra by either one of*

$$\{D_{H,\omega}(x_1^{(2)}),\, D_{H,\omega}(x_2^{(2)}),\, D_{H,\omega}(x_1^{(3)})\} \quad \textit{or}$$
$$\{D_{H,\omega}(x_1^{(2)}),\, D_{H,\omega}(x_2^{(2)}),\, D_{H,\omega}(x_2^{(3)})\};$$

(3) $\big(H(2;\underline{1};\Phi(\tau))^{(1)}{}_{(0)}\big)^{(1)} = \big(H(2;\underline{1};\Phi(\tau))^{(1)}{}_{(0)}\big)^{(2)}$.

Proof. (1) Set $I := \operatorname{span}\{D_{H,\omega}(x_1^{(a)}x_2^{(b)}) \mid 0 \le a,b \le p-1,\ 2 \le a+b \le 2p-3\}$. It is not hard to see that I is an ideal in $H(2;\underline{1};\Phi(\tau))^{(1)}{}_{(0)}$. As it has codimension 1, we get the inclusion "$\subset$".

Let us prove the reverse inclusion. If $a \ne b$, $a+b \ge 2$, then one has

$$D_{H,\omega}(x_1^{(a)}x_2^{(b)}) = \frac{1}{2(b-a)}[D_{H,\omega}(x_1x_2),\, D_{H,\omega}(x_1^{(a)}x_2^{(b)})] \in \big(H(2;\underline{1};\Phi(\tau))^{(1)}{}_{(0)}\big)^{(1)}.$$

If $a = b > 0$, then $a,b \le p-2$ and

$$2D_{H,\omega}(x_1^{(a)}x_2^{(b)}) = [D_{H,\omega}(x_1^{(a+1)}),\, D_{H,\omega}(x_2^{(b+1)})] \in \big(H(2;\underline{1};\Phi(\tau))^{(1)}{}_{(0)}\big)^{(1)}.$$

(2) Recall that (see Theorem 6.3.10(1))

$$H(2;\underline{1};\Phi(\tau))^{(1)}{}_{(0)} = H(2;\underline{1})^{(1)}{}_{(0)} = D_H(\mathcal{O}(2;\underline{1})_{(2)}).$$

We therefore may argue in $H(2;\underline{1})^{(1)}{}_{(0)}$ rather than in $H(2;\underline{1};\Phi(\tau))^{(1)}{}_{(0)}$.

The algebra G generated by $\{D_H(x_1^{(2)}), D_H(x_2^{(2)}), D_H(x_1^{(3)})\}$ contains the element $[D_H(x_1^{(2)}), D_H(x_2^{(2)})] = D_H(x_1x_2)$. Thus the graded degree 0 space $D_H(\mathcal{O}(2;\underline{1})_2)$ is contained in G. As $D_H(x_1^{(3)}) \in G$, Lemma 5.2.2(2) shows that $D_H(\mathcal{O}(2;\underline{1})_3) \subset G$.

Suppose inductively that G contains the graded degree $(s-2)$ space $D_H(\mathcal{O}(2;\underline{1})_s)$ with $s \ge 3$. If $s < p-1$ then G contains

$$0 \ne \binom{s+1}{2} D_H(x_1^{(s+1)}) = [D_H(x_1^{(s)}),\, D_H(x_1^{(2)}x_2)].$$

Suppose $s \geq p-1$. Then G contains $D_H(x_1^{(p-1)}x_2^{(b)})$ for some $0 \leq b \leq p-2$ (Lemma 5.2.2(2)). If $b < p-2$ then G contains

$$[D_H(x_1^{(p-1)}x_2^{(b)}), D_H(x_1x_2^{(2)})] = \left(\binom{p-1}{1}\binom{b+1}{1} - \binom{b+1}{2}\right)D_H(x_1^{(p-1)}x_2^{(b+1)}) \neq 0.$$

Apply Lemma 5.2.2(2). The claim follows directly if $b = p-2$.

The case when G is generated by $\{D_{H,\omega}(x_1^{(2)}), D_{H,\omega}(x_2^{(2)}), D_{H,\omega}(x_2^{(3)})\}$ is symmetric.

(3) is a direct consequence of (2). □

16.3.1 Central Extensions

In Volume 2, pp. 143–145 (Theorem 11.4.3, Equation (11.4.1), Lemma 11.4.4) we have described central extensions of a simple Lie algebra carrying an associative bilinear form.

Theorem 16.3.2. $H(2;\underline{1};\Phi(\tau))^{(1)}$ *carries a non-degenerate associative bilinear form given by*

$$\lambda\big(D_{H,\omega}(x_1^{(a)}x_2^{(b)}), D_{H,\omega}(x_1^{(c)}x_2^{(d)})\big) = \delta_{a+c,p-1}\delta_{b+d,p-1}(-1)^{a+b}. \tag{16.3.1}$$

Proof. Induce a Poisson multiplication on $\mathcal{O}(2;\underline{1})$ by setting

$$\{f,g\}_\omega := D_{H,\omega}(f)(g) = 2(1-x_1^{(p-1)}x_2^{(p-1)})\big(\partial_1(f)\partial_2(g) - \partial_2(f)\partial_1(g)\big). \tag{16.3.2}$$

As mentioned in Volume 1, p. 339, this turns $\mathcal{O}(2;\underline{1})$ into a Lie algebra (the required Jacobi identity holds trivially as there are only two generators x_1, x_2). Equation (6.5.5) in Volume 1, p. 339 shows that

$$D_{H,\omega} : \mathcal{O}(2;\underline{1}) \to H(2;\underline{1};\Phi(\tau))^{(1)}$$

defines a surjective Lie algebra homomorphism with kernel $F1$ when $\mathcal{O}(2;\underline{1})$ is equipped with the multiplication $\{\ ,\ \}_\omega$. From this and Equation (16.3.2) one easily concludes that the derived algebra of this Poisson algebra contains the set

$$\{x_1^{(a)}x_2^{(b)},\ a+b>0,\ (a,b) < (p-1,p-1)\} \cup \{1 - x^{(\tau(\underline{1}))}\},$$

and does not contain $F1$. Therefore the above set spans the derived algebra and is isomorphic to $H(2;\underline{1};\Phi(\tau))^{(1)}$. Put $\mathcal{H} := (\mathcal{O}(2;\underline{1}), \{\ ,\ \})^{(1)}$ which is an ideal of codimension 1 in $\mathcal{O}(2;\underline{1})$ and does not contain $x_1^{(p-1)}x_2^{(p-1)}$ as it does not contain $F1$ but contains $1-x_1^{(p-1)}x_2^{(p-1)}$. Observe that $\mathcal{O}(2;\underline{1}) = \mathcal{H} \oplus Fx_1^{(p-1)}x_2^{(p-1)}$. Define a non-denerate bilinear form $\lambda' : \mathcal{O}(2;\underline{1}) \times \mathcal{O}(2;\underline{1}) \to F$ by setting

$$fg \equiv \lambda'(f,g)x_1^{(p-1)}x_2^{(p-1)} \quad (\mathrm{mod}\ \mathcal{H}).$$

Since $x_1^{(a)}x_2^{(b)}x_1^{(c)}x_2^{(d)} = \binom{a+c}{a}\binom{b+d}{b}x_1^{(a+c)}x_2^{(b+d)}$ and $\binom{p-1}{a} = (-1)^a$, one obtains $\lambda'(x_1^{(a)}x_2^{(b)}, x_1^{(c)}x_2^{(d)}) = \delta_{a+c,p-1}\delta_{b+d,p-1}(-1)^{a+b}$. As

$$\begin{aligned}
&\{x_1^{(u)}x_2^{(v)}, x_1^{(a)}x_2^{(b)}\}_\omega\, x_1^{(c)}x_2^{(d)} + x_1^{(a)}x_2^{(b)}\, \{x_1^{(u)}x_2^{(v)}, x_1^{(c)}x_2^{(d)}\}_\omega \\
&= D_{H,\omega}(x_1^{(u)}x_2^{(v)})(x_1^{(a)}x_2^{(b)})\, x_1^{(c)}x_2^{(d)} + x_1^{(a)}x_2^{(b)}\, D_{H,\omega}(x_1^{(u)}x_2^{(v)})(x_1^{(c)}x_2^{(d)}) \\
&= D_{H,\omega}(x_1^{(u)}x_2^{(v)})(x_1^{(a)}x_2^{(b)}x_1^{(c)}x_2^{(d)}) = \{x_1^{(u)}x_2^{(v)}, x_1^{(a)}x_2^{(b)}x_1^{(c)}x_2^{(d)}\}_\omega \in \mathcal{H},
\end{aligned}$$

the invariance of λ' follows. The respective invariant form on $H(2;\underline{1};\Phi(\tau))^{(1)}$ is given as $\lambda(D_{H,\omega}(f), D_{H,\omega}(g)) = \lambda'(f,g)$ for all $f,g \in \mathcal{H}$. □

Theorem 16.3.3. *Up to algebra isomorphisms there are exactly two central extensions of $H(2;\underline{1};\Phi(\tau))^{(1)}$ by a 1-dimensional center, namely*

(1) *the split extension;*
(2) *the non-split extension $H(2;\underline{1};\Phi(\tau))^{(1)} \oplus Fw$, given by*

$$\begin{aligned}
&[D_{H,\omega}(x_1^{(a)}x_2^{(b)}) + \delta w, D_{H,\omega}(x_1^{(c)}x_2^{(d)}) + \delta' w]_w \\
&\quad = [D_{H,\omega}(x_1^{(a)}x_2^{(b)}), D_{H,\omega}(x_1^{(c)}x_2^{(d)})] + \delta_{a,0}\delta_{c,0}\delta_{b+d,p}(-1)^{b-1}w.
\end{aligned}$$

Proof. Let $\mathfrak{g}$ be a central extension, the multiplication denoted by $[\ ,\ ]_w$. Equation (11.4.1) states that the extension is given as $\mathfrak{g} = H(2;\underline{1};\Phi(\tau))^{(1)} \oplus Fw$, where

$$\begin{aligned}
&[D_{H,\omega}(x_1^{(a)}x_2^{(b)}) + \delta w, D_{H,\omega}(x_1^{(c)}x_2^{(d)}) + \delta' w]_w \\
&\quad = [D_{H,\omega}(x_1^{(a)}x_2^{(b)}), D_{H,\omega}(x_1^{(c)}x_2^{(d)})] \\
&\qquad + \lambda\big(D(D_{H,\omega}(x_1^{(a)}x_2^{(b)})), D_{H,\omega}(x_1^{(c)}x_2^{(d)})\big)w
\end{aligned}$$

for some derivation $D \in \operatorname{Der} H(2;\underline{1};\Phi(\tau))^{(1)}$. In the introductory remark we described the derivation algebra, so that $2D = r_1 D_{H,\omega}(x_1^{(p)}) + r_2 D_{H,\omega}(x_2^{(p)}) + d$ with $r_1, r_2 \in F$ and some inner derivation d. If $r_1 = r_2 = 0$, Theorem 11.4.3 implies that the extension is split. So assume one of these scalars is non-zero. Since $\{x_2, -x_1\}_\omega = \{x_1, x_2\}_\omega$, the mapping $x_1 \mapsto x_2$, $x_2 \mapsto -x_1$ induces an automorphism of the Poisson algebra (see Equation (16.3.2)), whence of $H(2;\underline{1};\Phi(\tau))^{(1)}$. Applying this automorphism if necessary gives $r_1 \neq 0$. Then apply the automorphism ϱ induced by the mapping $r_1^{1/p}x_1 + r_2^{1/p}x_2 \mapsto x_1$, $r_1^{-(1/p)}x_2 \mapsto x_2$, which brings us to the case

$\varrho(2D) = D_{H,\omega}(x_1^{(p)}) + d'$ with d' an inner derivation. Due to Theorem 11.4.3 we may assume $d' = 0$.

After this normalization, we compute using Equation (16.3.1) (observe in the following equations that the condition $b + d = p$ implies $b \neq 0$):

$$\begin{aligned}
&\lambda\big([\tfrac{1}{2}D_{H,\omega}(x_1^{(p)}), D_{H,\omega}(x_1^{(a)}x_2^{(b)})], D_{H,\omega}(x_1^{(c)}x_2^{(d)})\big)\\
&\quad = \delta_{a,0}\lambda\big(D_{H,\omega}(x_1^{(p-1)}x_2^{(b-1)}), D_{H,\omega}(x_1^{(c)}x_2^{(d)})\big)\\
&\quad = \delta_{a,0}\delta_{c,0}\delta_{b+d,p}(-1)^{b-1}.
\end{aligned}$$

□

16.3.2 Representations of dimension $\leq p^2$

For future reference we derive some rules on how elements in an associative algebra commute. Let A be an associative algebra and suppose that $z, x_1, .., x_n$, are elements of A. We use the multi-index notation $x^t := x_1^{t_1} \cdots x_n^{t_n}$, $s \leq t$ if and only if $s_i \leq t_i$ for all i, $|t| = \sum t_i$, $\binom{t}{s} = \prod \binom{t_i}{s_i}$. Lemma 2.1.5 shows that one has

$$zx^s = \sum_{0\leq t\leq s} (-1)^{|t|}\binom{s}{t} x^{s-t}\underbrace{[x_n,\ldots,[x_n,}_{t_n \text{ times}}\ldots,\underbrace{[x_1,\ldots,[x_1,}_{t_1 \text{ times}} z]\ldots].$$

Lemma 16.3.4. *Assume that $0 \leq a, b, i, j \leq p-1$, $0 < a + b$, and put*

$$g := (\operatorname{ad} D_{H,\omega}(x_2))^j (\operatorname{ad} D_{H,\omega}(x_1))^i (D_{H,\omega}(x_1^{(a)}x_2^{(b)})).$$

The following holds.

(1) $b > i \geq 0$:

$$g = \begin{cases} 0 & a < j \\ 2^{i+j}(-1)^j D_{H,\omega}(x_1^{(a-j)}x_2^{(b-i)}) & a \geq j. \end{cases}$$

(2) $b = i$:

$$g = \begin{cases} 2^{i+j}(-1)^{j+1} D_{H,\omega}(x_1^{(p-1+a-j)}x_2^{(p-1)}) & a \leq j \\ 2^{i+j}(-1)^j D_{H,\omega}(x_1^{(a-j)}) & a > j. \end{cases}$$

(3) $0 \leq b < i$:

$$g = \begin{cases} 2^{i+j}(-1)^{j+1} D_{H,\omega}(x_1^{(p-1-j)}x_2^{(p-1+b-i)}) & a = 0 \\ 0 & a > 0. \end{cases}$$

Proof. (1) Suppose $i = 0$: Since $b > 0$, we have for $a < j$

$$g = (\operatorname{ad} D_{H,\omega}(x_2))^j (D_{H,\omega}(x_1^{(a)}x_2^{(b)})) =$$
$$2^a(-1)^a(\operatorname{ad} D_{H,\omega}(x_2))^{j-a}(D_{H,\omega}(x_2^{(b)})) = 0$$

and $g = 2^j(-1)^j D_{H,\omega}(x_1^{(a-j)}x_2^{(b)})$ for $a \geq j$. Next, for $i > 0$ one obtains (observe that $b \geq 2$ in this case)

$$(\operatorname{ad} D_{H,\omega}(x_2))^j (\operatorname{ad} D_{H,\omega}(x_1))^i (D_{H,\omega}(x_1^{(a)}x_2^{(b)})) = \\ (\operatorname{ad} D_{H,\omega}(x_2))^j (\operatorname{ad} D_{H,\omega}(x_1))^{i-1}(2D_{H,\omega}(x_1^{(a)}x_2^{(b-1)})).$$

Induction on i yields the result.

(2) Suppose $a = 0$ (observe that one has, due to the general assumption, in this case $i = b > 0$, and recall from the definition that $(\operatorname{ad} D_{H,\omega}(x_1))(D_{H,\omega}(x_2)) = [D_{H,\omega}(x_1), D_{H,\omega}(x_2)] = -2D_{H,\omega}(x_1^{(p-1}x_2^{(p-1)}))$:

$$\begin{aligned} g &= 2^{i-1}(\operatorname{ad} D_{H,\omega}(x_2))^j (\operatorname{ad} D_{H,\omega}(x_1))(D_{H,\omega}(x_2)) \\ &= 2^{i-1}(\operatorname{ad} D_{H,\omega}(x_2))^j (-2D_{H,\omega}(x_1^{(p-1)}x_2^{(p-1)})) \\ &= 2^{i+j}(-1)^{j+1} D_{H,\omega}(x_1^{(p-1-j)}x_2^{(p-1)}). \end{aligned}$$

This is the claim.

Consider the case $a > 0$. One immediately gets

$$g = 2^i (\operatorname{ad} D_{H,\omega}(x_2))^j (D_{H,\omega}(x_1^{(a)})).$$

For $a > j$ the claim follows. If $a = j$, then

$$\begin{aligned} g &= 2^{i+j-1}(-1)^{j-1}(\operatorname{ad} D_{H,\omega}(x_2))(D_{H,\omega}(x_1)) \\ &= 2^{i+j}(-1)^{j-1} D_{H,\omega}(x_1^{(p-1)}x_2^{(p-1)}). \end{aligned}$$

In case $a < j$ one obtains

$$\begin{aligned} g &= 2^{i+a}(-1)^{a-1}(\operatorname{ad} D_{H,\omega}(x_2))^{j-a}(D_{H,\omega}(x_1^{(p-1)}x_2^{(p-1)})) \\ &= 2^{i+j}(-1)^{j-1} D_{H,\omega}(x_1^{(p-1-j+a)}x_2^{(p-1)}). \end{aligned}$$

(3) Finally let $0 \leq b < i$. Consider the case $a > 0$. Then

$$(\operatorname{ad} D_{H,\omega}(x_1))^i (D_{H,\omega}(x_1^{(a)}x_2^{(b)})) = 2^b (\operatorname{ad} D_{H,\omega}(x_1))^{i-b}(D_{H,\omega}(x_1^{(a)})) = 0.$$

So assume that $a = 0$. Then $b \neq 0$ and $(p-1)-(i-b) \neq 0$, and

$$\begin{aligned} g &= (\operatorname{ad} D_{H,\omega}(x_2))^j (\operatorname{ad} D_{H,\omega}(x_1))^i (D_{H,\omega}(x_2^{(b)})) \\ &= 2^b (\operatorname{ad} D_{H,\omega}(x_2))^j (\operatorname{ad} D_{H,\omega}(x_1))^{i-b}(D_{H,\omega}(-x_1^{(p-1)}x_2^{(p-1)})) \\ &= 2^{i+j}(-1)^{j+1} D_{H,\omega}(x_1^{(p-1-j)}x_2^{(p-1-i+b)}). \qquad \square \end{aligned}$$

Put in the following

$$\mathfrak{g} := H(2;\underline{1};\Phi(\tau))_{[p]} = \operatorname{Der} H(2;\underline{1};\Phi(\tau)) \subset W(2;\underline{1})$$

and $\mathfrak{g}_{(i)} := \mathfrak{g} \cap W(2;\underline{1})_{(i)}$. Let $\rho : \mathfrak{g} \to \mathfrak{gl}(V)$ be an irreducible representation of dimension $\leq p^2$ (in particular $\rho(\mathfrak{g})V \neq \{0\}$). Along with this representation, there comes a character χ given by

$$\rho(d)^p - \rho(d^{[p]}) = \chi(d)^p \mathrm{Id}_V \quad \forall d \in \mathfrak{g}.$$

Let V_0 denote an irreducible $\mathfrak{g}_{(0)}$-submodule of V. We abbreviate

$$X_i := \rho(D_{H,\omega}(x_i)),\ i = 1, 2.$$

Observe that

$$\begin{aligned} X_1^p &= \rho(D_{H,\omega}(x_1))^p = \rho(D_{H,\omega}(x_1)^{[p]}) + \chi(D_{H,\omega}(x_1))^p \mathrm{Id} \\ &= -\rho(D_{H,\omega}(x_1^{(p)})) + \chi(D_{H,\omega}(x_1))^p \mathrm{Id}, \\ X_2^p &= \rho(D_{H,\omega}(x_2))^p = \rho(D_{H,\omega}(x_2)^{[p]}) + \chi(D_{H,\omega}(x_2))^p \mathrm{Id} \\ &= \rho(D_{H,\omega}(x_2^{(p)})) + \chi(D_{H,\omega}(x_2))^p \mathrm{Id}. \end{aligned}$$

Applying the preceding lemma we obtain the following lemma.

Lemma 16.3.5. *Let $v \in V$ be such that $\rho(D_{H,\omega}(x_1^{(r)}x_2^{(s)}))v = 0$ for all $r, s \leq p-1$ with $3 \leq r+s \leq 2p-3$. Put for $3 \leq a+b \leq 2p-3$*

$$g := \rho(D_{H,\omega}(x_1^{(a)}x_2^{(b)}))X_1^i X_2^j v.$$

(1) *If $a+b \geq i+j+3$, then $g = 0$.*
(2) *If $a \geq j+3$, then $g = 0$.*
(3) *If $a = j$, $b = i+2$, then $g = 2^{a+b-2}(-1)^b \rho(D_{H,\omega}(x_2^{(2)}))v$.*
(4) *If $a = j+2$, $b = i$, then $g = 2^{a+b-2}(-1)^b \rho(D_{H,\omega}(x_1^{(2)}))v$.*

Proof. We mentioned at the beginning that

$$\begin{aligned} g = \sum_{\substack{0\leq r\leq i \\ 0\leq s\leq j}} (-1)^{r+s} \binom{i}{r}\binom{j}{s} X_1^{i-r} X_2^{j-s} \\ \times \rho\Big((\operatorname{ad} D_{H,\omega}(x_2))^s (\operatorname{ad} D_{H,\omega}(x_1))^r (D_{H,\omega}(x_1^{(a)}x_2^{(b)}))\Big)v. \end{aligned}$$

Observe that in all cases $a > j \geq s \geq 0$ or $b > i \geq r \geq 0$ holds. Lemma 16.3.4 yields

$$\rho\Big((\operatorname{ad} D_{H,\omega}(x_2))^s (\operatorname{ad} D_{H,\omega}(x_1))^r (D_{H,\omega}(x_1^{(a)}x_2^{(b)}))\Big) \in F\rho(D_{H,\omega}(x_1^{(a-s)}x_2^{(b-r)}))$$

where this expression vanishes if $a-s<0$ or $b-r<0$. The assumption on v yields that all summands with $(a-s)+(b-r)\geq 3$ vanish. Thus only the following pairs (r,s) contribute to the sum:

$$(r,s)\in\{(b,a),(b,a-1),(b,a-2),(b-1,a),(b-1,a-1),(b-2,a)\}=:\mathcal{S}$$

The conditions $i\geq r$ and $j\geq s$ yield in the respective cases of the lemma

(1) $a+b\geq i+j+3\geq r+s+3$, whence $\mathcal{S}=\emptyset$ and therefore $g=0$;

(2) $a\geq j+3\geq s+3$, whence $\mathcal{S}=\emptyset$ and therefore $g=0$;

(3) $a=j, b=i+2$: then $r\leq i=b-2$, whence $\mathcal{S}=\{(b-2,a)\}$; in this case $r=b-2=i, s=a=j$ holds, and we get

$$g=(-1)^{a+b}\rho\Big((\operatorname{ad} D_{H,\omega}(x_2))^a(\operatorname{ad} D_{H,\omega}(x_1))^{b-2}(D_{H,\omega}(x_1^{(a)}x_2^{(b)}))\Big)v;$$

(4) $a=j+2, b=i$: then $s\leq j=a-2$, whence $\mathcal{S}=\{(b,a-2)\}$; in this case $r=b=i, s=a-2=j$ holds, and we get

$$g=(-1)^{a+b}\rho\Big((\operatorname{ad} D_{H,\omega}(x_2))^{a-2}(\operatorname{ad} D_{H,\omega}(x_1))^{b}(D_{H,\omega}(x_1^{(a)}x_2^{(b)}))\Big)v.$$

The computation of these expressions is done by Lemma 16.3.4. □

If $a+b\geq 3$ or $(a,b)=(2,0)$ or $(a,b)=(0,2)$, then $D_{H,\omega}(x_1^{(a)}x_2^{(b)})^{[p]}=D_H(x_1^{(a)}x_2^{(b)})^{[p]}=0$ (Lemma 7.2.1(3)). Therefore $\rho(D_{H,\omega}(x_1^{(a)}x_2^{(b)}))$ has the only eigenvalue $\chi(D_{H,\omega}(x_1^{(a)}x_2^{(b)}))$. It is not hard to see that $D_{H,\omega}(x_1^{(p-1)}x_2^{(p-1)})$ centralizes $\mathfrak{g}_{(0)}$. Hence this element acts on V_0 as $\chi(D_{H,\omega}(x_1^{(p-1)}x_2^{(p-1)}))\mathrm{Id}_{V_0}$. Next observe that

$$\operatorname{span}\{D_{H,\omega}(x_1^{(a)}x_2^{(b)})\mid 0\leq a,b\leq p-1,\ 3\leq a+b\leq 2p-3\}$$

is an ideal of $\mathfrak{g}_{(0)}$. By the ENGEL–JACOBSON Theorem, if $\chi(D_{H,\omega}(x_1^{(a)}x_2^{(b)}))=0$ for all these elements, then these elements annihilate V_0.

Lemma 16.3.6. *One has*

$$\rho(D_{H,\omega}(x_1^{(a)}x_2^{(b)}))V_0=\{0\}\quad\textit{for}\quad 0\leq a,b\leq p-1,\quad 3\leq a+b,$$
$$\rho(D_{H,\omega}(x_i^{(p)}))|V_0=\chi(D_{H,\omega}(x_i^{(p)}))\mathrm{Id}_{V_0}\quad(i=1,2)$$

and

$$\rho(D_{H,\omega}(x_1^{(2)}))^2V_0=\rho(D_{H,\omega}(x_2^{(2)}))^2V_0=\{0\}.$$

Proof. (a) Suppose the set

$$\{i+j\mid 0\leq i,j\leq p-1,\ 3\leq i+j\leq 2p-3,\ \chi(D_{H,\omega}(x_1^{(i)}x_2^{(j)}))\neq 0\}$$

is non-empty and hence has a maximum k. Put

$$b := \max\{j \mid j \le k,\ \chi(D_{H,\omega}(x_1^{(k-j)}x_2^{(j)})) \ne 0\}, \quad a := k-b.$$

This definition implies that

$$\chi(D_{H,\omega}(x_1^{(i)}x_2^{(j)})) = 0 \text{ if } (2p-3 \ge i+j > a+b) \text{ or } (i+j = a+b \text{ and } j > b),$$

and $\chi(D_{H,\omega}(x_1^{(a)}x_2^{(b)})) \ne 0$ hold. We mentioned that $\rho(D_{H,\omega}(x_1^{(i)}x_2^{(j)}))$ has the single eigenvalue $\chi(D_{H,\omega}(x_1^{(i)}x_2^{(j)}))$. Put in Proposition 3.2.11

$$G := \mathfrak{g},\ M := \mathfrak{g}_{(0)},\ I := \operatorname{span}\{D_{H,\omega}(x_1^{(i)}x_2^{(j)}) \mid k+1 \le i+j \le 2p-3\}.$$

(i) Consider the case that $a, b < p-1$. Put

$$e_1 := D_{H,\omega}(x_2), \qquad e_2 := D_{H,\omega}(x_1),$$
$$f_1 := D_{H,\omega}(x_1^{(a+1)}x_2^{(b)}), \qquad f_2 := D_{H,\omega}(x_1^{(a)}x_2^{(b+1)}).$$

One has $[e_i, f_j], [e_i, [e_l, f_j]] \in D_{H,\omega}(\mathcal{O}(2;\underline{1})_{(k-1)})$. As $k \ge 3$, this space is contained in M. Next we observe that

$$[M, I] + R^{(1)} + F[e_1, f_2] \subset D_{H,\omega}(\mathcal{O}(2;\underline{1})_{(a+b+1)}) + FD_{H,\omega}(x_1^{(a-1)}x_2^{(b+1)}) \subset \ker\chi,$$

while $[e_i, f_i] \in F^* D_{H,\omega}(x_1^{(a)}x_2^{(b)}) \not\subset \ker\chi$. The proposition yields the estimate

$$\dim V \ge p^2 \dim V_0.$$

As $\dim V \le p^2$, we obtain $\dim V_0 = 1$. In particular, ${\mathfrak{g}_{(0)}}^{(1)}$ annihilates V_0. This, however, contradicts our assumption, since

$$2D_{H,\omega}(x_1^{(a)}x_2^{(b)}) = [D_{H,\omega}(x_1^{(a+1)}), D_{H,\omega}(x_2^{(b+1)})] \in {\mathfrak{g}_{(0)}}^{(1)}.$$

(ii) Consider the case $a = p-1$ (the case $b = p-1$ is similar). Then $b \le 2p-3-a = p-2$. Put in Proposition 3.2.11 $e_1 := D_{H,\omega}(x_1)$, $f_1 := D_{H,\omega}(x_1^{(p-1)}x_2^{(b+1)})$ to obtain the upper bound

$$\dim V_0 \le p.$$

We now apply the proposition with $G := \mathfrak{g}_{(0)}$,

$$M := I := \operatorname{span}\{D_{H,\omega}(x_1^{(i)}x_2^{(j)}) \mid k \le i+j \le 2p-3\},$$

and

$$\begin{aligned} e_1 &:= D_{H,\omega}(x_1^{(2)}), & e_2 &:= D_{H,\omega}(x_1 x_2), \\ f_1 &:= D_{H,\omega}(x_1^{(p-2)} x_2^{(b+1)}), & f_2 &:= D_{H,\omega}(x_1^{(p-1)} x_2^{(b)}). \end{aligned}$$

As M is an ideal in G, one gets $[e_i, f_j], [e_i, [e_l, f_j]] \in M$. Since $k \geq 3$, one has

$$\begin{aligned} [M, I] + R^{(1)} + F[e_1, f_2] &\subset D_{H,\omega}(\mathcal{O}(2; \underline{1})_{(k+1)}) + F D_{H,\omega}(x_1^{(p-2)} x_2^{(b+1)}) \\ &\subset \ker \chi. \end{aligned}$$

Finally observe that $[e_i, f_i] \in F^* D_{H,\omega}(x_1^{(p-1)} x_2^{(b)}) \not\subset \ker \chi$. The proposition yields the contradiction $\dim V_0 \geq p^2$.

We mentioned before that this result implies the statement that the space

$$\operatorname{span}\{D_{H,\omega}(x_1^{(i)} x_2^{(j)}) \mid 0 \leq a, b \leq p-1,\ 3 \leq i+j \leq 2p-3\}$$

annihilates V_0, and thereby proves the first statement if $(a, b) \neq (p-1, p-1)$.

(b) Let $u \in V_0$ be an eigenvector with respect to $\rho(D_{H,\omega}(x_1 x_2))$. The family

$$\left(X_1^i X_2^j u\right)_{(0,0) \leq (i,j) < (p-1,p-3)} \cup \left(X_1^i X_2^j \rho(D_{H,\omega}(x_1^{(2)}))u\right)_{(0,0) \leq (i,j) < (p-1,p-3)}$$

has $2p(p-2) - 2 > p^2$ elements and is therefore linearly dependent. There is a relation with not all coefficients vanishing

$$\sum_{(0,0) \leq (i,j) < (p-1,p-3)} \alpha_{i,j} X_1^i X_2^j u + \sum_{(0,0) \leq (i,j) < (p-1,p-3)} \beta_{i,j} X_1^i X_2^j \rho(D_{H,\omega}(x_1^{(2)}))u = 0.$$

Put

$$\begin{aligned} k &:= \max\{i + j \mid (0,0) \leq (i,j) < (p-1, p-3),\ \alpha_{i,j} \neq 0 \text{ or } \beta_{i,j} \neq 0\} \\ s &:= \max\{j \mid \alpha_{k-j,j} \neq 0 \text{ or } \beta_{k-j,j} \neq 0\} \leq p-3, \quad r := k - s \leq p-1. \end{aligned}$$

If $k = 0$, then $u, \rho(D_{H,\omega}(x_1^{(2)}))u$ are linearly dependent. Since they correspond to different eigenvalues with respect to $\rho(D_{H,\omega}(x_1 x_2))$, this is only possible if $\rho(D_{H,\omega}(x_1^{(2)}))u = 0$. So consider the case $k > 0$. Then

$$\begin{aligned} 0 &= \rho(D_{H,\omega}(x_1^{(s+2)} x_2^{(r)}))\times \\ &\times\left(\sum_{(0,0) \leq (i,j) < (p-1,p-3)} \alpha_{i,j} X_1^i X_2^j u + \sum_{(0,0) \leq (i,j) < (p-1,p-3)} \beta_{i,j} X_1^i X_2^j \rho(D_{H,\omega}(x_1^{(2)}))u\right). \end{aligned}$$

The definitions of k and s yield that only pairs (i, j) may contribute to the sum for which $i + j \leq k = r + s$ and $j \leq s$. The result of (a) gives that Lemma 16.3.5 applies (note that $r+s+2 \geq k+2 \geq 3$, $s+2 \leq p-1$ and $r+s+2 < 2(p-1)$). Parts 1 and 2 of that lemma show that we only need to consider pairs (i,j) with $r+s < i+j+1$ and $s < j+1$. As a result, all summands except those with $i = r$, $j = s$ vanish. Hence part 4 of that lemma gives

$$\begin{aligned}0 =&\rho(D_{H,\omega}(x_1^{(s+2)}x_2^{(r)}))\Big(\alpha_{r,s}X_1^rX_2^su + \beta_{r,s}X_1^rX_2^s\rho(D_{H,\omega}(x_1^{(2)}))u\Big)\\ =&2^{r+s}(-1)^r\alpha_{r,s}\rho(D_{H,\omega}(x_1^{(2)}))u + 2^{r+s}(-1)^r\beta_{r,s}\rho(D_{H,\omega}(x_1^{(2)}))^2u.\end{aligned}$$

As $\rho(D_{H,\omega}(x_1^{(2)}))u$ and $\rho(D_{H,\omega}(x_1^{(2)}))^2u$ correspond to different eigenvalues with respect to $\rho(D_{H,\omega}(x_1x_2))$, we obtain $\alpha_{r,s}\rho(D_{H,\omega}(x_1^{(2)}))u = \beta_{r,s}\rho(D_{H,\omega}(x_1^{(2)}))^2u = 0$. By definition, one of the coefficients is nonzero, which is possible only if $\rho(D_{H,\omega}(x_1^{(2)}))^2u = 0$. By symmetry we obtain the equation $\rho(D_{H,\omega}(x_2^{(2)}))^2u = 0$.

(c) By (b), the set

$$\mathcal{S} := \{v \in V_0 \mid \rho(D_{H,\omega}(x_2^{(2)}))v = 0,\ \rho(D_{H,\omega}(x_1x_2))v \in Fv\}$$

is nonzero. Since $[\rho(D_{H,\omega}(x_1^{(p)})), \rho(D_{H,\omega}(x_2^{(2)}))] = \rho(2D_{H,\omega}(x_1^{(p-1)}x_2))$ annihilates V_0 (by (a)), $\mathcal{S}$ is invariant under $\rho(D_{H,\omega}(x_1^{(p)}))$. Hence there is $u \in \mathcal{S}$ for which $\rho(D_{H,\omega}(x_1^{(p)}))u = \lambda u$ for some $\lambda \in F$. Consider the family

$$\Big(\rho(D_{H,\omega}(x_2^{(p)}))^iX_2^ju\Big)_{0\leq i,j\leq p-2} \cup \Big(\rho(D_{H,\omega}(x_2^{(p)}))^iX_2^jX_1u\Big)_{0\leq i,j\leq p-2}.$$

This family has $2(p-1)^2 > p^2$ elements and therefore is linearly dependent. Thus there is a relation with not all coefficients vanishing

$$\sum_{0\leq i,j\leq p-2}\alpha_{i,j}\rho(D_{H,\omega}(x_2^{(p)}))^iX_2^ju + \sum_{0\leq i,j\leq p-2}\beta_{i,j}\rho(D_{H,\omega}(x_2^{(p)}))^iX_2^jX_1u = 0.$$

If not all of the $\beta_{i,j}$ vanish, the application of $\rho(D_{H,\omega}(x_2^{(2)}))$ (in combination with $\rho(D_{H,\omega}(x_2^{(2)}))u = 0$) yields a non-trivial relation

$$\sum_{0\leq i,j\leq p-2}\beta_{i,j}\rho(D_{H,\omega}(x_2^{(p)}))^iX_2^{j+1}u = 0.$$

Considering eigenvectors with respect to $\rho(D_{H,\omega}(x_1x_2))$ we see in either case that there is $r \leq p-1$ and a relation

$$\sum_{0\leq i\leq k}\gamma_i\rho(D_{H,\omega}(x_2^{(p)}))^iX_2^ru = 0, \qquad k \leq p-2,\ \gamma_k \neq 0.$$

Multiplying by X_2^{p-r} if $r \neq 0$ (and observing that $X_2^p = \rho(D_{H,\omega}(x_2^{(p)})) + \chi(D_{H,\omega}(x_2))^p\mathrm{Id}$), we end up with a relation

$$\sum_{0\leq i\leq l}\delta_i\rho(D_{H,\omega}(x_2^{(p)}))^iu = 0, \qquad l \leq p-1,\ \delta_l \neq 0.$$

Choose l minimal and assume that $\chi(D_{H,\omega}(x_1^{(p-1)}x_2^{(p-1)})) \neq 0$. Then

$$\begin{aligned}
0 &= \big(\rho(D_{H,\omega}(x_1^{(p)})) - \lambda \mathrm{Id}\big)\Big(\sum_{0\le i\le l} \delta_i \rho(D_{H,\omega}(x_2^{(p)}))^i u\Big)\\
&= \sum_{0\le i\le l} \delta_i [\rho(D_{H,\omega}(x_1^{(p)})), \rho(D_{H,\omega}(x_2^{(p)}))^i] u\\
&= \sum_{0\le i\le l} i\delta_i \rho(D_{H,\omega}(x_2^{(p)}))^{i-1}\rho(D_{H,\omega}(2x_1^{(p-1)}x_2^{(p-1)}))u\\
&= 2\chi\big(D_{H,\omega}(x_1^{(p-1)}x_2^{(p-1)})\big)\sum_{0\le i\le l} i\delta_i \rho(D_{H,\omega}(x_2^{(p)}))^{i-1}u.
\end{aligned}$$

The minimality of l yields $l = 0$, whence $u = 0$, a contradiction.

(d) Note that $D_{H,\omega}(x_i^{(p)})^{[p]} = 0$ and, by the preceding results, $[D_{H,\omega}(x_i^{(p)}), \mathfrak{g}_{(0)}]$ acts nilpotently on V_0. Then $\rho(D_{H,\omega}(x_i^{(p)}))|V_0 = \chi(D_{H,\omega}(x_i^{(p)}))\mathrm{Id}_{V_0}$ holds. □

Lemma 16.3.7. *V contains a 1-dimensional $\big(H(2;\underline{1};\Phi(\tau))^{(1)}{}_{(0)}\big)^{(1)}$-submodule.*

Proof. (a) Assume that the lemma is not true. Let V_0' denote an irreducible $\big(H(2;\underline{1};\Phi(\tau))^{(1)}{}_{(0)}\big)^{(1)}$-submodule of V_0. Due to Lemma 16.3.6, this is an irreducible module for $FD_{H,\omega}(x_1^{(2)}) + FD_{H,\omega}(x_1x_2) + FD_{H,\omega}(x_2^{(2)}) \cong \mathfrak{sl}(2)$. Moreover, $\rho(D_{H,\omega}(x_i^{(2)}))^2 V_0 = \{0\}$ holds for $i = 1, 2$. By assumption, V_0' is not 1-dimensional. Then it can only be 2-dimensional. As $(D_{H,\omega}(x_2^{(2)}), \frac{1}{2}D_{H,\omega}(x_1x_2), \frac{-1}{4}D_{H,\omega}(x_1^{(2)}))$ is an $\mathfrak{sl}(2)$-triple, V_0' has a presentation

$$\begin{aligned}
&V_0' = Fu \bigoplus F\rho(D_{H,\omega}(x_1^{(2)}))u,\\
&\rho(D_{H,\omega}(x_2^{(2)}))u = 0,\ \rho(D_{H,\omega}(x_1x_2))u = 2u,\ \rho(D_{H,\omega}(x_1^{(2)}))^2u = 0.
\end{aligned}$$

(b) Put $w_1 := X_2\rho(D_{H,\omega}(x_1^{(2)}))u - 2X_1u$. Taking into account Lemma 16.3.6 we obtain

$$\begin{aligned}
\rho(D_{H,\omega}(x_1^{(3)}))w_1 &= 2\rho(D_{H,\omega}(x_1^{(2)}))^2u = 0,\\
\rho(D_{H,\omega}(x_1^{(2)}))w_1 &= 2X_1\rho(D_{H,\omega}(x_1^{(2)}))u\\
&\qquad + X_2\rho(D_{H,\omega}(x_1^{(2)}))^2u - 2\rho(D_{H,\omega}(x_1^{(2)}))X_1u = 0,\\
\rho(D_{H,\omega}(x_2^{(2)}))w_1 &= X_2\rho([D_{H,\omega}(x_2^{(2)}), D_{H,\omega}(x_1^{(2)})])u + 4X_2u\\
&= -2X_2\rho(D_{H,\omega}(x_1x_2))u + 4X_2u = 0.
\end{aligned}$$

Proposition 16.3.1 implies that Fw_1 is an $\big(H(2;\underline{1};\Phi(\tau))^{(1)}{}_{(0)}\big)^{(1)}$-submodule. Our present assumption yields $w_1 = 0$.

Set $w_2 := X_2^{p-1}u$. The general computation rule simplifies a bit since $(-1)^j\binom{p-1}{j} = 1$. Recall that $[X_2, X_1] = \rho(D_{H,\omega}(2x_1^{(p-1)}x_2^{(p-1)}))$. One computes

$$\begin{aligned}
\rho(D_{H,\omega}(x_2^{(3)}))w_2 &= X_2^{p-1}\rho(D_{H,\omega}(x_2^{(3)}))u = 0,\\
\rho(D_{H,\omega}(x_2^{(2)}))w_2 &= X_2^{p-1}\rho(D_{H,\omega}(x_2^{(2)}))u = 0,\\
\rho(D_{H,\omega}(x_1^{(2)}))w_2 &= X_2^{p-1}\rho(D_{H,\omega}(x_1^{(2)}))u - 2X_2^{p-2}X_1u\\
&\quad - 4X_2^{p-3}\rho(D_{H,\omega}(x_1^{(p-1)}x_2^{(p-1)}))u\\
&\quad - \sum_{3\le i\le p-1} 2^i(-1)^iX_2^{p-1-i}\rho(D_{H,\omega}(x_1^{((p-1)-(i-2))}x_2^{(p-1)}))u\\
&= X_2^{p-2}\big(X_2\rho(D_{H,\omega}(x_1^{(2)}))u - 2X_1u\big) = X_2^{p-2}w_1 = 0.
\end{aligned}$$

As before we conclude $w_2 = 0$.

Set $w_3 := X_1^{p-1}\rho(D_{H,\omega}(x_1^{(2)}))u$. One computes, using former results,

$$\begin{aligned}
\rho(D_{H,\omega}(x_1^{(2)}))w_3 &= X_1^{p-1}\rho(D_{H,\omega}(x_1^{(2)}))^2u = 0,\\
\rho(D_{H,\omega}(x_1^{(3)}))w_3 &= X_1^{p-1}\rho(D_{H,\omega}(x_1^{(2)}))\rho(D_{H,\omega}(x_1^{(3)}))u = 0,\\
\rho(D_{H,\omega}(x_2^{(2)}))w_3 &= X_1^{p-1}\rho(D_{H,\omega}(x_2^{(2)}))\rho(D_{H,\omega}(x_1^{(2)}))u\\
&\quad + 2X_1^{p-2}X_2\rho(D_{H,\omega}(x_1^{(2)}))u\\
&\quad - 4X_1^{p-3}\rho(D_{H,\omega}(x_1^{(p-1)}x_2^{(p-1)}))\rho(D_{H,\omega}(x_1^{(2)}))u\\
&\quad - \sum_{3\le i\le p-1} 2^iX_1^{p-1-i}\rho(D_{H,\omega}(x_1^{(p-1)}x_2^{((p-1)-(i-2))}))\\
&\qquad \times \rho(D_{H,\omega}(x_1^{(2)}))u\\
&= X_1^{p-1}\rho(D_{H,\omega}(x_2^{(2)}))\rho(D_{H,\omega}(x_1^{(2)}))u\\
&\quad + 2X_1^{p-2}X_2\rho(D_{H,\omega}(x_1^{(2)}))u\\
&= -2X_1^{p-1}\rho(D_{H,\omega}(x_1x_2))u + 2X_1^{p-2}X_2\rho(D_{H,\omega}(x_1^{(2)}))u\\
&= -4X_1^{p-1}u + 2X_1^{p-2}X_2\rho(D_{H,\omega}(x_1^{(2)}))u = 2X_1^{p-2}w_1 = 0.
\end{aligned}$$

As before we conclude $w_3 = 0$.

Recall that $X_1^p = -\rho(D_{H,\omega}(x_1^{(p)})) + \chi(D_{H,\omega}(x_1))^p\mathrm{Id}$ and $\rho(D_{H,\omega}(x_1^{(p)}))|V_0 = \chi(D_{H,\omega}(x_1^{(p)}))\mathrm{Id}_{V_0}$. As $X_1^{p-1}\rho(D_{H,\omega}(x_1^{(2)}))u = w_3 = 0$ and $\rho(D_{H,\omega}(x_1^{(2)}))u \in V_0$ is assumed to be nonzero by (a), we conclude

$$0 = X_1w_3 = \big(-\chi(D_{H,\omega}(x_1^{(p)})) + \chi(D_{H,\omega}(x_1))^p\big)\rho(D_{H,\omega}(x_1^{(2)}))u$$

and

$$(*) \qquad -\chi(D_{H,\omega}(x_1^{(p)})) + \chi(D_{H,\omega}(x_1))^p = 0.$$

As
$$[X_2^{p-3}, X_1]u \in \sum_{i\geq 1} FX_2^{p-3-i} D_{H,\omega}\big(x_1^{(p-i)} x_2^{(p-1)}\big)u = \{0\},$$
the former equation gives
$$\begin{aligned}X_1^{p-1}X_2^{p-3}X_1u = X_1^pX_2^{p-3}u &= -\rho(D_{H,\omega}(x_1^{(p)}))X_2^{p-3}u + \chi(D_{H,\omega}(x_1))^pX_2^{p-3}u\\ &= -[\rho(D_{H,\omega}(x_1^{(p)})), X_2^{p-3}]u = 0.\end{aligned}$$
Similar to former computations one gets
$$\begin{aligned}&\rho(D_{H,\omega}(x_1^{(2)}))X_2^{p-2}u - X_2^{p-2}\rho(D_{H,\omega}(x_1^{(2)}))u + 4X_2^{p-3}X_1u\\ &\quad\in \sum_{i\geq 2} FX_2^{p-2-i} D_{H,\omega}(x_1^{((p-1)-(i-2))}x_2^{(p-1)})u = \{0\},\\ &\rho(D_{H,\omega}(x_2^{(2)}))X_1^{p-1}X_2^{p-2}u - X_1^{p-1}\rho(D_{H,\omega}(x_2^{(2)}))X_2^{p-2}u - 2X_1^{p-2}X_2^{p-1}u\\ &\quad\in \sum_{2\leq i\leq p-1} FX_1^{p-1-i} D_{H,\omega}(x_1^{(p-1)}x_2^{((p-1)-(i-2))})X_2^{p-2}u\\ &\quad\subset \sum_{\substack{2\leq i\leq p-1\\ 0\leq j\leq p-2}} FX_1^{p-1-i}X_2^{p-2-j} D_{H,\omega}(x_1^{(p-1-j)}x_2^{((p-1)-(i-2))})u = \{0\}.\end{aligned}$$

Set $w_4 := X_1^{p-1}X_2^{p-2}u$. The above gives
$$\begin{aligned}\rho(D_{H,\omega}(x_1^{(2)}))w_4 &= X_1^{p-1}X_2^{p-2}\rho(D_{H,\omega}(x_1^{(2)}))u - 4X_1^{p-1}X_2^{p-3}X_1u\\ &= X_1^{p-1}X_2^{p-3}w_1 - 2X_1^{p-1}X_2^{p-3}X_1u = 0,\\ \rho(D_{H,\omega}(x_2^{(2)}))w_4 &= X_1^{p-1}\rho(D_{H,\omega}(x_2^{(2)}))X_2^{p-2}u + 2X_1^{p-2}X_2^{p-1}u\\ &= X_1^{p-1}X_2^{p-2}\rho(D_{H,\omega}(x_2^{(2)}))u + 2X_1^{p-2}w_2 = 0,\\ \rho(D_{H,\omega}(x_2^{(3)}))w_4 &\in 2X_1^{p-2}\rho(D_{H,\omega}(x_2^{(2)}))X_2^{p-2}u + 4X_1^{p-3}X_2^{p-1}u+\\ &\quad + \sum_{\substack{3\leq i\leq p-1\\ 0\leq j\leq p-2}} FX_1^{p-1-i}X_2^{p-2-j}\rho(D_{H,\omega}(x_1^{(p-1-j)}x_2^{((p-1)-(i-3))}))u\\ &= 2X_1^{p-2}X_2^{p-2}\rho(D_{H,\omega}(x_2^{(2)}))u + 4X_1^{p-3}w_2 + \{0\} = \{0\}.\end{aligned}$$
As before, $w_4 = 0$.

(c) From (b) we conclude (using $(*)$)
$$\begin{aligned}0 = X_1w_4 = X_1^pX_2^{p-2}u &= -\rho(D_{H,\omega}(x_1^{(p)}))X_2^{p-2}u + \chi(D_{H,\omega}(x_1))^pX_2^{p-2}u\\ &= -\rho(D_{H,\omega}(x_1^{(p)}))X_2^{p-2}u + X_2^{p-2}\rho(D_{H,\omega}(x_1^{(p)}))u\\ &= (\operatorname{ad} X_2)^{p-2}\big(\rho(D_{H,\omega}(x_1^{(p)}))\big)u\\ &= -2^{p-2}\rho(D_{H,\omega}(x_1^{(2)}))u.\end{aligned}$$

Thus $\rho(D_{H,\omega}(x_1^{(2)}))u = 0$, contradicting (a). □

Lemma 16.3.8. *V contains a 1-dimensional $\mathfrak{g}_{(0)}$-submodule.*

Proof. Lemma 16.3.7 in combination with Proposition 16.3.1(3) shows that the space

$$W := \{v \in V \mid \rho\Big(\big(H(2;\underline{1};\Phi(\tau))^{(1)}{}_{(0)}\big)^{(1)}\Big)v = 0\}$$

is nonzero. It is invariant under $\mathfrak{g}_{(0)}$. Choose an irreducible $\mathfrak{g}_{(0)}$-module $W_0 \subset W$. Lemma 16.3.6 shows that $D_{H,\omega}(x_1^{(p-1)}x_2^{(p-1)})$ annihilates every irreducible $\mathfrak{g}_{(0)}$-module. Therefore W_0 is annihilated by

$$\big(H(2;\underline{1};\Phi(\tau))^{(1)}{}_{(0)}\big)^{(1)} + FD_{H,\omega}(x_1^{(p-1)}x_2^{(p-1)}) = H(2;\underline{1};\Phi(\tau))^{(1)}{}_{(0)}.$$

Note that $\mathfrak{g}_{(0)}/H(2;\underline{1};\Phi(\tau))^{(1)}{}_{(0)}$ is a 2-dimensional commutative algebra spanned by the image of $FD_{H,\omega}(x_1^{(p)}) + FD_{H,\omega}(x_2^{(p)})$. But then $\mathfrak{g}_{(0)}$ acts as an abelian algebra, whence W_0 is 1-dimensional. □

Proposition 16.3.9. *Let $Fu \subset V$ be a 1-dimensional $\mathfrak{g}_{(0)}$-submodule. The following holds.*

(1) $\dim V \in \{1, p^2-1, p^2\}$;
(2) *If* $\dim V = p^2$, *then* $(X_1^i X_2^j u)_{0\le i,j\le p-1}$ *is a basis of* V;
(3) *If* $\dim V = p^2-1$, *then*
  (a) $(X_1^i X_2^j u)_{\substack{0\le i,j\le p-1\\ i+j\le 2p-3}}$ *is a basis of* V;
  (b) $X_1^p u = X_2^p u = 0$, $X_1^{p-1}X_2^{p-1}u = -u$.

Proof. (1) Suppose $\dim V < p^2-1$. Then the family $(X_1^i X_2^j u)_{\substack{0\le i,j\le p-1\\ i+j\le 2p-3}}$ is linearly dependent. Applying X_1, X_2 several times and $\rho(D_{H,\omega}(x_1^{(2)}))$ (if necessary) we obtain a relation

$$\sum \alpha_{i,j} X_1^i X_2^j u = 0, \qquad \alpha_{p-1,p-2} \neq 0,\ \alpha_{p-2,p-1} = 0.$$

Since $\rho(D_{H,\omega}(x_1^{(a)}x_2^{(b)}))u = 0$ if $a, b \le p-1$, $a+b \ge 2$, we obtain

$$\begin{aligned} 0 &= \rho(D_{H,\omega}(x_1^{(p-1)}x_2^{(p-1)}))\sum \alpha_{i,j} X_1^i X_2^j u \\ &= -\alpha_{p-1,p-2}(\operatorname{ad} X_1)^{p-1}(\operatorname{ad} X_2)^{p-2}\big(\rho(D_{H,\omega}(x_1^{(p-1)}x_2^{(p-1)}))\big)u \\ &\in F^* X_1 u. \end{aligned}$$

Then Fu is annihilated by $FD_{H,\omega}(x_1) + H(2;\underline{1};\Phi(\tau))^{(1)}{}_{(0)}$. It is not hard to see that then V is a trivial $H(2;\underline{1};\Phi(\tau))^{(1)}$-module. Thus $\dim V = 1$. This proves (1).

(2) follows from the fact that $\sum_{0\le i,j\le p-1} FX_1^iX_2^ju$ is a $\mathfrak{g}$-submodule.

(3) The proof of (1) shows that in the present situation the family $(X_1^iX_2^ju)_{\substack{0\le i,j\le p-1\\ i+j\le 2p-3}}$ is a basis of V. Thus there is a relation

$$X_1^{p-1}X_2^{p-1}u = \sum_{0\le i+j\le 2p-3} \alpha_{i,j}X_1^iX_2^ju.$$

Considering eigenvectors with respect to $\rho(D_{H,\omega}(x_1x_2))$ we may assume that $\alpha_{i,j} = 0$ for $i \neq j$. Thus

$$X_1^{p-1}X_2^{p-1}u = \sum_{0\le i\le p-2} \alpha_{i,i}X_1^iX_2^iu.$$

Then (as $\rho(D_{H,\omega}(x_1^{(p-i)}))u = 0$ if $0 < i < p-1$)

$$\begin{aligned}\sum_{0\le i\le p-2} \alpha_{i,i}X_1^{i+1}X_2^iu &= X_1^pX_2^{p-1}u\\ &= -\rho(D_{H,\omega}(x_1^{(p)}))X_2^{p-1}u + \chi(D_{H,\omega}(x_1))^pX_2^{p-1}u\\ &= -X_2^{p-1}\rho(D_{H,\omega}(x_1^{(p)}))u - (\operatorname{ad}X_2)^{p-1}\big(\rho(D_{H,\omega}(x_1^{(p)}))\big)u\\ &\quad + \chi(D_{H,\omega}(x_1))^pX_2^{p-1}u\\ &= -X_2^{p-1}\chi(D_{H,\omega}(x_1^{(p)}))u - 2^{p-1}X_1u + \chi(D_{H,\omega}(x_1))^pX_2^{p-1}u\\ &= -X_1u + X_2^{p-1}\big(-\chi(D_{H,\omega}(x_1^{(p)})) + \chi(D_{H,\omega}(x_1))^p\big)u.\end{aligned}$$

Since the family in question is linearly independent, this shows that $\alpha_{i,i} = 0$ for $i \neq 0$, $\alpha_{0,0} = -1$ and $-\chi(D_{H,\omega}(x_1^{(p)})) + \chi(D_{H,\omega}(x_1))^p = 0$. The former implies $X_1^{p-1}X_2^{p-1}u = -u$ and the latter implies $X_1^pu = 0$.

Finally, we observe

$$\begin{aligned}X_2^pu &= D_{H,\omega}(x_2^{(p)})u + \chi(D_{H,\omega}(x_2))^pu \in Fu,\\ X_1^{p-1}X_2^pu &= X_2X_1^{p-1}X_2^{p-1}u - [X_2, X_1^{p-1}]X_2^{p-1}u\\ &\in FX_2u + \sum_{i=1}^{p-1} FX_1^{p-1-i}D_{H,\omega}(x_1^{(p-1)}x_2^{((p-1)-(i-1))})X_2^{p-1}u\\ &\subset FX_2u + \sum_{\substack{1\le i\le p-1\\ 0\le j\le p-1}} FX_1^{p-1-i}X_2^{p-1-j}D_{H,\omega}(x_1^{(p-1-j)}x_2^{((p-1)-(i-1))})u\\ &\subset FX_2u.\end{aligned}$$

Set $X_2^pu =: \lambda u$. The preceding computation gives $\lambda X_1^{p-1}u \in FX_2u$. The vectors $X_1^{p-1}u$, X_2u are linearly independent, and hence we conclude $\lambda = 0$, whence $X_2^pu = 0$. □

We are now able to determine irreducible modules.

Theorem 16.3.10. *Let V be a non-trivial irreducible $H(2;\underline{1};\Phi(\tau))^{(1)}$-module of dimension $\leq p^2$. Extend the represention to make V a module for the minimal p-envelope $\mathfrak{g} :=$ Der $H(2;\underline{1};\Phi(\tau))^{(1)}$ with a character χ. One has* dim $V \in \{p^2 - 1, p^2\}$.

(1) *If* dim $V = p^2 - 1$, *then* $V \cong H(2;\underline{1};\Phi(\tau))^{(1)}$.
(2) *Suppose* dim $V = p^2$. *Set* $\mathfrak{g}_{(0)} := \mathfrak{g} \cap W(2;\underline{1})_{(0)}$. *Then V has a 1-dimensional $\mathfrak{g}_{(0)}$-submodul Fu, such that*

$$V \cong u(\mathfrak{g},\chi) \otimes_{u(\mathfrak{g}_{(0)},\chi|\mathfrak{g}_{(0)})} Fu.$$

In this case $\chi \neq 0$ holds.

In both cases one has $\chi(H(2;\underline{1};\Phi(\tau))^{(1)}{}_{(0)}) = 0$.

Proof. Due to [S-F, Theorem 5.1.1] V can be turned into a module for the p-envelope $\mathfrak{g}$. Therefore there exists a character χ on $\mathfrak{g}$ for this module. Proposition 16.3.9 shows that dim $V \in \{1, p^2 - 1, p^2\}$.

If dim $V = 1$, then, since $H(2;\underline{1};\Phi(\tau))^{(1)}$ is a perfect algebra, the module is trivial. So this case cannot happen.

If dim $V = p^2 - 1$, part (3) of the proposition proves that the module is uniquely determined. Hence it has to coincide with the natural module $H(2;\underline{1};\Phi(\tau))^{(1)}$.

Suppose dim $V = p^2$. Lemma 16.3.8 yields the existence of a 1-dimensional $\mathfrak{g}_{(0)}$-module Fu. Then V is the homomorphic image of the induced module $u(\mathfrak{g},\chi) \otimes_{u(\mathfrak{g}_{(0)},\chi|\mathfrak{g}_{(0)})} Fu$. Since $\mathfrak{g}/\mathfrak{g}_{(0)}$ is 2-dimensional, we obtain by a dimension argument that this homomorphism is injective.

Suppose $\chi = 0$. Then $\mathfrak{g}_{(0)}$ annihilates Fu, and one has in addition

$$X_1^p u = -\rho(D_{H,\omega}(x_1^{(p)}))u = 0, \; X_2^p u = \rho(D_{H,\omega}(x_2^{(p)}))u = 0.$$

Therefore $\sum_{0\leq i,j\leq p-1,\, i+j\neq 0} X_1^i X_2^j u$ is a proper submodule, which is not possible.

In both cases there exists a 1-dimensional $\mathfrak{g}_{(0)}$-submodule $FD_{H,\omega}(x_1^{(p-1)}x_2^{(p-1)})$ and Fu, respectively. Then ${\mathfrak{g}_{(0)}}^{(1)}$ annihilates this module. Observe that ${\mathfrak{g}_{(0)}}^{(1)} = H(2;\underline{1};\Phi(\tau))^{(1)}{}_{(0)}$ is a restricted subalgebra of $\mathfrak{g}$. Let $y \in {\mathfrak{g}_{(0)}}^{(1)}$. Since $\rho(y)u = \rho(y^{[p]})u = 0$, one has $\chi(y)^p\mathrm{Id} = \rho(y)^p - \rho(y^{[p]}) = 0$. Therefore $\chi(y) = 0$ holds. □

16.3.3 Splitting off the radical

In this subsection we derive some results which will be applied later on. Therefore some rather specific assumptions occur.

Proposition 16.3.11. *Let $\mathfrak{g}$ be a central extension of $H(2;\underline{1};\Phi(\tau))^{(1)}$ and V a finite dimensional $\mathfrak{g}$-module. If $\mathfrak{g}^{(1)} \cap C(\mathfrak{g})$ acts non-nilpotently on V, then* dim $V > p^{(p-1)/2}$.

Proof. Suppose the statement is not true. Decompose $\mathfrak{g} = H(2;\underline{1};\Phi(\tau))^{(1)} \oplus C$ as a vector space, where C is central. Considering composition factors, we may assume that V is a faithful irreducible module and $C = Fw$ is 1-dimensional. By assumption the central extension is non-split. Theorem 16.3.3(2) shows that we may take

$$[D_{H,\omega}(x_2^{(b)}), D_{H,\omega}(x_2^{(p-b)})] = (-1)^{b-1}w, \quad 0 \le b \le p-1.$$

Therefore

$$\sum_{1\le i\le p-1} FD_{H,\omega}(x_2^{(i)}) + Fw$$

is a p-dimensional Heisenberg algebra. The representation theory of this algebra (Volume 1, p. 149) shows that $\dim V \ge p^{(p-1)/2}$, whence equality holds. Moreover,

$$\operatorname{End} V = \operatorname{span}\{\prod_{j=1}^{p-1} \rho(D_{H,\omega}(x_2^{(j)}))^{k_j} \mid 0 \le k_j \le p-1\}$$

holds. In particular,

$$\rho(D_{H,\omega}(x_1^{(p-1)}x_2^{(p-1)})) = \sum \alpha_k \prod_{j=1}^{p-1} \rho(D_{H,\omega}(x_2^{(j)}))^{k_j}$$

for some scalars $\alpha_k = \alpha_{(k_1,\dots,k_{p-1})} \in F$. We conclude from Theorem 16.3.3 that $\rho(D_{H,\omega}(x_1^{(p-1)}x_2^{(p-1)}))$ commutes with all $\rho(D_{H,\omega}(x_2^{(r)}))$, $r \ge 2$. This gives $\alpha_k = 0$ except if $k = (0,\dots,0,k_{p-1})$. Consequently,

$$\rho(D_{H,\omega}(x_1^{(p-1)}x_2^{(p-1)})) = \sum_{l=0}^{p-1} \beta_l \rho(D_{H,\omega}(x_2^{(p-1)}))^l.$$

Next we conclude from Theorem 16.3.3 that $\rho(D_{H,\omega}(x_1^{(p-1)}x_2^{(p-1)}))$ commutes with $\rho(D_{H,\omega}(x_1x_2))$, and this gives $\beta_l = 0$ if $l \ne 0$. As a result, $\rho(D_{H,\omega}(x_1^{(p-1)}x_2^{(p-1)})) \in F\mathrm{Id}_V$ centralizes the whole algebra. This contradiction proves the claim. □

Let us refer to the notation

$$\gamma_1(\mathfrak{g}) = \mathfrak{g}, \qquad \gamma_{n+1}(\mathfrak{g}) = [\mathfrak{g}, \gamma_n(\mathfrak{g})],\ n \ge 1$$

introduced in Volume 2, p. 34.

Proposition 16.3.12. *Let $\mathfrak{g}$ be a restricted Lie algebra. Suppose $TR(\mathfrak{g}) = 2$, $\mathfrak{g}/\mathrm{rad}\,\mathfrak{g} \cong \mathrm{Der}\, H(2;\underline{1};\Phi(\tau))^{(1)}$, and $\mathrm{rad}\,\mathfrak{g} \ne C(\mathfrak{g})$. Let V be a faithful irreducible $\mathfrak{g}$-module of dimension $\le 2p^2$. The following holds.*

(1) *$V \cong u(\mathfrak{g},\chi) \otimes_{u(\mathfrak{g}_{(0)},\chi|\mathfrak{g}_{(0)})} V_0$, where χ is the character of the representation and $\mathfrak{g}_{(0)}$ is the inverse image of $\big(\mathrm{Der}\, H(2;\underline{1};\Phi(\tau))^{(1)}\big) \cap W(2;\underline{1})_{(0)}$ in $\mathfrak{g}$.*

(2) $\operatorname{rad}\mathfrak{g}$ *is abelian.*

(3) $\operatorname{rad}\mathfrak{g} \not\subset \mathfrak{g}_{(0)}{}^{(1)}$.

Proof. (1) We identify $\mathfrak{g}$ and its image in $\mathfrak{gl}(V)$. Set $\mathfrak{r} := \operatorname{rad}\mathfrak{g}$. Theorem 1.2.8(3) shows that $\mathfrak{r}$ is nilpotent. Then there is $n \geq 0$ such that $\gamma_{n+1}(\mathfrak{r}) \subset F\mathrm{Id}_V$. Choose n minimal with this condition. If $n = 0$, then $\mathfrak{r} \subset C(\mathfrak{g})$, which case we have excluded. Therefore $n > 0$. Choose an ideal A of $\mathfrak{g}$ minimal subject to the conditions

$$\mathfrak{g} \cap F\mathrm{Id}_V \subsetneq A \subset \mathfrak{r}, \quad A^{(1)} \subset F\mathrm{Id}_V.$$

Suppose $A^{(1)} \neq \{0\}$. Then $A^{(1)} = F\mathrm{Id}_V$. There is a pairing

$$f : A \times A \to F, \quad [a,b] = f(a,b)\mathrm{Id}_V \quad \forall a,b \in A.$$

Note that f is $\mathfrak{g}$-invariant. As the radical of f is $\mathfrak{g}$-invariant as well, the minimality of $A/F\mathrm{Id}_V$ implies $\operatorname{rad} f = F\mathrm{Id}_V$. Then A is a Heisenberg algebra acting faithfully on V. As $\dim V < p^3$, one has $\dim A/F\mathrm{Id}_V = 2s$ with $s \leq 2$ (Volume 1, p. 149). The nilpotency of $\mathfrak{r}$ in combination with the minimality of $A/F\mathrm{Id}_V$ and ENGEL's Theorem gives $[\mathfrak{r}, A] \subset F\mathrm{Id}_V$, and therefore $A/F\mathrm{Id}_V$ is a restricted $\mathfrak{g}/\mathfrak{r}$-module of dimension at most 4. A dimension argument proves $(\mathfrak{g}/\mathfrak{r}) \cdot (A/F\mathrm{Id}_V) = \{0\}$. The minimality of A shows that $A/F\mathrm{Id}_V$ is 1-dimensional. Hence $A^{(1)} = \{0\}$.

As V is $\mathfrak{g}$-irreducible, there is a linear function $\lambda \in A^*$ such that $x - \lambda(x)\mathrm{Id}_V$ is nilpotent for any $x \in A$. Let

$$V_0 := \{v \in V \mid y(v) = \lambda(y)v \quad \text{for all } y \in A\}$$

and

$$\mathfrak{g}^\lambda := \{x \in \mathfrak{g} \mid \lambda([x,A]) = 0\}.$$

Obviously $\mathfrak{g}^\lambda$ is a restricted subalgebra of $\mathfrak{g}$, and it is easy to see that $\mathfrak{g}^\lambda = \{x \in \mathfrak{g} \mid x(V_0) \subset V_0\}$. By Corollary 3.2.5 (setting $B = U(A)$), V_0 is $\mathfrak{g}^\lambda$-irreducible and

$$V \cong u(\mathfrak{g},\chi) \otimes_{u(\mathfrak{g}^\lambda, \chi|u(\mathfrak{g}^\lambda))} V_0$$

as $\mathfrak{g}$-modules. If $\mathfrak{g} = \mathfrak{g}^\lambda$, then the ideal $[\mathfrak{g}, A]$ acts nilpotently on V and hence annihilates this irreducible module. We obtain the contradiction $A \subset F\mathrm{Id}_V$. Therefore $\mathfrak{g} \neq \mathfrak{g}^\lambda$. Also,

$$\dim V = p^{\dim \mathfrak{g}/\mathfrak{g}^\lambda} \dim V_0.$$

As $\dim V \leq 2p^2$, one gets $\dim V_0 \leq 2p$ and $0 \neq \dim \mathfrak{g}/\mathfrak{g}^\lambda \leq 2$.

Suppose $\mathfrak{r} \subset \mathfrak{g}^\lambda$. Then $\mathfrak{g}^\lambda$ maps onto a restricted subalgebra of $\operatorname{Der} H(2;\underline{1};\Phi(\tau))^{(1)}$ of codimension 2 at most. The only such subalgebra is $\big(\operatorname{Der} H(2;\underline{1};\Phi(\tau))^{(1)}\big) \cap W(2;\underline{1})_{(0)}$. We then obtain $\dim V_0 \leq 2$. This is claim (1) if $\mathfrak{r} \subset \mathfrak{g}^\lambda$.

Suppose $\mathfrak{r} \not\subset \mathfrak{g}^\lambda$. Then $\mathfrak{r} + \mathfrak{g}^\lambda$ is a subalgebra of codimension at most 1, and the preceding deliberations show that even more $\mathfrak{r} + \mathfrak{g}^\lambda = \mathfrak{g}$ is true. We observe that

$\dim \mathfrak{r}/\mathfrak{r}\cap\mathfrak{g}^\lambda \le \dim \mathfrak{g}/\mathfrak{g}^\lambda \le 2$. Since $\mathfrak{r}$ is nilpotent, $[\mathfrak{r}, A] \ne A$ whence $[\mathfrak{r}, A] \subset F\mathrm{Id}_V$ and $[\mathfrak{r}^{(1)}, A] = \{0\}$. This gives $\mathfrak{r}^{(1)} \subset \mathfrak{g}^\lambda$. Consequently

$$\mathrm{Der}\, H(2;\underline{1};\Phi(\tau))^{(1)} \cong \mathfrak{g}/\mathfrak{r} \cong \mathfrak{g}^\lambda/\mathfrak{g}^\lambda \cap \mathfrak{r}$$

acts restrictedly on $\mathfrak{r}/\mathfrak{g}^\lambda\cap\mathfrak{r}$. As the latter is at most 2-dimensional, this action is trivial. Hence $[\mathfrak{g}, \mathfrak{r}] \subset \mathfrak{g}^\lambda \cap \mathfrak{r}$. Then

$$[\mathfrak{g}, \mathfrak{g}^\lambda] = [\mathfrak{r} + \mathfrak{g}^\lambda, \mathfrak{g}^\lambda] \subset \mathfrak{g}^\lambda,$$

whence $\mathfrak{g}^\lambda$ is an ideal of $\mathfrak{g}$. This ideal acts on V_0, which is at most $2p$-dimensional. Recall that $\mathfrak{g}^\lambda/\mathfrak{g}^\lambda \cap \mathfrak{r} \cong \mathfrak{g}/\mathfrak{r} \cong \mathrm{Der}\, H(2;\underline{1};\Phi(\tau))^{(1)}$.

Let $\rho_0 : \mathfrak{g}^\lambda \to \mathfrak{gl}(V_0)$ denote the representation, set $\mathfrak{g}' := \rho_0(\mathfrak{g}^\lambda)$, $\mathfrak{r}' := \rho_0(\mathfrak{g}^\lambda \cap \mathfrak{r})$. Note that

$$[\mathfrak{g}, \ker \rho_0] = [\mathfrak{r} + \mathfrak{g}^\lambda, \ker \rho_0] \subset [\mathfrak{r}, \mathfrak{g}^\lambda] + [\mathfrak{g}^\lambda, \ker \rho_0] \subset \mathfrak{g}^\lambda \cap \mathfrak{r} + (\ker \rho_0).$$

As a consequence, $\mathfrak{g}^\lambda \cap \mathfrak{r} + (\ker \rho_0)$ is an ideal of $\mathfrak{g}$. So we have $\mathfrak{g}^\lambda \cap \mathfrak{r} + (\ker \rho_0) = \mathfrak{g}^\lambda \cap \mathfrak{r}$ or $\mathfrak{g}^\lambda \cap \mathfrak{r} + (\ker \rho_0) = \mathfrak{g}^\lambda$.

If $\rho_0(\mathfrak{g}^\lambda)$ is solvable, then $(\mathfrak{g}^\lambda)^{(\infty)}$ annihilates V_0. But then $(\mathfrak{g}^\lambda)^{(\infty)}$ annihilates V (as it is an ideal), and therefore V is not faithful. So $\ker \rho_0 \subset \mathfrak{g}^\lambda \cap \mathfrak{r}$ holds, whence

$$\mathfrak{g}'/\mathfrak{r}' \cong \mathfrak{g}^\lambda/\mathfrak{g}^\lambda \cap \mathfrak{r} \cong \mathfrak{g}/\mathfrak{r} \cong \mathrm{Der}\, H(2;\underline{1};\Phi(\tau))^{(1)}.$$

Suppose $\mathfrak{r}'$ is central in $\mathfrak{g}'$. Proposition 16.3.11 yields $\mathfrak{g}'^{(1)} \cap \mathfrak{r}' = \{0\}$. Then $\mathfrak{g}'^{(1)} \cong H(2;\underline{1};\Phi(\tau))^{(1)}$ has a faithful modul of dimension $\le 2p$, which contradicts Theorem 16.3.10. So $\mathfrak{r}'$ is not central in $\mathfrak{g}'$. All the assumptions of this proposition are fulfilled. Inductively we obtain a contradiction. This contradiction proves (1).

(2), (3) As a result of (1), $\mathrm{rad}\, \mathfrak{g} \subset \mathfrak{g}_{(0)}$ and $\dim V_0 \le 2$ hold. Note that every element of the nilpotent ideal $\mathrm{rad}\, \mathfrak{g}$ has a unique eigenvalue on V. Computing traces, we obtain that every element of $(\mathrm{rad}\, \mathfrak{g}) \cap {\mathfrak{g}_{(0)}}^{(1)}$ has eigenvalue 0 on V_0. This means that $(\mathrm{rad}\, \mathfrak{g}) \cap {\mathfrak{g}_{(0)}}^{(1)}$ acts nilpotently on V. As $(\mathrm{rad}\, \mathfrak{g})^{(1)}$ is an ideal, it therefore annihilates V. The faithfulness of the representation gives claim (2). Similarly, as $\mathrm{rad}\, \mathfrak{g}$ is nonzero, it is not contained in ${\mathfrak{g}_{(0)}}^{(1)}$. This is claim (3). □

Proposition 16.3.13. *Let $\mathfrak{g}$ be a restricted Lie algebra with $TR(\mathfrak{g}) = 2$ such that*

(a) $\mathrm{rad}\, \mathfrak{g}$ *is abelian and* $\mathfrak{g}/\mathrm{rad}\, \mathfrak{g} \cong \mathrm{Der}\, H(2;\underline{1};\Phi(\tau))^{(1)}$*;*
(b) $\mathrm{rad}\, \mathfrak{g} \cong H(2;\underline{1};\Phi(\tau))^{(1)}$ *as* $(\mathfrak{g}/\mathrm{rad}\, \mathfrak{g})$*-modules.*

Let $\mathfrak{t}$ be a 2-dimensional torus in $\mathfrak{g}$, let $\mathfrak{g}_{(0)}$ denote the preimage of

$$\big(\mathrm{Der}\, H(2;\underline{1};\Phi(\tau))^{(1)}\big) \cap W(2;\underline{1})_{(0)}$$

in $\mathfrak{g}$, and suppose that

$$\mathrm{rad}\, \mathfrak{g} \not\subset {\mathfrak{g}_{(0)}}^{(1)}.$$

Then there exists a restricted subalgebra $\mathcal{M}$ *in* $\mathfrak{g}$ *with* $\mathfrak{t} \subset \mathcal{M}$ *and* $\mathfrak{g} = \mathcal{M} \oplus \operatorname{rad}\mathfrak{g}$. *In particular,* $\mathfrak{g}$ *is a split extension.*

Proof. (a) Since the standard maximal subalgebra of $\operatorname{Der} H(2;\underline{1};\Phi(\tau))^{(1)}$ is restricted, so is $\mathfrak{g}_{(0)}$. It is immediate from assumption (b) that the Lie algebra $\mathfrak{g}$ is centerless. By definition one has

$$\begin{aligned} \mathfrak{g}_{(0)} &= \big\{x \in \mathfrak{g}_{(0)} \mid x + \operatorname{rad}\mathfrak{g} \in \big(\operatorname{Der} H(2;\underline{1};\Phi(\tau))^{(1)}\big) \cap W(2;\underline{1})_{(0)}\big\},\\ \operatorname{rad}\mathfrak{g}_{(0)} &= \big\{x \in \mathfrak{g}_{(0)} \mid x + \operatorname{rad}\mathfrak{g} \in \big(\operatorname{Der} H(2;\underline{1};\Phi(\tau))^{(1)}\big) \cap W(2;\underline{1})_{(1)}\big\},\\ \mathfrak{g}_{(0)}/\operatorname{rad}\mathfrak{g}_{(0)} &\cong \mathfrak{sl}(2). \end{aligned}$$

Given $x \in \operatorname{rad}\mathfrak{g}_{(0)}$, one has $x^{[p]^e} \in \operatorname{rad}\mathfrak{g}$ for $e \gg 0$ forcing $x^{[p]^{e+1}} \in C(\mathfrak{g}) = \{0\}$. Thus $\operatorname{rad}\mathfrak{g}_{(0)}$ is $[p]$-nilpotent.

(b) By our assumption, $\operatorname{rad}\mathfrak{g} \not\subset [\mathfrak{g}_{(0)}, \mathfrak{g}_{(0)}]$. So there exist a subspace V in $\mathfrak{g}_{(0)}$ and a nonzero $c \in \operatorname{rad}\mathfrak{g}$ such that $[\mathfrak{g}_{(0)}, \mathfrak{g}_{(0)}] \subset V$ and $\mathfrak{g}_{(0)} = V \oplus Fc$. Clearly, V is an ideal of $\mathfrak{g}_{(0)}$. Let λ denote the linear function on $\mathfrak{g}_{(0)}$ with $\ker\lambda = V$ and $\lambda(c) = 1$ and let F_λ be the respective 1-dimensional representation of $\mathfrak{g}_{(0)}$. Define a linear function μ on $\mathfrak{g}$ such that

$$\mu(x)^p = \lambda(x)^p - \lambda(x^{[p]}) \qquad \forall x \in \mathfrak{g}_{(0)}.$$

Set

$$M := u(\mathfrak{g},\mu) \otimes_{u(\mathfrak{g}_{(0)},\mu|\mathfrak{g}_{(0)})} F_\lambda,$$

a p^2-dimensional induced $\mathfrak{g}$-module with p-character μ. Let M' be a composition factor of M. As $c^{[p]} \in C(\mathfrak{g}) = \{0\}$, the only eigenvalue of c on M' equals $\mu(c) = \lambda(c) = 1$. As $\operatorname{rad}\mathfrak{g}$ is the unique minimal ideal of $\mathfrak{g}$, we now deduce that M' is faithful. Proposition 16.3.12 then shows that $\dim M' = p^2 = \dim M$, i.e., $M = M'$ is irreducible. Theorem 3.1.1 in combination with Corollary 3.3.7 yields a simultaneous realization

$$\begin{aligned} &M \cong F_{\tilde\lambda} \otimes \mathcal{O}(2;\underline{1}),\\ &\sigma : \mathfrak{g} \hookrightarrow \big(\operatorname{Id}_{F_{\tilde\lambda}} \otimes \mathcal{O}(2;\underline{1})\big) \rtimes \big(\operatorname{Id}_{F_{\tilde\lambda}} \otimes W(2;\underline{1})\big), \end{aligned}$$

where $\tilde\lambda$ is obtained by λ and the respective Frobenius twist, σ is a Lie algebra homomorphism, and the projection π_2 into the second summand gives a transitive subalgebra of $W(2;\underline{1})$. Observe that $(\operatorname{Id}_{F_{\tilde\lambda}} \otimes \mathcal{O}(2;\underline{1})) \rtimes (\operatorname{Id}_{F_{\tilde\lambda}} \otimes W(2;\underline{1}))$ is restrictable. Let $u \mapsto u^p$ denote a p-mapping on this algebra. The homomorphism σ is not necessarily a restricted homomorphism. However, $\sigma(u)^p - \sigma(u^{[p]})$ centralizes $\sigma(\mathfrak{g})$ and hence acts on the irreducible module as a multiple of the identity,

$$\sigma(u)^p - \sigma(u^{[p]}) \in F\operatorname{Id}_{F_{\tilde\lambda}} \otimes 1 \qquad \forall u \in \mathfrak{g}.$$

Therefore $\mathfrak{t}' := \sigma(\mathfrak{t}) + F\,\mathrm{Id}_{F_{\tilde\lambda}} \otimes 1$ is a torus. Theorem 3.6.1 gives rise to an automorphism $\tau \in \mathrm{Aut}\,(\mathrm{Id}_{F_{\tilde\lambda}} \otimes \mathcal{O}(2;\underline{1}))$ such that $\tau \circ \mathfrak{t}' \circ \tau^{-1} \subset (\mathrm{Id}_{F_{\tilde\lambda}} \otimes F) \rtimes (\mathrm{Id}_{F_{\tilde\lambda}} \otimes W(2;\underline{1}))$. Observe that

$$\mathrm{Id}_{F_{\tilde\lambda}} \otimes W(2;\underline{1}) = \mathrm{Der}\left(F_{\tilde\lambda} \otimes \mathcal{O}(2;\underline{1})\right) = \tau \circ \left(\mathrm{Der}\left(F_{\tilde\lambda} \otimes \mathcal{O}(2;\underline{1})\right)\right) \circ \tau^{-1}.$$

Suppressing the notion of τ we may assume that the above inclusion gives an inclusion

$$\sigma : \mathfrak{t} \hookrightarrow \left(\mathrm{Id}_{F_{\tilde\lambda}} \otimes F\right) \rtimes \left(\mathrm{Id}_{F_{\tilde\lambda}} \otimes W(2;\underline{1})\right).$$

(c) Note that $\mathrm{rad}\,\mathfrak{g}$ is an abelian ideal and $\pi_2(\sigma(\mathfrak{g}))$ is a transitive subalgebra of $W(2;\underline{1})$. Then it can only be that $\pi_2(\sigma(\mathrm{rad}\,\mathfrak{g})) = \{0\}$. Since $C(\mathfrak{g}) = \{0\}$, we have $\sigma(\mathrm{rad}\,\mathfrak{g}) \cap ((\mathrm{Id}_{F_{\tilde\lambda}} \otimes F) \rtimes (\mathrm{Id}_{F_{\tilde\lambda}} \otimes W(2;\underline{1}))) = \{0\}$. Since $\mathrm{rad}\,\mathfrak{g}$ has dimension $p^2 - 1$, one now gets

$$\left(\mathrm{Id}_{F_{\tilde\lambda}} \otimes \mathcal{O}(2;\underline{1})\right) \rtimes \left(\mathrm{Id}_{F_{\tilde\lambda}} \otimes W(2;\underline{1})\right) = \sigma(\mathrm{rad}\,\mathfrak{g}) \oplus \left(\left(\mathrm{Id}_{F_{\tilde\lambda}} \otimes F\right) \rtimes \left(\mathrm{Id}_{F_{\tilde\lambda}} \otimes W(2;\underline{1})\right)\right).$$

Set $\mathcal{M} := \sigma^{-1}\left(\sigma(\mathfrak{g}) \cap \left((\mathrm{Id}_{F_{\tilde\lambda}} \otimes F) \rtimes (\mathrm{Id}_{F_{\tilde\lambda}} \otimes W(2;\underline{1}))\right)\right)$. By construction, $\mathfrak{g} = (\mathrm{rad}\,\mathfrak{g}) \oplus \mathcal{M}$ and $\mathfrak{t} \subset \mathcal{M}$. If $u \in \mathcal{M}$, then $\sigma(u)^p \in (\mathrm{Id}_{F_{\tilde\lambda}} \otimes F) \rtimes (\mathrm{Id}_{F_{\tilde\lambda}} \otimes W(2;\underline{1}))$. But then $\sigma(u^{[p]}) \in (\mathrm{Id}_{F_{\tilde\lambda}} \otimes F) \rtimes (\mathrm{Id}_{F_{\tilde\lambda}} \otimes W(2;\underline{1}))$ as well. Thus $\mathcal{M}$ is a restricted subalgebra of $\mathfrak{g}$ and our proof is complete. □

We conclude two results needed in later chapters.

Theorem 16.3.14. *Let $\mathfrak{g}$ be a restricted Lie algebra and $\mathfrak{t}$ a torus of maximal dimension in $\mathfrak{g}$. Suppose $\mathfrak{g}/\mathrm{rad}\,\mathfrak{g} \cong \mathrm{Der}\,H(2;\underline{1};\Phi(\tau))^{(1)}$, $TR(\mathfrak{g}) = 2$, $\mathrm{rad}\,\mathfrak{g} \neq C(\mathfrak{g})$, $\dim \mathrm{rad}\,\mathfrak{g} = p^2 - 1$. Let V be a faithful irreducible $\mathfrak{g}$-module of dimension $\leq 2p^2$. The following holds.*

(1) *$V \cong u(\mathfrak{g},\chi) \otimes_{u(\mathfrak{g}_{(0)},\chi|\mathfrak{g}_{(0)})} V_0$, where χ is the character of the representation and $\mathfrak{g}_{(0)}$ is the inverse image of $\left(\mathrm{Der}\,H(2;\underline{1};\Phi(\tau))^{(1)}\right) \cap W(2;\underline{1})_{(0)}$ in $\mathfrak{g}$.*

(2) *${\mathfrak{g}_{(0)}}^{(1)}$ is a restricted subalgebra of $\mathfrak{g}$ and $\chi({\mathfrak{g}_{(0)}}^{(1)}) = 0$ holds.*

(3) *There exists a subalgebra $\mathcal{M} \cong \mathrm{Der}\,H(2;\underline{1};\Phi(\tau))^{(1)}$ such that $\mathfrak{g} = \mathcal{M} \oplus \mathrm{rad}\,\mathfrak{g}$ and $\mathfrak{t} \subset \mathcal{M}$.*

Proof. Proposition 16.3.12 proves the statement on V.

It also shows that $\mathrm{rad}\,\mathfrak{g}$ is abelian. Let U be a composition factor of the $(\mathfrak{g}/\mathrm{rad}\,\mathfrak{g})$-module $\mathrm{rad}\,\mathfrak{g}$. If there is more than one factor, then all the factors have dimension less than $p^2 - 1$. Theorem 16.3.10 proves that all factors are trivial modules. In this case $\mathrm{rad}\,\mathfrak{g}$ is annihilated by $\mathfrak{g}^{(\infty)} + \mathrm{rad}\,\mathfrak{g}$. Note that $(\mathfrak{g}^{(\infty)} + \mathrm{rad}\,\mathfrak{g})/\mathrm{rad}\,\mathfrak{g} \cong H(2;\underline{1};\Phi(\tau))^{(1)}$. But then $\mathrm{rad}\,\mathfrak{g}$ is annihilated by the p-envelope of $\mathfrak{g}^{(\infty)} + \mathrm{rad}\,\mathfrak{g}$ which is $\mathfrak{g}$, whence the contradiction $\operatorname{rad}\mathfrak{g} = C(\mathfrak{g})$. Hence $\operatorname{rad}\mathfrak{g}$ is an irreducible $\mathfrak{g}/\operatorname{rad}\mathfrak{g}$-module. Due Theorem 16.3.10 it is isomorphic, as a module, to $H(2;\underline{1};\Phi(\tau))^{(1)}$. Proposition 16.3.12(2), (3) shows that Proposition 16.3.13 is applicable. Hence there is a decomposition

$$\mathfrak{g} = \mathcal{M} \oplus \operatorname{rad}\mathfrak{g}, \quad \mathcal{M} \cong \operatorname{Der} H(2;\underline{1};\Phi(\tau))^{(1)}.$$

It remains to prove assertion (2). Since $\mathfrak{g}$ is centerless, there exists only one mapping $[p]$ which makes $\mathfrak{g}$ restricted. This mapping then is the canonicial mapping on $\mathcal{M}$ and the trivial mapping on $\operatorname{rad}\mathfrak{g}$. In particular $\mathcal{M}_{(0)}{}^{(1)} \cong H(2;\underline{1};\Phi(\tau))^{(1)}{}_{(0)}$ is closed under the p-mapping. Then $\mathfrak{g}_{(0)}{}^{(1)} = \mathcal{M}_{(0)}{}^{(1)} \oplus (\mathfrak{g}_{(0)}{}^{(1)} \cap \operatorname{rad}\mathfrak{g})$ is a restricted subalgebra.

Let $x \in \mathfrak{g}_{(0)}{}^{(1)} \cap \operatorname{rad}\mathfrak{g}$. Since $x^{[p]} = 0$, then $\rho(x)^p = \chi(x)^p\operatorname{Id}$ (where ρ denotes the representation $\mathfrak{g} \to \mathfrak{gl}(V)$). Since $x \in \mathfrak{g}_{(0)}{}^{(1)}$, one also has $0 = \operatorname{trace}(\rho(x)|V_0) = \chi(x)\dim V_0$. As $\dim V_0 \leq 2$, this gives $\chi(x) = 0$.

Next observe that $\operatorname{rad}(\mathfrak{g}_{(0)}{}^{(1)} \cap \mathcal{M}) = \operatorname{rad}\mathcal{M}_{(0)}{}^{(1)} \cong H(2;\underline{1};\Phi(\tau))^{(1)}{}_{(1)}$ is a nilpotent and restricted subalgebra. One argues as before in order to obtain that every $x \in \operatorname{rad}\mathcal{M}_{(0)}{}^{(1)}$ is $[p]$-nilpotent, and hence we obtain inductively

$$\rho(x)^{p^e} = \sum_{i=0}^{e} \chi(x^{[p]^i})^{p^{e-i}}\operatorname{Id} = \chi(x)^{p^e}\operatorname{Id}, \quad e \gg 0,$$

and then $\chi(x) = 0$. As a result, $\operatorname{rad}\mathfrak{g}_{(0)}{}^{(1)}$ acts nilpotently on V_0. Let $V_0' \subset V_0$ be an irreducible $\mathfrak{g}_{(0)}{}^{(1)}$-module. Then $\operatorname{rad}\mathfrak{g}_{(0)}{}^{(1)}$ annihilates V_0', whence V_0' is irreducible as an $\mathfrak{sl}(2)$-module. As $\dim V_0'$ is at most 2-dimensional, it is a restricted module. Hence χ vanishes on all of $\mathfrak{g}_{(0)}{}^{(1)}$. □

Theorem 16.3.15. *Suppose $\mathcal{L}$ is a Lie algebra with $TR(\mathcal{L}) = 2$ and*

$$H(2;\underline{1};\Phi(\tau))^{(1)} \subset \mathcal{L}/\operatorname{rad}\mathcal{L} \subset H(2;\underline{1};\Phi(\tau))^{(1)}{}_{[p]}$$

holds. Let $\rho : \mathcal{L} \to V$ be an irreducible representation of dimension p^2 and let $\mathfrak{g}$ denote the p-envelope of $\rho(\mathcal{L})$ in $\mathfrak{gl}(V)$. Then there is a simultaneous realization

$$V \cong F \otimes \mathcal{O}(2;\underline{1}),$$
$$\mathfrak{g} \hookrightarrow \big(\operatorname{Id} \otimes \mathcal{O}(2;\underline{1})\big) \rtimes \big(\operatorname{Id} \otimes W(2;\underline{1})\big),$$

where the projection π_2 into the second summand gives a transitive subalgebra of $W(2;\underline{1})$.

Proof. If $\mathfrak{g}$ is solvable, then Corollary 3.3.9 applies and proves the claim. So assume in the following that $\mathfrak{g}$ is not solvable. This assumption implies that $\ker\rho$ is solvable and

$$\rho(\mathcal{L})/\operatorname{rad}\rho(\mathcal{L}) \cong \mathcal{L}/\operatorname{rad}\mathcal{L}.$$

Therefore we may suppress the notion of ρ.

Suppose $\operatorname{rad} \mathfrak{g} = C(\mathfrak{g})$. Then $\operatorname{rad} \mathcal{L} \subset (\operatorname{rad} \mathfrak{g}) \cap \mathcal{L} = C(\mathcal{L})$ as well. Proposition 16.3.11 shows that the central extension splits:

$$\mathcal{L} \cong H(2; \underline{1}; \Phi(\tau))^{(1)} \oplus C(\mathcal{L}).$$

Let $\mathcal{L}_{[p]}$ denote a minimal p-envelope of $\mathcal{L}$. Since both $(\operatorname{Der} H(2; \underline{1}; \Phi(\tau))^{(1)}) \oplus C(\mathcal{L})$ and $\mathcal{L}_{[p]}$ are restrictable, contain $\mathcal{L}$, and have equal centers $C(\mathcal{L})$, it is clear by Theorem 1.1.6(2) that

$$\mathcal{L}_{[p]} \cong \left(\operatorname{Der} H(2; \underline{1}; \Phi(\tau))^{(1)}\right) \oplus C(\mathcal{L})$$

as algebras. Since $\mathfrak{g}$ is a p-envelope of $\mathcal{L}$, one obtains (Theorem 1.1.6(2))

$$\mathfrak{g} \cong \left(\operatorname{Der} H(2; \underline{1}; \Phi(\tau))^{(1)}\right) \oplus C(\mathfrak{g}).$$

Theorem 16.3.10 applies. Set $\mathfrak{g}_{(0)} := \left(\left(\operatorname{Der} H(2; \underline{1}; \Phi(\tau))^{(1)}\right) \cap W(2; \underline{1})_{(0)}\right) \oplus C(\mathfrak{g})$. Then V has a 1-dimensional $\mathfrak{g}_{(0)}$-submodul Fu, such that

$$V \cong u(\mathfrak{g}, \chi) \otimes_{u(\mathfrak{g}_{(0)}, \chi|\mathfrak{g}_{(0)})} Fu.$$

Suppose $\operatorname{rad} \mathfrak{g} \neq C(\mathfrak{g})$. Since $\mathfrak{g}/\operatorname{rad} \mathfrak{g}$ is a p-envelope of $\mathcal{L}/\operatorname{rad} \mathcal{L}$ (Theorem 1.1.6(4)), one has $\mathfrak{g}/\operatorname{rad} \mathfrak{g} \cong H(2; \underline{1}; \Phi(\tau))^{(1)}{}_{[p]} \cong \operatorname{Der} H(2; \underline{1}; \Phi(\tau))^{(1)}$. Proposition 16.3.12 applies and shows that V is induced by a $\mathfrak{g}_{(0)}$-submodule V_0. As $\dim V = p^2$ and $\dim \mathfrak{g}/\mathfrak{g}_{(0)} = 2$, one has $\dim V_0 = 1$.

In both cases, Theorem 3.1.1, in combination with Corollary 3.3.7, yields a simultaneous realization

$$V \cong F \otimes \mathcal{O}(2; \underline{1}),$$
$$\mathfrak{g} \hookrightarrow \left(\operatorname{Id} \otimes \mathcal{O}(2; \underline{1})\right) \rtimes \left(\operatorname{Id} \otimes W(2; \underline{1})\right)$$

where the projection π_2 into the second summand gives a transitive subalgebra of $W(2; \underline{1})$. This is the claim. □

16.4 Some properties of Melikian algebras

In this section we will first show that all Melikian algebras $\mathcal{M}(\underline{n})$ carry a non-degenerate invariant symmetric bilinear form. As a consequence, central extensions of the restricted Melikian algebra $\mathcal{M}(1, 1)$ split. Secondly, we will derive information on the p-characters of irreducible representations of dimension ≤ 125 of $\mathcal{M}(1, 1)$. These results are subsidiary for Chapter 18.

Melikian algebras allow two different gradings. The *natural grading* is described in Volume 1, p. 200,

$$\mathcal{M}(\underline{n}) = \oplus_{i=-3}^{s} \mathcal{M}(\underline{n})_{[i]}.$$

Here $\mathcal{M}(\underline{n})_{[-3]} = F\partial_1 \oplus F\partial_2$, $\mathcal{M}(\underline{n})_{[0]} = \bigoplus_{i,j=1}^{2} Fx_i\partial_j \cong \mathfrak{gl}(2)$, and $\mathcal{M}(\underline{n})_{[s]} = Fx^{(\tau(\underline{n}))}\tilde{\partial}_1 \oplus Fx^{(\tau(\underline{n}))}\tilde{\partial}_2$, where $\tau(\underline{n}) = (5^{n_1}-1, 5^{n_2}-1)$ and $s = 3(5^{n_1}+5^{n_2})-7$. The *natural filtration* is given by $\mathcal{M}(\underline{n})_{(k)} := \sum_{i\geq k} \mathcal{M}(\underline{n})_{[i]}$.

The *Melikian grading* and *Melikian filtration* of depth 2 is described in Volume 2, p. 71. We have

$$\mathcal{M}(\underline{n})_{\langle k\rangle} := \operatorname{span}\{x^{(a)}\partial_j, x^{(b)}\tilde{\partial}_j, x^{(c)} \mid a_1+2a_2-1-\delta_{j,2} = b_1+2b_2-\delta_{j,2} = c_1+2c_2-1 = k\},$$

$$\mathcal{M}(\underline{n})_{\{k\}} := \sum_{i\geq k} \mathcal{M}(\underline{n})_{\langle i\rangle}.$$

Theorem 16.4.1. *Every Melikian algebra $\mathcal{M}(\underline{n})$ possesses a non-degenerate invariant symmetric bilinear form.*

Proof. (a) Abbreviate $\mathcal{M} := \mathcal{M}(\underline{n})$. The universal property of induced modules yields an $\mathcal{M}$-module homomorphism $\theta : U(\mathcal{M}) \otimes_{U(\mathcal{M}_{(0)})} \mathcal{M}_{[s]} \to \mathcal{M}$. This mapping is surjective since $\mathcal{M}$ is irreducible. Consequently,

$$\mathcal{M} \cong \Big(U(\mathcal{M}) \otimes_{U(\mathcal{M}_{(0)})} \mathcal{M}_{[s]}\Big)/\ker\theta.$$

Moreover, by definition this mapping has the property that $1\otimes u$ is mapped onto u for all $u \in \mathcal{M}_{[s]}$. Observe that both $U(\mathcal{M}) \otimes_{U(\mathcal{M}_{(0)})} \mathcal{M}_{[s]}$ and $\mathcal{M}$ are naturally graded and θ is homogeneous of degree 0. Therefore $\ker\theta$ is a graded submodule. It is now clear that $\ker\theta$ is the maximal submodule uniquely determined by the conditions that it is a graded submodule which meets $1\otimes \mathcal{M}_{[s]}$ trivially.

Regard every $f \in \mathcal{M}_{[i]}{}^*$ as a linear form on $\mathcal{M}$ by setting $f(u) = 0$ if $u \in \sum_{k\neq i} \mathcal{M}_{[k]}$. Then $\mathcal{M}^* = \oplus_{i=s}^{-3} \mathcal{M}_{[i]}{}^*$ is graded and $\mathcal{M}_{[-3]}{}^*$ is an $\mathcal{M}_{(0)}$-module. It is an irreducible module since $\mathcal{M}$ is so. Similarly to the preceding reasoning we obtain that there is a surjective homogeneous $\mathcal{M}$-module homomorphism

$$\theta^* : U(\mathcal{M}) \otimes_{U(\mathcal{M}_{(0)})} \mathcal{M}_{[-3]}{}^* \to \mathcal{M}^*$$

so that

$$\mathcal{M}^* \cong \Big(U(\mathcal{M}) \otimes_{U(\mathcal{M}_{(0)})} \mathcal{M}_{[-3]}{}^*\Big)/\ker\theta^*,$$

where $\ker\theta^*$ is the unique maximal graded submodule meeting $1\otimes\mathcal{M}_{[-3]}{}^*$ trivially.

(b) Both $\mathcal{M}_{[-3]}$ and $\mathcal{M}_{[s]}$ are 2-dimensional irreducible $\mathcal{M}_{[0]}$-modules. Using the multiplication table (4.3.1) in Volume 1, p. 199 it is easy to observe that

$$\big[x_1\partial_1, x^{(\tau(\underline{n}))}\tilde{\partial}_1\big] = (-2+2)x^{(\tau(\underline{n}))}\tilde{\partial}_1 = 0, \qquad \big[x_2\partial_1, x^{(\tau(\underline{n}))}\tilde{\partial}_1\big] = 0,$$
$$\big[x_2\partial_2, x^{(\tau(\underline{n}))}\tilde{\partial}_1\big] = (-1+2)x^{(\tau(\underline{n}))}\tilde{\partial}_1.$$

This shows that $x^{(\tau(\underline{n}))}\tilde{\partial}_1$ is a primitive vector of weight $(0,1)$ for the Borel subalgebra $\mathfrak{b} := Fx_1\partial_1 \oplus Fx_2\partial_2 \oplus Fx_2\partial_1$ of $\mathcal{M}_{[0]}$. Now let f be the linear function on $\mathcal{M}_{[-3]}$ such that $f(\partial_1) = 0$ and $f(\partial_2) = 1$. Then $(x_1\partial_1)(f) = -f \circ (x_1\partial_1) = 0$, $(x_2\partial_2)(f) = -f \circ (x_2\partial_2) = f$ and $(x_2\partial_1)(f) = -f \circ (x_2\partial_1) = 0$, showing that $f \in {\mathcal{M}_{[-3]}}^*$ is a primitive vector of weight $(0, 1)$ for the Borel subalgebra $\mathfrak{b}$. From this it is immediate that $\mathcal{M}_{[s]} \cong {\mathcal{M}_{[-3]}}^*$ as $\mathcal{M}_{[0]}$-modules. Then these are isomorphic modules for $\mathcal{M}_{(0)}$ as well. Consequently,

$$U(\mathcal{M}) \otimes_{U(\mathcal{M}_{(0)})} \mathcal{M}_{[s]} \cong U(\mathcal{M}) \otimes_{U(\mathcal{M}_{(0)})} {\mathcal{M}_{[-3]}}^*$$

under a homogenous isomorphism Ψ which extends the given one $\mathcal{M}_{[s]} \cong {\mathcal{M}_{[-3]}}^*$.

(c) Since $\ker \theta^*$ is the unique maximal graded submodule of $U(\mathcal{M}) \otimes_{U(\mathcal{M}_{(0)})} {\mathcal{M}_{[-3]}}^*$ meeting $1 \otimes {\mathcal{M}_{[-3]}}^*$ trivially, $\ker(\theta^* \circ \Psi)$ is the unique maximal graded submodule of $U(\mathcal{M}) \otimes_{U(\mathcal{M}_{(0)})} \mathcal{M}_{[s]}$ meeting $1 \otimes \mathcal{M}_{[s]}$ trivially. Consequently, $\ker \theta = \ker(\theta^* \circ \Psi)$. We obtain a composite homogenous module isomorphism $\psi : \mathcal{M} \to \mathcal{M}^*$.

(d) Define a bilinear form $b : \mathcal{M} \times \mathcal{M} \to F$ by setting $b(x, y) := \psi(x)(y)$ for all $x, y \in \mathcal{M}$. Since ψ is an isomorphism of $\mathcal{M}$-modules, the form b is non-degenerate and $\mathcal{M}$-invariant. Next we define a bilinear skew-symmetric form b' on $\mathcal{M}$ by setting $b'(x, y) := b(x, y) - b(y, x)$ for all $x, y \in \mathcal{M}$. As $\mathcal{M}$ is a simple Lie algebra, the invariant form b' is either non-degenerate or zero. As $\dim \mathcal{M} = 5^{n_1+n_2+1}$ is odd, it must be that $b' = 0$. Therefore, the form b is symmetric. □

Corollary 16.4.2. *If $\widetilde{\mathcal{M}}$ is a Lie algebra such that $\widetilde{\mathcal{M}}/C(\widetilde{\mathcal{M}}) \cong \mathcal{M}(1, 1)$, then $\widetilde{\mathcal{M}}^{(1)} \cong \mathcal{M}(1, 1)$ and $\widetilde{\mathcal{M}} = \widetilde{\mathcal{M}}^{(1)} \oplus C(\widetilde{\mathcal{M}})$.*

Proof. In view of Proposition 16.4.1, Theorem 11.4.3, and the fact that all derivations of $\mathcal{M}(1, 1)$ are inner, the result follows. □

From here on we denote by $\mathcal{M}$ the restricted Melikian algebra $\mathcal{M}(1, 1)$. The height of the natural filtration then is $s = 23$. For our constructions in Chapter 18 we will need some information on the p-characters of irreducible representations of dimension $\le 5^3$ of $\mathcal{M}$. Let in the following V denote an irreducible $\mathcal{M}$-module of dimension $\le 5^3$ with p-character χ and representation $\rho : \mathcal{M} \to \mathfrak{gl}(V)$. Thus we have

$$\rho(u)^5 = \rho(u^{[5]}) + \chi(u)^5 \mathrm{Id}_V \quad \forall u \in \mathcal{M}.$$

Set

$$\chi_l(u) := \delta_{i,l}\chi(u) \quad \forall u \in \mathcal{M}_{[i]},\ l \ge -3.$$

Then $\chi = \sum_{l=-3}^{23} \chi_l$. Let $d := \max\{l \mid \chi_l \neq 0\}$ if $\chi \neq 0$.

We intend to show that either $\chi = 0$ or $d = -3$. The original proof of this statement (see [P-S 08, Proposition 6.3]) uses [P-Sk 99] and its powerful general methods. In contrast, we have decided here to use the rather elementary Proposition 3.2.11. Clearly, a disadvantage of this procedure is the fact that one has to apply the multiplication details of the underlying algebra, but we hope that this disadvantage is offset by using no further prerequisites.

Lemma 16.4.3. $d \not\equiv 1 \pmod{(3)}$.

Proof. We assume $\chi \neq 0$ and $d \equiv 1 \pmod{(3)}$. Then there is an element $x^{(a)} := x_1^{(a_1)} x_2^{(a_2)}$ of maximal degree for which $\chi(x^{(a)}) \neq 0$. By symmetry we may assume $a_1 \geq a_2$, and then we may assume that a_1 is maximal with these properties. Note that $d = 3|a| - 2 < s$.

(a) Suppose $|a| := a_1 + a_2 \neq 0$. We have $a_1 \neq 0$ and $d \geq 1$. Set in Proposition 3.2.11

$$G := \mathcal{M}, \qquad M := \mathcal{M}_{(0)}, \qquad I := \mathcal{M}_{(d+1)},$$
$$e_1 := \tilde{\partial}_1, \qquad e_2 := \tilde{\partial}_2, \qquad f_1 := x^{(a)}\tilde{\partial}_2, \qquad f_2 := x^{(a)}\tilde{\partial}_1.$$

Then $[e_i, f_j], [e_i, [e_k, f_j]] \subset \mathcal{M}_{(d-1)} \subset \mathcal{M}_{(0)} = M$ and $[e_i, f_j] = (-1)^{i+1}\delta_{i,j}x^{(a)}$. The algebra R generated by the $[e_i, f_j]$ is $Fx^{(a)}$ whence abelian. Moreover, $[M, I] \subset \mathcal{M}_{(d+1)} \subset \ker \chi$. Proposition 3.2.11 yields that $\dim V_0 \leq 5$, where V_0 is an irreducible $\mathcal{M}_{(0)}$-submodule of V.

If $|a| > 2$, then $d > 4$. Put (recall that $a_1 \neq 0$)

$$G := \mathcal{M}_{(0)}, \qquad M := I := \mathcal{M}_{(d-2)},$$
$$e_1 := x_1\tilde{\partial}_1, \qquad e_2 := x_1\tilde{\partial}_2, \qquad f_1 := x^{(a-\epsilon_1)}\tilde{\partial}_2, \qquad f_2 := x^{(a-\epsilon_1)}\tilde{\partial}_1.$$

Then $[G, I], [G, [G, I]] \subset \mathcal{M}_{(d-2)} = M$, $[\mathcal{M}_{(d-2)}, \mathcal{M}_{(d-2)}] \subset \mathcal{M}_{(d+1)} \subset \ker \chi$, and $[e_i, f_j] = (-1)^{i+1}a_1\delta_{i,j}x^{(a)}$. Proposition 3.2.11 yields $\dim V_0 \geq 5^2$, a contradiction.

If $1 \leq |a| \leq 2$, then $1 \leq a_1 \leq 2$, $a_2 \leq 1$. Recall that $\chi(x^{(a+\epsilon_1-\epsilon_2)}) = 0$ by choice of a. Put

$$G := \mathcal{M}_{(0)}, \qquad M := I := \mathcal{M}_{(d)},$$
$$e_1 := x_1\partial_2, \qquad e_2 := x_2\partial_2, \qquad f_1 := x^{(a-\epsilon_1+\epsilon_2)}, \qquad f_2 := x^{(a)}.$$

Then $[G, I], [G, [G, I]] \subset \mathcal{M}_{(d)} = M$, $[\mathcal{M}_{(d)}, \mathcal{M}_{(d)}] \subset \mathcal{M}_{(d+1)} \subset \ker \chi$, and

$$[e_1, f_1] = a_1x^{(a)} \notin \ker \chi, \qquad [e_1, f_2] = (a_1 + 1)x^{(a+\epsilon_1-\epsilon_2)} \in \ker \chi,$$
$$[e_2, f_2] = (a_2 - 2)x^{(a)} \notin \ker \chi.$$

Proposition 3.2.11 yields $\dim V_0 \geq 5^2$, a contradiction.

(b) We are left with the case $|a| = 0$, i.e., $x^{(a)} = 1$ and $\chi(\mathcal{M}_{(-1)}) = 0$. Due to Theorem 10.5.8, there is for every $\alpha \in F$ an automorphism of the form $\mathrm{Id} + \alpha(\operatorname{ad} x_2) + \sum_{k>1}\varphi_k$, where the φ_k are homogenous linear mappings of degree k. Consequently, there is an automorphism θ which maps ∂_i to $\partial_i - \delta_{i,2}\chi(\partial_2)\chi(1)^{-1}1 + q_i$ with $q_i \in \mathcal{M}_{(-1)}$. This automorphism stabilizes all $\mathcal{M}_{(i)}$ and therefore we still have

$\chi(\mathcal{M}_{(-1)}) = 0$, $\chi(\theta(1)) \neq 0$. In addition $\chi(\theta(\partial_2)) = 0$ now holds. We suppress the notion of θ. In this case let us apply the Melikian grading. Set

$$G := \mathcal{M}, \qquad M := \sum_{i \leq 0} \mathcal{M}_{\langle i \rangle}, \qquad I := \mathcal{M}_{\langle -2 \rangle} = F\partial_2,$$

$$e_1 := x_2, \qquad f_1 := \partial_2 \in I.$$

Both $[e_1, f_1] = -1$ and $[e_1, [e_1, f_1]] = -2\tilde{\partial_1}$ are contained in M. Note that $[M, I] = F\partial_2 \subset \ker \chi$. Therefore Proposition 3.2.11 shows that there is a M-submodule W of dimension $\leq 5^2$.

Set

$$G := \sum_{i \leq 0} \mathcal{M}_{\langle i \rangle}, \quad M := I := \sum_{i \leq -1} \mathcal{M}_{\langle i \rangle},$$

and

$$e_1 := x_1,\ e_2 := x_1\partial_1,\ e_3 := \tilde{\partial_1},\ f_1 := \partial_1,\ f_2 := 1,\ f_3 := \tilde{\partial_2},\ .$$

Note that (see Volume 2, p. 71) $e_1, e_2, e_3 \in \mathcal{M}_{\langle 0 \rangle}$, $f_1, f_2, f_3 \in \mathcal{M}_{\langle -1 \rangle}$, and M is an ideal of G acting trigonalizably. Moreover,

$$[e_1, f_1] = -1, \qquad [e_1, f_2] = 2\tilde{\partial_2}, \qquad [e_1, f_3] = x_1\partial_2,$$

$$[e_2, f_2] = -2, \qquad [e_2, f_3] = 2\tilde{\partial_2}, \qquad [e_3, f_3] = 1.$$

Since $\tilde{\partial_2}, x_1\partial_2 \in \mathcal{M}_{(-1)} \subset \mathcal{M}_{(d+1)} \subset \ker \chi$, Proposition 3.2.11 shows that the module W has dimension $\geq 5^3$, a contradiction. □

Lemma 16.4.4. *If* $d \equiv 0 \pmod{(3)}$, *then* $d = -3$ *and* $\dim V = 5^3$.

Proof. We assume $\chi \neq 0$, $d \equiv 0 \pmod{(3)}$, and $d \neq -3$. There is an element $x^{(a)}\partial_i$ of maximal degree for which $\chi(x^{(a)}\partial_i) \neq 0$. By symmetry we may assume $i = 1$. Note that $d = 3|a| - 3$, whence $|a| \neq 0$ and $0 \leq d < s$.

(a) If $|a| \geq 2$, then $d \geq 3$. Set in Proposition 3.2.11

$$G := \mathcal{M}, \qquad M := \mathcal{M}_{(0)}, \qquad I := \mathcal{M}_{(d+1)},$$

$$e_1 := \tilde{\partial_1}, \qquad e_2 := 1, \qquad f_1 := x^{(a)}, \qquad f_2 := x^{(a)}\tilde{\partial_1}.$$

Since $e_1, e_2 \in \mathcal{M}_{(-2)}$ and $f_1, f_2 \in \mathcal{M}_{(d+1)}$, one has $[e_i, f_j], [e_i, [e_k, f_j]] \in \mathcal{M}_{(d+1-4)} \subset \mathcal{M}_{(0)} = M$. Moreover,

$$[e_1, f_1] = -x^{(a)}\partial_1, \quad [e_1, f_2] = 0, \quad [e_2, f_1] \in \mathcal{M}_{(d-1)}, \quad [e_2, f_2] = x^{(a)}\partial_1.$$

In particular, the subalgebra R generated by these elements is contained in $\mathcal{M}_{(d-1)}$, whence $R^{(1)} \subset \mathcal{M}_{(d+1)}$. Proposition 3.2.11 yields that $\dim V_0 \leq 5$, where V_0 is an irreducible $\mathcal{M}_{(0)}$-submodule of V.

Since $|a| \geq 2$, there are $r, s \in F$ such that $r(a_1 - 1) + sa_2 \neq 0$. Put

$$G := \mathcal{M}_{(0)}, \qquad M := \mathcal{M}_{(1)}, \qquad I := \mathcal{M}_{(d)},$$
$$e_1 := rx_1\partial_1 + sx_2\partial_2, \qquad f_1 := x^{(a)}\partial_1.$$

Then $[G, I], [G, [G, I]] \subset \mathcal{M}_{(d)} \subset M$, $[\mathcal{M}_{(1)}, \mathcal{M}_{(d)}] \subset \mathcal{M}_{(d+1)} \subset \ker \chi$, and $[e_1, f_1] = (r(a_1 - 1) + sa_2)x^{(a)}\partial_1 \neq 0$. Proposition 3.2.11 yields that there is a 1-dimensional $\mathcal{M}_{(1)}$-submodule. Then ${\mathcal{M}_{(1)}}^{(1)}$ acts nilpotently on V. However, choose k such that $a_k \neq 0$ and observe that

$$x_k, x^{(a-\epsilon_k)}\tilde{\partial_1} \in \mathcal{M}_{(1)}, \quad [x_k, x^{(a-\epsilon_k)}\tilde{\partial_1}] = a_k x^{(a)}\partial_1 \notin \ker \chi,$$

a contradiction.

(b) Suppose $|a| = 1$, i.e., $d = 0$. In this case we apply the Melikian filtration.

(i) If $\chi(x_2\partial_1) \neq 0$, then we set in Proposition 3.2.11

$$G := \mathcal{M}, \qquad M := \mathcal{M}_{\{0\}}, \qquad I := \mathcal{M}_{\{2\}},$$
$$e_1 := \partial_1, \qquad e_2 := 1, \qquad f_1 := x_1x_2\partial_1, \qquad f_2 := x_2\tilde{\partial_1}.$$

Since (see Volume 2, p. 71) $e_1, e_2 \in \mathcal{M}_{\langle -1\rangle}$ and $f_1, f_2 \in \mathcal{M}_{\langle 2\rangle}$, one has $[e_i, f_j]$, $[e_i, [e_k, f_j]] \in \mathcal{M}_{\{0\}} = M$. Moreover,

$$[e_1, f_1] = x_2\partial_1, \qquad [e_1, f_2] = 0, \qquad [e_2, f_1] = 2x_2, \qquad [e_2, f_2] = x_2\partial_1.$$

In particular, the subalgebra R generated by these elements is abelian. Moreover, $[M, I] \subset \mathcal{M}_{\{2\}} \subset \mathcal{M}_{(1)} \subset \ker \chi$. Proposition 3.2.11 yields that $\dim V_{\{0\}} \leq 5$, where $V_{\{0\}}$ is an irreducible $\mathcal{M}_{\{0\}}$-submodule of V.

Next, set

$$G := \mathcal{M}_{\{0\}}, \qquad M := I := \mathcal{M}_{\{1\}},$$
$$e_1 := \tilde{\partial_1}, \qquad e_2 := x_1\partial_1, \qquad f_1 := -x_2, \qquad f_2 := -x_2\partial_1.$$

Then

$$[e_1, f_1] = x_2\partial_1, \qquad [e_1, f_2] = 0, \qquad [e_2, f_1] = 2x_2, \qquad [e_2, f_2] = x_2\partial_1.$$

As before, we conclude $\dim V_{\{0\}} \geq 5^2$, a contradiction.

(ii) Suppose $\chi(x_1\partial_1) \neq 0$. As $\rho(x_1\partial_1)^5 = \rho(x_1\partial_1) + \chi(x_1\partial_1)^5 \mathrm{Id}_V$, this gives that all eigenvalues of $\rho(x_1\partial_1)$ are contained in $F \setminus \mathbb{F}_5$. In particular, $\rho(x_1\partial_1)$ is invertible. By the former results (and Volume 2, p. 71) one has $\mathcal{M}_{\{1\}} \subset Fx_2\partial_1 + \mathcal{M}_{(1)} \subset \ker \chi$. Let $V_{\{0\}}$ denote an irreducible $\mathcal{M}_{\{0\}}$-submodule of V. Observe that $\mathcal{M}_{\{1\}}$ annihilates this module. Suppose

$$\sum_{0 \leq i,j \leq 4} \rho(\partial_1)^i \rho(\partial_2)^j u_{i,j} = 0, \quad u_{i,j} \in V_{\{0\}}$$

is a non-trivial relation. Put

$$k := \max\{i + j \mid u_{i,j} \neq 0\}, \quad i_0 := \max\{i \mid u_{i,k-i} \neq 0\}.$$

Then (see Lemma 2.1.5, observe $\rho(x_1^{(r)} x_2^{(s)} \tilde{\partial}_1) V_{\{0\}} = \{0\}$ if $r + s > 0$)

$$0 = \rho(x_1^{(i_0)} x_2^{(k-i_0)} \tilde{\partial}_1) \cdot \sum_{0 \le i,j \le 4} \rho(\partial_1)^i \rho(\partial_2)^j u_{i,j} = (-1)^k \rho(\tilde{\partial}_1) u_{i_0,k-i_0}.$$

We conclude $0 = \rho(x_1)\rho(\tilde{\partial}_1) u_{i_0,k-i_0} = \rho(x_1\partial_1) u_{i_0,k-i_0}$, a contradiction. Consequently, $\dim V_{\{0\}} \le 5$. On the other hand, $V_{\{0\}}$ is a module for $\mathcal{M}_{(0)} \cong W(1;\underline{1})$ and e_0 as the image of $x_1\partial_1$ acts invertibly. Then $V_{\{0\}}$ is 5-dimensional and $V = \bigoplus_{0 \le i,j \le 4} \rho(\partial_1)^i \rho(\partial_2)^j V_{\{0\}}$. Write for arbitrary nonzero $u \in V_{\{0\}}$

$$\rho(\tilde{\partial}_1) u = \sum_{0 \le i,j \le 4} \rho(\partial_1)^i \rho(\partial_2)^j u_{i,j}.$$

Set k and i_0 as before. Since $\rho(x_1^{(r)} x_2^{(s)} \tilde{\partial}_1)\rho(\tilde{\partial}_1) u = 0$ whenever $r + s > 0$, we may proced as in the former computation to obtain $k = 0$, which means $\rho(\tilde{\partial}_1) u \in V_{\{0\}}$. We conclude $0 = \rho(x_1)\rho(\tilde{\partial}_1) u = \rho(x_1\partial_1) u$, a contradiction.

(c) As a consequence of the former deliberations we get $d = -3$. So we finally have to consider the case when $\chi(\partial_1) \neq 0$. Set (observe that we consider the natural grading here)

$$G := \sum_{i \le 0} \mathcal{M}_{[i]}, \quad M := I := \mathcal{M}_{[-3]} + \mathcal{M}_{[-2]}$$

and

$$e_1 := \tilde{\partial}_1,\ e_2 := x_2\partial_1,\ e_3 := x_1\partial_1,\ f_1 := 1,\ f_2 := \partial_2,\ f_3 := \partial_1.$$

Then M is an abelian ideal in G and

$$[e_1, f_1] = -\partial_1, \qquad [e_1, f_2] = 0, \qquad [e_1, f_3] = 0,$$
$$[e_2, f_2] = -\partial_1, \qquad [e_2, f_3] = 0, \qquad [e_3, f_3] = -\partial_1.$$

As before, we conclude $\dim V \ge 5^3$. □

Lemma 16.4.5. $d \not\equiv 2 \pmod{(3)}$.

Proof. We assume $\chi \neq 0$ and $d \equiv 2 \pmod{(3)}$.

(a) Assume $\chi(x^{(a)}\tilde{\partial}_1) \neq 0$ for some a with $3|a| - 1 = d$. Put

$$Q := \sum_{i=0}^{4} F x_1^{(i)} \partial_1 + \sum_{j=0}^{4} F x_2^{(j)} \partial_2 + \sum_{0 \le c_1, c_2 \le 4} F x^{(c)} \tilde{\partial}_1, \quad A := \sum_{0 \le c_1, c_2 \le 4} F x^{(c)} \tilde{\partial}_1.$$

One computes

$$[x_1^{(i)}\partial_1, x^{(c)}\tilde{\partial}_1] = \left(\binom{c_1+i-1}{i} - \binom{c_1+i-1}{i-1} + 2\binom{c_1+i-1}{i-1}\right) \times x^{(c+(i-1)\epsilon_1)}\tilde{\partial}_1$$
$$= \binom{c_1+i}{i} x^{(c+(i-1)\epsilon_1)}\tilde{\partial}_1,$$
$$[x_2^{(j)}\partial_2, x^{(c)}\tilde{\partial}_1] = \left(\binom{c_2+j-1}{j} + 2\binom{c_2+j-1}{j-1}\right) x^{(c+(j-1)\epsilon_2)}\tilde{\partial}_1.$$

Consequently, Q is a subalgebra and A is an abelian ideal of Q. Each element $u \in A$ satisfies $u^{[5]} = 0$ (Lemma 10.5.5(1)), hence acts on V with single eigenvalue $\chi(u)$. Therefore

$$\dim V \geq 5^{\dim Q/A^{\perp}}, \quad A^{\perp} := \{u \in Q \mid \chi([u, A]) = 0\},$$

and this gives $\dim Q/A^{\perp} \leq 3$.

Suppose $a_2 = 4$. Put

$$e_j := x_2^{(j)}\partial_2 \in Q, \quad f_j := x_1^{(a_1)}x_2^{(5-j)}\tilde{\partial}_1 \in A, \qquad j = 1, \ldots, 4.$$

The former computations give $[e_j, f_j] = (-1)^{j-1}x^{(a)}\tilde{\partial}_1$. Since $[e_i, f_j] = 0$ for $i > j$, the matrix $(\chi([e_i, f_j]))$ is triangular. So this case is impossible.

Suppose $a_2 = 3$. Put

$$e_j := x_2^{(j)}\partial_2, \quad f_j := x_1^{(a_1)}x_2^{(4-j)}\tilde{\partial}_1, \qquad j = 0, 2, 3, 4.$$

The former computations give $[e_j, f_j] = \alpha_j x^{(a)}\tilde{\partial}_1$ with $\alpha_0 = 1, \alpha_2 = -1, \alpha_3 = 2, \alpha_4 = 2$. Derive a contradiction as before.

As the result, $\chi(x^{(c)}\tilde{\partial}_1) = 0$, whenever $3|c| - 1 = d$ and $c_2 = 4, 3$.

(b) Assume $\chi(x^{(a)}\tilde{\partial}_1) \neq 0$ for some a with $3|a| - 1 = d$ and $a_1 = 4$. Note that $d = 3|a| - 1 \geq 11$.

The symmetric version of (a) gives $\chi(x_1^{(3)}x_2^{(a_2+1)}\tilde{\partial}_2) = 0$. Set in Proposition 3.2.11

$$G := \mathcal{M}, \qquad M := \mathcal{M}_{(0)}, \qquad I := \mathcal{M}_{(d+2)},$$
$$e_1 := 1, \qquad e_2 := \partial_2, \qquad f_1 := x_1^{(4)}x_2^{(a_2+1)}, \qquad f_2 := x_1^{(4)}x_2^{(a_2+1)}\tilde{\partial}_1.$$

Since $f_1, f_2 \in \mathcal{M}_{(d+2)}$, one has $[e_i, f_j] \in \mathcal{M}_{(d-1)} \subset M$, $[e_i, [e_k, f_j]] \in \mathcal{M}_{(d-4)} \subset M$. In particular, the subalgebra R generated by all $[e_i, f_j]$ has the property $R^{(1)} \subset \mathcal{M}_{(d+1)}$. Moreover, $[e_1, f_2] \in \mathcal{M}_{(d+1)} \subset \ker \chi$, $[e_2, f_2] = x_1^{(4)}x_2^{(a_2)}\tilde{\partial}_1$, and

$$\chi([e_1, f_1]) = \chi\big(-2x_1^{(4)}x_2^{(a_2)}\tilde{\partial}_1 + 2x_1^{(3)}x_2^{(a_2+1)}\tilde{\partial}_2\big) = -2\chi\big(x_1^{(4)}x_2^{(a_2)}\tilde{\partial}_1\big) \neq 0.$$

Therefore there is a $\mathcal{M}_{(0)}$-submodule V_0 of V of dimension ≤ 5.

Set in Proposition 3.2.11

$$G := \mathcal{M}_{(0)}, \qquad M := I := \mathcal{M}_{(d-4)},$$

$$e_1 := x_1^{(2)}, \qquad e_2 := x_1, \qquad f_1 := x_1^{(2)} x_2^{(a_2+1)}, \qquad f_2 := x_1^{(3)} x_2^{(a_2+1)}.$$

Observe that

$$\begin{aligned}
[e_1, f_1] &= [x_1^{(2)}, x_1^{(2)} x_2^{(a_2+1)}] \\
&= -2x_1^{(2)} x_1^{(2)} x_2^{(a_2)} \tilde{\partial}_1 + 2(x_1^{(2)} x_1 x_2^{(a_2+1)} - x_1^{(2)} x_2^{(a_2+1)} x_1) \tilde{\partial}_2 \\
&= -2x_1^{(4)} x_2^{(a_2)} \tilde{\partial}_1, \\
[e_2, f_2] &= [x_1, x_1^{(3)} x_2^{(a_2+1)}] \\
&= -2x_1 x_1^{(3)} x_2^{(a_2)} \tilde{\partial}_1 + 2(x_1 x_1^{(2)} x_2^{(a_2+1)} - x_1^{(3)} x_2^{(a_2+1)}) \tilde{\partial}_2 \\
&= -8x_1^{(4)} x_2^{(a_2)} \tilde{\partial}_1 + 4x_1^{(3)} x_2^{(a_2+1)} \tilde{\partial}_2.
\end{aligned}$$

Argue as before to derive a contradiction.

(c) By definition of d there is $x^{(a)}\tilde{\partial}_i$ for which $\chi(x^{(a)}\tilde{\partial}_i) \neq 0$ for some a with $3|a| - 1 = d$. By symmetry we may assume $i = 1$. In this case we assume $a \neq 0$. Take a_1 maximal under the present conditions and observe that $a_2 \leq 2$ and $a_1 \neq 4$ hold.

If $a_2 = 2$, then set

$$\begin{aligned}
&e_1 := \partial_1, && e_2 := \partial_2, && e_3 := x_2\partial_2, && e_4 := x_2^{(3)}\partial_2, \\
&f_1 := x_1^{(a_1+1)} x_2^{(2)} \tilde{\partial}_1, && f_2 := x_1^{(a_1)} x_2^{(3)} \tilde{\partial}_1, && f_3 := x_1^{(a_1)} x_2^{(2)} \tilde{\partial}_1, && f_4 := x_1^{(a_1)} \tilde{\partial}_1;
\end{aligned}$$

if $a_2 = 1$, then set

$$\begin{aligned}
&e_1 := \partial_1, && e_2 := \partial_2, && e_3 := x_2\partial_2, && e_4 := x_2^{(2)}\partial_2, \\
&f_1 := x_1^{(a_1+1)} x_2 \tilde{\partial}_1, && f_2 := x_1^{(a_1)} x_2^{(2)} \tilde{\partial}_1, && f_3 := x_1^{(a_1)} x_2 \tilde{\partial}_1, && f_4 := x_1^{(a_1)} \tilde{\partial}_1;
\end{aligned}$$

if $a_2 = 0$, (which implies $a_1 \neq 0$), then set

$$\begin{aligned}
&e_1 := \partial_1, && e_2 := \partial_2, && e_3 := x_2\partial_2, && e_4 := x_1^{(2)}\partial_1, \\
&f_1 := x_1^{(a_1+1)} \tilde{\partial}_1, && f_2 := x_1^{(a_1)} x_2 \tilde{\partial}_1, && f_3 := x_1^{(a_1)} \tilde{\partial}_1, && f_4 := x_1^{(a_1-1)} \tilde{\partial}_1.
\end{aligned}$$

It is not hard to see that $\chi([e_i, f_i]) \neq 0$ and $\chi([e_i, f_j]) = 0$ if $i > j$. In view of (a) this gives the contradiction $\dim Q/A^{\perp} \geq 4$.

(d) We end up with the case $|a| = 0$, whence $\chi(\tilde{\partial}_1) \neq 0$. Let V_0 be an irreducible $\mathcal{M}_{(0)}$-submodule of V. Observe that $\mathcal{M}_{(1)}$ annihilates this module and χ vanishes on $\mathcal{M}_{(0)}$, and therefore V_0 is an irreducible and restricted $\mathfrak{sl}(2)$-module.

Suppose

$$\sum_{0\le i,j,k\le 4} \rho(\partial_1)^i \rho(\partial_2)^j \rho(\tilde{\partial}_1)^k u_{i,j,k} = 0, \quad u_{i,j,k} \in V_0$$

is a non-trivial relation. Put

$$l := \max\{i+j \mid u_{i,j,k} \neq 0 \text{ for some } k\}, \quad i_0 := \max\{i \mid u_{i,l-i,k} \neq 0 \text{ for some } k\}.$$

Then (see Lemma 2.1.5, observe $\rho(x_1^{(r)} x_2^{(s)} \tilde{\partial}_1) V_0 = \{0\}$ if $r+s>0$)

$$\begin{aligned} 0 &= \rho(x_1^{(i_0)} x_2^{(l-i_0)} \tilde{\partial}_1) \cdot \sum_{0\le i,j,k\le 4} \rho(\partial_1)^i \rho(\partial_2)^j \rho(\tilde{\partial}_1)^k u_{i,j,k} \\ &= (-1)^l \sum_{k=0}^{4} \rho(\tilde{\partial}_1)^{k+1} u_{i_0,l-i_0,k}. \end{aligned}$$

Note that $\rho(\tilde{\partial}_1)^5 u_{i_0,l-i_0,k} = \chi(\tilde{\partial}_1)^5 u_{i_0,l-i_0,k}$ and $\chi(\tilde{\partial}_1) \neq 0$. Thus, there is a non-trivial relation $\sum_{r=0}^{4} \rho(\tilde{\partial}_1)^r u_r = 0$. Apply $\rho(x_2 \tilde{\partial}_1)$ several times if necessary to obtain that $\rho(x_2\partial_1) u_r = 0$ whenever $\rho(\tilde{\partial}_1)^r u_r \neq 0$. Then every such u_r is a multiple of a distinguished $x_1\partial_1$-eigenvector $u \neq 0$. Considering eigenvalues with respect to $\rho(x_1\partial_1)$ we arrive at the contradiction $\rho(\tilde{\partial}_1)^r u = 0$ for some r. Consequently, $\dim V_0 = 1$ and

$$V = \oplus_{0\le i,j,k\le 4} F\rho(\partial_1)^i \rho(\partial_2)^j \rho(\tilde{\partial}_1)^k u, \quad V_0 = Fu.$$

Write

$$\rho(\tilde{\partial}_2) u = \sum_{0\le i,j,k\le 4} \alpha_{i,j,k} \rho(\partial_1)^i \rho(\partial_2)^j \rho(\tilde{\partial}_1)^k u.$$

Set l and i_0 as before. Suppose $l > 0$. Since $\rho(x_1^{(r)} x_2^{(s)} \tilde{\partial}_1) \rho(\tilde{\partial}_2) u = 0$ whenever $r+s>0$, we may proceed as in the former computation to obtain $l=0$, which means $\rho(\tilde{\partial}_2) u \in V_0$. However, $F\tilde{\partial}_2 + \mathcal{M}_{(0)}$ generate $\mathcal{M}$ as an algebra, whence $V = V_0$, a contradiction. □

We summarize:

Theorem 16.4.6. *If V is an irreducible $\mathcal{M}(1,1)$-module of dimension $\le 5^3$, then the p-character of V vanishes on the subspace $\bigoplus_{i\ge -2} \mathcal{M}(1,1)_{[i]}$. If V has a nonzero p-character, then* $\dim V = 5^3$.

Chapter 17

Sections

In this chapter we will be studying 2-sections of L relative to a torus $\mathfrak{t} \subset L_{[p]}$. Many properties of these are ruled by properties of simple Lie algebras of absolute toral rank 2. The latter algebras are known by the main result Theorem 14.3.15 of Volume 2.

17.1 On trigonalizability

For every subalgebra Q of $L_{[p]}$ we denote by $Q_{[p]}$ the p-envelope of Q in $L_{[p]}$ and by $\operatorname{rad}_p Q_{[p]}$ the maximal ideal of $Q_{[p]}$ consisting of $[p]$-nilpotent elements. We set $\operatorname{nil} Q := Q \cap \operatorname{rad}_p Q_{[p]}$. In particular, we have the twofold notation $\operatorname{rad}_p Q_{[p]} = \operatorname{nil} Q_{[p]}$.

Let $\mathfrak{t}$ denote a maximal torus of $L_{[p]}$ and $u \in L_\alpha(\mathfrak{t})$ a root vector. Write $u = u_s + u_n$ where u_s, u_n are the semisimple and $[p]$-nilpotent parts of u. The maximality of $\mathfrak{t}$ ensures that $u_s \in \mathfrak{t}$, one has $u^{[p]^e} = u_s{}^{[p]^e}$ for $e \gg 0$. In §1.3 we defined the notion of extended roots by setting $\mu(u) := \mu(u_s) = \mu(u^{[p]^e})^{p^{-e}}$ for $\mu \in \Gamma(L, \mathfrak{t})$.

In order to reduce the investigations of 2-sections to investigations of simple algebras of absolute toral rank 2, the subsequent proposition will be used several times.

Proposition 17.1.1. *Let $\mathfrak{t}_0 \subset L_{[p]}$ be a torus such that $\mathfrak{h}_0 := C_L(\mathfrak{t}_0)$ is nilpotent, let $\mathfrak{t}$ be the unique maximal torus in $\mathfrak{t}_0 + \mathfrak{h}_{0,[p]} \subset L_{[p]}$, and suppose one of the following conditions holds for $\alpha, \beta \in \Gamma(L, \mathfrak{t})$:*

(a) *there are $h_1 \in [L_\alpha, L_{-\alpha}]$ and $h_2 \in [L_\beta, L_{-\beta}]$ such that $\alpha(h_1) = 0$, $\beta(h_1) \neq 0$, and $\alpha(h_2) \neq 0$;*
(b) *the torus $\mathfrak{t}$ is maximal in $L_{[p]}$ and there are $u \in L_\alpha$ and $h_2 \in [L_\beta, L_{-\beta}]$ such that $\beta(u) \neq 0$ and $\alpha(h_2) \neq 0$;*
(c) *the torus $\mathfrak{t}$ is maximal in $L_{[p]}$ and there are $u \in L_\alpha$ and $v \in L_\beta$ such that $\beta(u) \neq 0$ and $\alpha(v) \neq 0$.*

Set $\mathfrak{g} := \sum_{\gamma \in (\mathbb{F}_p\alpha + \mathbb{F}_p\beta)\setminus\{0\}} (L_\gamma + [L_\gamma, L_{-\gamma}])$. Then the following holds.

(1) *Every ideal of $\mathfrak{g}$ is $\mathfrak{t}$-invariant.*
(2) *If I is a maximal ideal of $\mathfrak{g}$ and $\pi : \mathfrak{g} \to \mathfrak{g}/I$ denotes the canonical homomorphism, then $h_1, h_2 \notin I$ ($u, h_2 \notin I$, $u, v \notin I$, respectively) and $\pi(\mathfrak{g})$ is simple with*

$TR(\pi(\mathfrak{g})) \geq 2$. Moreover, if (a) holds, then $\pi(\mathfrak{h}_0 \cap \mathfrak{g})$ is a Cartan subalgebra in $\pi(\mathfrak{g})$ satisfying

$$\dim \pi(\mathfrak{h}_0 \cap \mathfrak{g})/\mathrm{nil}\, \pi(\mathfrak{h}_0 \cap \mathfrak{g}) \geq 2 \quad \text{and} \quad TR\big(\pi(\mathfrak{h}_0 \cap \mathfrak{g}), \pi(\mathfrak{g})\big) = 2.$$

(3) *Suppose $\mathfrak{t}$ is a maximal torus in $L_{[p]}$. Then* $\mathrm{rad}\,\mathfrak{g}$ *is nilpotent and, moreover, is the unique maximal ideal in $\mathfrak{g}$. If $\mathfrak{t}$ is a torus of maximal toral rank, then $TR(\mathfrak{g}/\mathrm{rad}\,\mathfrak{g}) = 2$.*

(4) *Suppose $\mathfrak{t}$ is a maximal torus in $L_{[p]}$ and let $\mathfrak{h}'$ be a subalgebra in $C_{L_{[p]}}(\mathfrak{t})$ containing $\mathfrak{h}_0$ such that all elements in $\bigcup_{h \in \mathfrak{h}'} (\mathrm{ad}\, h)^{p-1}(\mathfrak{h}' \cap \mathfrak{g})$ act nilpotently on $\mathfrak{g}/\mathrm{rad}\,\mathfrak{g}$. In cases (a) or (b)* $\mathrm{rad}\,\mathfrak{g}$ *is $\mathfrak{h}'$-invariant.*

Proof. (i) Let $t_1,\ t_2$ denote the semisimple parts of h_1, h_2 in case (a), of u, h_2 in case (b) and of u, v in case (c). In case (a) one has $t_1,\ t_2 \in \mathfrak{h}_{0,[p]}$, whence $t_1,\ t_2 \in \mathfrak{t}$ by definition of $\mathfrak{t}$, in cases (b) and (c) the last inclusion follows from the maximality of $\mathfrak{t}$. In case (a), $\alpha(h_1) = 0$ by our assumption, while in cases (b) and (c) the maximality of $\mathfrak{t}$ implies that $\alpha(u) = 0$. Thus we have in all cases under consideration

$$\alpha(t_1) = 0, \quad \beta(t_1) \neq 0, \quad \alpha(t_2) \neq 0.$$

Consequently, $\mathfrak{t} = Ft_1 \oplus Ft_2 \oplus \big(\mathfrak{t} \cap (\ker\alpha) \cap (\ker\beta)\big)$. Since $\mathfrak{t} \cap (\ker\alpha) \cap (\ker\beta)$ annihilates $\mathfrak{g}$ and $t_1, t_2 \in \mathfrak{g}_{[p]}$, every ideal of $\mathfrak{g}$ is $\mathfrak{t}$-invariant.

(ii) Let I be any ideal of $\mathfrak{g}$. Suppose $I_{[p]} \cap \mathfrak{t} \not\subset (\ker\alpha) \cap (\ker\beta)$. Then there is $t \in I_{[p]} \cap \mathfrak{t}$ with $\alpha(t) \neq 0$ or $\beta(t) \neq 0$. Suppose $\alpha(t) \neq 0$. Then $L_{i\alpha} = [t,\ L_{i\alpha}] \subset I$ for all $i \in \mathbb{F}_p^*$. In particular, $h_1 \in I$ in case (a) and $u \in I$ in cases (b) or (c). Then $t_1 \in I_{[p]} \cap \mathfrak{t}$ in all cases showing that $L_\gamma \subset I$ for all $\gamma \in (\mathbb{F}_p\alpha + \mathbb{F}_p\beta) \setminus \mathbb{F}_p\alpha$. This gives $I = \mathfrak{g}$.

Suppose $\beta(t) \neq 0$. As before we obtain $L_\beta \in I$ and therefore $h_2 \in I$ and $v \in I$, respectively. But then we are in the former case. Thus, if $I \neq \mathfrak{g}$, then $I_{[p]} \cap \mathfrak{t}$ centralizes $\mathfrak{g}$.

(iii) Let I be a maximal ideal of $\mathfrak{g}$. Since $t_1,\ t_2 \in \mathfrak{g}_{[p]}$, one has $\mathfrak{g}^{(1)} = \mathfrak{g}$. So $\mathfrak{g}/I$ is simple. Since $I_{[p]} \cap \mathfrak{t} \subset \mathfrak{t} \cap (\ker\,\alpha) \cap (\ker\,\beta)$ by part (ii), it is clear that $h_1,\ h_2 \notin I$ in case (a), $u,\ h_2 \notin I$ in case (b), and $u, v \notin I$ in case (c). This implies that $L_\alpha \not\subset I$ and $L_\beta \not\subset I$. So, $\pi(h_1)$ and $\pi(h_2)$ (respectively, $\pi(u)$ and $\pi(h_2)$, $\pi(u)$ and $\pi(v)$) generate a torus in $\pi(\mathfrak{g})_{[p]} \subset \mathrm{Der}\,\mathfrak{g}$, which distinguishes $\pi(\mathfrak{g}_\alpha) \neq \{0\}$ and $\pi(\mathfrak{g}_\beta) \neq \{0\}$. From this it is immediate that $TR(\pi(\mathfrak{g})) \geq 2$. If (a) holds, then $\pi(\mathfrak{h}_0 \cap \mathfrak{g})$ is self-normalizing, hence is a CSA in $\pi(\mathfrak{g})$. Moreover, $\pi(h_1)$ and $\pi(h_2)$ are linearly independent modulo $\mathrm{nil}\,\pi(\mathfrak{h}_0 \cap \mathfrak{g})$, so that

$$2 \leq TR\big(\pi(\mathfrak{h}_0 \cap \mathfrak{g}), \pi(\mathfrak{g})\big) \leq TR(\mathfrak{h}_0, \mathfrak{g}) \leq TR(\mathfrak{t}, \mathfrak{g}_{[p]}) = \dim\,\mathfrak{t}/\mathfrak{t} \cap (\ker\,\alpha) \cap (\ker\,\beta) \leq 2.$$

(iv) Now suppose that $\mathfrak{t}$ is a maximal torus in $L_{[p]}$. Let $I \subset \mathfrak{g}$ be a proper ideal of $\mathfrak{g}$ and $x \in I_\gamma := \mathfrak{g}_\gamma \cap I$ where $\gamma \in \mathbb{F}_p\alpha + \mathbb{F}_p\beta$. As $\mathfrak{t}$ is maximal, $x^{[p]^e} \in \mathfrak{t}$ for $e \gg 0$. We have shown in (ii) that $I_{[p]} \cap \mathfrak{t}$ centralizes $\mathfrak{g}$. It follows that $\bigcup_{\gamma \in \mathbb{F}_p\alpha+\mathbb{F}_p\beta} \operatorname{ad}_{\mathfrak{g}} I_\gamma$ is a Lie set consisting of nilpotent endomorphisms. Thus, the ENGEL–JACOBSON Theorem yields that I acts nilpotently on $\mathfrak{g}$. Therefore, $I \subset \operatorname{rad}\mathfrak{g}$. Moreover, $\mathfrak{g} \neq \operatorname{rad}\mathfrak{g}$, for $\mathfrak{g}^{(1)} = \mathfrak{g}$. Then $\operatorname{rad}\mathfrak{g}$ is nilpotent and is the unique maximal ideal of $\mathfrak{g}$.

If $\mathfrak{t}$ is even more a torus of maximal toral rank, then $TR(\mathfrak{g}) \leq 2$ by Theorem 1.3.11(3). Therefore by part (iii),

$$2 \leq TR(\mathfrak{g}/\operatorname{rad}\mathfrak{g}) \leq TR(\mathfrak{g}) \leq 2.$$

Then $TR(\mathfrak{g}/\operatorname{rad}\mathfrak{g}) = 2$.

(v) Now assume that $\mathfrak{t}$ is a maximal torus in $L_{[p]}$ and consider cases (a) or (b) of our proposition. Let $\mathfrak{h}'$ be a subalgebra in $C_{L_{[p]}}(\mathfrak{t})$ containing $\mathfrak{h}_0$. Clearly, $[\mathfrak{h}', L_\gamma] \subset L_\gamma$ for all γ. Then $[\mathfrak{h}', \mathfrak{g}] \subset \mathfrak{g}$ forcing $[\mathfrak{h}', \mathfrak{h}_0 \cap \mathfrak{g}] \subset \mathfrak{h}_0 \cap \mathfrak{g}$.

Let $\mathfrak{r}$ be the maximal $\mathfrak{h}'$-invariant solvable ideal in $\mathfrak{g}$, and let

$$\phi : \mathfrak{h}' + \mathfrak{g} \longrightarrow (\mathfrak{h}' + \mathfrak{g})/\mathfrak{r}$$

denote the canonical homomorphism. Let $\mathcal{J}$ be any nonzero $\phi(\mathfrak{h}')$-invariant ideal of $\phi(\mathfrak{g})$ and set $I := \phi^{-1}(\mathcal{J})$. Then I is an $\mathfrak{h}'$-invariant ideal of $\mathfrak{g}$ containing $\mathfrak{r}$ properly. In particular I is non-solvable. And so part (iv) of this proof shows that $I = \mathfrak{g}$ and, as a consequence, $\mathcal{J} = \phi(\mathfrak{g})$. This means that $\phi(\mathfrak{g})$ is $\phi(\mathfrak{h}')$-simple, whence $\phi(\mathfrak{h}' + \mathfrak{g})$ is semisimple with unique minimal ideal $\phi(\mathfrak{g})$. By Corollary 3.3.5, there are a simple algebra S, $m \in \mathbb{N}$, and $\underline{n} \in \mathbb{N}^m$ such that

$$\phi(\mathfrak{g}) \cong S \otimes \mathcal{O}(m;\underline{n}) \subset \phi(\mathfrak{h}' + \mathfrak{g}) \subset \big((\operatorname{Der} S) \otimes \mathcal{O}(m;\underline{n})\big) \rtimes \big(\operatorname{Id} \otimes W(m;\underline{n})\big),$$

such that $\pi_2(\mathfrak{h}') \subset W(m;\underline{n})$ is a transitive subalgebra.

Note that the semisimple parts of $\phi(h_1), \phi(h_2)$, ($\phi(u), \phi(h_2)$, respectively) generate a torus $\tilde{\mathfrak{t}}$, which by construction is contained in $(\operatorname{Der} S) \otimes \mathcal{O}(m;\underline{n})$. Thus Theorem 10.6.4(3) applies for $\phi(\mathfrak{h}' + \mathfrak{g}\big)$ and $\tilde{\mathfrak{t}}$. Case (a) of that theorem means that $\phi(\mathfrak{g})$ is simple and, consequently, $\operatorname{rad}\mathfrak{g} = \ker\phi$ is $\mathfrak{h}'$-invariant. This is the claim. Case (b) of that theorem is impossible, since $\phi(h_2) \in C_{S\otimes\mathcal{O}(m;\underline{n})}(\tilde{\mathfrak{t}})$. We have to exclude case (c). In order to do this, we observe the following. In part (i) of this proof it is mentioned that t_1, t_2 denote the semisimple parts of h_1, h_2 ($u^{[p]}, h_2$, respectively) and $\mathfrak{t} = Ft_1 \oplus Ft_2 \oplus \big(\mathfrak{t} \cap (\ker\alpha) \cap (\ker\beta)\big)$. Then $\mathfrak{h}'$ is the Fitting-0-component of $F(\operatorname{ad}_{\mathfrak{h}'+\mathfrak{g}} h_1) + F(\operatorname{ad}_{\mathfrak{h}'+\mathfrak{g}} h_2)$, whence $\phi(\mathfrak{h}')$ is the Fitting-0-component of $F(\operatorname{ad}_{\phi(\mathfrak{h}'+\mathfrak{g})} \phi(h_1)) + F(\operatorname{ad}_{\phi(\mathfrak{h}'+\mathfrak{g})} \phi(h_2))$, or, respectively, $\mathfrak{h}'$ is the Fitting-0-component of $F(\operatorname{ad}_{\mathfrak{h}'+\mathfrak{g}} u)^p + F(\operatorname{ad}_{\mathfrak{h}'+\mathfrak{g}} h_2)$, whence $\phi(\mathfrak{h}')$ is the Fitting-0-component of $F(\operatorname{ad}_{\phi(\mathfrak{h}'+\mathfrak{g})} \phi(u))^p + F(\operatorname{ad}_{\phi(\mathfrak{h}'+\mathfrak{g})} \phi(h_2))$. Consequently, we have in both cases $\phi(\mathfrak{h}') = C_{\phi(\mathfrak{h}'+\mathfrak{g})}(\tilde{\mathfrak{t}})$. So, case (c) is excluded by the explicit assumption. □

Theorem 17.1.2. *Let $\mathfrak{t}_0 \subset L_{[p]}$ be a torus and $\alpha_1, \dots, \alpha_s \in \Gamma(L, \mathfrak{t}_0)$. Assume that $\mathfrak{h} := L(\alpha_1, \dots, \alpha_s)$ is nilpotent. Then $\mathfrak{h}$ acts trigonalizably on L if and only if $\mathfrak{h}_0 := C_L(\mathfrak{t}_0)$ also does. If $\mathfrak{h}$ acts non-trigonalizably on L, then one has $\gamma_3(\mathfrak{h}_0) \not\subset \operatorname{nil} \mathfrak{h}_0$. In any case $\gamma_4(\mathfrak{h}) \subset \operatorname{nil} \mathfrak{h}$ holds.*

Proof. By assumption $\mathfrak{t}_0 \cap \bigcap_{i=1}^s (\ker \alpha_i) + \mathfrak{h}_{[p]}$ is nilpotent. Let $\mathfrak{t}$ denote its uniquely determined maximal torus. As $\mathfrak{t}_0 \cap \bigcap_{i=1}^s (\ker \alpha_i) \subset \mathfrak{t}$ and $C_L(\mathfrak{t}_0 \cap \bigcap_{i=1}^s (\ker \alpha_i)) = L(\alpha_1, \dots, \alpha_s) = \mathfrak{h}$, one has $\mathfrak{h} \supset C_L(\mathfrak{t})$. Hence

$$C_L(\mathfrak{t}) = \mathfrak{h} \supset \mathfrak{h}_0.$$

We may assume that $\mathfrak{h}$ acts non-trigonalizably on L, because otherwise $\mathfrak{h}_0$ acts trigonalizably. Therefore we take the assumptions of *Setting* 12.5 (Volume 2, p. 210) as fulfilled for $\mathfrak{g} = L, \mathfrak{t}, \mathfrak{h}$.

(a) Since $\mathfrak{h}$ is assumed to act non-trigonalizably, there is $k \geq 2$ maximal under the condition $\gamma_k(\mathfrak{h}) \not\subset \operatorname{nil} \mathfrak{h}$. Note that $[\mathfrak{t}+\gamma_{k-1}(\mathfrak{h}), \mathfrak{h}] = \gamma_k(\mathfrak{h}) \not\subset \operatorname{nil} \mathfrak{h}$ and $[\mathfrak{t}+\operatorname{rad}_p \mathfrak{h}_{[p]}, \mathfrak{h}] \subset \operatorname{nil} \mathfrak{h}$. Therefore $\mathfrak{t}+\gamma_{k-1}(\mathfrak{h}) \not\subset \mathfrak{t}+\operatorname{rad}_p \mathfrak{h}_{[p]}$ holds. As $\mathfrak{h}$ acts nilpotently on $\mathfrak{t}+\gamma_{k-1}(\mathfrak{h})$, there is $b \in \mathfrak{t} + \gamma_{k-1}(\mathfrak{h}) \setminus \mathfrak{t} + \operatorname{rad}_p \mathfrak{h}_{[p]}$ such that $[b, \mathfrak{h}] \subset \mathfrak{t} + \operatorname{rad}_p \mathfrak{h}_{[p]}$. Decompose $b = b_s + b_n$, where b_s is the semisimple and b_n is the $[p]$-nilpotent part of b. Note that according to *Setting* 12.5(4) $b_s \in \mathfrak{t}$. Then b_n also has the above properties. Thus we may assume that $b = b_n$ is $[p]$-nilpotent. If $[b, \mathfrak{h}] \subset \operatorname{nil} \mathfrak{h}$, then $Fb + \operatorname{rad}_p \mathfrak{h}_{[p]}$ is an ideal of $\mathfrak{h}_{[p]}$ consisting of $[p]$-nilpotent elements, whence $b \in \operatorname{rad}_p \mathfrak{h}_{[p]}$, which is not true by assumption. It is now easy to see that b satisfies the assertions of Lemma 12.5.1. Lemmas 12.5.2–12.5.5 apply with this b. Find accordingly roots

$$\alpha, \beta \in \Gamma(L, \mathfrak{t}) \text{ satisfying } \alpha([b, [L_\beta, L_{-\beta}]]) \neq 0, \ \beta([b, [L_\alpha, L_{-\alpha}]]) \neq 0.$$

Define $\mathfrak{r}(\alpha, \beta)$ as in Volume 2, p. 216, and π, $\mathcal{L}$, $\mathcal{H}$, $\mathcal{T}$ as in Lemma 12.5.5. Theorem 12.5.7 shows that $p = 5$, $\mathcal{L}^{(1)}$ is simple, and $\mathcal{T}$ is a 2-dimensional torus contained in $\mathcal{L}^{(1)}$. Moreover, Theorem 12.5.8 applies and gives rise to an isomorphism

$$\mathcal{L}^{(1)} \cong \mathcal{M}(1, 1), \text{ which maps } \mathcal{T} \text{ onto } F(1 + x_1)\partial_1 \oplus F(1 + x_2)\partial_2.$$

Since $\mathcal{M}(1, 1) \cong \operatorname{Der} \mathcal{M}(1, 1)$ (Theorem 7.1.4), we obtain even more

$$\mathcal{L} = \mathcal{L}^{(1)} = \pi(\mathfrak{g}(\alpha, \beta)).$$

We compute (see the remark in Volume 1, p. 209)

$$\mathcal{H} = F(1 + x_1)^4(1 + x_2)^3\tilde{\partial}_1 \oplus F(1 + x_1)^3(1 + x_2)^4\tilde{\partial}_2 \oplus F(1 + x_1)^2(1 + x_2)^2 \oplus \mathcal{T},$$

$$[(1 + x_1)^4(1 + x_2)^3\tilde{\partial}_1, (1 + x_1)^3(1 + x_2)^4\tilde{\partial}_2] = (1 + x_1)^2(1 + x_2)^2,$$

$$[(1 + x_1)^2(1 + x_2)^2, (1 + x_1)^4(1 + x_2)^3\tilde{\partial}_1] = (1 + x_1)\partial_1,$$

$$[(1 + x_1)^2(1 + x_2)^2, (1 + x_1)^3(1 + x_2)^4\tilde{\partial}_2] = (1 + x_2)\partial_2.$$

This shows that

$$\gamma_3(C_{\mathcal{L}}(\mathcal{T})) = \mathcal{T}, \quad \gamma_4(C_{\mathcal{L}}(\mathcal{T})) = \{0\}.$$

These equations give by construction (see Lemma 12.5.5(3))

$$\gamma_3(\pi(\mathfrak{h})) = \gamma_3(C_{\mathcal{L}}(\mathcal{T})) \not\subset \operatorname{nil} \pi(\mathfrak{h}), \quad \gamma_4(\mathfrak{h}) \subset \ker \pi.$$

By choice of $b \in \mathfrak{t}+\gamma_{k-1}(\mathfrak{h})$ and α, one has $\alpha(\gamma_k(\mathfrak{h})) \neq 0$. If $k \geq 4$, then $\alpha(\ker \pi) \neq 0$ and α were not a root on $\mathcal{L}$. But $\mathcal{M}(1,1)$ has p^2 roots with respect to $F(1+x_1)\partial_1 \oplus F(1+x_2)\partial_2$. Therefore $k \leq 3$. This proves $\gamma_4(\mathfrak{h}) \subset \operatorname{nil} \mathfrak{h}$.

(b) Next we will show that $\gamma_3(\mathfrak{h}_0) \not\subset \operatorname{nil} \mathfrak{h}_0$. Put

$$\tilde{\mathfrak{h}} := \mathfrak{t} + \mathfrak{h}_{[p]}, \qquad \tilde{L} := \mathfrak{t} + \mathfrak{h}_{[p]} + L.$$

Note that $\tilde{\mathfrak{h}}$ is restricted, nilpotent, and has the unique maximal torus $\mathfrak{t}$. Recall the notation of Lemma 12.5.5. Due to Lemma 12.5.5(1), $\mathfrak{r}(\alpha,\beta)$ is $\tilde{L}(\alpha,\beta)$-invariant. Therefore the p-envelope $\tilde{L}(\alpha,\beta)_{[p]}$ of $\tilde{L}(\alpha,\beta)$ in $L_{[p]}$ preserves $\mathfrak{r}(\alpha,\beta)$, hence acts on $\mathcal{L} = \tilde{L}(\alpha,\beta)/\mathfrak{r}(\alpha,\beta)$ as derivations. Since $\mathcal{L} \cong \mathcal{M}(1,1) \cong \operatorname{Der} \mathcal{M}(1,1)$ by the previous result, π extends to an epimorphism of restricted Lie algebras

$$\pi' : \tilde{L}(\alpha,\beta)_{[p]} \longrightarrow \mathcal{M}(1,1)$$

where

$$\ker \pi' = \{x \in \tilde{L}(\alpha,\beta)_{[p]} \mid [x, \tilde{L}(\alpha,\beta)_{[p]}] \subset \mathfrak{r}(\alpha,\beta)\}.$$

We identify $\pi'(\mathfrak{t}) = \mathcal{T} = F(1+x_1)\partial_1 \oplus F(1+x_2)\partial_2$.

Suppose $\tilde{\mathfrak{h}} \cap \mathfrak{r}(\alpha,\beta)$ contains an element h acting non-nilpotently on $\tilde{L}(\alpha,\beta)$ and let $h_s \in \mathfrak{t}$ denote the semisimple part of h. There exists a nonzero $\nu \in \mathbb{F}_5\alpha + \mathbb{F}_5\beta$ such that $\nu(h_s) \neq 0$. But then $\tilde{L}_\nu \subset \mathfrak{r}(\alpha,\beta)$, so that ν is not a $\mathcal{T}$-root of $\mathcal{M}(1,1)$. However, $\mathcal{M}(1,1)$ has p^2 $\mathcal{T}$-roots. This contradiction shows that $\tilde{\mathfrak{h}} \cap \mathfrak{r}(\alpha,\beta)$ acts nilpotently on $\tilde{L}(\alpha,\beta)$.

Let I be any ideal of $\tilde{L}(\alpha,\beta)_{[p]}$ which is not contained in $\ker \pi'$. Since $\tilde{L}(\alpha,\beta)/\mathfrak{r}(\alpha,\beta) \cong \mathcal{M}(1,1)$ is simple, it can only be that $(\ker \pi') \cap \tilde{L}(\alpha,\beta) = \mathfrak{r}(\alpha,\beta)$ and $\tilde{L}(\alpha,\beta) = I \cap \tilde{L}(\alpha,\beta) + \mathfrak{r}(\alpha,\beta)$. As $\mathfrak{t} \subset \tilde{L}(\alpha,\beta)$, both I and $\mathfrak{r}(\alpha,\beta)$ are $\mathfrak{t}$-stable. Then $\tilde{\mathfrak{h}} = \tilde{\mathfrak{h}} \cap I + \tilde{\mathfrak{h}} \cap \mathfrak{r}(\alpha,\beta)$. Thus, any $t \in \mathfrak{t}$ can be written as $t = h_1 + h_2$ with $h_1 \in \tilde{\mathfrak{h}} \cap I$ and $h_2 \in \tilde{\mathfrak{h}} \cap \mathfrak{r}(\alpha,\beta)$. By our discussion above, h_2 acts nilpotently on $\tilde{L}(\alpha,\beta)$. Also $0 = [t,h_2] = [h_1,h_2]$. Hence, for $e \gg 0$, $t^{[p]^e} = h_1^{[p]^e} + h_2^{[p]^e} \in h_1^{[p]^e} + \mathfrak{t} \cap (\ker \alpha) \cap (\ker \beta)$. But then $\mathfrak{t} = \mathfrak{t} \cap I_{[p]} + \mathfrak{t} \cap (\ker \alpha) \cap (\ker \beta)$. As a conseqence, $\tilde{L}_\gamma \subset I$ for any $\gamma \in (\mathbb{F}_5\alpha + \mathbb{F}_5\beta) \setminus \{0\}$. In other words, $\tilde{L}(\alpha,\beta) \subset I + \tilde{\mathfrak{h}} \cap \mathfrak{r}(\alpha,\beta)$. Since $\tilde{L}(\alpha,\beta)_{[p]}/\tilde{L}(\alpha,\beta)$ is abelian and $\tilde{\mathfrak{h}}$ is nilpotent, we conclude

$$\tilde{L}(\alpha,\beta)_{[p]}{}^{(\infty)} \subset I.$$

(c) Note that $\mathfrak{h}$ (and hence $\mathfrak{h}_{[p]}$ and the center $C(\mathfrak{h}_{[p]})$) are $\mathfrak{t}_0$-invariant. Moreover, $\mathfrak{t}_0 \cap_{i=1}^{s} (\ker \alpha_i)$ centralizes $\mathfrak{h}$. Therefore,

$$\big(\mathfrak{t}_0 \cap_{i=1}^{s} (\ker \alpha_i) + C(\mathfrak{h}_{[p]})\big)^{[p]} = \\ \mathfrak{t}_0 \cap_{i=1}^{s} (\ker \alpha_i) + \Big(\sum_{\delta\in\mathfrak{t}_0^*} C(\mathfrak{h}_{[p]})_\delta\Big)^{[p]} \subset \mathfrak{t}_0 + \sum_{\delta\in\mathfrak{t}_0^*} C(\mathfrak{h}_{[p]})_{p\delta}$$

is centralized by $\mathfrak{t}_0$. Recall that by definition $\mathfrak{t} \subset \mathfrak{t}_0 \cap_{i=1}^{s} (\ker \alpha_i) + \mathfrak{h}_{[p]}$ and $[\mathfrak{t}, \mathfrak{h}] = \{0\}$. Then we have, moreover, $\mathfrak{t} \subset \mathfrak{t}_0 \cap_{i=1}^{s} (\ker \alpha_i) + C(\mathfrak{h}_{[p]})$ and

$$[\mathfrak{t}_0, \mathfrak{t}] = [\mathfrak{t}_0, \mathfrak{t}^{[p]}] \subset [\mathfrak{t}_0, \big(\mathfrak{t}_0 \cap_{i=1}^{s} (\ker \alpha_i) + C(\mathfrak{h}_{[p]})\big)^{[p]}] = \{0\},$$

so that $\mathfrak{t}_0$ respects the root space decomposition of $\tilde{L}$ relative to $\mathfrak{t}$. In other words, $[\mathfrak{t}_0, \tilde{L}_\gamma] \subset \tilde{L}_\gamma$ for all $\gamma \in \Gamma(\tilde{L}, \mathfrak{t})$. In particular, $\tilde{L}(\alpha, \beta)$ is $\mathfrak{t}_0$-invariant.

Next, let J be the sum of all $\mathfrak{t}_0$-invariant ideals of $\tilde{L}(\alpha, \beta)_{[p]}$ contained in $\ker \pi'$ and

$$\Phi : \tilde{L}(\alpha, \beta)_{[p]} \longrightarrow \tilde{L}(\alpha, \beta)_{[p]}/J$$

be the canonical restricted homomorphism. Since J is $\mathfrak{t}_0$-invariant, the torus $\mathfrak{t}_0$ acts on $\Phi(\tilde{L}(\alpha, \beta)_{[p]})$. Let $\{0\} \neq \mathcal{J} \subset \Phi(\tilde{L}(\alpha, \beta)_{[p]})$ be a minimal $\mathfrak{t}_0$-invariant ideal, set $I := \Phi^{-1}(\mathcal{J})$, which is a $\mathfrak{t}_0$-invariant ideal of $\tilde{L}(\alpha, \beta)_{[p]}$. In particular, $I \not\subset \ker \pi'$. Then (b) applies to I, whence $\tilde{L}(\alpha, \beta)_{[p]}{}^{(\infty)} \subset I$. The minimality of $\mathcal{J}$ implies that

$$\mathcal{J} = \mathcal{J}^{(\infty)} = \Phi(\tilde{L}(\alpha, \beta)_{[p]})^{(\infty)}$$

is the unique minimal $\mathfrak{t}_0$-invariant ideal of $\Phi(\tilde{L}(\alpha, \beta)_{[p]})$. By Corollary 3.3.6 there are a simple Lie algebra S and $m \in \mathbb{N}$ such that

$$\Phi(\tilde{L}(\alpha, \beta)_{[p]})^{(\infty)} \cong S \otimes \mathcal{O}(m; \underline{1})$$

as Lie algebras. Since $J \subset \ker \pi'$, there exists a restricted Lie algebra epimorphism

$$\Psi : \Phi(\tilde{L}(\alpha, \beta)_{[p]}) = \tilde{L}(\alpha, \beta)_{[p]}/J \twoheadrightarrow \tilde{L}(\alpha, \beta)_{[p]}/\ker \pi' \cong \mathcal{M}(1, 1)$$

such that $\pi' = \Psi \circ \Phi$.

Since $\pi'\big(\tilde{L}(\alpha, \beta)_{[p]}{}^{(\infty)}\big)$ is a nonzero ideal in the simple algebra $\pi'(\tilde{L}(\alpha, \beta)_{[p]}) \cong \mathcal{M}(1, 1)$ and $S \otimes \mathcal{O}(m; \underline{1})_{(1)}$ is the unique maximal ideal of $S \otimes \mathcal{O}(m; \underline{1})$, we obtain

$$\begin{aligned} \mathcal{M}(1, 1) &\cong \tilde{L}(\alpha, \beta)_{[p]}{}^{(\infty)}/\tilde{L}(\alpha, \beta)_{[p]}{}^{(\infty)} \cap (\ker \pi') \\ &\cong \Phi\big(\tilde{L}(\alpha, \beta)_{[p]}{}^{(\infty)}\big)/\Phi\big(\tilde{L}(\alpha, \beta)_{[p]}{}^{(\infty)}\big) \cap (\ker \Psi) \\ &\cong S \otimes \mathcal{O}(m; \underline{1})/S \otimes \mathcal{O}(m; \underline{1})_{(1)} \cong S. \end{aligned}$$

(d) Since the ideal $\mathcal{J}$ is $\mathfrak{t}_0$-invariant, Φ extends to a homomorphism of restricted Lie algebras $\mathfrak{t}_0 + \tilde{L}(\alpha,\beta)_{[p]} \to \operatorname{Der}\mathcal{J}$. Since $\mathcal{J} \cong S \otimes \mathcal{O}(m;\underline{1})$, the latter induces a restricted homomorphism

$$\Phi' : \mathfrak{t}_0 + \tilde{L}(\alpha,\beta)_{[p]} \longrightarrow \big((\operatorname{Der} S) \otimes \mathcal{O}(m;\underline{1})\big) \rtimes \big(\mathrm{Id} \otimes W(m;\underline{1})\big)$$

such that $S \otimes \mathcal{O}(m;\underline{1}) \subset \Phi'(\tilde{L}(\alpha,\beta)_{[p]})$ and $\pi_2(\mathfrak{t}_0 + \Phi'(\tilde{L}(\alpha,\beta)_{[p]}))$ is a transitive subalgebra in $W(m;\underline{1})$ (we denote by π_2 the canonical projection onto the second summand). Since $S \cong \mathcal{M}(1,1) \cong \operatorname{Der} S$, there exists a restricted transitive Lie subalgebra $\mathcal{D} \subset W(m;\underline{1})$ such that

$$\Phi'\big(\mathfrak{t}_0 + \tilde{L}(\alpha,\beta)_{[p]}\big) = \big(S \otimes \mathcal{O}(m;\underline{1})\big) \rtimes \big(\mathrm{Id} \otimes \mathcal{D}\big).$$

Write

$$\Phi'\big(\tilde{L}(\alpha,\beta)_{[p]}\big) = \big(S \otimes \mathcal{O}(m;\underline{1})\big) \rtimes \big(\mathrm{Id} \otimes \mathcal{D}_0\big)$$

where $\mathcal{D}_0$ is a subalgebra of $\mathcal{D}$. By (c),

$$S \otimes \mathcal{O}(m;\underline{1})_{(1)} = \Phi'\big(\tilde{L}(\alpha,\beta)_{[p]}{}^{(\infty)} \cap (\ker \pi')\big),$$

whence it is invariant under $\mathcal{D}_0$. Therefore we have $\mathcal{D}_0 \subset W(m;\underline{1})_{(0)}$. On the other hand, $\Phi'(\tilde{L}(\alpha,\beta)_{[p]})$ is a $\mathfrak{t}_0$-invariant ideal, whence $\mathcal{D}_0$ is $\mathcal{D}$-invariant. However, $\mathcal{D}$ is a transitive subalgebra of $W(m;\underline{1})$, and this shows that $\mathcal{D}_0 = \{0\}$. Consequently,

$$\Phi'\big(\tilde{L}(\alpha,\beta)_{[p]}\big) = S \otimes \mathcal{O}(m;\underline{1}) \text{ and } \mathcal{D} = (\pi_2 \circ \Phi')(\mathfrak{t}_0).$$

From the first we conclude

$$\begin{aligned}\Phi'\big(\tilde{L}(\alpha,\beta)_{[p]}\big) &= \Phi'\big(\tilde{L}(\alpha,\beta)_{[p]}\big)^{(\infty)} = \Phi'(L(\alpha,\beta))^{(\infty)} = \Phi'(L(\alpha,\beta))\\ &= S \otimes \mathcal{O}(m;\underline{1}) \cong \mathcal{M}(1,1) \otimes \mathcal{O}(m;\underline{1}).\end{aligned}$$

There is a commutative diagram

$$\begin{array}{ccc} \tilde{L}(\alpha,\beta)_{[p]} & \overset{\pi'}{\twoheadrightarrow} & \mathcal{M}(1,1) \\ \searrow \Phi' & & \nearrow \Psi \\ & S \otimes \mathcal{O}(m;\underline{1}) & \end{array}$$

and $\Phi'(\mathfrak{t}_0) \subset ((\operatorname{Der} S) \otimes \mathcal{O}(m;\underline{1})) \rtimes (\mathrm{Id} \otimes W(m;\underline{1}))$ holds.

(e) As a further consequence, $\Phi'(\mathfrak{t}_0)$ acts transitively. Thanks to Theorem 3.6.1 the isomorphism Φ' can be chosen such that there exist toral elements $t_1,\dots,t_m \in \mathfrak{t}_0$ and a subtorus $\mathfrak{t}_0' \subset \mathfrak{t}_0$ such that $\mathfrak{t}_0 = (\oplus_{i=1}^m F t_i) \oplus \mathfrak{t}_0'$ and

$$\begin{aligned} \Phi'(t_i) &= \mathrm{Id} \otimes (1 + x_i)\partial_i && \forall\, i \le m,\\ \Phi'(t) &= \omega(t) \otimes 1 + \mathrm{Id} \otimes \lambda(t) && \forall\, t \in \mathfrak{t}_0', \end{aligned}$$

where ω and λ are restricted homomorphisms from $\mathfrak{t}_0'$ into $\operatorname{Der} S = S$ and $\sum_{i=1}^m Fx_i\partial_i$, respectively. As $[\lambda(\mathfrak{t}_0'), (1+x_i)\partial_i] \subset (\pi_2 \circ \Phi')(\mathfrak{t}_0^{(1)}) = \{0\}$ for all $i \le m$, it must be that $\lambda = 0$. So

$$\Phi'(\mathfrak{t}_0) = \Big(\sum_{i=1}^m F \operatorname{Id} \otimes (1+x_i)\partial_i\Big) \oplus \big(\omega(\mathfrak{t}_0') \otimes F\big).$$

Recall that $[\mathfrak{t}_0, \mathfrak{t}] = \{0\}$. Then

$$[\Phi'(\mathfrak{t}_0), \Phi'(\mathfrak{t})] = \{0\}, \text{ in particular } \Phi'(\mathfrak{t}) \subset C_S(\omega(\mathfrak{t}_0')) \otimes F.$$

We mentioned earlier that $\pi'(\mathfrak{t}) = \mathcal{T}$ is 2-dimensional. Then $\dim \Phi'(\mathfrak{t}) \ge 2$. Since $TR(\mathcal{M}(1,1)) = 2$ (see Corollary 10.5.14) and $\mathcal{M}(1,1)$ is centerless, then $\Phi'(\mathfrak{t})$ is a torus of maximal dimension in $S \otimes F$. Moreover, $\omega(\mathfrak{t}_0') \otimes F$ is a torus contained in $C_{S\otimes F}(\Phi'(\mathfrak{t}))$. Therefore $\Phi'(\mathfrak{t}) \supset \omega(\mathfrak{t}_0') \otimes F$, whence

$$\Phi'(\mathfrak{h}) = \Phi'\big(C_{L(\alpha,\beta)}(\mathfrak{t})\big) \subset C_S(\omega(\mathfrak{t}_0')) \otimes \mathcal{O}(m;\underline{1}).$$

We mentioned that $\Phi'(\tilde{L}(\alpha,\beta)_{[p]}) = \Phi'(L(\alpha,\beta))$ holds. Hence

$$\Phi'(\mathfrak{h}_0) = \Phi'\big(C_{L(\alpha,\beta)}(\mathfrak{t}_0)\big) = C_S(\omega(\mathfrak{t}_0')) \otimes F.$$

(f) We proved in (a) that $\gamma_3(\pi(\mathfrak{h})) \not\subset \operatorname{nil} \pi(\mathfrak{h})$. Since $\pi = \Psi \circ \Phi|_{\tilde{L}(\alpha,\beta)}$, we obtain $\gamma_3(\Phi(\mathfrak{h})) \not\subset \operatorname{nil} \Phi(\mathfrak{h})$ and then $\gamma_3(\Phi'(\mathfrak{h})) \not\subset \operatorname{nil} \Phi'(\mathfrak{h})$. Due to (e) this can only be if $\gamma_3(C_S(\omega(\mathfrak{t}_0')))$ acts non-nilpotently on S. Then $\gamma_3(\Phi'(\mathfrak{h}_0)) \not\subset \operatorname{nil} \Phi'(\mathfrak{h}_0)$, whence (again by (e)) $\gamma_3(\mathfrak{h}_0) \not\subset \operatorname{nil} \mathfrak{h}_0$. □

Let us apply this result for a particular setting. Recall that $\tilde{H} := C_{L_{[p]}}(T)$ and $\tilde{H}^{(1)} \subset H$. Set $\mathfrak{n} := \{h \in \tilde{H} \mid [H,h] \subset \operatorname{nil} H\}$. Note that $[\mathfrak{n} \cap L, \mathfrak{n}] \subset [H,\mathfrak{n}] \subset \operatorname{nil} H$ and $T \subset \mathfrak{n}$.

Lemma 17.1.3. *For $\alpha \in \Gamma(L,T)$ and $x \in L_\alpha$ one has $\alpha([x^{[p]}, \mathfrak{n}]) = 0$.*

Proof. (a) Suppose α is solvable. Then $\alpha([L_\alpha, L_{-\alpha}]) = 0$ and hence $[x^{[p]}, \mathfrak{n}] = (\operatorname{ad} x)^p(\mathfrak{n}) \subset [L_\alpha, L_{-\alpha}] \subset \ker \alpha$.

(b) Suppose α is not solvable. Note that $TR(\tilde{H}, L_{[p]}(\alpha)) = 1$ (see Theorem 1.3.11(2)). Apply Theorem 11.1.1 with $\mathfrak{g} = L_{[p]}(\alpha)$, $\mathfrak{h} = \tilde{H}$ and $\mathfrak{t}_0 = T$. Let $\pi : L_{[p]}(\alpha) \to L_{[p]}(\alpha)/\operatorname{rad} L_{[p]}(\alpha)$ denote the canonical homomorphism, which is a homomorphism of restricted algebras. That theorem shows that $\pi(L_{[p]}(\alpha))$ has a unique minimal ideal

$$\mathfrak{s} = \pi(L_{[p]}(\alpha)^{(\infty)}) = \pi(L(\alpha)^{(\infty)})$$

and there is a realization $\mathfrak{s} \cong \mathfrak{s}' \otimes \mathcal{O}(m;\underline{1})$ such that simultaneously one may take the image of T as $Ft_0 \otimes 1$. Moreover, since

$$0 \ne TR(\mathfrak{s}') \le TR(L_{[p]}(\alpha)) \le 1,$$

$\mathfrak{s}'$ is one of $\mathfrak{sl}(2)$, $W(1;\underline{1})$, $H2;\underline{1})^{(2)}$. In particular, $\mathfrak{s}'$ is restricted, and hence $\mathfrak{s}$ is a restricted subalgebra of the semisimple algebra $\pi(L_{[p]}(\alpha))$. Since $\pi(x) \in \mathfrak{s}$, there is $h \in \tilde{H} \cap L(\alpha,\beta) = H$ for which $\pi(h) = \pi(x)^{[p]}$. Consequently,

$$\pi([x^{[p]},\mathfrak{n}]) = \pi\big((\operatorname{ad} x)^p(\mathfrak{n})\big) = \big(\operatorname{ad}\pi(x)\big)^p(\pi(\mathfrak{n})) = [\pi(h),\pi(\mathfrak{n})] \subset \pi(\operatorname{nil} H).$$

As α is a root on $\pi(L_{[p]}(\alpha))$, one obtains $[x^{[p]},\mathfrak{n}] \subset \ker\alpha$. □

Lemma 17.1.4. *For $\alpha \in \Gamma(L,T)$ and $x \in L_\alpha$ one has $[x^{[p]},\mathfrak{n}] \subset \operatorname{nil} H$.*

Proof. (a) Suppose $\alpha(H) = 0$. Then $L(\alpha)$ is nilpotent. Theorem 17.1.2 implies that $\gamma_4(L(\alpha)) \subset \operatorname{nil} L(\alpha)$ and thus consists of $[p]$-nilpotent elements. In particular, $\gamma_p(L(\alpha)) \cap \tilde{H} \subset H$ is an ideal of H consisting of $[p]$-nilpotent elements. Then $[x^{[p]},\mathfrak{n}] \subset \gamma_p(L(\alpha)) \cap \tilde{H} \subset \operatorname{nil} H$.

(b) Now assume that $\alpha(H) \neq 0$, and suppose that $[x^{[p]},\mathfrak{n}] \not\subset \operatorname{nil} H$. Put

$$\Omega := \{\gamma \in \Gamma(L,T) \mid \gamma([x^{[p]},\mathfrak{n}]) \neq 0\} \neq \emptyset.$$

Since L is simple, we then have $H = \sum_{\gamma\in\Omega} [L_\gamma, L_{-\gamma}]$. Since $\alpha(H) \neq 0$ and $\cup_{\gamma\in\Omega} [L_\gamma, L_{-\gamma}]$ is a Lie set, the ENGEL–JACOBSON Theorem shows that there is $\beta \in \Omega$ such that $\alpha([L_\beta, L_{-\beta}]) \neq 0$. By Lemma 17.1.3 and the definition of β we have

$$\alpha([x^{[p]},\mathfrak{n}]) = 0, \quad \beta([x^{[p]},\mathfrak{n}]) \neq 0, \quad \alpha([L_\beta, L_{-\beta}]) \neq 0.$$

As $[x^{[p]},\mathfrak{n}] \subset [L_\alpha, L_{-\alpha}]$, assumption (a) of Proposition 17.1.1 with

$$\mathfrak{t}_0 = T, \quad \mathfrak{h}_0 = H, \quad \mathfrak{t} = T$$

is satisfied. Then all parts of this proposition are applicable. Accordingly, we set

$$\mathfrak{g} := \sum_{(\gamma\in\mathbb{F}_p\alpha+\mathbb{F}_p\beta)\setminus\{0\}} (L_\gamma + [L_\gamma, L_{-\gamma}])$$

and $\pi(\mathfrak{g}) := \mathfrak{g}/\operatorname{rad}\mathfrak{g}$. The proposition (part (3)) shows that $\ker\pi$ is nilpotent and $\pi(\mathfrak{g})$ is simple of absolute toral rank 2.

Suppose $\pi([x^{[p]},\mathfrak{n}]) \subset \pi(\operatorname{nil} H)$. Since $\beta([x^{[p]},\mathfrak{n}]) \neq 0$, the choice of β yields $L_\beta \subset \ker\pi$. But then $[L_\beta, L_{-\beta}] \subset \ker\pi$ acts nilpotently, which is not true as $\alpha([L_\beta, L_{-\beta}]) \neq 0$. In order to derive a contradiction, we intend to show that $\pi([x^{[p]},\mathfrak{n}]) \subset \pi(\operatorname{nil} H)$ holds.

Put $\mathfrak{h}' := \mathfrak{n} + H$. Then $\mathfrak{h}' \cap \mathfrak{g} \subset H$ and for all $h \in \mathfrak{h}'$ one has (by Theorem 17.1.2 with $s = 0$)

$$(\operatorname{ad} h)^{p-1}(\mathfrak{h}' \cap \mathfrak{g}) \subset [\mathfrak{n}, H] + \gamma_p(H) \subset \operatorname{nil} H.$$

The proposition (part (4)) then shows that $\mathfrak{n}$ acts on $\pi(\mathfrak{g})$ as derivations. Denote by $\pi' : \mathfrak{n} + \mathfrak{g} \to \operatorname{Der}\pi(\mathfrak{g})$ the canonical homomorphism (which may considered an extension of π).

We mentioned above that the choice of β yields elements $h_1 \in [L_\alpha, L_{-\alpha}]$, $h_2 \in [L_\beta, L_{-\beta}]$ with $\alpha(h_1) = 0$, $\beta(h_1) \neq 0$, $\alpha(h_2) \neq 0$. Let q_1, q_2 denote the semisimple parts of $\pi(h_1)$, $\pi(h_2)$, respectively. Note that $q_1, q_2 \in \pi'(T)$. If q_1, q_2 were linearly dependent, then some non-trivial linear combination $rh_1 + sh_2$ were contained in $\ker \pi$. It is not hard to see that then $h_1 \in \ker \pi$ and then $L_\beta \subset \ker \pi$. As before, this is impossible. Consequently, $\pi'(T)$ is a torus of dimension 2 (which is the maximal possible dimension) in the minimal p-envelope $\mathfrak{g}_{[p]}$ of $\mathfrak{g}$.

Theorem 14.3.15 in combination with Theorem 10.6.3 gives a complete list of algebras for $\pi(\mathfrak{g})$. Suppose $\pi(\mathfrak{g})$ is restricted. Then there is $h \in H \cap \mathfrak{g}$ such that $\pi(h) = \pi(x)^{[p]}$. Then

$$\pi([x^{[p]}, \mathfrak{n}]) = \pi\big((\operatorname{ad} x)^p(\mathfrak{n})\big) = \big(\operatorname{ad} \pi(x)\big)^p(\pi'(\mathfrak{n})) = [\pi(h), \pi'(\mathfrak{n})] \subset \pi(\operatorname{nil} H).$$

If $\pi(\mathfrak{g})$ is one of $W(1; \underline{2})$, $H(2; \underline{1}; \Phi(\tau))^{(1)}$, $H(2; \underline{1}; \Phi(1))$, then all root spaces of $\pi(\mathfrak{g})$ are 1-dimensional (Theorems 7.6.2, 7.6.5, 10.3.2, and 10.4.6). Since both $\mathfrak{n}$ and $x^{[p]}$ act on root spaces, $[x^{[p]}, \mathfrak{n}]$ annihilates all these. Consequently, $[x^{[p]}, \mathfrak{n}] \subset \ker \pi$.

Now we are left with the case where $\pi(\mathfrak{g}) \cong H(2; (1,2))^{(2)}$. By Lemma 10.2.3, $\pi(\mathfrak{g})_{[p]} = \pi'(T) + \pi(\mathfrak{g})$. Thus there exist $t \in T$ and $h \in H \cap \mathfrak{g}$ such that $\pi(x)^{[p]} = \pi(t) + \pi(h)$. As before,

$$\pi([x^{[p]}, \mathfrak{n}]) = [\pi(t) + \pi(h), \pi'(\mathfrak{n})] = [\pi(h), \pi'(\mathfrak{n})] \subset \pi(\operatorname{nil} H). \qquad \square$$

Lemma 17.1.5. *One has* $[\tilde{H}, \mathfrak{n}] \subset \operatorname{nil} H$ *and* $\operatorname{nil} H = H \cap \operatorname{rad}_p \tilde{H}$.

Proof. By definition we have $\operatorname{nil} H \subset \mathfrak{n}$. Therefore the set $\{h \in \tilde{H} \mid [h, \mathfrak{n}] \subset \operatorname{nil} H\}$ is closed under $[p]$-powers. As it contains H by definition, and all $x^{[p]}$, $x \in L_\alpha, \alpha \neq 0$, by Lemma 17.1.4, it contains $\tilde{H}$.

In particular, $[\tilde{H}, \operatorname{nil} H] \subset \operatorname{nil} H$ holds. Consequently, $\operatorname{nil} H$ is an ideal of $\tilde{H}$ consisting of $[p]$-nilpotent elements. Therefore one has $\operatorname{nil} H \subset H \cap \operatorname{rad}_p \tilde{H}$. Conversely, $H \cap \operatorname{rad}_p \tilde{H}$ is an ideal of $H_{[p]}$ consisting of $[p]$-nilpotent elements. Hence $H \cap \operatorname{rad}_p \tilde{H} \subset H \cap \operatorname{rad}_p H_{[p]} = \operatorname{nil} H$. $\square$

Definition 17.1.6. *Let* $\mathfrak{g}$ *be any Lie algebra and* $\mathfrak{t}$ *a torus of maximal dimension in a minimal* p*-envelope of* $\mathfrak{g}$. *Then* $\mathfrak{t}$ *is called* standard *if* $C_{\mathfrak{g}}(\mathfrak{t})$ *acts trigonalizably on* $\mathfrak{g}$, *otherwise* non-standard.

Theorem 17.1.7.

(1) *One has* $\gamma_4(\tilde{H}) \subset \operatorname{rad}_p \tilde{H}$.
(2) *If* T *is standard, then* $\tilde{H}^{(1)} \subset \operatorname{rad}_p \tilde{H}$ *holds.*
(3) *If* T *is non-standard, then* $\gamma_3(H) \not\subset \operatorname{nil} H$.

Proof. (a) Suppose T is standard. Then we have $H \subset \mathfrak{n}$, and therefore Lemma 17.1.5 shows that $[\tilde{H}, H] \subset \operatorname{nil} H$. Then $\tilde{H} \subset \mathfrak{n}$ holds by definition. As before, this lemma gives $[\tilde{H}, \tilde{H}] \subset \operatorname{nil} H \subset \operatorname{rad}_p \tilde{H}$. We obtain a proof of (2), and in the case of a standard torus also of (1).

(b) Suppose T is non-standard.

(i) Set in Theorem 17.1.2 $\mathfrak{t}_0 = T$ and $s = 0$. As a result, $\gamma_3(H) \not\subset \operatorname{nil} H$ and $\gamma_4(H) \subset \operatorname{nil} H$.

(ii) It remains to show that $\gamma_4(\tilde{H}) \subset \operatorname{rad}_p \tilde{H}$. As $\tilde{H}$ is nilpotent, there is $k \geq 1$ minimal such that $[\gamma_k(\tilde{H}), H] \subset \operatorname{nil} H$. By (i) and the fact that $\gamma_3(H) \subset [\gamma_2(\tilde{H}), H]$, one concludes $k \geq 3$. As $\operatorname{nil} H = H \cap \operatorname{rad}_p H_{[p]}$ and therefore $[T + \operatorname{rad}_p H_{[p]}, H] \subset \operatorname{nil} H$, the definition of k gives

$$\gamma_{k-1}(\tilde{H}) \subset \tilde{H}^{(1)} \subset H, \quad \gamma_{k-1}(\tilde{H}) \not\subset T + \operatorname{rad}_p H_{[p]}.$$

Take any $h \in \gamma_k(\tilde{H})_{[p]}$ and decompose $h = h_s + h_n$, where $h_s \in T \cap \gamma_k(\tilde{H})_{[p]}$ is the semisimple and $h_n \in \gamma_k(\tilde{H})_{[p]}$ is the $[p]$-nilpotent part of h. By definition of k, $[h_n, H] = [h, H] \subset \operatorname{nil} H$. Consequently, the set of all $[p]$-nilpotent elements of $\gamma_k(\tilde{H})_{[p]}$ is an ideal of $H_{[p]}$ whence is contained in $\operatorname{rad}_p H_{[p]}$. This shows that $\gamma_k(\tilde{H}) \subset T + \operatorname{rad}_p H_{[p]}$. We conclude

$$[\gamma_{k-1}(\tilde{H}), H] \subset \gamma_k(\tilde{H}) \subset T + \operatorname{rad}_p H_{[p]}.$$

(iii) Choose according to the definition of k an element $b' \in \gamma_{k-1}(\tilde{H})$ for which $[b', H] \not\subset \operatorname{nil} H$. Decompose $b' = b'_s + b'_n$, where b'_s is the semisimple and b'_n is the $[p]$-nilpotent part of b'. Then

$$b'_s \in T \cap \gamma_{k-1}(\tilde{H})_{[p]}, \; b'_n \in \gamma_{k-1}(\tilde{H})_{[p]}, \; [b'_n, H] = [b', H] \not\subset \operatorname{nil} H.$$

It is not hard to prove (using the arguments in the proof of Lemma 12.5.1) that $b := b'_n$ satisfies the assertions of Lemma 12.5.1. Lemmas 12.5.2–12.5.5 apply with this b. Find accordingly roots

$$\alpha, \beta \in \Gamma(L, T) \text{ satisfying } \alpha([b, [L_\beta, L_{-\beta}]]) \neq 0, \; \beta([b, [L_\alpha, L_{-\alpha}]]) \neq 0.$$

Define $\mathfrak{r}(\alpha, \beta)$ as in Volume 2, p. 216 and π, $\mathcal{L}$, $\mathcal{H}$, $\mathcal{T}$ as in Lemma 12.5.5. Theorem 12.5.7 shows that $p = 5$, $\mathcal{L}^{(1)}$ is simple, and $\mathcal{T}$ is a 2-dimensional torus in $\mathcal{L}^{(1)}$. Moreover, Theorem 12.5.8 applies and gives rise to an isomorphism

$$\mathcal{L}^{(1)} \cong \mathcal{M}(1, 1), \text{ which maps } \mathcal{T} \text{ onto } F(1 + x_1)\partial_1 \oplus F(1 + x_2)\partial_2.$$

Since $\mathcal{M}(1, 1) \cong \operatorname{Der} \mathcal{M}(1, 1)$ (Theorem 7.1.4), we obtain even more $\mathcal{L}^{(1)} = \mathcal{L}$. Using the remark from Volume 1, p. 209, it is easy to compute that $\gamma_4(C_{\mathcal{L}}(\mathcal{T})) = \{0\}$.

Since $\operatorname{rad}_p H_{[p]} \subset \mathfrak{n}$, Lemma 17.1.5 shows that $[\tilde{H}, \operatorname{rad}_p H_{[p]}] \subset \operatorname{nil} H$. Therefore

$$[b, [\tilde{H}, H]] \subset [\gamma_k(\tilde{H}), H] + [\tilde{H}, T + \operatorname{rad}_p H_{[p]}] \subset \operatorname{nil} H.$$

This shows that the ideal $\mathfrak{r}(\alpha, \beta)$ is invariant under $\tilde{H}$. Therefore $\tilde{H}$ acts on $\mathcal{M}(1,1)$ by derivations. As all derivations are inner, this subalgebra is mapped into $C_{\mathcal{L}}(\mathcal{T})$. As a consequence, $\gamma_4(\tilde{H}) \subset \ker \pi$. On the other hand, by choice of b and α one has $\alpha(\gamma_k(\tilde{H})) \supset \alpha([b, H]) \neq \{0\}$. If $k \geq 4$, then α were not a root on $\mathcal{L}$. But $\mathcal{M}(1,1)$ has p^2 roots with respect to $F(1+x_1)\partial_1 \oplus F(1+x_2)\partial_2$. Therefore $k \leq 3$. This proves $[\gamma_3(\tilde{H}), H] \subset \operatorname{nil} H$ and thereby shows that $\gamma_3(\tilde{H}) \subset \mathfrak{n}$. Now apply Lemma 17.1.5. □

As a consequence we determine 1-sections.

Theorem 17.1.8. *Let $\alpha \in \Gamma(L, T)$.*

(1) *Either $L_{[p]}[\alpha]$ is isomorphic to one of*

$$\{0\},\ \mathfrak{sl}(2),\ W(1;\underline{1}),\ H(2;\underline{1})^{(2)},\ H(2;\underline{1})^{(1)}$$

or $p = 5$, $L_{[p]}$ possesses a torus of maximal dimension which is non-standard, and

$$L_{[p]}[\alpha] \cong H(2;\underline{1})^{(2)} \oplus F D_H(x_1^{(5)}).$$

(2) $\operatorname{rad} L(\alpha) = L(\alpha) \cap \operatorname{rad} L_{[p]}(\alpha)$.

(3) *$L[\alpha] = L_{[p]}[\alpha]$ holds except if*

$$L_{[p]}[\alpha] = H(2;\underline{1})^{(1)} \text{ or } L_{[p]}[\alpha] = H(2;\underline{1})^{(2)} + F D_H(x_1^{(5)})$$

and $L[\alpha] = \left(L_{[p]}[\alpha]\right)^{(1)} \cong H(2;\underline{1})^{(2)}$.

Proof. The theorem is obviously true if α is solvable. Suppose α is non-solvable.

(1) As T is a torus of maximal dimension, $TR(L_{[p]}(\alpha)) = 1$ (by Theorem 1.3.11(3)). Combining Theorems 11.1.2 and 17.1.7 one obtains that $L_{[p]}[\alpha]$ is one of $\mathfrak{sl}(2)$, $W(1;\underline{1})$, or $H(2;\underline{1})^{(2)} \subset L_{[p]}[\alpha] \subset H(2;\underline{1})$.

Suppose $H(2;\underline{1})^{(2)} \subset L_{[p]}[\alpha] \subset H(2;\underline{1})$. If $L_{[p]}[\alpha] \subset H(2;\underline{1})^{(1)}$, then $L_{[p]}[\alpha] = H(2;\underline{1})^{(2)}$ or $L_{[p]}[\alpha] = H(2;\underline{1})^{(1)}$ because $\dim H(2;\underline{1})^{(1)}/H(2;\underline{1})^{(2)} = 1$. These cases are listed in the theorem.

So from here on we may assume that $L_{[p]}[\alpha]$ contains a nonzero element

$$D = a D_H(x_1^{(p)}) + b D_H(x_2^{(p)}) + c D_H(x_1^{p-1} x_2^{p-1})$$

with $a, b \in F$, not both of a, b vanishing. Applying the automorphism σ of $H(2;\underline{1})$ induced by the rule $\sigma(x_1) = x_2, \sigma(x_2) = -x_1$ if necessary we may assume that $a \neq 0$. Applying Lemma 11.4.6, we may assume further that $b = c = 0$. In other words, it can be assumed that $D = x_1^{p-1}\,\partial_2 = -D_H(x_1^{(p)})$. Note that

$$x_1^{p-1}\,\partial_2 \equiv (1+x_1)^{p-1}\,\partial_2 \pmod{H(2;\underline{1})^{(2)}}.$$

It follows that $C := C_{L_{[p]}[\alpha]}(D_H((1+x_1)x_2))$ contains $v_{-1} := (1+x_1)^{p-1}\,\partial_2$. For $0 \le i \le p-2$ put $v_i := D_H((1+x_1)^{i+1}x_2^{i+1})$. It is easy to check that $v_0, \dots, v_{p-2}$ pairwise commute and

$$[v_{-1}, v_i] = (i+1)v_{i-1}, \qquad 0 \le i \le p-2.$$

Moreover, $L_{[p]}[\alpha]$ contains all v_i's with $0 \le i \le p-3$. Since v_0 is a toral element, this implies that C is a non-trigonalizable CSA in $L_{[p]}[\alpha]$. Choose a toral element $t \in L_{[p]}(\alpha)$, which acts on $L_{[p]}[\alpha]$ as v_0. Then $T' := Ft \oplus (T \cap (\ker \alpha))$ is a torus of maximal dimension in $L_{[p]}$. It is non-standard, due to Theorem 17.1.7(2). It is easy to check that the *Setting* 12.5 holds for $\mathfrak{g} = L$ and $\mathfrak{t} = T'$. Theorem 12.5.7(1) yields $p = 5$.

Finally, suppose $L_{[p]}[\alpha] \supsetneq H(2;\underline{1})^{(2)} \oplus Fx_1^4\partial_2$. Then C contains $\lambda x_2^4\partial_1 + \mu v_3$ with $\lambda \ne 0$ or $\mu \ne 0$. Observe that

$$[v_{-1}, \lambda x_2^4\partial_1 + \mu v_3] \equiv -\lambda v_3 \pmod{H(2;\underline{1})^{(2)}}.$$

From this it is immediate that $v_3 \in C$. It follows that $(\operatorname{ad} v_{-1})^3(v_3) \in F^*v_0 \cap \gamma_4(C)$ is not $[p]$-nilpotent. This contradicts Theorem 17.1.7 and thereby shows that the case we are considering is impossible.

(2) In any case $\big(L_{[p]}[\alpha]\big)^{(1)} \in \{\mathfrak{sl}(2), W(1;\underline{1}), H(2;\underline{1})^{(2)}\}$ is a simple algebra of co-dimension ≤ 1 in $L_{[p]}[\alpha]$ and is the unique minimal ideal. Then the image of $L(\alpha)$ (as it is a non-solvable ideal) in $L_{[p]}[\alpha]$ is semisimple. This gives $L(\alpha) \cap \operatorname{rad} L_{[p]}(\alpha) = \operatorname{rad} L(\alpha)$.

(3) By (2) there are injections $\big(L_{[p]}[\alpha]\big)^{(1)} \hookrightarrow L[\alpha] \hookrightarrow L_{[p]}[\alpha]$. Therefore there are only the listed cases when these algebras do not coincide. □

We introduce the homomorphism

$$\Psi_\alpha : L_{[p]}(\alpha) \longrightarrow L_{[p]}[\alpha] = L_{[p]}(\alpha)/\operatorname{rad} L_{[p]}(\alpha) \tag{17.1.1}$$

and recall that (by Theorem 17.1.8(2)) there is a natural restriction of Ψ_α to $L(\alpha)$. In order to avoid ambiguities we sometimes indicate the underlying algebra L by writing $\Psi_{L_{[p]},\alpha}$, $\Psi_{L,\alpha}$, etc. Next define a subalgebra of $L(\alpha)$

$$Q(\alpha) := \begin{cases} L(\alpha) & \text{if } \alpha \text{ is solvable or classical,} \\ L(\alpha) \cap \Psi_\alpha^{-1}(W(1;\underline{1})_{(0)}) & \text{if } \alpha \text{ is Witt,} \\ L(\alpha) \cap \Psi_\alpha^{-1}(H(2;\underline{1})_{(0)}) & \text{if } \alpha \text{ is Hamiltonian.} \end{cases} \tag{17.1.2}$$

One defines similarly a subalgebra with respect to $L_{[p]}(\alpha)$. If necessary we indicate the underlying algebra as $Q(L_{[p]}(\alpha))$, $Q(L(\alpha))$, etc. We also define the following solvable ideal of $Q(\alpha)$

$$Q(\alpha)_{(1)} := \begin{cases} \sum_{i \in \mathbb{F}_p^*} (L_{i\alpha} + [L_{i\alpha}, L_{-i\alpha}]) & \text{if } \alpha \text{ is solvable,} \\ \operatorname{rad} L(\alpha) & \text{if } \alpha \text{ is classical,} \\ L(\alpha) \cap \Psi_\alpha^{-1}(W(1;\underline{1})_{(1)}) & \text{if } \alpha \text{ is Witt,} \\ L(\alpha) \cap \Psi_\alpha^{-1}(H(2;\underline{1})_{(1)}) & \text{if } \alpha \text{ is Hamiltonian.} \end{cases} \tag{17.1.3}$$

Note that $Q(\alpha)$ is solvable if α is solvable or Witt, and $Q(\alpha)/Q(\alpha)_{(1)} \cong \mathfrak{sl}(2)$ if α is classical or Hamiltonian. Since the subalgebras $W(1;\underline{1})_{(0)}$ and $H(2;\underline{1})_{(0)}$ are invariant under all automorphisms of the respective algebra, the definition of $Q(\alpha)$ is independent of any isomorphism as mentioned in Theorem 17.1.8(1). Also, $Q(\alpha)_{(1)}$ is nilpotent as ker Ψ_α is nilpotent.

Definition 17.1.9. *The algebra $Q(\alpha)$ is called* the maximum compositionally classical *subalgebra of $L(\alpha)$.*

Recall the definition of proper and improper roots from Definition 11.2.2. In the present context a root $\alpha \in \Gamma(L, T)$ is proper if and only if $[T, Q(\alpha)] \subset Q(\alpha)$. In order to unify notation *we call the 0 root proper* as well. The torus T is a proper torus if and only if all roots are proper.

17.2 1-sections in simple Lie algebras of absolute toral rank 2

For further applications we will need to know how 1-sections fit into 2-sections. As a first step we investigate the 1-sections of the simple Lie algebras of absolute toral rank 2. We are particularly interested how the subalgebras $Q(\alpha)$ sits in these algebras.

Lemma 17.2.1. *Let $\mathfrak{g}$ be one of the simple Cartan type algebras $W(2;\underline{1})$, $S(3;\underline{1})^{(1)}$, $H(4;\underline{1})^{(1)}$, $K(3;\underline{1})$, $W(1;\underline{2})$, $H(2;(1,2))^{(2)}$, and $\mathfrak{g}_{(0)}$ the natural subalgebra of maximal dimension. Suppose $\mathfrak{t} \subset \mathfrak{g}_{[p]}$ is a 2-dimensional torus and $\alpha \in \Gamma(\mathfrak{g}, \mathfrak{t})$ is a nonzero root. Then*

$$Q(\mathfrak{g}(\alpha)) = \mathfrak{g}_{(0)} \cap \mathfrak{g}(\alpha) + (\mathfrak{g} \cap \mathfrak{t}) \cap (\ker \alpha)$$

holds provided

$$\dim \mathfrak{g}(\alpha)/\big(\mathfrak{g}_{(0)} \cap \mathfrak{g}(\alpha) + (\mathfrak{g} \cap \mathfrak{t}) \cap (\ker \alpha)\big) \leq r$$

with $r = 0$, or $r = 1$ if α is Witt, or $r = 2$ if α is Hamiltonian.

Proof. Set $\mathfrak{g}_{(0)}(\alpha) = \mathfrak{g}(\alpha) \cap \mathfrak{g}_{(0)}$. As $(\mathfrak{g} \cap \mathfrak{t}) \cap (\ker \alpha) \subset \ker \Psi_{\mathfrak{g},\alpha} \subset Q(\mathfrak{g}(\alpha))$ by definition, one has

$$Q(\mathfrak{g}(\alpha)) \cap \big(\mathfrak{g}_{(0)} + (\mathfrak{g} \cap \mathfrak{t}) \cap (\ker \alpha)\big) = Q(\mathfrak{g}(\alpha)) \cap \mathfrak{g}_{(0)} + (\mathfrak{g} \cap \mathfrak{t}) \cap (\ker \alpha).$$

Recall that $\mathfrak{t}$ is standard (Corollary 12.5.10), and α is solvable, classical, Witt or Hamiltonian (Definition 11.2.2, Corollary 11.2.4). If $\mathfrak{g}(\alpha) \subset \mathfrak{g}_{(0)} + (\mathfrak{g} \cap \mathfrak{t}) \cap (\ker \alpha)$, properties of $\mathfrak{g}_{(0)}/\operatorname{rad} \mathfrak{g}_{(0)}$ imply that α can only be solvable or classical. In this case $Q(\mathfrak{g}(\alpha)) = \mathfrak{g}(\alpha)$ holds. This is the claim for $r = 0$.

Suppose α is Witt and $\dim \mathfrak{g}(\alpha)/\big(\mathfrak{g}_{(0)}(\alpha) + (\mathfrak{g} \cap \mathfrak{t}) \cap (\ker \alpha)\big) \leq 1$. The homomorphism $\Psi_{\mathfrak{g},\alpha} : \mathfrak{g}(\alpha) \to \mathfrak{g}(\alpha)/\operatorname{rad} \mathfrak{g}(\alpha) \cong W(1; \underline{1})$ maps $\mathfrak{g}_{(0)}(\alpha) + (\mathfrak{g} \cap \mathfrak{t}) \cap (\ker \alpha)$ onto a subalgebra of codimension ≤ 1. It cannot be mapped onto $W(1; \underline{1})$ because $\mathfrak{g}_{(0)}$ is solvable or $\mathfrak{g}_{(0)}/\operatorname{rad} \mathfrak{g}_{(0)}$ is classical. So the image has to be the unique subalgebra of codimension 1, and this shows that $\Psi_{\mathfrak{g},\alpha}\big(\mathfrak{g}_{(0)}(\alpha) + (\mathfrak{g} \cap \mathfrak{t}) \cap (\ker \alpha)\big) = W(1; \underline{1})_{(0)}$. Therefore $\mathfrak{g}_{(0)}(\alpha) + (\mathfrak{g} \cap \mathfrak{t}) \cap (\ker \alpha) \subset Q(\mathfrak{g}(\alpha))$. Since both these spaces have codimension 1 in $\mathfrak{g}(\alpha)$, the claim holds for $r = 0$.

Suppose α is Hamiltonian and $\dim \mathfrak{g}(\alpha)/\big(\mathfrak{g}_{(0)}(\alpha) + (\mathfrak{g} \cap \mathfrak{t}) \cap (\ker \alpha)\big) \leq 2$ holds. Then the homomorphism $\Psi_{\mathfrak{g},\alpha} : \mathfrak{g}(\alpha) \to \mathfrak{g}(\alpha)/\operatorname{rad} \mathfrak{g}(\alpha) \hookrightarrow H(2; \underline{1})$ maps $\mathfrak{g}_{(0)}(\alpha) + (\mathfrak{g} \cap \mathfrak{t}) \cap (\ker \alpha)$ onto a subalgebra of codimension ≤ 2 in $\Psi_\alpha(\mathfrak{g}(\alpha))$. Then $\Psi_{\mathfrak{g},\alpha}\big(\mathfrak{g}_{(0)}(\alpha) + (\mathfrak{g} \cap \mathfrak{t}) \cap (\ker \alpha)\big) \cap H(2; \underline{1})^{(2)}$ has codimension ≤ 2 in $H(2; \underline{1})^{(2)}$. Since $H(2; \underline{1})^{(2)}{}_{(0)}$ is the only proper subalgebra of codimension ≤ 2 in $H(2; \underline{1})^{(2)}$ and $\mathfrak{g}_{(0)}$ is solvable or $\mathfrak{g}_{(0)}/\operatorname{rad} \mathfrak{g}_{(0)}$ is classical, it happens that $\Psi_{\mathfrak{g},\alpha}\big(\mathfrak{g}_{(0)}(\alpha) + (\mathfrak{g} \cap \mathfrak{t}) \cap (\ker \alpha)\big) \cap H(2; \underline{1})^{(2)} = H(2; \underline{1})^{(2)}{}_{(0)}$. This shows that $\Psi_\alpha\big(\mathfrak{g}_{(0)}(\alpha) + (\mathfrak{g} \cap \mathfrak{t}) \cap (\ker \alpha)\big) \subset H(2; \underline{1})_{(0)}$. As before, the claim holds. □

Theorem 17.2.2. *Let $\mathfrak{g} = W(2; \underline{1})$, take $z_1 \in \{x_1, 1 + x_1\}$.*

(1) *If $t = z_1 \partial_1$, then*

$$C_{\mathfrak{g}}(t) = \sum_{j=0}^{p-1} F x_2^j \partial_2 + \sum_{j=0}^{p-1} z_1 x_2^j \partial_1,$$

$$\operatorname{rad} C_{\mathfrak{g}}(t) = \sum_{j=0}^{p-1} z_1 x_2^j \partial_1 \subset W(2; \underline{1})_{(0)} + Ft,$$

$$C_{\mathfrak{g}}(t)/\operatorname{rad} C_{\mathfrak{g}}(t) \cong \sum_{j=0}^{p-1} F x_2^j \partial_2 \cong W(1; \underline{1}).$$

(2) *If $t = x_1\partial_1 + x_2\partial_2$, then $C_{\mathfrak{g}}(t) \subset W(2; \underline{1})_{(0)}$, $\operatorname{rad} C_{\mathfrak{g}}(t) \subset W(2; \underline{1})_{(1)} + Ft$, and $C_{\mathfrak{g}}(t)/\operatorname{rad} C_{\mathfrak{g}}(t) \cong \mathfrak{sl}(2)$.*

(3) *If $t = ax_1\partial_1 + bx_2\partial_2$ with $a, b, a - b \in \mathbb{F}^*$, then $C_{\mathfrak{g}}(t) \subset W(2; \underline{1})_{(1)} + Fx_1\partial_1 + Fx_2\partial_2$.*

Proof. For a general element $t = az_1\partial_1 + bx_2\partial_2$ it is easy to check that

$$[t, z_1^i x_2^j \partial_k] = \big(a(i - \delta_{k,1}) + b(j - \delta_{k,2})\big) z_1^i x_2^j \partial_k.$$

All assertions follow. □

Corollary 17.2.3. *Let $\mathfrak{t}$ be a 2-dimensional torus in $W(2;\underline{1})$. For any nonzero root $\alpha \in \Gamma(W(2;\underline{1}), \mathfrak{t})$ one has*

$$Q(W(2;\underline{1})(\alpha)) = W(2;\underline{1})_{(0)} \cap W(2;\underline{1})(\alpha) + \mathfrak{t} \cap (\ker \alpha).$$

Proof. Choose a toral element t_α such that $\mathfrak{t} \cap (\ker \alpha) = Ft_\alpha$. By definition $\mathfrak{t} \cap (\ker \alpha) \subset W(2;\underline{1})(\alpha)$ holds. Theorem 7.5.1 shows that t_α is conjugate by an automorphism ψ to one of $(1 + x_1)\partial_1$ or $ax_1\partial_1 + bx_2\partial_2$ with $a, b \in \mathbb{F}_p$. Note that $\psi(W(2;\underline{1})_{(i)}) = W(2;\underline{1})_{(i)}$ for all i. We suppress the notion of this automorphism.

Suppose $t_\alpha = z_1\partial_1$ with $z_1 \in \{x_1, 1 + x_1\}$. Theorem 17.2.2(1) applies and shows

$$W(2;\underline{1})(\alpha) = \sum_{j=0}^{p-1} Fx_2^j\partial_2 + \sum_{j=0}^{p-1} z_1x_2^j\partial_1,$$

and that α is Witt. Therefore

$$W(2;\underline{1})_{(0)} \cap W(2;\underline{1})(\alpha) + Ft_\alpha = \sum_{j=1}^{p-1} Fx_2^j\partial_2 + \sum_{j=0}^{p-1} Fz_1x_2^j\partial_2$$

has codimension 1. Lemma 17.2.1 proves the claim. One argues similarly in the case $a = 0$.

In the case $a \neq 0$, $b \neq 0$, $z_1 = x_1$, $z_2 = x_2$ one has (see Theorem 17.2.2(2),(3)) $W(2;\underline{1})(\alpha) \subset W(2;\underline{1})_{(0)}$. As before, this is the claim. □

Theorem 17.2.4. *Let $\mathfrak{g} = S(3;\underline{1})^{(1)}$, take $z_1 \in \{x_1, 1 + x_1\}$.*

(1) *If $t = z_1\partial_1 - x_3\partial_3 = D_{1,3}(z_1x_3)$, then*

$$\begin{aligned} C_{\mathfrak{g}}(t) = &\sum_{i=1}^{p-1}\sum_{j=0}^{p-1} FD_{1,3}(z_1^i x_2^j x_3^i) \\ &+ \sum_{i=0}^{p-2}\sum_{j=0}^{p-1} FD_{1,2}(z_1^{i+1} x_2^j x_3^i) + \sum_{j=1}^{p-1} FD_{1,2}(x_2^j x_3^{p-1}) \\ &+ \sum_{i=0}^{p-2}\sum_{j=0}^{p-1} FD_{2,3}(z_1^i x_2^j x_3^{i+1}) + \sum_{j=1}^{p-1} FD_{2,3}(z_1^{p-1} x_2^j). \end{aligned}$$

If $z_1 = x_1$, then $C_{\mathfrak{g}}(t)/\operatorname{rad} C_{\mathfrak{g}}(t)$ is of Witt type. If $z_1 = 1 + x_1$, then $C_{\mathfrak{g}}(t)/\operatorname{rad} C_{\mathfrak{g}}(t)$ is of Hamiltonian type.

(2) *If* $t = x_1\partial_1 + x_2\partial_2 - 2x_3\partial_3$, *then* $C_{\mathfrak{g}}(t) \subset S(3;\underline{1})^{(1)}{}_{(0)}$, $\operatorname{rad} C_{\mathfrak{g}}(t) \subset S(3;\underline{1})^{(1)}{}_{(1)} + Ft$, *and* $C_{\mathfrak{g}}(t)/\operatorname{rad} C_{\mathfrak{g}}(t) \cong \mathfrak{sl}(2)$.

(3) *If* $t = ax_1\partial_1 + bx_2\partial_2 - (a+b)x_3\partial_3$ *with* $a, b, a \pm b \in \mathbb{F}^*$, *then* $C_{\mathfrak{g}}(t) \subset S(3;\underline{1})^{(1)}{}_{(1)} + F(x_1\partial_1 - x_3\partial_3) + F(x_2\partial_2 - x_3\partial_3)$.

Proof. (1) Note that

$$S(3;\underline{1})^{(1)} = \operatorname{span}\{D_{r,s}(f) \mid 1 \le r \le s \le 3,\ f \in \mathcal{O}(3;\underline{1})\}$$

(see Volume 1, p. 187). Since

$$[z_1\partial_1 - x_3\partial_3, D_{r,s}(z_1^i x_2^j x_3^k)] = (i - k - \delta_{r,1} - \delta_{s,1} + \delta_{r,3} + \delta_{s,3}) D_{r,s}(z_1^i x_2^j x_3^k),$$

one has

$$\begin{aligned} C_{\mathfrak{g}}(z_1\partial_1 - x_3\partial_3) = &\sum_{i=0}^{p-1}\sum_{j=0}^{p-1} FD_{1,3}(z_1^i x_2^j x_3^i) \\ &+ \sum_{i=0}^{p-2}\sum_{j=0}^{p-1} FD_{1,2}(z_1^{i+1} x_2^j x_3^i) + \sum_{j=0}^{p-1} FD_{1,2}(x_2^j x_3^{p-1}) \\ &+ \sum_{i=0}^{p-2}\sum_{j=0}^{p-1} FD_{2,3}(z_1^i x_2^j x_3^{i+1}) + \sum_{j=0}^{p-1} FD_{2,3}(z_1^{p-1} x_2^j). \end{aligned}$$

Note that $D_{1,3}(x_2^j) = D_{1,2}(x_3^{p-1}) = D_{2,3}(z_1^{p-1}) = 0$.

The following congruence modulo $S(3;\underline{1})^{(1)} \cap W(3;\underline{1})_{(0)} = S(3;\underline{1})^{(1)}{}_{(0)}$ holds:

$$\begin{aligned} C_{\mathfrak{g}}(z_1\partial_1 - x_3\partial_3) &\equiv FD_{1,3}(z_1x_3) + FD_{1,2}(z_1) + FD_{1,2}(z_1x_2) \\ &\quad + FD_{2,3}(x_3) + FD_{2,3}(z_1^{p-1}x_2) \\ &= F(z_1\partial_1 - x_3\partial_3) + F\partial_2 + F(z_1\partial_1 - x_2\partial_2) + Fz_1^{p-1}\partial_3 \\ &\equiv F(z_1\partial_1 - x_3\partial_3) + F\partial_2 + Fz_1^{p-1}\partial_3 \\ &= Ft + F\partial_2 + Fz_1^{p-1}\partial_3. \end{aligned}$$

(i) Suppose $z_1 = x_1$. Direct computation establishes that $\sum_{j=0}^{p-1} FD_{1,2}(x_1x_2^j)$ is an algebra isomorphic to $W(1;\underline{1})$ under the mapping

$$D_{1,2}(x_1x_2^j) = jx_1x_2^{j-1}\partial_1 - x_2^j\partial_2 \mapsto -x^j\partial.$$

Then $C_{\mathfrak{g}}(t)$ can only be Witt or Hamiltonian. Moreover, $S(3;\underline{1})^{(1)}{}_{(0)}$ is t-invariant and the above congruence shows in this case that $\partial_2 = -D_{1,2}(x_1)$ spans the complement of $C_{S(3;\underline{1})^{(1)}{}_{(0)}}(t)$ in $C_{S(3;\underline{1})^{(1)}}(t)$. This proves the claim for this case.

(ii) Suppose $z_1 = 1 + x_1$. A direct computation establishes that the algebra $\sum_{\substack{0\le i,j\le p-1\\(i,j)\neq(0,0)}} FD_{2,3}(z_1^{i-1}x_2^j x_3^i)$ is isomorphic to $H(2;\underline{1})^{(1)}$ under the mapping

$$\begin{aligned} D_{2,3}(z_1^{i-1}x_2^j x_3^i) &= iz_1^{i-1}x_2^j x_3^{i-1}\partial_2 - jz_1^{i-1}x_2^{j-1}x_3^i\partial_3 \\ &\mapsto ix_1^{i-1}x_2^j\partial_2 - jx_1^i x_2^{j-1}\partial_1 = D_H(x_1^i x_2^j). \end{aligned}$$

Then $C_{\mathfrak g}(t)$ can only be Hamiltonian. This proves the claim for this case.

(2), (3) If $t = ax_1\partial_1 + bx_2\partial_2 - (a+b)x_3\partial_3$ with $a, b, a+b \neq 0$, then easy calculation shows that $C_{\mathfrak g}(t) \subset S(3;\underline{1})^{(1)}{}_{(0)}$. Moreover, if $a = b$ then t centralizes $S(3;\underline{1})^{(1)}{}_{(0)}/S(3;\underline{1})^{(1)}{}_{(1)}$. If $a \neq b$, then $C_{\mathfrak g}(t) \subset S(3;\underline{1})^{(1)}{}_{(1)} + F(x_1\partial_1 - x_3\partial_3) + F(x_2\partial_2 - x_3\partial_3)$. □

Corollary 17.2.5. *Let $\mathfrak t$ be a 2-dimensional torus in $S(3;\underline{1})^{(1)}$. For any nonzero root $\alpha \in \Gamma(S(3;\underline{1})^{(1)}, \mathfrak t)$ one has*

$$Q(S(3;\underline{1})^{(1)}(\alpha)) = S(3;\underline{1})^{(1)}{}_{(0)} \cap S(3;\underline{1})^{(1)}(\alpha) + \mathfrak t \cap (\ker \alpha).$$

Proof. Choose a toral element t_α such that $\mathfrak t \cap (\ker\alpha) = Ft_\alpha$. Note that $\mathfrak t \cap (\ker\alpha) \subset S(3;\underline{1})^{(1)}(\alpha)$. Using Theorems 7.5.5 and 16.1.1, one may normalize t_α by an automorphism ψ, so that $\psi(t_\alpha) = a(z_1\partial_1 - x_3\partial_3) + b(x_2\partial_2 - x_3\partial_3)$, where $z_1 \in \{1 + x_1, x_1\}$ and $b = 0$ if $t_\alpha \notin S(3;\underline{1})^{(1)}{}_{(0)}$. Note that all $S(3;\underline{1})^{(1)}{}_{(i)}$ are invariant under ψ. We suppress the notion of ψ.

Suppose $b = 0$. Theorem 17.2.4(1) shows

$$\begin{aligned} S(3;\underline{1})^{(1)}(\alpha) = &\sum_{i=1}^{p-1}\sum_{j=0}^{p-1} FD_{1,3}(z_1^i x_2^j x_3^i) \\ &+ \sum_{i=0}^{p-2}\sum_{j=0}^{p-1} FD_{1,2}(z_1^{i+1}x_2^j x_3^i) + \sum_{j=1}^{p-1} FD_{1,2}(x_2^j x_3^{p-1}) \\ &+ \sum_{i=0}^{p-2}\sum_{j=0}^{p-1} FD_{2,3}(z_1^i x_2^j x_3^{i+1}) + \sum_{j=1}^{p-1} FD_{2,3}(z_1^{p-1}x_2^j). \end{aligned}$$

Moreover, if $z_1 = x_1$ then $S(3;\underline{1})^{(1)}(\alpha)$ is of Witt type, and if $z_1 = 1 + x_1$ then $S(3;\underline{1})^{(1)}(\alpha)$ is of Hamiltonian type. In the related proof we also computed

$$S(3;\underline{1})^{(1)}(\alpha) \equiv Ft_\alpha + F\partial_2 + Fz_1^{p-1}\partial_3 \pmod{S(3;\underline{1})^{(1)}{}_{(0)}}.$$

If $z_1 = x_1$, then $S(3;\underline{1})^{(1)}(\alpha) \cap S(3;\underline{1})^{(1)}{}_{(0)} + Ft_\alpha$ has codimension 1. If $z_1 = 1 + x_1$, then $S(3;\underline{1})^{(1)}(\alpha) \cap S(3;\underline{1})^{(1)}{}_{(0)} + Ft_\alpha$ is a subalgebra of codimension 2. Lemma 17.2.1 yields the claim. One argues similarly in the cases $a = 0$ and $a+b = 0$.

In the case $a \neq 0, b \neq 0, a+b \neq 0, z_1 = x_1$ (see Theorem 17.2.4(2),(3)) one has $S(3;\underline{1})^{(1)}(\alpha)) \subset S(3;\underline{1})^{(1)}{}_{(0)}$. As before, the claim follows. □

Theorem 17.2.6. *Let $\mathfrak{g} = H(4;\underline{1})^{(1)}$, take $z_1 \in \{x_1, 1+x_1\}$.*

(1) *If $t = D_H(z_1x_3)$, then*

$$C_{\mathfrak{g}}(t) = \sum_{\substack{0\leq i,j,l\leq p-1\\ (i,j,l)\neq(0,0,0),(p-1,p-1,p-1)}} FD_H(z_1^i x_2^j x_3^i x_4^l),$$

$$\operatorname{rad} C_{\mathfrak{g}}(t) = \sum_{\substack{0\leq i,j,l\leq p-1\\ i\neq 0,\ (i,j,l)\neq(p-1,p-1,p-1)}} FD_H(z_1^i x_2^j x_3^i x_4^l),$$

$$C_{\mathfrak{g}}(t)/\operatorname{rad} C_{\mathfrak{g}}(t) \cong \sum_{\substack{0\leq j,l\leq p-1\\ (j,l)\neq(0,0)}} FD_H(x_2^j x_4^l) \cong H(2;\underline{1})^{(1)}.$$

(2) *If $t = aD_H(x_1x_3) + bD_H(x_2x_4)$ with $b = \pm a \neq 0$, then $C_{\mathfrak{g}}(t) \subset H(4;\underline{1})^{(1)}{}_{(0)}$ and $C_{\mathfrak{g}}(t)/\operatorname{rad} C_{\mathfrak{g}}(t) \cong \mathfrak{sl}(2)$.*

(3) *If $t = aD_H(x_1x_3) + bD_H(x_2x_4)$ with $a, b, a \pm b \in F^*$, then $C_{\mathfrak{g}}(t) \subset H(4;\underline{1})^{(1)}{}_{(1)} + FD_H(x_1x_3) + FD_H(x_2x_4)$.*

Proof. (1) Since

$$[D_H(z_1x_3), D_H(z_1^i x_2^j x_3^k x_4^l)] = (k-i)D_H(z_1^i x_2^j x_3^k x_4^l),$$

we have that

$$C_{\mathfrak{g}}(D_H(z_1x_3)) = \sum_{\substack{0\leq j,l\leq p-1\\ (j,l)\neq(0,0)}} FD_H(x_2^j x_4^l) + \sum_{\substack{0\leq i,j,l\leq p-1\\ i\neq 0,\ (i,j,l)\neq(p-1,p-1,p-1)}} FD_H(z_1^i x_2^j x_3^i x_4^l).$$

Obviously, the first sum is an algebra isomorphic to $H(2;\underline{1})^{(1)}$ and the second sum is a nilpotent ideal, whence the radical of the whole algebra.

(2), (3) Suppose $t = aD_H(x_1x_3) + bD_H(x_2x_4)$ and $a, b \neq 0$. Then t is homogeneous, and one therefore computes $C_{\mathfrak{g}}(t)$ in the homogeneous components. Observe that the homogeneous degree -1 part $H(4;\underline{1})^{(1)}{}_{-1}$ is spanned by

$$\{D_H(x_1), D_H(x_2), D_H(x_3), D_H(x_4)\}$$

and these are eigenvectors with respect to t of respective eigenvalues

$$-a, -b, a, b.$$

This gives $C_{\mathfrak{g}}(t) \subset H(4;\underline{1})^{(1)}{}_{(0)}$. The homogeneous degree 0 part $H(4;\underline{1})^{(1)}{}_0$ is spanned by

$$\{D_H(x_1^2), D_H(x_1x_2), D_H(x_1x_3), D_H(x_1x_4), D_H(x_2^2), D_H(x_2x_3), D_H(x_2x_4), \\ D_H(x_3^2), D_H(x_3x_4), D_H(x_4^2)\},$$

where all these elements are eigenvectors with respect to t of the respective eigenvalues

$$-2a, -a-b, 0, -a+b, -2b, -b+a, 0, 2a, a+b, 2b.$$

Thus, if $a = b \neq 0$, then

$$C_{\mathfrak{g}}(t) \subset FD_H(x_1x_3) + FD_H(x_1x_4) + FD_H(x_2x_3) + FD_H(x_2x_4) + H(4;\underline{1})^{(1)}{}_{(1)},$$

and in case $a = -b \neq 0$

$$C_{\mathfrak{g}}(t) \subset FD_H(x_1x_2) + FD_H(x_1x_3) + FD_H(x_2x_4) + FD_H(x_3x_4) + H(4;\underline{1})^{(1)}{}_{(1)}.$$

These sections are of the classical type. In all other cases, $C_{\mathfrak{g}}(t) \subset FD_H(x_1x_3) + FD_H(x_2x_4) + H(4;\underline{1})^{(1)}{}_{(1)}$. □

Corollary 17.2.7. *Let* $\mathfrak{t}$ *be a 2-dimensional torus in* $H(4;\underline{1})^{(1)}$. *For any nonzero root* $\alpha \in \Gamma(H(4;\underline{1})^{(1)}, \mathfrak{t})$ *one has*

$$Q(H(4;\underline{1})^{(1)}(\alpha)) = H(4;\underline{1})^{(1)}{}_{(0)} \cap H(4;\underline{1})^{(1)}(\alpha) + \mathfrak{t} \cap (\ker \alpha).$$

Proof. Choose a toral element t_α such that $\mathfrak{t} \cap (\ker \alpha) = Ft_\alpha$. Note that $\mathfrak{t} \cap (\ker \alpha) \subset H(4;\underline{1})^{(1)}(\alpha)$. Using Theorems 7.5.8 and 16.1.2, one may normalize t_α by an automorphism ψ of $H(4;\underline{1})^{(1)}$, so that $\psi(t_\alpha) = aD_H(z_1x_3) + bD_H(x_2x_4)$, where $z_1 \in \{1+x_1, x_1\}$ and $b = 0$ if $t_\alpha \notin H(4;\underline{1})^{(1)}{}_{(0)}$. Note that all $H(4;\underline{1})^{(1)}{}_{(i)}$ are invariant under ψ. We suppress the notion of ψ.

Suppose $b = 0$. We stated in Theorem 17.2.6(1)

$$H(4;\underline{1})^{(1)}(\alpha) = \sum_{\substack{0 \le j,l \le p-1 \\ (j,l) \neq (0,0)}} FD_H(x_2^j x_4^l) + \sum_{\substack{0 \le i,j,l \le p-1 \\ i \neq 0,\ (i,j,l) \neq (p-1,p-1,p-1)}} FD_H(z_1^i x_2^j x_3^i x_4^l),$$

where the first sum $\sum_{\substack{0 \le j,l \le p-1 \\ (j,l) \neq (0,0)}} FD_H(x_2^j x_4^l)$ is an algebra isomorphic to $H(2;\underline{1})^{(1)}$ and the second sum is the radical of the whole algebra. Observe that the following congruence modulo $H(4;\underline{1})^{(1)}{}_{(0)}$ holds

$$\begin{aligned} H(4;\underline{1})^{(1)}(\alpha) &\equiv FD_H(x_2) + FD_H(x_4) + FD_H(z_1x_3) \\ &\equiv FD_H(x_2) + FD_H(x_4) + Ft_\alpha. \end{aligned}$$

Lemma 17.2.1 yields the claim. One argues similarly if $a = 0$. Finally, if $a \neq 0$, $b \neq 0$, $z_1 = x_1$, then the theorem proves $H(4;\underline{1})^{(1)}(\alpha) \subset H(4;\underline{1})^{(1)}{}_{(0)}$. Again the claim follows. □

Theorem 17.2.8. *Let* $\mathfrak{g} = K(3;\underline{1})$.

(1) *If* $t = D_K(1+x_3)$, *then*

$$C_{\mathfrak{g}}(t) = \sum_{0\le i,j\le p-1} F D_K\big(x_1^i x_2^j (1+x_3)^{\frac{2-(i+j)}{2}}\big) \cong \mathcal{P}(2;\underline{1})$$

is isomorphic to the Poisson algebra described in Volume 1, pp. 402/403 under the isomorphism Γ *as in Theorem 7.5.15.*

(2) *If* $t = D_K((1+x_1)x_2)$, *then*

$$C_{\mathfrak{g}}(t) = \sum_{0\le i,j\le p-1} F D_K\big((1+x_1)^i x_2^i (x_3+x_1x_2)^j\big),$$

$$\operatorname{rad} C_{\mathfrak{g}}(t) = \sum_{1\le i\le p-1,\ 0\le j\le p-1} F D_K\big((1+x_1)^i x_2^i (x_3+x_1x_2)^j\big),$$

$$C_{\mathfrak{g}}(t)/\operatorname{rad} C_{\mathfrak{g}}(t) \cong \sum_{0\le j\le p-1} F D_K\big((x_3+x_1x_2)^j\big) \cong W(1;\underline{1}),$$

where the last isomorphism is given by the mapping $D_K((x_3+x_1x_2)^j) \mapsto 2x^j\partial$.

(3) *If* $t = aD_K(x_1x_2) + bD_K(x_3)$ *with* $a, b \in \mathbb{F}_p$, *then the following is true:*

(a) *if* $b = 0$, $a \neq 0$, *then*

$$C_{\mathfrak{g}}(t) = \sum_{0\le i,k\le p-1} F D_K(x_1^i x_2^i x_3^k),$$

$$\operatorname{rad} C_{\mathfrak{g}}(t) = \sum_{1\le i\le p-1,\ 0\le k\le p-1} F D_K(x_1^i x_2^i x_3^k),$$

$$C_{\mathfrak{g}}(t)/\operatorname{rad} C_{\mathfrak{g}}(t) \cong \sum_{0\le k\le p-1} F D_K(x_3^k) \cong W(1;\underline{1});$$

(b) *if* $a = b \neq 0$, *then*

$$C_{\mathfrak{g}}(t) = \sum_{0\le i,j,k\le p-1,\ j+k\equiv 1\,(p)} F D_K(x_1^i x_2^j x_3^k),$$

$$\operatorname{rad} C_{\mathfrak{g}}(t) = \operatorname{span}\{D_K(x_1^i x_2^j x_3^{p+1-j}) \mid 0\le i\le p-1,\ 2\le j\le p-1\} + \operatorname{span}\{D_K(x_1^i(x_3+x_1x_2)) \mid 0\le i\le p-1\},$$

$$C_{\mathfrak{g}}(t)/\operatorname{rad} C_{\mathfrak{g}}(t) \cong \sum_{0\le i\le p-1} F D_K(x_1^i x_2) \cong W(1;\underline{1});$$

(c) *if* $a = 0$, $b \neq 0$, *then*

$$C_{\mathfrak{g}}(t) = \sum_{0\le i,j,k\le p-1,\; i+j+2k\equiv 2\,(p)} FD_K(x_1^i x_2^j x_3^k) \subset K(3;\underline{1})_{(0)},$$

$$\operatorname{rad} C_{\mathfrak{g}}(t) = C_{\mathfrak{g}}(t) \cap K(3;\underline{1})_{(1)} + Ft,$$

$$C_{\mathfrak{g}}(t)/\operatorname{rad} C_{\mathfrak{g}}(t) \cong \mathfrak{sl}(2);$$

(d) *if* $a \neq 0$, $b \neq 0$, $a \pm b \neq 0$, *then*

$$C_{\mathfrak{g}}(t) \subset K(3;\underline{1})_{(1)} + FD_K(x_1x_2) + FD_K(x_3).$$

Proof. (1) If $t = D_K(1 + x_3)$, then Equation (7.5.1) and Theorem 7.5.15 apply and give the claim.

(2) Suppose $t = D_K((1 + x_1)x_2)$. Equation (4.2.10) allows (via the mapping D_K) to make the computations in $\mathcal{O}(3;\underline{1})$ with the product $\langle\, ,\, \rangle$. Moreover, $\mathcal{O}(3;\underline{1})$ is spanned by the set $\{(1 + x_1)^i x_2^j (x_3 + x_1x_2)^k \mid 0 \le i, j, k \le p - 1\}$. Note that (see Equations (4.2.12)–(4.2.16))

$$\langle 1 + x_1, x_2\rangle = 1, \quad \langle 1 + x_1, x_3 + x_1x_2\rangle = 2(1 + x_1), \quad \langle x_2, x_3 + x_1x_2\rangle = 0.$$

Equation (4.2.18) gives

$$\langle f, \prod_{l=1}^{r} g_l\rangle = \sum_{j=1}^{r} \langle f, g_j\rangle \prod_{l\neq j} g_l + 2(r-1)\partial_3(f) \prod_{l=1}^{r} g_l,$$

and thereby shows

$$\begin{aligned}
&\big\langle (1 + x_1)^i x_2^j (x_3 + x_1x_2)^k, (1 + x_1)^q x_2^r (x_3 + x_1x_2)^s\big\rangle \\
&= q\big\langle (1 + x_1)^i x_2^j (x_3 + x_1x_2)^k, (1 + x_1)\big\rangle (1 + x_1)^{q-1} x_2^r (x_3 + x_1x_2)^s \\
&\quad + r\big\langle (1 + x_1)^i x_2^j (x_3 + x_1x_2)^k, x_2\big\rangle (1 + x_1)^q x_2^{r-1} (x_3 + x_1x_2)^s \\
&\quad + s\big\langle (1 + x_1)^i x_2^j (x_3 + x_1x_2)^k, x_3 + x_1x_2\big\rangle (1 + x_1)^q x_2^r (x_3 + x_1x_2)^{s-1} \\
&\quad + 2k(q + r + s - 1)(1 + x_1)^{i+q} x_2^{j+r} (x_3 + x_1x_2)^{k+s-1} \\
&= qj\big\langle x_2, 1 + x_1\big\rangle (1 + x_1)^{i+q-1} x_2^{j+r-1} (x_3 + x_1x_2)^{k+s} \\
&\quad + qk\big\langle x_3 + x_1x_2, 1 + x_1\big\rangle (1 + x_1)^{i+q-1} x_2^{j+r} (x_3 + x_1x_2)^{k+s-1} \\
&\quad + ri\big\langle 1 + x_1, x_2\big\rangle (1 + x_1)^{i+q-1} x_2^{j+r-1} (x_3 + x_1x_2)^{k+s} \\
&\quad + rk\big\langle x_3 + x_1x_2, x_2\big\rangle (1 + x_1)^{i+q} x_2^{j+r-1} (x_3 + x_1x_2)^{k+s-1} \\
&\quad + si\big\langle 1 + x_1, x_3 + x_1x_2\big\rangle (1 + x_1)^{i+q-1} x_2^{j+r} (x_3 + x_1x_2)^{k+s-1} \\
&\quad + sj\big\langle x_2, x_3 + x_1x_2\big\rangle (1 + x_1)^{i+q} x_2^{j+r-1} (x_3 + x_1x_2)^{k+s-1} \\
&\quad + \big(2k(q + r + s - 1) - 2s(i + j + k - 1)\big)(1 + x_1)^{i+q} x_2^{j+r} (x_3 + x_1x_2)^{k+s-1}
\end{aligned}$$

$$\begin{aligned}&= (ri-qj)(1+x_1)^{i+q-1}x_2^{j+r-1}(x_3+x_1x_2)^{k+s}\\ &\quad + \big(-2qk+2si+2kq+2kr+2ks-2k-2si-2sj-2sk+2s\big)\\ &\quad \times (1+x_1)^{i+q}x_2^{j+r}(x_3+x_1x_2)^{k+s-1}\\ &= (ri-qj)(1+x_1)^{i+q-1}x_2^{j+r-1}(x_3+x_1x_2)^{k+s}\\ &\quad + \big(2k(r-1)-2s(j-1)\big)(1+x_1)^{i+q}x_2^{j+r}(x_3+x_1x_2)^{k+s-1}.\end{aligned}$$

This proves the claim on $C_{\mathfrak{g}}(t)$.

Next we consider the case $i = j$ and $q = r$,

$$\begin{aligned}&\langle (1+x_1)^i x_2^i (x_3+x_1x_2)^k, (1+x_1)^q x_2^q (x_3+x_1x_2)^s\rangle =\\ &\qquad\qquad (2k(q-1)-2s(j-1))(1+x_1)^{i+q}x_2^{i+q}(x_3+x_1x_2)^{k+s-1}.\end{aligned}$$

This proves that $\sum_{q>0}\sum_{s=0}^{p-1} D_K\big((1+x_1)^q x_2^q (x_3+x_1x_2)^s\big)$ is a nilpotent ideal in $C_{\mathfrak{g}}(t)$, and if $i = j = q = r = 0$, then

$$\langle (x_3+x_1x_2)^k, (x_3+x_1x_2)^s\rangle = 2(s-k)(x_3+x_1x_2)^{k+s-1}.$$

This shows that the mapping

$$C_{\mathfrak{g}}(t)/\operatorname{rad} C_{\mathfrak{g}}(t) \to W(1;\underline{1}),\ D_K((x_3+x_1x_2)^k) \mapsto 2x^k\partial$$

is an isomorphism.

(3) Suppose $t = aD_K(x_1x_2) + bD_K(x_3)$ and $a, b \in \mathbb{F}_p$. Equations (4.2.14) and (4.2.15) show that

$$C_{\mathfrak{g}}(t) = \operatorname{span}\{D_K(x_1^i x_2^j x_3^k) \mid a(j-i) + b(i+j+2k-2) = 0\}.$$

This proves the claim on $C_{\mathfrak{g}}(t)$ in cases (a)–(c). Now suppose $a \neq 0, b \neq 0, a\pm b \neq 0$. Then the triples

$$(i,j,k) \in \{(0,0,0),(1,0,0),(0,1,0),(2,0,0),(0,2,0)\}$$

do not occur as exponents in $C_{\mathfrak{g}}(t)$. Therefore $C_{\mathfrak{g}}(t) \subset K(3;\underline{1})_{(1)} + FD_K(x_1x_2) + FD_K(x_3)$. This is the claim in case (d).

Next we are going to prove the statements in (a)–(c) on the radical. From that, one easily concludes on the semisimple quotients. Using the reformulation of Equation (4.2.8) for ordinary powers instead of divided powers, one obtains

$$\begin{aligned}\langle x_1^i x_2^j x_3^k, x_1^q x_2^r x_3^s\rangle &= (ir-jq)x_1^{i+q-1}x_2^{j+r-1}x_3^{k+s}\\ &\quad + \big((q+r+2s-2)k-(i+j+2k-2)s\big)x_1^{i+q}x_2^{j+r}x_3^{k+s-1}.\end{aligned}$$

In case (a) we have $i = j$ and $q = r$. Then $\langle x_1^i x_2^i x_3^k, x_1^q x_2^q x_3^s\rangle = ((2q-2)k - (2i-2)s)x_1^{i+q} x_2^{i+q} x_3^{k+s-1}$. Thus the elements with $i > 0$ span a nilpotent ideal.

Suppose $j + k \equiv 1\ (p)$, $r + s \equiv 1\ (p)$, which is case (b). Set

$$V := \sum_{q=0}^{p-1} \sum_{r \geq 2} F x_1^q x_2^r x_3^{p+1-r}.$$

If $j + r > 2$, then $\langle x_1^i x_2^j x_3^k, x_1^q x_2^r x_3^s\rangle \in V$. If $j = 0$ and $r = 2$, then $k = 1$, $s = p-1$ and

$$\begin{aligned}\langle x_1^i x_3, x_1^q x_2^2 x_3^{p-1}\rangle &= 2i x_1^{i+q-1} x_2 x_3^p + (q+i-2) x_1^{i+q} x_2^2 x_3^{p-1}\\ &= (q+i-2) x_1^{i+q} x_2^2 x_3^{p-1} \in V.\end{aligned}$$

These computations show that V is an ideal. The complementary space in $C_{\mathfrak{g}}(t)$ is represented by $\sum_{i=0}^{p-1} F x_1^i x_2 + \sum_{i=0}^{p-1} F x_1^i x_3$. Put

$$W := \sum_{q=0}^{p-1} F x_1^q (x_3 + x_1 x_2).$$

One has

$$\begin{aligned}\langle x_1^i x_2, x_1^q(x_3 + x_1x_2)\rangle &= -q x_1^{q+i-1} x_3 - (i-1) x_1^{q+i} x_2 + (i-(q+1)) x_1^{q+i} x_2\\ &= -q x_1^{q+i-1}(x_3 + x_1 x_2) \in W,\\ \langle x_1^i x_3, x_1^q(x_3 + x_1x_2)\rangle &= ((q-2)-(i-2)) x_1^{i+q} x_3 + i x_1^{i+q} x_3 + q x_1^{i+q+1} x_2\\ &= q x_1^{i+q}(x_3 + x_1 x_2) \in W.\end{aligned}$$

Consequently, $W + V$ is an ideal which is contained in $K(3;\underline{1})_{(1)} + Ft$. Therefore it is solvable. It is complemented by the space $\sum_{i=0}^{p-1} F D_K(x_1^i x_2) \cong W(1;\underline{1})$.

In case (c) we have $a = 0$, which means $t = bD_K(x_3)$. Then $C_{\mathfrak{g}}(t) \subset K(3;\underline{1})_{(0)}$ and $D_K(x_1^2), D_K(x_2^2) \in C_{\mathfrak{g}}(t)$. Consequently, $\operatorname{rad} C_{\mathfrak{g}}(t) = C_{\mathfrak{g}}(t) \cap K(3;\underline{1})_{(1)} + Ft$. □

Corollary 17.2.9. *Let $\mathfrak{t}$ be a 2-dimensional torus in $K(3;\underline{1})$. For any nonzero root $\alpha \in \Gamma(K(3;\underline{1}), \mathfrak{t})$ one has*

$$Q(K(3;\underline{1})(\alpha)) = K(3;\underline{1})_{(0)} \cap K(3;\underline{1})(\alpha) + \mathfrak{t} \cap (\ker \alpha).$$

Proof. Choose a toral element t_α such that $\mathfrak{t} \cap (\ker \alpha) = Ft_\alpha$. Note that $\mathfrak{t} \cap (\ker \alpha) \subset K(3;\underline{1})(\alpha)$. We have to distinguish several cases.

(i) Suppose $t_\alpha \notin K(3;\underline{1})_{(-1)}$. Lemma 7.5.14 states that one may normalize $\psi(t_\alpha) = D_K(1 + x_3)$, while Theorem 7.5.15 establishes an isomorphism of Lie algebras

$$\Gamma : C_{K(3;\underline{1})}(D_K(1+x_3)) \to \mathcal{P}(2;\underline{1}),\ D_K\big(x_1^a x_2^b (1+x_3)^{\frac{2-(a+b))}{2}}\big) \mapsto x_1^a x_2^b.$$

Here $\mathcal{P}(2;\underline{1})$ is the Poisson central extension of $H(2;\underline{1})^{(1)}$ (Volume 1, p. 403). Therefore α is Hamiltonian and, since $\Gamma(FD_K(1+x_3)) = F$, we obtain that $K(3;\underline{1})_{(0)} \cap K(3;\underline{1})(\alpha) + F\psi(t_\alpha)$ contains $\Gamma^{-1}(F + \sum_{a+b\geq 2} Fx_1^a x_2^b) = \psi\big(Q(K(3;\underline{1})(\alpha))\big)$ as a subalgebra of codimension 2. Since ψ stabilizes $K(3;\underline{1})_{(0)}$, Lemma 17.2.1 proves the claim.

(ii) Suppose $t_\alpha \in K(3;\underline{1})_{(-1)} \setminus K(3;\underline{1})_{(0)}$. Due to Theorem 16.1.3 we may normalize $\psi(t_\alpha) = D_K((1+x_1)x_2)$. Theorem 17.2.8(2) gives that

$$\psi(K(3;\underline{1})(\alpha)) = \sum_{0\leq i,j\leq p-1} FD_K((1+x_1)^i x_2^i (x_3 + x_1 x_2)^j)$$

and that α is of the Witt type. One concludes

$$\begin{aligned}\psi(K(3;\underline{1})(\alpha)) &= \psi(K(3;\underline{1})(\alpha)) \cap K(3;\underline{1})_{(0)} + FD_K((1+x_1)x_2) + FD_K(1)\\ &= \psi(K(3;\underline{1})(\alpha)) \cap K(3;\underline{1})_{(0)} + F\psi(t_\alpha) + FD_K(1).\end{aligned}$$

Since ψ stabilizes $K(3;\underline{1})_{(0)}$, Lemma 17.2.1 proves the claim.

(iii) Suppose $t_\alpha \in K(3;\underline{1})_{(0)}$. Theorem 7.5.13 shows that we may assume $\psi(t_\alpha) = aD_K(x_1x_2) + bD_K(x_3)$, $a, b \in F$. Since $\psi(t_\alpha)$ is toral, one has $a, b \in \mathbb{F}_p$. If $K(3;\underline{1})(\alpha) \subset K(3;\underline{1})_{(0)}$ then Lemma 17.2.1 gives the claim. So assume that $\psi(t_\alpha)$ does annihilate one of $FD_K(1)$, $FD_K(x_1)$, $FD_K(x_2)$. Then $b = 0$, $a+b = 0$, or $a-b = 0$, respectively. In either case $\dim K(3;\underline{1})(\alpha)/K(3;\underline{1})_{(0)}(\alpha) = 1$. Theorem 17.2.8(3) shows that in the first and third case α is Witt. The second case is transformed into the third one by the automorphism $x_1 \mapsto x_2$, $x_2 \mapsto -x_1$, $x_3 \mapsto x_3$. Hence, also in this case α is of the Witt type. Since ψ stabilizes $K(3;\underline{1})_{(0)}$, Lemma 17.2.1 proves the claim. □

Corollary 17.2.10.

(1) *Let $\mathfrak{g}$ be one of the simple Cartan type algebras $W(2;\underline{1})$, $S(3;\underline{1})^{(1)}$, $H(4;\underline{1})^{(1)}$, $K(3;\underline{1})$, $W(1;\underline{2})$, $H(2;(1,2))^{(2)}$, and $\mathfrak{g}_{(0)}$ the natural subalgebra of maximal dimension. Suppose $\mathfrak{t} \subset \mathfrak{g}_{[p]}$ is a 2-dimensional torus and $\alpha \in \Gamma(\mathfrak{g},\mathfrak{t})$ is a nonzero root. The following holds.*

$$Q(\mathfrak{g}(\alpha)) = \mathfrak{g}_{(0)} \cap \mathfrak{g}(\alpha) + (\mathfrak{g} \cap \mathfrak{t}) \cap (\ker \alpha).$$

(2) *Let $\mathfrak{g} = H(2;\underline{1};\Phi(1))$. Suppose $\mathfrak{t} \subset \mathfrak{g}_{[p]}$ is a 2-dimensional torus and $\alpha \in \Gamma(\mathfrak{g},\mathfrak{t})$ is a nonzero root. Then*

$$Q(\mathfrak{g}(\alpha)) = \mathfrak{g}(\alpha) \cap \big(W(2;\underline{1})_{(0)} + \mathfrak{t} \cap (\ker \alpha)\big).$$

Proof. (a) The statement follows from Corollaries 17.2.3, 17.2.5, 17.2.7 and 17.2.9 if $\mathfrak{g}$ is one of $W(2;\underline{1})$, $S(3;\underline{1})^{(1)}$, $H(4;\underline{1})^{(1)}$, $K(3;\underline{1})$.

(b) Suppose $\mathfrak{g} = W(1;\underline{2})$: If Theorem 7.6.2 or Theorem 7.6.5(3) applies, then $\mathfrak{g}(\alpha) \cong W(1;\underline{1})$. In this case $\mathfrak{g}(\alpha) \cap W(1;\underline{2})_{(0)}$ is a subalgebra of codimension at most 1 in $\mathfrak{g}(\alpha) \cap W(1;\underline{2})$. Lemma 17.2.1 proves the claim.

In the remaining case, Theorem 7.6.5(4) applies. Then α is solvable. In the proof of that part of the theorem in Volume 1, p. 426, we mentioned that $\mathfrak{g}(\alpha) = \sum_{j=0}^{p-1} F(x\partial + w_{1,j\alpha}) \subset W(1;\underline{2})_{(0)}$. This is the claim.

(c) Suppose $\mathfrak{g} = H(2;(1,2))^{(2)}$: If Corollary 10.2.9 or Theorem 10.2.14(4) or Theorem 10.2.16(3) applies, then $\mathfrak{g}(\alpha) \cong H(2;\underline{1})^{(2)}$. As $\mathfrak{g}(\alpha) \cap H(2;(1,2))_{(0)}$ has codimension at most 2, Lemma 17.2.1 proves the claim in this case.

Consider the case of Theorem 10.2.14(3). Then α is solvable. In the proof of that part of the theorem in Volume 2, p. 33, we mentioned that $\mathfrak{g}(\alpha) \subset \sum_{i,k=0}^{p-1} FD_H(u_0^i(1+u_2)^{kj})$ for some $j \in \mathbb{F}_p^*$. Here u_0 and u_2 are described in Lemma 10.2.13. In particular, $u_0 \in \mathcal{O}(1;(1,2))_{(1)}$, $u_2 \in \mathcal{O}(1;(1,2))_{(p)}$. One obtains that $(\mathfrak{g} \cap \mathfrak{t}) \cap (\ker \alpha) = FD_H(u_0)$, and we conclude

$$\mathfrak{g}(\alpha) \subset \sum_{i,k=0}^{p-1} FD_H(u_0^i(1+u_2)^{kj}) \subset FD_H(u_0) + \mathfrak{g}_{(0)}.$$

This is the claim.

Finally, consider the case of Theorem 10.2.16(2). Then α is solvable. In the proof of that part of the theorem in Volume 2, p. 38, we mentioned that

$$H(2;(1,2))^{(2)}(\alpha) \subset Ft_1 + C_{H(2;(1,2))^{(2)}{}_{(1)}}(t_1).$$

Here t_1 spans $H(2;(1,2))^{(2)} \cap \mathfrak{t} = \mathfrak{t} \cap (\ker \alpha)$. This is the claim.

(d) Suppose $\mathfrak{g} = H(2;\underline{1};\Phi(1))$: In the cases of Theorems 16.2.3 and 16.2.4, the claim follows. In the case of Theorem 16.2.5 one obtains

$$Q(H(2;\underline{1};\Phi(1))(\alpha)) = H(2;\underline{1};\Phi(1))(\alpha) \cap H(2;\underline{1};\Phi(1))_{(0)},$$
$$\mathfrak{t} \cap (\ker \alpha) \subset W(2;\underline{1})_{(0)}.$$

This is the claim. □

17.3 On the $[p]$-nilpotency of elements

Next we want to show that some distinguished elements are $[p]$-nilpotent.

Theorem 17.3.1. *Let $\alpha \in \Gamma(L,T)$ satisfy $\alpha(H) \neq 0$.*

(1) *Suppose α is solvable. Then $\gamma_4(L_{[p]}(\alpha)) \subset \operatorname{nil} L(\alpha)$ holds.*
(2) *Suppose α is Witt or $L[\alpha,\beta] \not\cong \mathcal{M}(1,1)$ for every $\beta \in \Gamma(L,T)$. Then $\big(\operatorname{ad} Q(L_{[p]}(\alpha))\big)^3\big(Q(L_{[p]}(\alpha))_{(1)}\big)$ acts nilpotently on L. Moreover, if α is proper, then in addition every $x \in Q(L_{[p]}(\alpha)) \cap L_\alpha$ acts nilpotently on L.*

(3) *Suppose α is Hamiltonian and there is $\beta \in \Gamma(L, T)$ with $L[\alpha, \beta] \cong \mathcal{M}(1, 1)$. In this case $\big(\mathrm{ad}\, Q(L_{[p]}(\alpha))\big)^3\big(\mathrm{rad}\, L_{[p]}(\alpha)\big)$ acts nilpotently on L.*

(4) *Suppose α is classial and there is $\beta \in \Gamma(L, T)$ with $L[\alpha, \beta] \cong \mathcal{M}(1, 1)$. In this case*

$$\sum_{i,j \in \mathbb{F}_p^*} [(\mathrm{rad}\, L_{[p]}(\alpha))_{i\alpha}, (\mathrm{rad}\, L_{[p]}(\alpha))_{j\alpha}]$$

consists of $[p]$-nilpotent elements.

Proof. (a) If α is solvable or classical, then α is proper. Otherwise there is a toral element $h \in Q(L_{[p]}(\alpha))$, which image spans the proper torus of $L_{[p]}[\alpha]$. Then $T' := Fh \oplus (T \cap (\ker \alpha))$ is a torus of maximal dimension. Define a T'-root α' by setting $\alpha'(h) = 1$, $\alpha'\big(T \cap (\ker \alpha)\big) = 0$. Then $L_{[p]}(\alpha) = C_{L_{[p]}}\big(T \cap (\ker \alpha)\big) = L_{[p]}(\alpha')$ is a 1-section with respect to T', and T' acts as a proper torus on it. Moreover, $L_{[p]}(\alpha, \beta) = C_{L_{[p]}}(T \cap (\ker \alpha) \cap (\ker \beta)) = C_{L_{[p]}}(T' \cap (\ker \alpha') \cap (\ker \beta))$, and both T and T' act naturally on this T-section as tori of the same dimension, which is at most 2. So $L_{[p]}(\alpha, \beta)$ can be considered a section $L_{[p]}(\alpha', \beta')$ with respect to T'. In other words, we may assume that α is a proper root. Therefore the torus $\Psi_\alpha(T)$ respects the natural subalgebra of maximal dimension in $L_{[p]}[\alpha]$ (cf. Theorem 17.1.8(1)). Then $\Psi_\alpha(T)$ is conjugate to $Fx\partial$ or $FD_H(x_1x_2)$, respectively. In particular, $\Psi_\alpha(\tilde{H})$ is contained in the natural subalgebra of maximal dimension. This means $\tilde{H} \subset Q(L_{[p]}(\alpha))$. We abbreviate $Q := Q(L_{[p]}(\alpha))$.

(b) Set

$$\mathcal{S}_1 := Q_\alpha \cup_{i \in \mathbb{F}_p^*} Q_{(1), i\alpha} \cup_{i \in \mathbb{F}_p^*} [Q_{i\alpha}, Q_{(1), -i\alpha}].$$

If α is solvable or Witt, then $Q = \tilde{H} + Q_{(1)}$, and if α is classical or Hamiltonian, then $Q/Q_{(1)} \cong \mathfrak{sl}(2)$. In the first case $Q_\alpha \in Q_{(1)}$, and in the second case Q_α is at most 1-dimensional modulo $Q_{(1)}$. Consequently, $\mathcal{S}_1$ is a Lie set. Note that

$$\gamma_4(L_{[p]}(\alpha)) \subset (\mathrm{span}\, \mathcal{S}_1) + \gamma_4(\tilde{H}) \qquad \text{if } \alpha \text{ is solvable,}$$
$$Q_\alpha + (\mathrm{ad}\, Q)^3(Q_{(1)}) \subset (\mathrm{span}\, \mathcal{S}_1) + \gamma_4(\tilde{H}), \qquad \text{otherwise.}$$

Theorem 17.1.7(1) shows that $\gamma_4(\tilde{H}) \subset \mathrm{rad}_p\, \tilde{H}$. Since $\tilde{H} \subset Q$, $\mathcal{S}_1 \cup (\mathrm{rad}_p\, \tilde{H})$ is also a Lie set.

If $\mathcal{S}_1$ consists of $[p]$-nilpotent elements, then the ENGEL–JACOBSON Theorem proves all statements. So we may assume that the set

$$\Omega_1 := \{\mu \in \Gamma(L, T) \mid \mu(\mathcal{S}_1) \neq 0\}$$

is non-void (for the notion of extended roots see §1.3). Since L is simple, we obtain $H = \sum_{\mu \in \Omega_1} [L_\mu, L_{-\mu}]$. As $\alpha(H) \neq 0$, there is $\beta_1 \in \Omega_1$ such that

$\alpha([L_{\beta_1}, L_{-\beta_1}]) \neq 0$. Clearly $\alpha \notin \Omega_1$. Proposition 17.1.1(3) applies and shows the following. Set

$$\mathfrak{g}_1 := \sum_{\gamma \in (\mathbb{F}_p\alpha + \mathbb{F}_p\beta_1)\setminus\{0\}} (L_\gamma + [L_\gamma, L_{-\gamma}]), \quad \bar{\mathfrak{g}}_1 := \mathfrak{g}_1/\operatorname{rad}\mathfrak{g}_1.$$

Then $\operatorname{rad}\mathfrak{g}_1$ is nilpotent and $\bar{\mathfrak{g}}_1$ is simple of absolute toral rank 2. More exactly, by construction there are root vectors $u_1 \in \mathcal{S}_1 \subset \mathfrak{g}_1$ and $u_2 \in [\mathfrak{g}_{1,\beta_1}, \mathfrak{g}_{1,-\beta_1}]$ which semisimple parts $u_1^{[p]^e}, u_2^{[p]^e}$ span $T/T \cap (\ker\alpha) \cap (\ker\beta_1)$. Therefore $\mathfrak{t} := T/T \cap (\ker\alpha) \cap (\ker\beta_1)$ acts as a 2-dimensional torus on $\mathfrak{g}_1/\operatorname{rad}\mathfrak{g}_1$ and may be regarded as a torus of maximal dimension in $\bar{\mathfrak{g}}_{1,[p]}$. We may also consider all roots in $\mathbb{F}_p\alpha + \mathbb{F}_p\beta_1$ as $\mathfrak{t}$-roots. Since $\operatorname{rad}\mathfrak{g}_1$ acts nilpotently on $\mathfrak{g}_1$, one has $u_2 \notin \operatorname{rad}\mathfrak{g}_1$. Therefore $\alpha(C_{\bar{\mathfrak{g}}_1}(\mathfrak{t})) \neq 0$ holds.

Let $\pi : \mathfrak{g}_1 \to \bar{\mathfrak{g}}_1$ denote the homomorphism. Note that $\mathcal{S}_1 \subset \mathfrak{g}_1$. Suppose $\pi(\mathcal{S}_1)$ acts nilpotently on $\bar{\mathfrak{g}}_1$. By choice of $\beta_1 \in \Omega_1$, this assumption implies $\mathfrak{g}_{\beta_1} \subset \operatorname{rad}\mathfrak{g}_1$. But then $u_2 \in \operatorname{rad}\mathfrak{g}_1$, which contradicts a former statement. Consequently, $\pi(\mathcal{S}_1)$ contains an element which acts non-nilpotently on $\bar{\mathfrak{g}}_1$.

Since $(\operatorname{rad}\mathfrak{g}_1) \cap \mathfrak{g}_1(\alpha) \subset \operatorname{rad}\mathfrak{g}_1(\alpha)$, there exists the composition of homomorphisms

$$\mathfrak{g}_1(\alpha) \overset{\pi}{\twoheadrightarrow} \bar{\mathfrak{g}}_1(\alpha) \twoheadrightarrow \bar{\mathfrak{g}}_1(\alpha)/\operatorname{rad}\bar{\mathfrak{g}}_1(\alpha) \cong \mathfrak{g}_1(\alpha)/\operatorname{rad}\mathfrak{g}_1(\alpha).$$

If α is non-solvable, then Theorem 11.1.2 with $\mathfrak{g} = L_{[p]}(\alpha)$ and $\mathfrak{h} = \tilde{H}$ shows that every ideal of $L_{[p]}(\alpha)/\operatorname{rad}L_{[p]}(\alpha)$ is semisimple in its own right. Since the image of $\mathfrak{g}_1(\alpha)$ in $L_{[p]}(\alpha)/\operatorname{rad}L_{[p]}(\alpha)$ is an ideal, the above picture can be completed by adding a canonical embedding

$$\Psi_\alpha : \mathfrak{g}_1(\alpha) \overset{\pi}{\twoheadrightarrow} \bar{\mathfrak{g}}_1(\alpha) \twoheadrightarrow \mathfrak{g}_1(\alpha)/\operatorname{rad}\mathfrak{g}_1(\alpha) \hookrightarrow L_{[p]}(\alpha)/\operatorname{rad}L_{[p]}(\alpha) = L_{[p]}[\alpha].$$

So, for non-solvable α, one has

$$\pi\big(Q(\mathfrak{g}_1(\alpha))\big) = Q(\bar{\mathfrak{g}}_1(\alpha)), \ \pi\big(Q(\mathfrak{g}_1(\alpha))_{(1)}\big) = Q(\bar{\mathfrak{g}}_1(\alpha))_{(1)}.$$

These equations are trivially true if α is solvable (as in that case $Q(\mathfrak{g}_1(\alpha)) = \mathfrak{g}_1(\alpha)$ and $Q(\bar{\mathfrak{g}}_1(\alpha)) = \bar{\mathfrak{g}}_1(\alpha)$). Consequently,

$$\pi(\mathcal{S}_1) = Q(\bar{\mathfrak{g}}_1(\alpha))_\alpha \cup_{i\in\mathbb{F}_p^*} Q(\bar{\mathfrak{g}}_1(\alpha))_{(1),i\alpha} \cup_{i\in\mathbb{F}_p^*} [Q(\bar{\mathfrak{g}}_1(\alpha))_{i\alpha}, Q(\bar{\mathfrak{g}}_1(\alpha))_{(1),-i\alpha}].$$

Set $G := \bar{\mathfrak{g}}_1$, $S := \pi(\mathcal{S}_1)$. We have now reduced our problem to the following:

- G is simple of absolute toral rank 2;
- $\mathfrak{t}$ is a 2-dimensional torus in $G_{[p]}$ and α is a proper $\mathfrak{t}$-root;
- $S = Q(G(\alpha))_\alpha \cup_{i\in\mathbb{F}_p^*} Q(G(\alpha))_{(1),i\alpha} \cup_{i\in\mathbb{F}_p^*} [Q(G(\alpha))_{i\alpha}, Q(G(\alpha))_{(1),-i\alpha}]$ acts non-nilpotently on G;
- $\alpha([G_{\beta_1}, G_{-\beta_1}]) \neq 0$.

Since α is a proper root, one has $[G_{\beta_1}, G_{-\beta_1}] \subset Q(G(\alpha))$. So there is $h \in C_{Q(G(\alpha))}(\mathfrak{t})$ for which $\alpha(h) \neq 0$.

(c) Theorem 10.6.3 gives the list of possible algebras for G.

If G is of classical type, then $Q(G(\alpha)) = G(\alpha) \cong \mathfrak{sl}(2) \oplus (\mathfrak{t} \cap (\ker \alpha))$. Consequently, $Q(G(\alpha))_{(1)} = \mathfrak{t} \cap (\ker \alpha)$. As all root vectors for nonzero roots act nilpotently in classical algebras, we conclude that every element of S acts nilpotently, a contradiction.

Suppose G is of Cartan type $W(2;\underline{1})$, $S(3;\underline{1})^{(1)}$, $H(4;\underline{1})^{(1)}$, $K(3;\underline{1})$, $W(1;\underline{2})$ or $H(2;(1,2))^{(2)}$. Corollary 17.2.10(1) shows that

$$Q(G(\alpha)) = G_{(0)} \cap G(\alpha) + (\mathfrak{g} \cap \mathfrak{t}) \cap (\ker \alpha).$$

Choose $h_0 \in G_{(0)} \cap G(\alpha)$ such that $h - h_0 \in (\mathfrak{g} \cap \mathfrak{t}) \cap (\ker \alpha)$. Set $\mathfrak{q} := F h_0 + \operatorname{span} S$. Since $\alpha(h_0) = \alpha(h) \neq 0$, one has

$$Q(G(\alpha))_{i\alpha} = [h_0, Q(G(\alpha))_{i\alpha}] \subset G_{(0)} \cap G(\alpha) \text{ if } i \in \mathbb{F}_p^*.$$

Therefore $\mathfrak{q} \subset G_{(0)}$ and $S \subset \mathfrak{q}^{(1)}$ hold.

If $G = H(2;\underline{1};\Phi(1))$, then $Q(G(\alpha)) = G(\alpha) \cap (W(2;\underline{1})_{(0)} + \mathfrak{t} \cap (\ker \alpha))$ (Corollary 17.2.10(2)). In this case we choose $h_0 \in W(2;\underline{1})_{(0)}$ such that $h - h_0 \in \mathfrak{t} \cap (\ker \alpha)$. Then $\mathfrak{q} \subset W(2;\underline{1})_{(0)}$ and $S \subset \mathfrak{q}^{(1)}$ hold.

Note that $\mathfrak{q} \subset G_{(0)}$ and $\mathfrak{q} \subset W(2;\underline{1})_{(0)}$, resp., is solvable and acts on the space $G/G_{(0)}$ or $W(2;\underline{1})/W(2;\underline{1})_{(0)}$ (either one of dimension $\leq 4 < p$). The representation theory of solvable algebras implies that the derived algebra acts nilpotently on this space. Then every element of S acts nilpotently on the respective space. Let $v \in S$ be an arbitrary element, and set V the Fitting-1-component of $\operatorname{ad} v$ in G. Since v stabilizes the space $G_{(0)}$ ($W(2;\underline{1})_{(0)}$, respectively) and acts nilpotently on the respective quotient space, we conclude that $V \subset G_{(0)}$ or $V \subset W(2;\underline{1})_{(0)} \cap H(2;\underline{1};\Phi(1)) = G_{(0)}$. Due to Proposition 1.3.5, then $V + [V,V] \subset G_{(0)}$ is an ideal of the simple algebra G. So $V = \{0\}$, and this shows that S acts nilpotently on G, a contradiction.

Suppose $G \cong H(2;\underline{1};\Phi(\tau))^{(1)}$. Then $C_G(\mathfrak{t}) = \{0\}$ (Theorem 10.3.2(2)). This contradicts the existence of h. So this case cannot occur at all.

(d) As a consequence, $G \cong \mathcal{M}(1,1)$ is Melikian. Theorem 10.5.13 shows that no root is solvable. In that theorem we consider the 1-dimensional torus $\mathfrak{t} \cap (\ker \alpha)$ and $C := G(\alpha)$. Suppose α is Witt. Then we are in cases (3) or (4) of that theorem. Let as in part (3a) in Volume 2, pp. 100/101 be $\mathfrak{t} \cap (\ker \alpha) = F t_1$ where $t_1 = \tilde{\partial}_1 + x_1 \partial_1$. There we computed

$$C(t_1) = \sum_{i=0}^{4} F(x_2^i \partial_2 - 2i x_2^{i-1} x_1 \partial_1) \oplus J_1,$$

$$J_1 = F t_1 \oplus (J_1 \cap \mathcal{M}(1,1)_{\{1\}}) = \operatorname{rad} C(t_1).$$

Here $\mathcal{M}(1,1)_{\{\cdot\}}$ means the Melikian filtration. Also, $\sum_{i=0}^{4} F(x_2^i\partial_2 - 2ix_2^{i-1}x_1\partial_1) \cong W(1;\underline{1})$ holds under the isomorphism $x_2^i\partial_2 - 2ix_2^{i-1}x_1\partial_1 \mapsto x^i\partial$. From this it is not difficult to conclude that

$$Q(C(t_1)) = \sum_{i=1}^{4} F(x_2^i\partial_2 - 2ix_2^{i-1}x_1\partial_1) \oplus J_1 = C(t_1) \cap \mathcal{M}(1,1)_{\{0\}},$$

$$Q(C(t_1))^{(1)} \subset \sum_{i=2}^{4} F(x_2^i\partial_2 - 2ix_2^{i-1}x_1\partial_1) \oplus \big(J_1 \cap \mathcal{M}(1,1)_{\{1\}}\big) \subset \mathcal{M}(1,1)_{\{1\}}.$$

Then (recall the existence of $h \in Q(C(t_1))$ satisfying $\alpha(h) \neq 0$) $S \subset Q(C(t_1))^{(1)} \subset \mathcal{M}(1,1)_{\{1\}}$ acts nilpotently on $\mathcal{M}(1,1)$. This contradicts our assumption.

Let, as in part (3b) in Volume 2, p. 101, be $\mathfrak{t} \cap (\ker\alpha) = Ft_2$ where $t_2 = \tilde{\partial}_1 + x_2\partial_2$. There we computed

$$C(t_2) = \sum_{i=0}^{4} F(x_1^i\partial_1 - 2ix_1^{i-1}\tilde{\partial}_1) \oplus J_2,$$

$$J_2 = Ft_2 \oplus (J_2 \cap \mathcal{M}(1,1)_{\{1\}}) = \operatorname{rad} C(t_2).$$

Also, $\sum_{i=0}^{4} F(x_1^i\partial_1 - 2ix_1^{i-1}\tilde{\partial}_1) \cong W(1;\underline{1})$ holds under the isomorphism $x_1^i\partial_1 - 2ix_1^{i-1}\tilde{\partial}_1 \mapsto x^i\partial$. From this it is not difficult to compute that

$$Q(C(t_2)) = \sum_{i=1}^{4} F(x_1^i\partial_1 - 2ix_1^{i-1}\tilde{\partial}_1) \oplus J_2 = C(t_2) \cap \mathcal{M}(1,1)_{\{0\}},$$

$$Q(C(t_2))^{(1)} \subset \sum_{i=2}^{4} F(x_1^i\partial_1 - 2ix_1^{i-1}\tilde{\partial}_1) \oplus \big(J_2 \cap \mathcal{M}(1,1)_{\{1\}}\big) \subset \mathcal{M}(1,1)_{\{1\}}.$$

Then $S \subset Q(C(t_2))^{(1)} \subset \mathcal{M}(1,1)_{\{1\}}$ acts nilpotently on $\mathcal{M}(1,1)$. This contradicts our assumption.

Let as in part (4a) in Volume 2, pp. 101/102 be $\mathfrak{t} \cap (\ker\alpha) = Ft_1$ where $t_1 = x_1\partial_1$. We computed there

$$C(t_1) = \sum_{i=0}^{4} Fx_2^i\partial_2 \oplus J_1,$$

$$J_1 = Ft_1 \oplus (J_1 \cap \mathcal{M}(1,1)_{(1)}) = \operatorname{rad} C(t_1).$$

Here $\mathcal{M}(1,1)_{(\cdot)}$ means the natural filtration. Also, $\sum_{i=0}^{4} Fx_2^i\partial_2 \cong W(1;\underline{1})$ holds under the isomorphism $x_2^i\partial_2 \mapsto x^i\partial$. From this it is not hard to compute that

$$Q(C(t_1)) = \sum_{i=1}^{4} Fx_2^i\partial_2 \oplus J_1 = C(t_1) \cap \mathcal{M}(1,1)_{(0)},$$

$$Q(C(t_1))^{(1)} \subset \sum_{i=2}^{4} F x_2^i \partial_2 \oplus \big(J_1 \cap \mathcal{M}(1,1)_{(1)}\big) \subset \mathcal{M}(1,1)_{(1)}.$$

Then $S \subset Q(C(t_1))^{(1)} \subset \mathcal{M}(1,1)_{(1)}$ acts nilpotently on $\mathcal{M}(1,1)$. This contradicts our assumption.

Let as in part (4c) in Volume 2, p. 102 be $\mathfrak{t} \cap (\ker \alpha) = Ft_3$ where $t_3 = x_1\partial_1 + 2x_2\partial_2$. We computed there

$$C(t_3) = \mathcal{M}(1,1)_{\langle 0\rangle} + C(t_3) \cap \mathcal{M}(1,1)_{\{5\}},$$
$$\operatorname{rad} C(t_3) = Ft_3 + C(t_3) \cap \mathcal{M}(1,1)_{\{5\}}.$$

Here $\mathcal{M}(1,1)_{\langle 0\rangle} \cong W(1;\underline{1})$ means the homogeneous degree 0 component in the Melikian grading. Observing the isomorphism mentioned in Volume 2, p. 71, one easily computes

$$Q(C(t_3)) = \big(Fx_1\partial_1 + Fx_1 + Fx_1\tilde{\partial_2} + Fx_1^2\partial_2\big) \oplus \operatorname{rad} C(t_3) \subset Ft_3 + \mathcal{M}(1,1)_{\{0\}},$$
$$Q(C(t_3))^{(1)} = \big(Fx_1 + Fx_1\tilde{\partial_2} + Fx_1^2\partial_2\big) \oplus \big(C(t_3) \cap \mathcal{M}(1,1)_{\{5\}}\big) \subset \mathcal{M}(1,1)_{(1)}.$$

Then $S \subset Q(C(t_1))^{(1)} \subset \mathcal{M}(1,1)_{(1)}$ acts nilpotently on $\mathcal{M}(1,1)$. This contradicts our assumption.

(e) We summarize to prove assertions (1) and (2). If α is solvable, then part (d) does not occur (as mentioned at the beginning of part (d)). Apply (c) to obtain a contradiction. If α is Witt, then the deliberations in (c) exclude the cases where $\bar{\mathfrak{g}}_1$ is of classical or Cartan type. Theorem 10.5.13 shows that part (d) of this proof exhausts all possible cases when $\bar{\mathfrak{g}}_1$ is Melikian. Finally, if no 2-section is of the Melikian type, then part (d) of this proof does not occur. We obtain the claim by part (c) of the proof.

(f) Next, let α be Hamiltonian. Set

$$\mathcal{S}_2 := \cup_{i \in \mathbb{F}_p^*} (\operatorname{rad} L_{[p]}(\alpha))_{i\alpha} \cup_{i \in \mathbb{F}_p^*} [Q_{i\alpha}, (\operatorname{rad} L_{[p]}(\alpha))_{-i\alpha}] \subset \mathcal{S}_1$$

and assume that $\mathcal{S}_2$ acts non-nilpotently. Choose a root β_2 as in (b), and define $\mathfrak{g}_2$ accordingly. Applying (c), we end up with $\bar{\mathfrak{g}}_2 \cong \mathcal{M}(1,1)$. Consider in Theorem 10.5.13 the 1-dimensional torus $\mathfrak{t} \cap (\ker \alpha)$ and $C := \bar{\mathfrak{g}}_2(\alpha)$. We are in cases (1) and (2) of that theorem, in which cases $\operatorname{rad} \bar{\mathfrak{g}}_2(\alpha) \subset \mathfrak{t} \cap (\ker \alpha)$. Consequently, $\pi(\mathcal{S}_2) = \{0\}$, a contradiction. This proves assertion (3).

(g) Finally, let α be classical. Set

$$\mathcal{S}_3 := \cup_{i,j \in \mathbb{F}_p^*} [(\operatorname{rad} L_{[p]}(\alpha))_{i\alpha}, (\operatorname{rad} L_{[p]}(\alpha))_{j\alpha}] \subset \mathcal{S}_1.$$

Choose a root β_3 as in (b) and define $\mathfrak{g}_3$ accordingly. Applying (c), we end up with $\bar{\mathfrak{g}}_3 \cong \mathcal{M}(1,1)$. Consider in Theorem 10.5.13 the 1-dimensional torus $\mathfrak{t} \cap (\ker \alpha)$ and $C := \bar{\mathfrak{g}}_3(\alpha)$. We are in case (4) of that theorem. One has $\mathfrak{t} \cap (\ker \alpha) \subset \mathcal{M}(1,1)_{(0)}$ and $\operatorname{rad} \bar{\mathfrak{g}}_3(\alpha) \subset \mathfrak{t} \cap (\ker \alpha) + \mathcal{M}(1,1)_{(1)}$. Consequently, $\pi(\mathcal{S}_3) \subset (\operatorname{rad} \bar{\mathfrak{g}}_3(\alpha))^{(1)} \subset \mathcal{M}(1,1)_{(1)}$ acts nilpotently, a contradiction. This proves assertion (4). □

Theorem 17.3.2. *If $\alpha \in \Gamma(L,T)$ satisfies $\alpha \neq 0$ and $\alpha(H) = 0$, then*

$$\sum_{i,j\in\mathbb{F}_p^*} [L_{i\alpha}, L_{j\alpha}] + \sum_{i,j\in\mathbb{F}_p^*} [[L_{i\alpha}, L_{-i\alpha}], L_{j\alpha}]$$

consists of $[p]$-nilpotent elements.

Proof. Set $\mathcal{S} := \bigcup_{i,j\in\mathbb{F}_p^*} ([L_{i\alpha}, L_{j\alpha}] \cup [[L_{i\alpha}, L_{-i\alpha}], L_{j\alpha}])$. Assume that the statement is not true. Then the set

$$\Omega := \{\mu \in \Gamma(L,T) \mid \mu(\mathcal{S}) \neq 0\}$$

is not empty. Note that $\alpha \notin \Omega$, because $\alpha(H) = 0$ and $\alpha(L_{i\alpha}{}^{[p]}) = 0$ if $i \neq 0$. The subalgebra $\sum_{\mu\in\Gamma(L,T)} \sum_{i\geq 0} L_\mu{}^{[p]^i}$ of $L_{[p]}$ is closed under $[p]$-th powers and hence coincides with $L_{[p]}$. In particular, this subalgebra contains T. Since $\alpha \neq 0$, there is $\beta \in \Gamma(L,T)$ such that $\alpha(L_\beta{}^{[p]}) \neq 0$.

(a) Suppose $\beta \in \Omega$. Set $\mathfrak{g} := \sum_{\mu\in(\mathbb{F}_p\alpha+\mathbb{F}_p\beta)\setminus\{0\}} (L_\mu + [L_\mu, L_{-\mu}])$. Proposition 17.1.1(3) implies that $\bar{\mathfrak{g}} := \mathfrak{g}/\mathrm{rad}\,\mathfrak{g}$ is simple of absolute toral rank 2. Then $\bar{\mathfrak{g}}(\alpha)$ is a trigonalizable CSA of $\bar{\mathfrak{g}}$ of toral rank 1 in $\bar{\mathfrak{g}}$ (Proposition 11.2.6). This contradicts the choice of $\beta \in \Omega$.

Consequently, $\beta \notin \Omega$. Proposition 1.3.6(1) yields $L_\beta = \sum_{\delta\in\Omega} [L_\delta, L_{\beta-\delta}]$. Then there is $d > 0$, and for $i = 1, \dots, d$ there are $\delta_i \in \Omega$ and $x_i \in [L_{\delta_i}, L_{\beta-\delta_i}]$ such that $\alpha\big((\sum_{j=1}^d x_j)^{[p]}\big) \neq 0$. By JACOBSON's formula,

$$\Big(\sum_{j=1}^d x_j\Big)^{[p]} \equiv \sum_{j=1}^d x_j{}^{[p]} \pmod{H}.$$

In addition, $\sum_{j=1}^d Fx_j{}^{[p]} + H/H \subset \tilde{H}/H$ is abelian. Since α vanishes on H, it is linear on the space $\sum_{j=1}^d Fx_j{}^{[p]}$. Therefore there is $\gamma \in \Omega$ such that $\alpha([L_\gamma, L_{\beta-\gamma}]^{[p]}) \neq 0$.

(b) Let $\mathrm{rad}_T\, L(\alpha,\beta,\gamma)$ denote the maximal T-invariant solvable ideal of $L(\alpha,\beta,\gamma)$. Assume inductively $L_\gamma \in (\mathrm{rad}_T\, L(\alpha,\beta,\gamma))^{(r)}$. Then $[L_\gamma, L_{\beta-\gamma}] \subset (\mathrm{rad}_T\, L(\alpha,\beta,\gamma))^{(r)}$ and hence $\sum_{i\in\mathbb{F}_p^*} L_{i\alpha} \subset (\mathrm{rad}_T\, L(\alpha,\beta,\gamma))^{(r)}$. But then $L_\gamma \in (\mathrm{rad}_T\, L(\alpha,\beta,\gamma))^{(r+1)}$ also holds (recall that $\gamma \in \Omega$). As $L_\gamma \neq \{0\}$, this is impossible.

As a consequence, $L(\alpha,\beta,\gamma)$ is not solvable. Set

$$L[\alpha,\beta,\gamma] := L(\alpha,\beta,\gamma)/\mathrm{rad}_T\, L(\alpha,\beta,\gamma) \neq \{0\},$$

and let

$$\Psi : L(\alpha,\beta,\gamma) + T \to \mathrm{Der}\, L[\alpha,\beta,\gamma]$$

denote the canonical homomorphism. Identify $L[\alpha,\beta,\gamma]$ and $\Psi(L(\alpha,\beta,\gamma))$. Arguing as before, we obtain that γ, β and some $i\alpha$, $i \in \mathbb{F}_p^*$, are $\Psi(T)$-roots on $L[\alpha,\beta,\gamma]$. Therefore $\Psi(T)$ has toral rank 3 in $L[\alpha,\beta,\gamma] + \Psi(T)$, and hence is a torus of maximal toral rank in this algebra (as $TR(L[\alpha,\beta,\gamma]) \leq TR(L(\alpha,\beta,\gamma)) \leq 3$ by Theorem 1.3.11(3)).

(c) Let S be a minimal $\Psi(T)$-invariant ideal of $L[\alpha,\beta,\gamma]$ and $S_{[p]}$ its p-envelope in $\operatorname{Der} L[\alpha,\beta,\gamma]$. Put

$$\overline{T}_1 := \Psi(T) \cap \big(S_{[p]} + C(\Psi(T) + S_{[p]})\big).$$

It follows from Theorem 1.2.9 (taking in that theorem $L[\alpha,\beta,\gamma] + \Psi(T)$ for L and its p-envelope in $\operatorname{Der} L[\alpha,\beta,\gamma]$) that $\overline{T}_1$ is a torus of maximal dimension in $S_{[p]} + C(\Psi(T) + S_{[p]})$. Note that $\overline{T}_1 \neq \{0\}$ because S is not nilpotent.

Suppose $\alpha(\overline{T}_1) = 0$. Then either $\gamma(\overline{T}_1) \neq 0$ or $(\beta-\gamma)(\overline{T}_1) \neq 0$, for $\overline{T}_1 \neq \{0\}$. In any event, $[\Psi(L_\gamma), \Psi(L_{\beta-\gamma})] \subset S$ and therefore $\alpha({S_\beta}^{[p]}) \neq 0$. But then $\alpha(\overline{T}_1) \neq 0$, a contradiction. Thus, $\alpha(\overline{T}_1) \neq 0$, forcing $\sum_{i\in\mathbb{F}_p^*} \Psi(L_{i\alpha}) \subset S$. As $\gamma \in \Omega$, there is $t \in \overline{T}_1$ which satisfies $\alpha(t) = 0$, $\gamma(t) \neq 0$. We obtain that

$$\alpha,\gamma \text{ are independent on } \overline{T}_1 \text{ and } \quad \sum_{i,j\in\mathbb{F}_p,\ (i,j)\neq(0.0)} \Psi(L_{i\alpha+j\gamma}) \subset S.$$

In particular, $TR(S) \geq 2$. We also have $TR(S) \leq TR(L[\alpha,\beta,\gamma]) = 3$.

Since S is a minimal $\Psi(T)$-invariant ideal of $L[\alpha,\beta,\gamma]$, there is a realization $S = \mathfrak{s} \otimes \mathcal{O}(m;\underline{1})$ where $\mathfrak{s}$ is a simple algebra. Recall that $S_{[p]} = \mathfrak{s} \otimes \mathcal{O}(m;\underline{1}) + \mathfrak{s}_{[p]} \otimes F$ and $TR(\mathfrak{s}) = TR(S)$ (Theorem 10.6.4(1c), (1d)).

(d) Suppose $TR(\mathfrak{s}) = 3$. Then $\dim \overline{T}_1 = TR(S) = TR(\mathfrak{s}) = 3$. This shows that there is a 3-dimensional torus in $S_{[p]}$, which acts as $\Psi(T)$ on S. In particular, every ideal of S is $\Psi(T)$-invariant. Since $\operatorname{rad} S = \mathfrak{s} \otimes \mathcal{O}(m;\underline{1})_{(1)}$ is $[p]$-nilpotent, the mapping

$$\varphi : S_{[p]} \to S_{[p]}/\operatorname{rad} S \cong \mathfrak{s}_{[p]}$$

is a restricted homomorphism, and there is a torus $\mathfrak{t}_0 \subset \mathfrak{s}_{[p]}$, which acts as $\Psi(T)$ on $\mathfrak{s}$. Arguing as in (b), one concludes that γ, β and some $i\alpha$, $i \in \mathbb{F}_p^*$ are $\mathfrak{t}_0$-roots on $\mathfrak{s}$. Set $\mathfrak{t} := \mathfrak{t}_0 \cap (\ker \alpha)$ and $\mathfrak{h} := C_{\mathfrak{s}}(\mathfrak{t}) = \mathfrak{s}(\alpha)$. The construction shows that "Setting 12.5" is fulfilled for $\mathfrak{g} = \mathfrak{s}$. Since $\dim \mathfrak{t} = 2$, the deliberations of Chapter 12 imply that $\mathfrak{s} \cong \mathcal{M}(1,1)$. But $TR(\mathcal{M}(1,1)) = 2 \neq TR(\mathfrak{s})$, a contradiction.

(e) As a consequence, $TR(S) = 2$.

(i) Suppose $\Psi(T) \not\subset ((\operatorname{Der} \mathfrak{s}) \otimes \mathcal{O}(m;\underline{1})) \rtimes (\operatorname{Id} \otimes W(m;\underline{1})_{(0)})$. Recall that α, γ are independent on $\overline{T}_1$ and $\overline{T}_1 \subset (\operatorname{Der} \mathfrak{s}) \otimes \mathcal{O}(m;\underline{1})$. Hence

$$\Psi(T) = \overline{T}_1 \oplus \big(\Psi(T) \cap (\ker \alpha) \cap (\ker \gamma)\big),$$

$$\pi_2\big(\Psi(T) \cap (\ker \alpha) \cap (\ker \gamma)\big) \not\subset W(m;\underline{1})_{(0)}.$$

Since $\dim \Psi(T)/\overline{T}_1 = 1$, Theorem 3.6.1 (where we take the trivial grading of $A := S$) shows that we can choose the realization of S in such a way that

$$\Psi(T) \cap (\ker \alpha) \cap (\ker \gamma) = F\mathrm{Id} \otimes (1 + x_1)\partial_1.$$

Then $S(\alpha, \gamma) = \mathfrak{s} \otimes F[x_2, \dots, x_m]$ and the evaluation map ev: $S(\alpha, \gamma) \twoheadrightarrow \mathfrak{s}$, taking $s \otimes f \in \mathfrak{s} \otimes F[x_2, \dots, x_m]$ to $f(0)s \in \mathfrak{s}$, is $\Psi(T)$-equivariant. Then $\mathfrak{s}(\alpha)$ is a trigonalizable CSA of $\mathfrak{s}$ of toral rank 1 in $\mathfrak{s}$ (Proposition 11.2.6). As a consequence, $S(\alpha)^{(1)}$ acts nilpotently on all $\mathfrak{s} \otimes \mathcal{O}(m; \underline{1})_{(i)}/\mathfrak{s} \otimes \mathcal{O}(m; \underline{1})_{(i+1)}$ for $i \geq 0$. Therefore $S(\alpha)^{(1)}$ acts nilpotently on S, which contradicts the choice of $\gamma \in \Omega$.

(ii) Suppose $\Psi(T) \subset ((\mathrm{Der}\, \mathfrak{s}) \otimes \mathcal{O}(m; \underline{1})) \rtimes (\mathrm{Id} \otimes W(m; \underline{1})_{(0)})$. Because S and $\mathfrak{s} \otimes \mathcal{O}(m; \underline{1})_{(1)}$ are both $\Psi(T)$-invariant, $\Psi(T)$ acts on $\mathfrak{s} \cong S/\mathfrak{s} \otimes \mathcal{O}(m; \underline{1})_{(1)}$ as a torus $\mathfrak{t}_1 \subset \mathrm{Der}\, \mathfrak{s}$. Set $\mathfrak{t} := \mathfrak{t}_1 \cap (\ker \alpha)$ and $\mathfrak{h} := C_{\mathfrak{s}}(\mathfrak{t}) = \mathfrak{s}(\alpha)$. Arguing as in (i), we obtain that $\mathfrak{h}$ acts non-trigonalizably on $\mathfrak{s}$. Therefore the "Setting 12.5" is fulfilled for $\mathfrak{g} = \mathfrak{s}$. Since $\dim \mathfrak{t} \leq 2$, the deliberations of Chapter 12 imply that $\mathfrak{s} \cong \mathcal{M}(1, 1)$. Then $\mathrm{Der}\, \mathfrak{s} \cong \mathfrak{s}$, which implies that $\dim \mathfrak{t}_1 = TR(\mathcal{M}(1, 1)) = 2$. But then Proposition 11.2.6 shows that $\mathfrak{h}$ is a trigonalizable CSA of $\mathfrak{s}$. This contradiction completes the proof of the theorem. □

For the sake of citation we collect some consequences.

Corollary 17.3.3.

(1) *If $\alpha \in \Gamma(L, T)$ is solvable and $\alpha(H) \neq 0$, then*

$$\sum_{i \in \mathbb{F}_p^*} \big(L_{i\alpha} + [L_{i\alpha}, L_{-i\alpha}]\big) \subset \mathrm{nil}\, L(\alpha).$$

(2) *If $\alpha \in \Gamma(L, T)$ is a proper root, then*

$$\sum_{i,j \in \mathbb{F}_p^*} [(\mathrm{rad}\, L(\alpha))_{i\alpha}, (\mathrm{rad}\, L(\alpha))_{j\alpha}]$$

consists of $[p]$-nilpotent elements.

(3) *Suppose $\alpha \in \Gamma(L, T)$ is non-classical, or α is classical and $L[\alpha, \beta] \not\cong \mathcal{M}(1, 1)$ for every $\beta \in \Gamma(L, T)$. Then $(\mathrm{ad}\, Q(\alpha))^3(\mathrm{rad}\, L(\alpha))$ consists of $[p]$-nilpotent elements and $\gamma_4(\mathrm{rad}\, L(\alpha)) \subset \mathrm{nil}\, L(\alpha)$ holds.*

(4) *Suppose T is standard and $\alpha \in \Gamma(L, T)$ is proper. If $\alpha(H) \neq 0$ and α does not fit into a Melikian 2-section, then*

$$[Q(\alpha), Q(\alpha)_{(1)}] \cup_{i \in \mathbb{F}_p^*} Q_{i\alpha} \cup H^{(1)}$$

consists of $[p]$-nilpotent elements.

Proof. (1) Suppose α is solvable and $\alpha(H) \neq 0$. Then $\sum_{i\in\mathbb{F}_p^*}(L_{i\alpha}+[L_{i\alpha},L_{-i\alpha}]) \subset \gamma_4(L_{[p]}(\alpha))$. Therefore Theorem 17.3.1(1) proves (1).

(2) If $\alpha(H) = 0$, then Theorem 17.3.2 proves the statement.

Suppose α is proper and $\alpha(H) \neq 0$. If α is solvable, claim (1) proves (2). So assume that α is non-solvable.

If α is classical, then $H \subset L_{[p]}(\alpha) = Q(L_{[p]}(\alpha))$. Suppose α is Witt or Hamiltonian. Definition 11.2.2 then says that, as α is proper, the torus $\Psi_\alpha(T)$ respects the natural subalgebra of maximal dimension in $L_{[p]}[\alpha]$ (cf. Theorem 17.1.8(1)). Then $\Psi_\alpha(T)$ is conjugate to $Fx\partial$ or $FD_H(x_1x_2)$, respectively. In particular, $\Psi_\alpha(H)$ is contained in this subalgebra of maximal dimension. This gives $H \subset Q(L_{[p]}(\alpha))$. Next Theorem 17.1.8(2) states that $\operatorname{rad} L(\alpha) \subset \operatorname{rad} L_{[p]}(\alpha) \subset Q(L_{[p]}(\alpha))_{(1)}$. Therefore we have in all cases

$$\begin{aligned}\sum_{i,j\in\mathbb{F}_p^*}[(\operatorname{rad} L(\alpha))_{i\alpha},(\operatorname{rad} L(\alpha))_{j\alpha}] &\subset \sum_{i,j\in\mathbb{F}_p^*}[(\operatorname{rad} L_{[p]}(\alpha))_{i\alpha},(\operatorname{rad} L_{[p]}(\alpha))_{j\alpha}]\\ = \sum_{i,j\in\mathbb{F}_p^*}&[(\operatorname{rad} L_{[p]}(\alpha))_{i\alpha},[H,[H,(\operatorname{rad} L_{[p]}(\alpha))_{j\alpha}]]]\\ \subset \big(\operatorname{ad} Q(L_{[p]}(\alpha))\big)^3&\big(\operatorname{rad} L_{[p]}(\alpha)\big) \subset \big(\operatorname{ad} Q(L_{[p]}(\alpha))\big)^3\big(Q(L_{[p]}(\alpha))_{(1)}\big).\end{aligned}$$

Claim (2) follows from Theorem 17.3.1.

(3) If $\alpha(H) = 0$, then $L(\alpha)$ is nilpotent. Theorem 17.1.2 proves claim (3). So assume $\alpha(H) \neq 0$. If α is solvable, then Theorem 17.3.1(1) gives the claim. So assume that α is non-solvable. Note that $\operatorname{rad} L(\alpha) \subset \operatorname{rad} L_{[p]}(\alpha) \subset Q(L_{[p]}(\alpha))_{(1)}$ and $Q(\alpha) \subset Q(L_{[p]}(\alpha))$ by Theorem 17.1.8. Due to our assumptions, we are in cases (2) or (3) of Theorem 17.3.1, which gives the claim.

(4) Note that $H \subset Q(\alpha)$ (as α is proper) and

$$\begin{aligned}[Q(\alpha),Q(\alpha)_{(1)}] + H^{(1)} &= \sum_{i\in\mathbb{F}_p^*} Q(\alpha)_{(1),i\alpha} + \sum_{i\in\mathbb{F}_p^*}[Q(\alpha)_{i\alpha},Q(\alpha)_{(1),-i\alpha}] + H^{(1)}\\ &\subset \big(\operatorname{ad} Q(\alpha)\big)^3\big(Q(\alpha)_{(1)}\big) + H^{(1)}.\end{aligned}$$

The present assumption on T implies that $H^{(1)}$ consists of $[p]$-nilpotent elements. Observe that every $i\alpha$ with $i \in \mathbb{F}_p^*$ is proper as α is so. Since no 2-section $L(\alpha,\beta)$ is of Melikian type, we are in case (2) of Theorem 17.3.1. This theorem proves the claim. □

17.4 2-sections

In the previous section we have seen that the reduction to simple algebras of absolut toral rank 2 yields strong information inside L. A slightly weaker but still fruitful method is to study 2-sections of L relative to T.

For roots $\alpha_i \in \Gamma(L, T)$ we set

$$\begin{aligned}&\mathrm{rad}_T\, L(\alpha_1, \dots, \alpha_s) \text{ the maximal } T\text{-invariant solvable ideal of } L(\alpha_1, \dots, \alpha_s),\\ &L[\alpha_1, \dots, \alpha_s] := L(\alpha_1, \dots, \alpha_s)/\mathrm{rad}_T\, L(\alpha_1, \dots, \alpha_s).\end{aligned}$$

Because of Theorem 17.1.8(2), this new notation does not conflict with an older one introduced in Equation (17.1.1). Let $\tilde{S} = \tilde{S}[\alpha_1, \dots, \alpha_s]$ be the T-socle of $L[\alpha_1, \dots, \alpha_s]$, the sum of all minimal T-invariant ideals of $L[\alpha_1, \dots, \alpha_s]$. The case $\tilde{S} = \{0\}$ is possible and occurs if and only if $L(\alpha_1, \dots, \alpha_s)$ is solvable. Next suppose $\tilde{S} \neq \{0\}$. Then

$$\tilde{S} = \oplus_{i=1}^{r} \tilde{S}_i,$$

where each $\tilde{S}_i$ is a minimal T-invariant ideal of $L[\alpha_1, \dots, \alpha_s]$. Recall that $L(\alpha_1, \dots, \alpha_s)_{[p]}$ denotes the restricted subalgebra of $L_{[p]}$ generated by $L(\alpha_1, \dots, \alpha_s)$. Since $\mathrm{Der}\,\tilde{S}$ is centerless, the natural mapping $L(\alpha_1, \dots, \alpha_s) \to L[\alpha_1, \dots, \alpha_s] \hookrightarrow \mathrm{Der}\,\tilde{S}$ can naturally be extended to a restricted homomorphism

$$\Psi_{\alpha_1,\dots,\alpha_s} : L(\alpha_1, \dots, \alpha_s)_{[p]} + T \longrightarrow \mathrm{Der}\,\tilde{S}, \tag{17.4.1}$$

where we take the associative p-power in $\mathrm{Der}\,\tilde{S}$. One may identitify $L[\alpha_1, \dots, \alpha_s]$ and its image in $\mathrm{Der}\,\tilde{S}$. The p-envelope $L[\alpha_1, \dots, \alpha_s]_{[p]} + \Psi_{\alpha_1,\dots,\alpha_s}(T)$ of $L[\alpha_1, \dots, \alpha_s] + \Psi_{\alpha_1,\dots,\alpha_s}(T)$ in $\mathrm{Der}\,\tilde{S}$ is centerless. Then

$$\Psi_{\alpha_1,\dots,\alpha_s}(L(\alpha_1, \dots, \alpha_s)_{[p]} + T) = L[\alpha_1, \dots, \alpha_s]_{[p]} + \Psi_{\alpha_1,\dots,\alpha_s}(T) \subset \mathrm{Der}\,\tilde{S}$$

is the minimal p-envelope.

Consequently, every torus of maximal toral rank has maximal dimension. Applying Lemma 1.2.6(3) to $L[\alpha_1, \dots, \alpha_s]_{[p]} + \Psi_{\alpha_1,\dots,\alpha_s}(T)$ and $I = \tilde{S}_i$ we obtain that $\Psi_{\alpha_1,\dots,\alpha_s}(T)$ is a torus of maximal dimension in $L[\alpha_1, \dots, \alpha_s]_{[p]} + \Psi_{\alpha_1,\dots,\alpha_s}(T)$ and

$$\dim \left(\Psi_{\alpha_1,\dots,\alpha_s}(T) \cap \tilde{S}_{i,[p]}\right) = TR(\tilde{S}_i) \tag{17.4.2}$$

holds.

In what follows we take $s = 2$, fix roots $\alpha_1 := \alpha, \alpha_2 := \beta$, and set $\overline{T} := \Psi_{\alpha,\beta}(T)$. Note that $r \leq TR(\tilde{S}) \leq TR(L[\alpha, \beta]) \leq TR(L(\alpha, \beta)) \leq 2$ (Theorems 1.2.8(5), 1.2.7(1), 1.2.8(3), 1.3.11(3)). If $r > 1$, then $r = 2 = TR(\tilde{S})$ and $TR(\tilde{S}_i) = 1$ for $i = 1, 2$.

By a similar reasoning we obtain that $\dim(\overline{T} \cap \tilde{S}_{[p]}) = TR(\tilde{S})$. Note that T acts as an at most 2-dimensional torus on the 2-section $L(\alpha, \beta)$. Therefore $\dim \overline{T} \leq 2$ holds. So if $TR(\tilde{S}) = 2$ then $\overline{T} \subset \tilde{S}_{[p]}$ is a torus of maximal dimension.

Theorem 17.4.1. *If $r = 2$, then there are $\mu_1, \mu_2 \in \Gamma(L, T)$ such that*

$$L[\mu_1]^{(1)} \oplus L[\mu_2]^{(1)} \subset L[\alpha, \beta] + \Psi_{\alpha,\beta}(T) \subset L[\mu_1] \oplus L[\mu_2].$$

Moreover, there are $h_i \in L[\mu_i]^{(1)}$ such that $\Psi_{\alpha,\beta}(T) = Fh_1 \oplus Fh_2$.

Proof. The classification of simple Lie algebras of absolute toral rank 1 shows that each $\tilde{S}_i$ is restricted. By Equation (17.4.2) we get $\dim(\overline{T} \cap \tilde{S}_i) = 1$ and hence

$$\overline{T} = (\overline{T} \cap \tilde{S}_1) \oplus (\overline{T} \cap \tilde{S}_2).$$

Pick $\mu_i \in \Gamma(L, T)$ with $\mu_i(\overline{T} \cap \tilde{S}_i) \neq 0$. Since $[\overline{T} \cap \tilde{S}_i, \tilde{S}_j] \subset \tilde{S}_i \cap \tilde{S}_j = \{0\}$ for $i \neq j$, it can only be that

$$\Gamma(L[\alpha, \beta], \overline{T}) \subset \mathbb{F}_p\mu_1 \cup \mathbb{F}_p\mu_2 \text{ and } L[\alpha, \beta] = \big(\tilde{S}_1(\mu_1) \oplus \tilde{S}_2(\mu_2)\big) + \Psi_{\alpha,\beta}(H).$$

Next observe that

$$L[\alpha, \beta](\mu_i) = \tilde{S}_i + \Psi_{\alpha,\beta}(H), \quad (\mathrm{rad}_T\, L(\alpha, \beta)) \cap L(\mu_i) \subset \mathrm{rad}\, L(\mu_i)$$

and let $\pi_i : L[\alpha, \beta](\mu_i) \twoheadrightarrow L[\mu_i]$ denote the canonical homomorphism. Note that $[(\ker \pi_i) \cap \tilde{S}_i, \tilde{S}_j] = \{0\}$ if $i \neq j$. Therefore $(\ker \pi_i) \cap \tilde{S}_i$ is a solvable ideal even of $\tilde{S}_i + \Psi_{\alpha,\beta}(H) + \tilde{S}_j = L[\alpha, \beta] + \overline{T}$. This gives $(\ker \pi_i) \cap \tilde{S}_i = \{0\}$. Then $\pi_i(\tilde{S}_i) \cong \tilde{S}_i$ is a nonzero ideal of $L[\mu_i]$ satisfying $\pi_i(\tilde{S}_i) = \pi_i(\tilde{S}_i)^{(1)}$. According to Theorem 17.1.8, $\pi_i(\tilde{S}_i) = L[\mu_i]^{(1)}$. Now let

$$\pi = \pi_1 \oplus \pi_2 : L[\alpha, \beta] \to L[\mu_1] \oplus L[\mu_2].$$

If $x \in \ker \pi$, then $[x, \tilde{S}_i] \subset (\ker \pi) \cap \tilde{S}_i = \{0\}$. Since $L[\alpha, \beta]$ is isomorphic to a subalgebra of $\mathrm{Der}(\tilde{S}_1 \oplus \tilde{S}_2)$, we have $x = 0$. So π is injective.

Take h_i any nonzero element of $\overline{T} \cap \tilde{S}_i$. □

The semisimple quotients $L[\mu_i]$ are described in Theorem 17.1.8.

A very strong tool in the classification proofs is to switch tori. This will become particularly important in Chapter 20. A first application will be used here.

Lemma 17.4.2. *Let $\alpha, \beta \in \Gamma(L, T)$ be such that $\Psi_{\alpha,\beta}(T)$ is a 2-dimensional torus. Suppose $\mathfrak{t} \subset \Psi_{\alpha,\beta}(L(\alpha, \beta)_{[p]} + T)$ is any 2-dimensional torus. The following holds.*

(1) *$\gamma_4(C_{L[\alpha,\beta]}(\mathfrak{t}))$ acts nilpotently on $L[\alpha, \beta]$.*

(2) *Let $\mu \in \Gamma(L[\alpha, \beta], \mathfrak{t})$ be a solvable $\mathfrak{t}$-root. If there is $x \in L[\alpha, \beta]_\mu$ which acts non-nilpotently on $L[\alpha, \beta]$, then $\mu(C_{L[\alpha,\beta]}(\mathfrak{t})) = 0$.*

(3) *Let $\mu \in \Gamma(L[\alpha, \beta], \mathfrak{t})$ be a solvable $\mathfrak{t}$-root. Then $\sum_{i,j \in \mathbb{F}_p^*} [L[\alpha, \beta]_{i\mu}, L[\alpha, \beta]_{j\mu}]$ acts nilpotently on $L[\alpha, \beta]$.*

Proof. Recall that the restricted homomorphism

$$\Psi_{\alpha,\beta} : L(\alpha, \beta)_{[p]} + T \to L[\alpha, \beta]_{[p]} + \overline{T} \subset \mathrm{Der}\, \tilde{S}$$

is surjective. Choose a torus $T_0 \subset L(\alpha, \beta)_{[p]} + T$ which is mapped onto $\mathfrak{t}$ ([S-F,

Theorem 2.4.5]) and set $T' := T_0 + T \cap (\ker \alpha) \cap (\ker \beta)$. Then

$$\dim T' \geq (\dim \mathfrak{t}) + (\dim T \cap (\ker \alpha) \cap (\ker \beta)) = \dim T.$$

Therefore T' is a torus of maximal dimension in $L_{[p]}$.

Note that $L(\alpha, \beta) = C_L(T \cap (\ker \alpha) \cap (\ker , \beta))$ is a T'-section. Moreover, the torus T' acts on it as a 2-dimensional torus (as even more the image $\mathfrak{t}$ is 2-dimensional). Thus there are $\alpha', \beta' \in \Gamma(L, T')$ such that $L(\alpha, \beta) = L(\alpha', \beta')$. Lemma 1.2.6(2a) shows that $\big(L(\alpha, \beta)_{[p]} + T\big)/\big(L(\alpha, \beta)_{[p]} + T'\big)$ is $[p]$-nilpotent. As this quotient also is toral, $L(\alpha, \beta)_{[p]} + T' = L(\alpha, \beta)_{[p]} + T$ holds. In particular, $\mathrm{rad}_{T'} L(\alpha, \beta) = \mathrm{rad}_T L(\alpha, \beta)$. Hence $L[\alpha, \beta] = L[\alpha', \beta']$ still is the T'-semisimple quotient of a T'-section.

Obviously, $C_{L[\alpha,\beta]}(\mathfrak{t}) = \Psi_{\alpha',\beta'}(C_L(T'))$ holds, and for $\mu \in \Gamma(L[\alpha, \beta], \mathfrak{t})$ there is $\mu' \in \Gamma(L, T')$ such that $L[\alpha, \beta]_{i\mu} = \Psi_{\alpha',\beta'}\big(L(\alpha, \beta)_{i\mu'}\big)$ for all $i \in \mathbb{F}_p^*$. Theorem 17.1.7(1) proves (1), while Corollary 17.3.3(1) proves (2) and Corollary 17.3.3(2) proves (3). □

Theorem 17.4.3. *If $r = 1$ and $TR(\tilde{S}) = 2$, then one of the following holds.*

(1) *$L[\alpha, \beta] = \tilde{S}$ is one of the simple classical algebras A_2, B_2, G_2;*

(2) *$L[\alpha, \beta] = \tilde{S}$ is one of the simple Cartan type algebras $W(2; \underline{1})$, $S(3; \underline{1})^{(1)}$, $K(3; \underline{1})$, $W(1; \underline{2})$, $H(2; \underline{1}; \Phi(\tau))^{(1)}$, $H(2; \underline{1}; \Phi(1))$;*

(3) *$L[\alpha, \beta] = \tilde{S} \cong \mathcal{M}(1, 1)$;*

(4) *$H(4; \underline{1})^{(1)} = \tilde{S} \subset L[\alpha, \beta] \subset H(4; \underline{1})^{(1)} + FD_H(x_1^{p-1} \cdots x_4^{p-1})$;*

(5) *$H(2; (1, 2))^{(2)} = \tilde{S} \subset L[\alpha, \beta] \subset H(2; (1, 2))$.*

In any case, $\Psi_{\alpha,\beta}(T) \subset \tilde{S}_{[p]}$ holds.

Proof. Due to Theorem 17.1.7, Corollary 10.6.5 applies for $\mathfrak{g} := L[\alpha, \beta]$. Our assumption $r = 1$ excludes case (a), Equation (17.4.2) excludes case (b), and the assumption $TR(\tilde{S}) = 2$ excludes case (c). Hence only cases (d) and (e) of that corollary can occur here.

(a) Consider case (d), in which $\tilde{S} = H(2; \underline{1}; \Phi(\tau))^{(1)} \otimes \mathcal{O}(m; \underline{n})$ and $m \neq 0$. Equation (17.4.2) shows that $\overline{T} \subset \tilde{S}_{[p]}$. This gives

$$H(2; \underline{1}; \Phi(\tau))^{(1)} \otimes \mathcal{O}(m; \underline{n}) \subset L[\alpha, \beta] \subset H(2; \underline{1}; \Phi(\tau))^{(1)} \otimes \mathcal{O}(m; \underline{n}) + \Psi_{\alpha,\beta}(H).$$

Moreover, Theorem 10.6.4(3) applies. Since $m \neq 0$ and $\gamma_4(H) \subset \mathrm{nil}\, H$, only part (b) can occur. This means that the subalgebra $C_{\tilde{S}}(\overline{T})$ acts nilpotently on $\tilde{S}$. By the ENGEL–JACOBSON Theorem, each 1-section $\tilde{S}(\mu)$ relative to $\overline{T}$ must be solvable. From this it is immediate that μ is solvable for any $\mu \in (\mathbb{F}_p\alpha + \mathbb{F}_p\beta) \setminus \{0\}$.

Suppose there is $\mu \in (\mathbb{F}_p\alpha + \mathbb{F}_p\beta) \setminus \{0\}$ with $\mu(H) = 0$. As T is a maximal torus in $L_{[p]}$, the 1-section $\mathfrak{h} := L(\mu)$ is nilpotent. Theorem 17.1.2 shows that $\gamma_4(\mathfrak{h}) \subset \mathrm{nil}\, \mathfrak{h}$. If $\mathfrak{h}$ acts nilpotently on $L(\alpha, \beta)$, then $L(\alpha, \beta)$ were solvable by Proposition 1.3.7. As this is not true, one has $TR(\mathfrak{h}, L(\alpha,\beta)) \neq 0$. Next observe that $T \cap \big(\mathfrak{h}_{[p]} + C(L(\alpha,\beta)_{[p]} + T)\big) \subset T \cap (\ker \mu)$. Theorem 1.2.9 yields (setting $K = \mathfrak{h}$)

$$\begin{aligned} TR(\mathfrak{h}, L(\alpha,\beta)) &\leq TR(\mathfrak{h}, L(\alpha,\beta) + T) \\ &\leq TR(L(\alpha,\beta) + T) - \dim T/T \cap \big(\mathfrak{h}_{[p]} + C(L(\alpha,\beta)_{[p]} + T)\big) \\ &\leq TR(L(\alpha,\beta) + T) - \dim T/T \cap (\ker \mu) \\ &\leq 2 - \dim T + \dim T \cap (\ker \mu) = 1, \end{aligned}$$

and $T \cap \big(\mathfrak{h}_{[p]} + C(L(\alpha,\beta)_{[p]} + T)\big) = T \cap (\ker \mu)$. Therefore $TR(\mathfrak{h}, L(\alpha,\beta)) = 1$ holds and $\mathfrak{h}$ is self-normalizing. Set in Theorem 11.1.1 $\mathfrak{g} := L(\alpha,\beta)$. Parts (2), (6), and (3) of that theorem state that $\mathfrak{g}/\mathrm{rad}\,\mathfrak{g}$ has a unique minimal ideal $(\mathfrak{g}/\mathrm{rad}\,\mathfrak{g})^{(\infty)}$ which is simple in its own right. We now recall that $\Psi_{\alpha,\beta}(\mathrm{rad}\,\mathfrak{g})$ is $\overline{T}$-invariant, since $\overline{T} \subset \tilde{S}_{[p]}$. Then $\mathrm{rad}\,\mathfrak{g}$ is T-invariant, whence $\mathfrak{g}/\mathrm{rad}\,\mathfrak{g} = L[\alpha,\beta]$. This in turn implies that $\tilde{S} = (L[\alpha,\beta])^{(\infty)}$ is simple. But then $m = 0$ contrary to our assumption.

Therefore, $\mu(H) \neq 0$ for all $\mu \in (\mathbb{F}_p\alpha + \mathbb{F}_p\beta) \setminus \{0\}$. According to Corollary 17.3.3(1) all elements in the union

$$\bigcup_{\gamma \in (\mathbb{F}_p\alpha + \mathbb{F}_p\beta)\setminus\{0\}} \big(L_\gamma \cup [L_\gamma, L_{-\gamma}]\big)$$

are $[p]$-nilpotent. But then $L(\alpha,\beta)$ is solvable (by the ENGEL–JACOBSON Theorem). This contradiction shows that case (d) does not occur.

We are now in case (e) of Corollary 10.6.5. It follows that $\tilde{S}$ is simple with $TR(\tilde{S}) = 2$. Then $\tilde{S}$ is listed in Theorem 10.6.3 (cf. Theorem 14.3.15).

(b) If $\tilde{S}$ is classical or one of $W(2;\underline{1})$, $K(3;\underline{1})$, $\mathcal{M}(1,1)$, then $\mathrm{Der}\,\tilde{S} \cong \tilde{S}$ (Theorems 7.1.2 and 7.1.4) holds. As $\tilde{S} \subset L[\alpha,\beta] \subset \mathrm{Der}\,\tilde{S}$, the claim follows.

(c) Suppose $\tilde{S} \cong W(1;\underline{2})$. There is a 2-dimensional torus $\mathfrak{t} \subset W(1;\underline{2})_{[p]}$ which is ruled by Theorem 7.6.5. Theorem 7.6.5(4) tells us that there is a $\mathfrak{t}$-root μ which is solvable and there are root vectors in $W(1;\underline{2})_\mu$ which act non-nilpotently on $W(1;\underline{2})$. Due to Lemma 17.4.2, this is only possible if $\mu(C_{L[\alpha,\beta]}(\mathfrak{t})) = 0$. Consequently, $W(1;\underline{2}) \subset L[\alpha,\beta] \subsetneq \mathrm{Der}\,W(1;\underline{2})$. Finally we mention that $\mathrm{Der}\,W(1;\underline{2}) = W(1;\underline{2}) + F\partial^p$ (Theorem 7.2.2). Then $\tilde{S} = L[\alpha,\beta] = W(1;\underline{2})$.

(d) If $\tilde{S} \cong H(2;\underline{1};\Phi(\tau))^{(1)}$, then one argues similarly. Applying Theorem 10.3.2(5) in combination with Lemma 17.4.2(2) one obtains first that

$$\mu(C_{L[\alpha,\beta]}(\overline{T})) = 0 \text{ for all } \mu \in \mathbb{F}_p\alpha + \mathbb{F}_p\beta.$$

This gives $L[\alpha,\beta] \cap \overline{T} = \{0\}$. Next, $\mathrm{Der}\,H(2;\underline{1};\Phi(\tau))^{(1)} = H(2;\underline{1};\Phi(\tau))^{(1)} + \overline{T}$ (Theorems 10.3.1 and 10.3.2(2)). We conclude $L[\alpha,\beta] = H(2;\underline{1};\Phi(\tau))^{(1)}$.

(e) Suppose $\tilde{S} \cong H(2;\underline{1};\Phi(1))$. Then $\mathrm{Der}\,\tilde{S} = \tilde{S} + F(x_1\partial_1 + x_2\partial_2)$ (Theorem 7.1.3(4)). Take in Theorem 16.2.1(2) $\delta_1 = \delta_2 = 1$, $r_1 = -1$, $r_2 \notin \mathbb{F}_p$. We obtain an isomorphic copy $\mathfrak{g} := \mathcal{D}(\mathcal{O}(2;\underline{1})) = \Phi_\sigma(H(2;\underline{1};\Phi(1)))$ where $\mathcal{D}$ is given by Equation (16.2.4). The theorem says that $F(1+x_1)\partial_1 \oplus F(1+x_2)\partial_2 \subset \mathfrak{g}_{[p]}$ is a 2-dimensional torus for which $\mathcal{D}((1+x_1)(1+x_2)) = -(r_2+1)(1+x_1)\partial_1$ holds. Then Theorem 16.2.2(1) shows that the 1-section $\mathfrak{g}(\epsilon_2)$ is abelian and every root space is spanned by a non-nilpotent root vector. Due to Lemma 17.4.2, this is only possible if $\epsilon_2\big(C_{\Phi_\sigma(L[\alpha,\beta])}(\mathfrak{t})\big) = 0$. Consequently, $\tilde{S} \subset L[\alpha,\beta] \subsetneq \operatorname{Der}\tilde{S}$. Then $\tilde{S} = L[\alpha,\beta]$.

(f) Suppose $\tilde{S} \cong S(3;\underline{1})^{(1)}$ and $\Psi_{\alpha,\beta}(H) \not\subset \tilde{S}$. We have

$$\operatorname{Der}\tilde{S} = \tilde{S} \oplus Ft_0 \oplus Fx_1^{p-1}x_2^{p-1}\partial_3 \oplus Fx_1^{p-1}x_3^{p-1}\partial_2 \oplus Fx_2^{p-1}x_3^{p-1}\partial_1,$$

where $t_0 := x_1\partial_1 + x_2\partial_2 + x_3\partial_3$ (Theorem 7.1.2(2)). If $\Psi_{\alpha,\beta}(H)$ contains an element $t_0 + r_1x_2^{p-1}x_3^{p-1}\partial_1 + r_2x_1^{p-1}x_3^{p-1}\partial_2 + r_3x_1^{p-1}x_2^{p-1}\partial_3 + E$ for some $r_i \in F$ and $E \in \tilde{S}$, then $TR(L[\alpha,\beta]) \geq 3$, which is not true. Thus,

$$L[\alpha,\beta] \subset \tilde{S} + Fx_2^{p-1}x_3^{p-1}\partial_1 + Fx_1^{p-1}x_3^{p-1}\partial_2 + Fx_1^{p-1}x_2^{p-1}\partial_3.$$

We may assume by symmetry that

$$L[\alpha,\beta] \not\subset \tilde{S} \oplus Fx_2^{p-1}x_3^{p-1}\partial_1 \oplus Fx_1^{p-1}x_3^{p-1}\partial_2 =: \tilde{S}'.$$

Let $z_1 := 1+x_1$, $z_2 := 1+x_2$, and $\mathfrak{t} := F(z_1\partial_1 - x_3\partial_3) \oplus F(z_2\partial_2 - x_3\partial_3)$, a 2-dimensional torus in $\tilde{S}$. Set $\mathfrak{h} := C_{L[\alpha,\beta]}(\mathfrak{t})$. Since $L[\alpha,\beta]/\tilde{S}$ is a trivial $\tilde{S}$-module, we have $L[\alpha,\beta] = \tilde{S} + \mathfrak{h}$. As $L[\alpha,\beta]$ is not contained in $\tilde{S}'$, it follows that $\mathfrak{h} \not\subset \tilde{S}'$. Therefore there is an element

$$h := z_1^{p-1}z_2^{p-1}\partial_3 + r_1\, z_2^{p-1}x_3^{p-1}\partial_1 + r_2\, z_1^{p-1}x_3^{p-1}\partial_2 + E'$$

contained in $\mathfrak{h}$, where $r_i \in F$ and $E' \in \mathfrak{h} \cap \tilde{S}$. As for all $1 \leq i \leq p-2$ $D_{1,3}(z_1^{i+1}z_2^i x_3^{i+1}) = (i+1)(z_1z_2x_3)^i(z_1\partial_1 - x_3\partial_3)$ is an element of $\mathfrak{h} \cap \tilde{S}$, it is immediate that the following holds

$$\begin{aligned}
&[h, (z_1z_2x_3)^i(z_1\partial_1 - x_3\partial_3)] \\
&\quad = i(z_1z_2x_3)^{i-1}(z_1\partial_1 - x_3\partial_3) + [E', (z_1z_2x_3)^i(z_1\partial_1 - x_3\partial_3)] \\
&\quad \equiv i(z_1z_2x_3)^{i-1}(z_1\partial_1 - x_3\partial_3) \pmod{(\mathfrak{h} \cap \tilde{S})^{(1)}}.
\end{aligned}$$

Therefore, $(\operatorname{ad} h)^{p-2}(D_{1,3}(z_1^{p-1}z_2^{p-2}x_3^{p-1})) \equiv (p-2)!(z_1\partial_1 - x_3\partial_3)$. Since $(\mathfrak{h}\cap\tilde{S})^{(1)}$ acts nilpotently on $\tilde{S}$ by Corollary 12.5.10 and the element on the right is toral in $\tilde{S}$, we have $\gamma_{p-1}(\mathfrak{h}) \not\subset \operatorname{nil}\mathfrak{h}$. Since $p \geq 5$, this contradicts Lemma 17.4.2(1).

(g) Suppose $\tilde{S} = H(4;\underline{1})^{(1)}$ and $H \not\subset \tilde{S}$. Put $\tilde{S}' := \tilde{S} + FD_H(x_1^{p-1}\cdots x_4^{p-1})$. One has

$$\operatorname{Der}\tilde{S} = \tilde{S}' \oplus Ft_0 \oplus \sum_{i=1}^{4} FD_H(x_i^{(p)})$$

with $t_0 := \sum_{i=1}^4 x_i\partial_i$ (Theorem 7.1.2(3)). Since $TR(\tilde{S}) = 2$, it must be that $L[\alpha,\beta] \subset \tilde{S}' \oplus \sum_{i=1}^4 FD_H(x_i^{(p)})$. No generality will be lost by assuming that $L[\alpha,\beta] \not\subset \tilde{S}' + \sum_{i=2}^4 FD_H(x_i^{(p)})$. This time we set

$$\mathfrak{t} := F((1+x_1)\partial_1 - x_3\partial_3) \oplus F(x_2\partial_2 - x_4\partial_4).$$

As before, we set $\mathfrak{h} := C_{L[\alpha,\beta]}(\mathfrak{t})$ and find an element

$$h := (1+x_1)^{p-1}\partial_3 + r_2 x_2^{p-1}\partial_4 + r_3 x_3^{p-1}\partial_1 + r_4 x_4^{p-1}\partial_2 + E$$

in $\mathfrak{h}$, where $r_i \in F$ and $E \in \mathfrak{h} \cap \tilde{S}'$. Applying $(\operatorname{ad} h)^{p-2}$ to the element

$$D_H((1+x_1)^{p-1}x_3^{p-1}) = -(1+x_1)^{p-2}x_3^{p-1}\partial_3 + (1+x_1)^{p-1}x_3^{p-2}\partial_1 \in \mathfrak{h} \cap \tilde{S}'$$

we obtain an element of $F^*((1+x_1)\partial_1 - x_3\partial_3) + (\mathfrak{h} \cap \tilde{S}')^{(1)}$. As before, this contradicts Lemma 17.4.2(1).

(h) Suppose $\tilde{S} = H(2;(1,2))^{(2)}$. By Theorems 7.2.2(5) and 7.1.2(3) one has

$$H(2;(1,2))_{[p]} = H(2;(1,2)) \oplus F\partial_2^p$$

and

$$\operatorname{Der} H(2;(1,2))^{(2)} = H(2;(1,2))_{[p]} \oplus F(x_1\partial_1 + x_2\partial_2).$$

Since $TR(L[\alpha,\beta]) = TR(H(2;(1,2))^{(2)}) = 2$, the restricted quotient

$$L[\alpha,\beta]_{[p]}/H(2;(1,2))^{(2)}{}_{[p]}$$

must be $[p]$-nilpotent (Lemma 1.2.6(2a)). On the other hand, the restricted algebra $\operatorname{Der} H(2;(1,2))^{(2)}/H(2;(1,2))_{[p]}$ is mapped naturally into the torus $F(x_1\partial_1 + x_2\partial_2)$, so one gets $L[\alpha,\beta] \subset H(2;(1,2))_{[p]}$.

Due to the deliberations made in Volume 2, p. 39 there is a torus

$$\mathfrak{t} \subset H(2;(1,2))^{(2)}{}_{[p]} \subset L[\alpha,\beta]_{[p]}$$

which is ruled by Theorem 10.2.16. Then $\mathfrak{t} = Ft_1 \oplus Ft_2$, where $t_1 \in H(2;(1,2))_{(0)}$ and $t_2 = s\partial_2^p + v_2$, $s \neq 0$, $v_2 \in H(2;(1,2))^{(2)}$ (cf. Volume 2, p. 37). Note that $L[\alpha,\beta]_{[p]} \subset H(2;(1,2))_{[p]} = H(2;(1,2)) + Ft_2$. Set $\mathfrak{h} := C_{L[\alpha,\beta]}(\mathfrak{t})$. Looking at the associated graded algebra it is easy to compute that

$$\mathfrak{h} \subset C_{H(2;(1,2))+\mathfrak{t}}(t_1) \subset C_{H(2;(1,2))_{(1)}}(t_1) + \mathfrak{t}.$$

Take any element $h + r_1t_1 + r_2t_2 \in \mathfrak{h}$, where $h \in H(2;(1,2))_{(1)}$ and $r_i \in F$. Clearly, $0 = [\mathfrak{t}, h + r_1t_1 + r_2t_2] = [\mathfrak{t}, h]$. This shows that

$$\mathfrak{h} = \mathfrak{h} \cap H(2;(1,2))_{(1)} + \mathfrak{h} \cap \mathfrak{t}.$$

There is a root $\gamma \neq 0$ such that $\gamma(t_1) = 0$. Then $\tilde{S}(\gamma)$ is nilpotent, hence $L[\alpha,\beta](\gamma)$ is solvable. Moreover, according to Theorem 10.2.16(2) there is an element $d_\gamma \in H(2;(1,2))^{(2)}{}_\gamma$ which acts non-nilpotently. Lemma 17.4.2(2) shows that $\gamma(\mathfrak{h}) = 0$. As a consequence, $\gamma(\mathfrak{h} \cap \mathfrak{t}) = 0$. Since $\gamma(t_1) = 0$ and $\gamma \neq 0$, we get $\mathfrak{h} \cap \mathfrak{t} = Ft_1$. This shows that $\mathfrak{h} \subset H(2;(1,2))$, whence $L[\alpha,\beta] \subset H(2;(1,2))$.

(i) Equation (17.4.2) gives $\overline{T} \subset \tilde{S}_{[p]}$ in all cases. □

Theorem 17.4.4. *If $r = 1$ and $TR(\tilde{S}) = 1$, then one of the following holds.*

(1) $L[\alpha, \beta] = L[\kappa]$ *for some* $\kappa \in (\mathbb{F}_p\alpha + \mathbb{F}_p\beta) \setminus \{0\}$. *Moreover,* $\tilde{S} = L[\kappa]^{(1)}$ *and* $\dim \Psi_{\alpha,\beta}(T) = 1$.

(2) $H(2; \underline{1})^{(2)} \subset L[\alpha, \beta] \subset H(2; \underline{1})^{(1)}$. *Moreover,*

$$\Psi_{\alpha,\beta}(T) = F(\delta_1 + x_1)\partial_1 \oplus F(\delta_2 + x_2)\partial_2,\ \delta_i \in \{0, 1\}.$$

(3) $\tilde{S} = \mathfrak{s} \otimes \mathcal{O}(1; \underline{1})$, *where* $\mathfrak{s}$ *is one of* $\mathfrak{sl}(2)$, $W(1; \underline{1})$, $H(2; \underline{1})^{(2)}$. *Moreover,*

$$L[\alpha, \beta] \subset (\operatorname{Der} \mathfrak{s})^{(1)} \otimes \mathcal{O}(1; \underline{1}),$$

and $\Psi_{\alpha,\beta}(T) = (Fh \otimes 1) \oplus (F\mathrm{Id} \otimes (1 + x_1)\partial_1)$ *where* h *is a nonzero toral element in* $\mathfrak{s}$.

(4) $\tilde{S} = \mathfrak{s} \otimes \mathcal{O}(m; \underline{1})$, *where* $\mathfrak{s}$ *is one of* $\mathfrak{sl}(2)$, $W(1; \underline{1})$, $H(2; \underline{1})^{(2)}$ *and* $m > 0$. *There exists a classical root* $\kappa \in \mathbb{F}_p\alpha + \mathbb{F}_p\beta$ *such that*

$$L[\alpha, \beta] = \mathfrak{s} \otimes \mathcal{O}(m; \underline{1}) + L[\alpha, \beta](\kappa),$$

$$L[\kappa, \nu] \cong \mathcal{M}(1, 1) \text{ for some } \nu \in \Gamma(L, T).$$

(5) $\tilde{S} = \mathfrak{s} \otimes \mathcal{O}(1; \underline{1})$, *where* $\mathfrak{s}$ *is one of* $\mathfrak{sl}(2)$, $W(1; \underline{1})$, $H(2; \underline{1})^{(2)}$, *and* $L[\alpha, \beta]$ *is a subalgebra in* $((\operatorname{Der} \mathfrak{s}) \otimes \mathcal{O}(1; \underline{1})) \rtimes (\mathrm{Id} \otimes W(1; \underline{1}))$ *such that*

$$\pi_2\big(L[\alpha, \beta]\big) = \pi_2(L[\alpha, \beta](\kappa)) \cong L[\kappa] \cong W(1; \underline{1})$$

for some Witt root $\kappa \in \mathbb{F}_p\alpha + \mathbb{F}_p\beta$ *satisfying* $\kappa(\Psi_{\alpha,\beta}(T) \cap \tilde{S}) = 0$. *Moreover,*

$$\Psi_{\alpha,\beta}(T) = F(h \otimes 1) \oplus F(d \otimes 1 + \mathrm{Id} \otimes t_0),$$

where $h \in \mathfrak{s}$, $d \in \operatorname{Der} \mathfrak{s}$, $t_0 \in W(1; \underline{1})$ *are toral; in addition,* $d = 0$ *and* $t_0 = (1 + x_1)\partial_1$, *or* $t_0 = x_1\partial_1$.

(6) $\tilde{S} = \mathfrak{s} \otimes \mathcal{O}(2; \underline{1})$, *where* $\mathfrak{s}$ *is one of* $\mathfrak{sl}(2)$, $W(1; \underline{1})$, $H(2; \underline{1})^{(2)}$, *and* $L[\alpha, \beta]$ *is a subalgebra in* $((\operatorname{Der} \mathfrak{s}) \otimes \mathcal{O}(2; \underline{1})) \rtimes (\mathrm{Id} \otimes W(2; \underline{1}))$ *such that*

$$H(2; 1)^{(2)} \subset \pi_2(L[\alpha, \beta]) = \pi_2(L[\alpha, \beta](\kappa)) \cong L[\kappa] \subset H(2; \underline{1})$$

for some Hamiltonian root $\kappa \in \mathbb{F}_p\alpha + \mathbb{F}_p\beta$ *satisfying* $\kappa(\Psi_{\alpha,\beta}(T) \cap \tilde{S}) = 0$. *Moreover,*

$$\Psi_{\alpha,\beta}(T) = F(h \otimes 1) \oplus F(d \otimes 1 + \mathrm{Id} \otimes t_0),$$

where $h \in \mathfrak{s}$, $d \in \operatorname{Der} \mathfrak{s}$, $t_0 \in W(2; \underline{1})$ *are toral; in addition,* $d = 0$ *and* $t_0 = D_H((1 + x_1)x_2)$, *or* $t_0 = D_H(x_1x_2)$.

Proof. Since $r = 1$, one has that $\tilde{S}$ is a minimal $\overline{T}$-invariant ideal of $L[\alpha, \beta]$.

(a) Suppose $TR(L[\alpha, \beta] + \overline{T}) = 1$. Set, according to Equation (17.4.2), $\overline{T} \cap \tilde{S}_{[p]} = Fh$, where h is toral. By the assumption of this case Fh is a maximal torus of $L[\alpha, \beta]_{[p]} + \overline{T}$. Consequently, $\overline{T} = Fh$ is 1-dimensional. This then shows that $L[\alpha, \beta]$ is not only $\overline{T}$-semisimple but semisimple.

Let $\kappa \in \Gamma(\tilde{S}, \overline{T}) \setminus \{0\}$. In this case $\kappa(h) \neq 0$, $\tilde{S} = \tilde{S}(\kappa)$ and $L[\alpha, \beta] = \tilde{S}(\kappa) + \Psi_{\alpha,\beta}(H)$ hold. Then there is an isomorphism

$$L[\alpha, \beta] = \tilde{S}(\kappa) + \Psi_{\alpha,\beta}(H) \cong L(\kappa)/L(\kappa) \cap \mathrm{rad}_T\, L(\alpha, \beta).$$

Since the former algebra is semisimple, it coincides with $L[\kappa]$. Theorem 17.1.8 shows that $L[\kappa]^{(1)}$ is simple. This is case (1) of the theorem.

(b) From now on suppose that $TR(L[\alpha, \beta] + \overline{T}) = 2$. Set in Corollary 10.6.6 $\mathfrak{g} := L[\alpha, \beta] + \overline{T}$ and $V := \tilde{S}$. In the first case of that corollary we consider

$$H(2; \underline{1})^{(2)} \subset L[\alpha, \beta] \subset CH(2; \underline{1}), \quad \overline{T} = F(\delta_1 + x_1)\partial_1 \oplus F(\delta_2 + x_2)\partial_2,\ \delta_i \in \{0, 1\}.$$

Clearly $L[\alpha, \beta] + \overline{T}$ contains $\mathfrak{t}_1 := F(1 + x_1)\partial_1 \oplus F(1 + x_2)\partial_2$. Set $z_i := 1 + x_i$ for $i = 1, 2$ and define linear forms $(\epsilon_i)_{i=1,2}$ dual to $(z_j\partial_j)_{j=1,2}$. It is not hard to see that $D_H(z_1^i z_2^j)$ is a root vector for the root $(i-1)\epsilon_1 + (j-1)\epsilon_2$. Consider $\mu := \epsilon_1 + \epsilon_2$. As

$$\mu(C_{H(2;\underline{1})^{(2)}}(\mathfrak{t}_1)) = \mu(FD_H(z_1 z_2)) = 0,$$

one has that $L[\alpha, \beta](\mu)$ is solvable. In addition, $H(2; \underline{1})^{(2)}{}_{\mu}$ contains the element $D_H(z_1^2 z_2^2)$, and this element acts non-nilpotently. Therefore Lemma 17.4.2(2) gives $\mu(C_{L[\alpha,\beta]}(\mathfrak{t}_1)) = 0$. We now observe that $\mathrm{Der}\, H(2; \underline{1})^{(2)} = H(2; \underline{1})^{(2)} + FD_H(z_1^{p-1} z_2^{p-1}) + Fz_1^{p-1}\partial_2 + Fz_2^{p-1}\partial_1 + \mathfrak{t}_1$. Note that $D_H(z_1^{p-1} z_2^{p-1})$, $z_1^{p-1}\partial_2$, $z_2^{p-1}\partial_1$ are root vectors for respective roots $-2(\epsilon_1 + \epsilon_2)$, $-(\epsilon_1 + \epsilon_2)$, $-(\epsilon_1 + \epsilon_2)$. This shows that $C_{\mathrm{Der}\, H(2;\underline{1})^{(2)}}(\mathfrak{t}_1) = \mathfrak{t}_1$, whence

$$C_{L[\alpha,\beta]}(\mathfrak{t}_1) \subset \mathfrak{t}_1 \cap (\ker \mu) = FD_H(z_1 z_2) \subset H(2; \underline{1})^{(2)}.$$

Next we conclude (since the spaces $FD_H(z_1^{p-1} z_2^{p-1})$, $Fz_1^{p-1}\partial_2 + Fz_2^{p-1}\partial_1$, $\mathfrak{t}_1$ sticking out of $H(2; \underline{1})^{(2)}$ belong to different roots),

$$L[\alpha, \beta] \subset H(2; \underline{1})^{(2)} + FD_H(z_1^{p-1} z_2^{p-1}) + Fz_1^{p-1}\partial_2 + Fz_2^{p-1}\partial_1 = H(2; \underline{1}).$$

The case $L[\alpha, \beta] \subset H(2; \underline{1})^{(1)}$ is listed as case (2) of the theorem. So we may assume that $L[\alpha, \beta]$ contains a nonzero element

$$az_1^{p-1}\partial_2 + bz_2^{p-1}\partial_1 + cD_H(z_1^{p-1} z_2^{p-1})$$

with $a, b \in F$, not both of a, b vanishing. Recall that $az_1^{p-1}\partial_2 + bz_2^{p-1}\partial_1$ and $D_H(z_1^{p-1} z_2^{p-1})$ are $\mathfrak{t}_1$-root vectors for different roots. So we may assume $c = 0$. By symmetry we may assume that $a \neq 0$, so that $L[\alpha, \beta]$ contains an element

$$(1 + x_1)^{p-1}\partial_2 + b_1 x_2^{p-1}\partial_1, \quad b_1 \in F.$$

Put

$$v_{-1} := (1+x_1)^{p-1}\partial_2 + b_1 x_2^{p-1}\partial_1,$$
$$v_i := D_H((1+x_1)^{i+1}x_2^{i+1}) = (i+1)\big((1+x_1)^i x_2^{i+1}\partial_2 - (1+x_1)^{i+1}x_2^i\partial_1\big),$$
$$0 \le i \le p-2.$$

It is easy to check that $v_0, \ldots, v_{p-2}$ pairwise commute and

$$[v_{-1}, v_1] = 2D_H((1+x_1)x_2)$$

holds. In the present case, $L[\alpha,\beta] + \overline{T}$ also contains the torus

$$\mathfrak{t}_2 := F(1+x_1)\partial_1 \oplus F x_2\partial_2.$$

Define a $\mathfrak{t}_2$-root μ by $\mu((1+x_1)\partial_1) = \mu(x_2\partial_2) = 1$. It follows that $v_{\pm 1} \in L[\alpha,\beta]_{\pm\mu}$ and $L[\alpha,\beta](\mu) \cap H(2;\underline{1})^{(2)} = \sum_{i=0}^{p-3} F v_i$. Since the latter is abelian, then μ is solvable. Due to Lemma 17.4.2(3) this is not possible.

(c) In the other case of Corollary 10.6.6 there is a realization $\tilde{S} \cong \mathfrak{s} \otimes \mathcal{O}(m;\underline{1})$, where $\mathfrak{s} \in \{\mathfrak{sl}(2), W1;\underline{1}), H(2;\underline{1})^{(2)}\}$, $m \neq 0$, and the above isomorphism gives rise to a restricted injective homomorphism

$$\Phi : L[\alpha,\beta]_{[p]} + \overline{T} \hookrightarrow \operatorname{Der}\big(\mathfrak{s} \otimes \mathcal{O}(m;\underline{1})\big) = \big((\operatorname{Der}\mathfrak{s}) \otimes \mathcal{O}(m;\underline{1})\big) \rtimes \big(\mathrm{Id} \otimes W(m;\underline{1})\big)$$

such that $\pi_2(L[\alpha,\beta]_{[p]} + \overline{T})$ is a transitive subalgebra of $W(m;\underline{1})$ and

$$\Phi(\overline{T}) = (Fh \otimes 1) \oplus F(d \otimes 1 + \mathrm{Id} \otimes t_0),$$

where $h \in \mathfrak{s}$, $d \in \operatorname{Der}\mathfrak{s}$, $t_0 \in W(m;\underline{1})$ are toral elements. In addition, $d = 0$ and $t_0 = (1+x_1)\partial_1$, or $t_0 \in \sum_{i=1}^m F x_i\partial_i$ is nonzero. In what follows we identify $L[\alpha,\beta] + \overline{T}$ with with is image under Φ.

(d) Suppose there is $\nu \in (\mathbb{F}_p\alpha + \mathbb{F}_p\beta) \setminus \{0\}$ with $\nu(H) = 0$. Set $\mathfrak{t}_0 := T \cap (\ker \nu)$ and $\mathfrak{h} := L(\nu)$. Then $\mathfrak{h}$ is nilpotent. As $h \otimes 1 \in (\mathfrak{s} \otimes F) \cap \Psi_{\alpha,\beta}(H) \subset \Psi_{\alpha,\beta}(\mathfrak{h})$, then $\mathfrak{h}$ does not act nilpotently on $L(\alpha,\beta)$. But then it is self-normalizing, whence a CSA of $L(\alpha,\beta)$. Observing that $T \cap \big(L(\nu)_{[p]} + C(L(\alpha,\beta)_{[p]} + T)\big) \subset T \cap (\ker \nu)$ and applying Theorem 1.2.9 we get

$$0 \neq TR(L(\mu), L(\alpha,\beta)) \le TR(L(\mu), L(\alpha,\beta) + T)$$
$$\le TR(L(\alpha,\beta) + T) - \dim T/T \cap (\ker \nu) \le 1.$$

Consequently, $\mathfrak{h}$ has toral rank 1 in $L(\alpha,\beta)$. On the other hand, Theorem 17.1.2 shows that $\gamma_4(\mathfrak{h})$ acts nilpotently on L. We then derive from Theorem 11.1.1 that $(L(\alpha,\beta)/\mathrm{rad}\, L(\alpha,\beta))^{(\infty)} \cong L(\alpha,\beta)^{(\infty)}/\mathrm{rad}\, L(\alpha,\beta)^{(\infty)}$ is the unique minimal ideal of $L(\alpha,\beta)/\operatorname{rad} L(\alpha,\beta)$ and is simple in its own right. There is a surjective homomorphism

$$\varphi : L[\alpha,\beta] \twoheadrightarrow L(\alpha,\beta)/\operatorname{rad} L(\alpha,\beta).$$

Since $\tilde{S}$ is perfect and $(L(\alpha,\beta)/\operatorname{rad} L(\alpha,\beta))^{(\infty)}$ is the unique minimal ideal in $L(\alpha,\beta)/\operatorname{rad} L(\alpha,\beta)$, φ maps $\tilde{S}$ surjectively onto this simple ideal. Consequently,

$$\mathfrak{s} \otimes \mathcal{O}(m;\underline{1})_{(1)} = \big(\mathfrak{s} \otimes \mathcal{O}(m;\underline{1})\big) \cap (\ker \varphi)$$

is an ideal of $L[\alpha,\beta]$. This yields $\pi_2(L[\alpha,\beta]) \subset W(m;\underline{1})_{(0)}$. On the other hand, $\pi_2(L[\alpha,\beta])$ is an ideal of $\pi_2(L[\alpha,\beta]_{[p]} + \overline{T})$ and the latter subalgebra is transitive in $W(m;\underline{1})$. So it must be that $\pi_2(L[\alpha,\beta]) = \{0\}$. As π_2 is a homomorphism of restricted algebras, $\pi_2(L[\alpha,\beta]_{[p]}) = \{0\}$ holds as well.

As a consequence, $L[\alpha,\beta]_{[p]} \subset (\operatorname{Der}\mathfrak{s}) \otimes \mathcal{O}(m;\underline{1})$ and $\pi_2(L[\alpha,\beta]_{[p]} + \overline{T}) = Ft_0$ is a transitive subalgebra of $W(m;\underline{1})$. The latter means that $t_0 \notin W(m;\underline{1})_{(0)}$. By the above characterization of $\overline{T}$ one gets $d = 0$ and $t_0 = (1 + x_1)\partial_1$. We conclude that $m = 1$.

Since $TR(L[\alpha,\beta]) \le TR(L(\alpha,\beta)) \le 2$ and $\overline{T}$ acts on $L[\alpha,\beta]$ as a 2-dimensional torus, one has that $\overline{T}$ is a torus of maximal dimension in $L[\alpha,\beta]_{[p]} + \overline{T}$. Theorem 1.2.9 now yields that $\overline{T} \cap L[\alpha,\beta]_{[p]} = Fh \otimes 1$ is a torus of maximal dimension in $L[\alpha,\beta]_{[p]}$. In particular,

$$0 \neq TR(\mathfrak{s}) \le TR(\varphi(L[\alpha,\beta])) \le TR(L[\alpha,\beta]_{[p]}) = 1.$$

But then $\varphi(L[\alpha,\beta])/\mathfrak{s}$ is $[p]$-nilpotent. We conclude $\mathfrak{s} \in \{\mathfrak{sl}(2), W(1;\underline{1})\}$ or else $\mathfrak{s} = H(2;\underline{1})^{(2)}$ and $\varphi(L[\alpha,\beta]) \subset H(2;\underline{1}) = (\operatorname{Der} H(2;\underline{1})^{(2)})^{(1)}$. This is case (3) of the theorem.

(e) From now on suppose that $\gamma(H) \neq 0$ for all $\gamma \in (\mathbb{F}_p\alpha + \mathbb{F}_p\beta) \setminus \{0\}$. Recall that we are still in the case of Corollary 10.6.6 with $m \neq 0$. The present assumption implies that $T \subset H_{[p]}$. If $C_{L[\alpha,\beta]}(h \otimes 1) \subset \overline{H}$, then $L[\alpha,\beta] = \tilde{S} + \overline{H}$. As in former cases Theorem 10.6.4(3) yields the contradiction $m = 0$.

Fix acccordingly a root $\kappa \in (\mathbb{F}_p\alpha + \mathbb{F}_p\beta) \setminus \{0\}$ with $\kappa(h \otimes 1) = 0$. Then

$$\begin{aligned} L[\alpha,\beta] &= \tilde{S} + L[\alpha,\beta](\kappa), \\ L[\alpha,\beta](\kappa) &\subset C_{\operatorname{Der}\mathfrak{s}}(h) \otimes \mathcal{O}(m;\underline{1}) + \operatorname{Id} \otimes \pi_2(L[\alpha,\beta](\kappa)), \\ Fh \otimes \mathcal{O}(m;\underline{1}) &\subset \operatorname{rad}(L[\alpha,\beta](\kappa)). \end{aligned}$$

If κ is classical and $L[\kappa,\nu] \cong \mathcal{M}(1,1)$ for some $\nu \in \Gamma(L,T)$, then we are in case (4).

(f) Assume in the following that this case does not happen. We intend to show that cases (5) or (6) occur. Corollary 17.3.3(3) applies and shows that $(\operatorname{ad} Q(\kappa))^3(\operatorname{rad} L(\kappa))$ acts nilpotently on L. Take $y \in Q(\kappa)$. Then the subspace

$$(\operatorname{ad} y)^3\big(\Psi_{\alpha,\beta}^{-1}(Fh \otimes \mathcal{O}(m;\underline{1})) \cap L(\kappa)\big)$$

consists of elements acting nilpotently on L. Hence all elements in $(\operatorname{ad} \bar{y})^3(Fh \otimes \mathcal{O}(m; \underline{1}))$ act nilpotently on $\tilde{S}$. This forces

$$\pi_2\big(\Psi_{\alpha,\beta}(Q(\kappa))\big) \subset \pi_2\big(\Psi_{\alpha,\beta}(L(\kappa))\big) \cap W(m; \underline{1})_{(0)}. \qquad (*)$$

Since $T \subset H_{[p]}$, we have that

$$\pi_2(L[\alpha, \beta]_{[p]} + \overline{T}) = \pi_2\big(\Psi_{\alpha,\beta}(L(\kappa))_{[p]}\big) = \pi_2\big(\Psi_{\alpha,\beta}(L(\kappa))\big)_{[p]}.$$

Consequently, $\pi_2(\Psi_{\alpha,\beta}(L(\kappa)))_{[p]}$ is a transitive subalgebra of $W(m; \underline{1})$. On the other hand, as $\operatorname{rad} L(\kappa) \subset Q(\kappa)$, we conclude by $(*)$ that $\pi_2(\Psi_{\alpha,\beta}(\operatorname{rad} L(\kappa))) \subset W(m; \underline{1})_{(0)}$. As $\pi_2\big(\Psi_{\alpha,\beta}(\operatorname{rad} L(\kappa))\big)$ is an ideal in $\pi_2\big(\Psi_{\alpha,\beta}(L(\kappa))\big)_{[p]}$ and the latter is a transitive subalgebra, we then conclude $\pi_2(\Psi_{\alpha,\beta}(\operatorname{rad} L(\kappa))) = \{0\}$. Since $(\ker \pi_2) \cap L[\alpha, \beta](\kappa) \subset C_{\operatorname{Der} \mathfrak{s}}(h) \otimes \mathcal{O}(m; \underline{1})$, and $C_{\operatorname{Der} \mathfrak{s}}(h)$ is solvable, we also obtain $\big(\ker(\pi_2 \circ \Psi_{\alpha,\beta})\big) \cap L(\kappa) \subset \operatorname{rad} L(\kappa)$, whence

$$\big(\ker(\pi_2 \circ \Psi_{\alpha,\beta})\big) \cap L(\kappa) = \operatorname{rad} L(\kappa) \subset Q(\kappa).$$

As a consequence,

$$\pi_2(L[\alpha, \beta]) = \pi_2\big(L[\alpha, \beta](\kappa)\big) \cong L(\kappa)/\operatorname{rad} L(\kappa) = L[\kappa].$$

Recall that $\pi_2\big(\Psi_{\alpha,\beta}(L(\kappa))\big) = L[\kappa]$ is a restricted algebra (Theorem 17.1.8(3)). Then $\pi_2(\Psi_{\alpha,\beta}(L(\kappa)))$ is transitive in $W(m; \underline{1})$. As a conclusion from this result and $(*)$ we derive

$$\begin{aligned}
\dim L(\kappa)/Q(\kappa) &= \dim \pi_2\big(\Psi_{\alpha,\beta}(L(\kappa))\big)/\pi_2\big(\Psi_{\alpha,\beta}(Q(\kappa))\big) \\
&\geq \dim \pi_2\big(\Psi_{\alpha,\beta}(L(\kappa))\big)/\pi_2\big(\Psi_{\alpha,\beta}(L(\kappa))\big) \cap W(m; \underline{1})_{(0)} \\
&\geq \dim W(m; \underline{1})/W(m; \underline{1})_{(0)} = m \neq 0.
\end{aligned}$$

So κ is not solvable or classical, if κ is Witt then $m = 1$, and if κ is Hamiltonian then $m \leq 2$. Conversely, if κ were Hamiltonian and $m = 1$ then

$$p = \dim W(m; \underline{1}) \geq \dim L[\kappa] \geq \dim H(2; \underline{1})^{(2)} = p^2 - 2,$$

a contradiction.

Since $\pi_2(L[\alpha, \beta]) \cong L[\kappa]$ is Witt or Hamiltonian, one has $\overline{T} \cap \tilde{S} = \overline{T} \cap (\ker \pi_2) \subset \ker \kappa$. Due to Theorem 17.1.8, $L[\kappa]$ has the desired form, namely $L[\kappa] \cong W(1; \underline{1})$ if κ is Witt, and $H(2; \underline{1})^{(2)} \subset L[\kappa] \subset H(2; \underline{1})$ if κ is Hamiltonian.

We verify the statement on t_0. Recall that $t_0 \in W(m; \underline{1})$ is toral and nonzero, and $d = 0$ and $t_0 = (1 + x_1)\partial_1$, or $t_0 \in \sum F x_i \partial_i$. This is the claim if $m = 1$. So consider the final case that $m = 2$, $H(2; \underline{1})^{(2)} \subset L[\kappa] \subset H(2; \underline{1})$. By the previous deliberations, the torus $\overline{T}$ is mapped into $L[\kappa]$. Hence we have $t_0 \in H(2; \underline{1})$. Since $H(2; \underline{1})/H(2; \underline{1})^{(2)}$ is $[p]$-nilpotent, we get even better $t_0 \in H(2; \underline{1})^{(2)}$. There is an automorphism σ of $\mathcal{O}(2;\underline{1})$ which stabilizes $H(2;\underline{1})^{(2)}$ and maps Ft_0 onto $FD_H((1+x_1)x_2)$ or $FD_H(x_1x_2)$. More exactly, if $t_0 \notin H(2;\underline{1})^{(2)}{}_{(0)}$, then one has $d = 0$ and may obtain $\sigma \circ t_0 \circ \sigma^{-1} = D_H((1+x_1)x_2)$ by Theorem 16.1.2. Then $\mathrm{Id}_{\mathfrak{s}} \otimes \sigma$ induces an automorphism of $\mathrm{Der}\big(\mathfrak{s} \otimes \mathcal{O}(2;\underline{1})\big)$ which normalizes $L[\kappa] + \overline{T}$ in the desired manner. □

Chapter 18

Solving the case when T is non-standard

In this chapter we assume that T is a torus of maximal dimension in the semisimple p-envelope $L_{[p]}$ of L and T **is non-standard**. Due to Theorem 12.5.7 this implies $p = 5$.

18.1 2-sections revisited

In this section we aim to reduce the list of 2-sections of L relative to T described in §17.4 by using the particular assumption on T. Theorem 17.1.7 shows that for $h_i \in \tilde{H}$ one has $(\sum h_i)^{[p]} \equiv \sum h_i^{[p]} \pmod{\mathrm{rad}_p \tilde{H}}$. Therefore the roots on T can be regarded as *linear* functions on the nilpotent restricted Lie algebra $\tilde{H}$ (cf. §1.3). Define

$$\Omega := \Omega(L,T) := \{\delta \in \Gamma(L,T) \mid \delta(H^{(1)}) \neq 0\}.$$

The general assumption of this chapter means $\Omega \neq \emptyset$. For given roots α, β, set in the following $\overline{T} := \Psi_{\alpha,\beta}(T)$ and $\overline{H} := \Psi_{\alpha,\beta}(H)$.

Lemma 18.1.1. *Let $\alpha \in \Omega$ be non-solvable. Then $L_{[p]}[\alpha] = L[\alpha] \cong H(2;\underline{1})^{(2)} + FD_H(x_1^{(5)})$, and Ψ_α (cf. Equation* (17.1.1)*) can be chosen such that it maps T onto $FD_H((1+x_1)x_2)$. In particular, $\gamma_3(\Psi_\alpha(H)) = \Psi_\alpha(T)$ and $\alpha(\gamma_3(\Psi_\alpha(H))) \neq 0$ hold.*

Proof. If α does not vanish on $H^{(1)} \cap \mathrm{rad}\, L(\alpha)$, then $L(\alpha) = H + \mathrm{rad}\, L(\alpha)$ is solvable. This contradicts our assumption. Hence, since $\alpha(H^{(1)}) \neq 0$, $\Psi_\alpha(H)$ acts non-trigonalizably. Then $\Psi_\alpha(\tilde{H})$ also acts non-trigonalizably.

Due to Theorem 17.1.8, $L_{[p]}[\alpha]$ is one of the simple algebras $\mathfrak{sl}(2)$, $W(1;\underline{1})$, $H(2;\underline{1})^{(2)}$, or $H(2;\underline{1})^{(1)}$, or $p = 5$ and $H(2;\underline{1})^{(2)} + FD_H(x_1^{(5)})$. The first three cases cannot occur, since in these cases every CSA is abelian.

Consider the fourth and fifth case. Since $L_{[p]}[\alpha]/H(2;\underline{1})^{(2)}$ is $[p]$-nilpotent, every torus is contained in $H(2;\underline{1})^{(2)}$. Recall that such a torus is conjugate to $FD_H(x_1x_2)$ or $FD_H((1+x_1)x_2)$. If $\Psi_\alpha(T) = FD_H(x_1x_2)$, then $\Psi_\alpha(\tilde{H}) \subset FD_H(x_1x_2) + \Psi_\alpha(\tilde{H}) \cap H(2;\underline{1})_{(1)}$. In this case $\Psi_\alpha(\tilde{H})$ would act trigonalizably.

We thus have $\Psi_\alpha(T) = FD_H((1+x_1)x_2)$. If $L_{[p]}[\alpha] \cong H(2;\underline{1})^{(1)}$, then $\Psi_\alpha(\tilde{H}) = \sum_{i=1}^{p-1} FD_H((1+x_1)^i x_2)^i)$, which is abelian. Therefore the only possible case is

$L_{[p]}[\alpha] \cong H(2;\underline{1})^{(2)} + FD_H(x_1^{(5)})$. Due to Theorem 17.1.8(3) either $L[\alpha] = L_{[p]}[\alpha]$ holds (which is the claim) or $L[\alpha] \cong H(2;\underline{1})^{(2)}$. In the latter case $\Psi_\alpha(H)$ would be abelian, a contradiction.

Under this normalization we have $\Psi_\alpha(H) = \sum_{i=1}^{3} FD_H((1+x_1)^i x_2^i) + F(1+x_1)^4\partial_2$. It is not hard to compute that $\gamma_3(\Psi_\alpha(H)) = FD_H((1+x_1)x_2) = \Psi_\alpha(T)$. Therefore $\alpha(\gamma_3(\Psi_\alpha(H))) \neq 0$. □

Lemma 18.1.2. *Suppose $\alpha \in \Omega$ and $\beta \in \Gamma(L,T)$ are such that $L(\alpha,\beta)$ is not solvable and $\alpha\big(\Psi_{\alpha,\beta}(H)^{(1)}\big) = 0$ holds. Then $L[\alpha,\beta]$ is as in case (1) of Theorem 17.4.4, i.e., $r = 1$, $TR(\tilde{S}) = 1$, and $L[\alpha,\beta] = L[\kappa]$ for some $\kappa \in (\mathbb{F}_p\alpha + \mathbb{F}_p\beta)\setminus \mathbb{F}_p\alpha$.*

Proof. Since $\alpha(H^{(1)}) \neq 0$ by definition of α but $\alpha\big(\Psi_{\alpha,\beta}(H)^{(1)}\big) = 0$, there is $h \in H^{(1)} \cap \operatorname{rad}_T L(\alpha,\beta)$ for which $\alpha(h) \neq 0$. Then one has $L(\alpha,\beta) = \operatorname{rad}_T L(\alpha,\beta) + L(\kappa)$ for some $\kappa \in \Gamma(L,T)\setminus\mathbb{F}_p^*\alpha$ satisfying $\kappa(h) = 0$. The case $\kappa = 0$ is not possible because $L(\alpha,\beta)$ is not solvable. In particular, $\operatorname{rad} L(\alpha,\beta) = \operatorname{rad}_T L(\alpha,\beta) + \operatorname{rad} L(\kappa)$, whence $\operatorname{rad} L(\alpha,\beta)$ is T-invariant (Theorem 17.1.8(2)). This result implies $L[\alpha,\beta] = L[\kappa]$, and then $r = 1$ and $TR(\tilde{S}) = 1$ because $L(\alpha,\beta)$ is assumed not to be solvable. □

Instead of Theorem 17.4.3 we have the stronger result

Lemma 18.1.3. *Suppose $\alpha \in \Omega$ and $\beta \in \Gamma(L,T)$. If $r = 1$ and $TR(\tilde{S}) = 2$, then one of the following occurs:*

- *$L[\alpha,\beta] \cong \mathcal{M}(1,1)$;*
- *$H(2;(1,2))^{(2)} \cong \tilde{S} \subset L[\alpha,\beta] \subset H(2;(1,2))$; furthermore, $\Psi_{\alpha,\beta}(T) \cap H(2;(1,2))^{(2)}$ is not a torus.*

In particular, every nonzero root μ of $\mathbb{F}_p\alpha + \mathbb{F}_p\beta$ is contained in Ω; it is Hamiltonian, and

$$(\operatorname{rad} L(\mu))_\mu \subset \operatorname{rad}_T L(\alpha,\beta), \quad [H, \operatorname{rad} L(\mu)] \subset \operatorname{rad}_T L(\alpha,\beta)$$

holds.

Proof. (a) In the present case $L[\alpha,\beta]$ is described in Theorem 17.4.3. Due to Lemma 18.1.2, $\overline{H}^{(1)}$ acts non-nilpotently on $L[\alpha,\beta]$. In cases (1), (2), and (4) of Theorem 17.4.3 we have $\overline{H}^{(1)} = \{0\}$ (Corollary 12.5.10). These cases do not occur here.

(b) Suppose $L[\alpha,\beta] \cong \mathcal{M}(1,1)$. This case is listed in the present lemma. By Lemma 18.1.2, $\overline{T}$ is a non-standard torus. Corollary 12.5.10 shows that we may normalize $\overline{T} = F(1+x_1)\partial_1 \oplus F(1+x_2)\partial_2$. Our computations in Volume 1, p. 209 yield $\overline{H}^{(1)} \supset \overline{T}$. Therefore no nonzero root in $F_p\alpha + F_p\beta$ vanishes on $H^{(1)}$, whence all these roots are contained in Ω.

One has $\overline{T} \cap \mathcal{M}(1,1)_{(-2)} = \{0\}$. Theorem 10.5.13(1) applies and gives the final claim for this case.

(c) We are left with the case $\tilde{S} \cong H(2;(1,2))^{(2)}$, $L[\alpha,\beta] \subset H(2;(1,2))$, and $\overline{T} \subset \tilde{S}_{[p]}$. The 1-sections are described by Corollary 10.2.9 and Theorems 10.2.14 and 10.2.16.

Suppose Theorem 10.2.14 applies. Part (1) shows that there is a root γ satisfying $\gamma(\overline{T} \cap H(2;(1,2))^{(2)}) = 0$. By part (3) this root is solvable. Due to part (2) there is a root vector $d_\gamma \in H(2;(1,2))^{(2)}{}_\gamma$ which acts non-nilpotently. Therefore Lemma 17.4.2(2) gives $\gamma(\overline{H}) = 0$. However, Corollary 10.2.15 yields that no nonzero root vanishes on $\overline{H}$. So this case is impossible.

Next, consider the case that Theorem 10.2.16 applies. Then $\overline{T} \cap \tilde{S} = Ft$, where t is a toral element in $\tilde{S}_{(0)}$. As $\operatorname{gr} t \in H(2;(1,2))_{(0)}/H(2;(1,2))_{(1)} \cong \mathfrak{sl}(2)$ acts invertibly on $H(2;(1,2))/H(2;(1,2))_{(0)}$, this implies that $\overline{H} \subset \overline{T} + C_{H(2;(1,2))_{(1)}}(\overline{T})$. But then $\overline{H}^{(1)} \subset H(2;(1,2))_{(1)}$ acts nilpotently on $H(2;(1,2))$, a contradiction.

As a consequence, Corollary 10.2.9 applies. In this case $\overline{T} \cap H(2;(1,2))^{(2)} \neq \{0\}$ is not a torus. Let μ be any nonzero root. The corollary shows that there is an isomorphism

$$\varphi_\mu : H(2;(1,2))^{(2)}(\mu) \cong H(2;\underline{1})^{(2)},$$
$$\varphi_\mu(\overline{T} \cap H(2;(1,2))^{(2)}) = FD_H((1+x_1)x_2).$$

This isomorphism maps $\overline{H}' := \overline{H} \cap H(2;(1,2))^{(2)}$ onto $\sum_{i=1}^{3} FD_H((1+x_1)^i x_2^i)$. Consequently, μ is Hamiltonian and $\overline{H}'$ is abelian.

In the present case $\overline{T} \cap H(2;(1,2))^{(2)}$ is 1-dimensional and is not a torus. On the other hand, $H(2;(1,2))^{(2)}{}_{(0)}$ is closed under $[p]$-powers (Lemma 7.1.1(3)), and this gives that $\overline{T}$ meets $H(2;(1,2))^{(2)}{}_{(0)}$ trivially. In particular, $\overline{H}' \cap H(2;(1,2))^{(2)}{}_{(0)}$ must be $[p]$-nilpotent. Therefore $\overline{H}' \not\subset H(2;(1,2))^{(2)}{}_{(0)}$ holds. Suppose that $\overline{H}' + H(2;(1,2))^{(2)}{}_{(0)} = H(2;(1,2))^{(2)}$. As $\overline{H}'$ is abelian, then $\overline{H}' \cap H(2;(1,2))^{(2)}{}_{(0)} = \{0\}$. But this gives $\dim \overline{H}' = 2$ contrary to Corollary 10.2.9(1). As a consequence,

$$\dim \overline{H}'/\overline{H}' \cap H(2;(1,2))^{(2)}{}_{(0)} = \dim \overline{T} \cap H(2;(1,2))^{(2)} = 1.$$

As a result of these deliberations we obtain

$$\overline{H}' = \left(\overline{T} \cap H(2;(1,2))^{(2)}\right) \oplus \left(\overline{H}' \cap H(2;(1,2))^{(2)}{}_{(0)}\right),$$

and $\overline{H}' \cap H(2;(1,2))^{(2)}{}_{(0)} = \operatorname{nil} \overline{H}'$.

Note that $\gamma_3(\overline{H}) \subset \gamma_3(H(2;(1,2))) = H(2;(1,2))^{(2)}$. Then $\gamma_3(\overline{H}) \subset \overline{H}'$. As $\alpha(\gamma_3(\overline{H})) \neq 0$ (Lemma 18.1.1), one concludes $\gamma_3(\overline{H}) \not\subset \operatorname{nil} \overline{H}'$. Therefore

$$\overline{T} \cap H(2;(1,2))^{(2)} \equiv \gamma_3(\overline{H}) \pmod{\operatorname{nil} \overline{H}'}.$$

Since no root vanishes on $\overline{T} \cap H(2;(1,2))^{(2)}$ (Corollary 10.2.9(3)), we conclude $\mu(\gamma_3(\overline{H})) \neq 0$. Therefore $\mu \in \Omega$. The existence of φ_μ also implies that $\Psi_{\alpha,\beta}(\operatorname{rad} L(\mu)) \cap H(2;(1,2))^{(2)}(\mu) = \{0\}$. Take any $u \in \Psi_{\alpha,\beta}(\operatorname{rad} L(\mu))$. Then $[u, H(2;(1,2))^{(2)}(\mu)] = \{0\}$. Since $H(2;(1,2))^{(2)}(\mu) \cong H(2;\underline{1})^{(2)}$, we have

$$H(2;(1,2))^{(2)} = H(2;(1,2))^{(2)}(\mu) + H(2;(1,2))_{(1)}.$$

This gives $u = 0$. So $\Psi_{\alpha,\beta}(\operatorname{rad} L(\mu)) = \{0\}$ and the final statement of the lemma holds even in the stronger version $\operatorname{rad} L(\mu) \subset \operatorname{rad}_T L(\alpha,\beta)$. □

Next we turn our attention to Theorem 17.4.4.

Lemma 18.1.4. *Suppose $\alpha \in \Omega$ and $\beta \in \Gamma(L,T)$. Case (2) of Theorem 17.4.4 does not occur.*

Proof. In the case under consideration one has $L[\alpha,\beta] \subset H(2;\underline{1})^{(1)}$ and $\overline{T} = F(\delta_1 + x_1)\partial_1 \oplus F(\delta_2 + x_2)\partial_2$, where $\delta_i \in \{0,1\}$. Set $z_i := \delta_i + x_i$ for $i = 1,2$. Since $[z_i\partial_i, D_H(z_1^a z_2^b)] = (\delta_{i,1}(a-1) + \delta_{i,2}(b-1))D_H(z_1^a z_2^b)$, one has $\overline{H} \subset \overline{T}$. Then $\overline{H}^{(1)} = \{0\}$, which contradicts Lemma 18.1.2. □

Lemma 18.1.5. *For any non-solvable $\alpha \in \Omega$ there exists $\kappa \in \Gamma(L,T)$ such that $L[\alpha,\kappa] \cong \mathcal{M}(1,1)$ and $\alpha([[L_{i\alpha}, L_{-i\alpha}],[L_\kappa, L_{-\kappa}]]) \neq 0$ for some $i \in \mathbb{F}_p^*$.*

Proof. Lemma 18.1.1 implies that $L[\alpha] \cong H(2;\underline{1})^{(2)} \oplus F(1+x_1)^4\partial_2$, $\Psi_\alpha(T) = FD_H((1+x_1)x_2)$, and $\alpha(\gamma_3(H)) \neq 0$ hold. Choose $k \in \mathbb{F}_p^*$ for which $\alpha([L_{k\alpha}, L_{-k\alpha}]) \neq 0$. The set $\Omega_1 := \{\delta \in \Gamma(L,T) \mid \delta([L_{k\alpha}, L_{-k\alpha}]) \neq 0\}$ is non-empty. Since $\Psi_\alpha(H) \cap H(2;\underline{1})^{(2)}$ has codimension 1 in $\Psi_\alpha(H)$, Proposition 1.3.6(1) implies that there exists $\kappa \in \Omega_1$ with the property that

$$\Psi_\alpha(H) = \Psi_\alpha(H) \cap H(2;\underline{1})^{(2)} + \Psi_\alpha([L_\kappa, L_{-\kappa}]).$$

Hence there exist $h \in [L_{i\alpha}, L_{-i\alpha}]$ for some $i \in \mathbb{F}_p^*$ and $h' \in [L_\kappa, L_{-\kappa}]$ for which

$$\Psi_\alpha(h) = [D_H((1+x_1)^4), D_H(x_2^4)] = D_H((1+x_1)^3 x_2^3),$$
$$\Psi_\alpha(h') \equiv (1+x_1)^4\partial_2 \pmod{\Psi_\alpha(H) \cap H(2;\underline{1})^{(2)}}.$$

Consequently, these elements satisfy $\alpha([h',[h',h]]) \neq 0$. Note that $\kappa([h',[h',h]]) \in \kappa([[L_\kappa, L_{-\kappa}],[L_\kappa, L_{-\kappa}]]) = 0$ by Theorem 17.1.8. Set $h_1 := [h',[h',h]]$. Next, choose $h_2 \in [L_{k\alpha}, L_{-k\alpha}]$ for which $\kappa(h_2) \neq 0$. We obtain that assumption (a) of Proposition 17.1.1 holds (taking $\mathfrak{t}_0 = T$, κ for α, $k\alpha$ for β). Set $\mathfrak{g} := \sum_{\gamma \in (\mathbb{F}_p\alpha + \mathbb{F}_p\kappa)\setminus\{0\}} (L_\gamma + [L_\gamma, L_{-\gamma}])$. The first part of this proposition yields that $\operatorname{rad}\mathfrak{g}$ is T-invariant, the forth part implies that $\operatorname{rad}\mathfrak{g}$ is H-invariant. This means that $\operatorname{rad}\mathfrak{g}$ is a T-invariant ideal of $L(\alpha,\kappa)$. Then $\mathfrak{g} \cap \operatorname{rad}_T L(\alpha,\kappa) = \operatorname{rad}\mathfrak{g}$ and $L[\alpha,\kappa]$ contains an ideal $\tilde{S}$ which is isomorphic to $\mathfrak{g}/\mathrm{rad}\,\mathfrak{g}$. The third part of the proposition shows that $\tilde{S} \cong \mathfrak{g}/\mathrm{rad}\,\mathfrak{g}$ is simple of absolute toral rank 2.

Equation (17.4.2) now says that $\tilde{S}_{[p]}$ contains $\Psi_{\alpha,\kappa}(T)$, whence $\Psi_{\alpha,\kappa}(L_\gamma) \subset \tilde{S}$ for all nonzero $\gamma \in \mathbb{F}_p\alpha + \mathbb{F}_p\kappa$. Since $h \in [L_{i\alpha}, L_{-i\alpha}]$, $h' \in [L_\kappa, L_{-\kappa}]$ and $\alpha([h',[h',h]]) \neq 0$, the torus $\Psi_{\alpha,\kappa}(T)$ is non-standard for $\tilde{S}$. Applying Corollary 12.5.10 we conclude that $\tilde{S} \cong \mathcal{M}(1,1)$. Then $L[\alpha,\kappa] \subset \mathrm{Der}\,\mathcal{M}(1,1) \cong \mathcal{M}(1,1)$, finishing the proof. □

Lemma 18.1.6. *Suppose $\alpha \in \Omega$ and $\beta \in \Gamma(L,T)$. Case* (3) *of Theorem 17.4.4 does not occur.*

Proof. (a) The case under consideration is given by

$$\mathfrak{s} \otimes \mathcal{O}(1;\underline{1}) \subset L[\alpha,\beta] \subset (\mathrm{Der}\,\mathfrak{s})^{(1)} \otimes \mathcal{O}(1;\underline{1})$$

and

$$\overline{T} = \big(Fh \otimes 1\big) \oplus \big(F\mathrm{Id} \otimes (1+x_1)\partial_1\big),$$

where $\mathfrak{s}$ is one of $\mathfrak{sl}(2)$, $W(1;\underline{1})$, $H(2;\underline{1})^{(2)}$ and h is a nonzero toral element of $\mathfrak{s}$. Then $\overline{H} \subset C_{(\mathrm{Der}\,\mathfrak{s})^{(1)}}(h) \otimes F$. Note that in all cases $(\mathrm{Der}\,\mathfrak{s})^{(1)}$ is closed under the associative p-power mapping, and $TR((\mathrm{Der}\,\mathfrak{s})^{(1)}) = 1$ holds. This gives $\overline{H}_{[p]} \cap \overline{T} = Fh \otimes 1$. Lemma 18.1.2 proves that $\alpha\big(\overline{H}^{(1)}{}_{[p]} \cap \overline{T}\big) \neq 0$. From this we conclude that $\overline{H}^{(1)}{}_{[p]} \cap \overline{T} = Fh \otimes 1$ as well. We conclude that any $\lambda \in \mathbb{F}_p\alpha + \mathbb{F}_p\beta$ is contained in Ω if and only if $\lambda(h \otimes 1) \neq 0$. In particular, $\alpha(h \otimes 1) \in \mathbb{F}_p^*$.

Define $\mu \in \mathbb{F}_p\alpha + \mathbb{F}_p\beta$ by the choice

$$\mu(h \otimes 1) = 0, \quad \mu(\mathrm{Id} \otimes (1+x_1)\partial_1) = 1.$$

Then $\mu(\overline{H}) = 0$, even more $\mu(H) = 0$ holds since $\mu(H \cap (\ker \Psi_{\alpha,\beta})) = 0$. Therefore $L(\mu)$ is a nilpotent 1-section. Next, we observe

$$h \otimes (1+x_1) \in \Psi_{\alpha,\beta}(L_\mu) \text{ and } h \otimes 1 = \big(h \otimes (1+x_1)\big)^{[p]} \in \Psi_{\alpha,\beta}(L_\mu)^{[p]}.$$

In particular, α does not vanish on $\Psi_{\alpha,\beta}(L_\mu{}^{[p]}) = \Psi_{\alpha,\beta}(L_\mu)^{[p]}$, whence $\alpha(L_\mu{}^{[p]}) \neq 0$.

To simplify notation we substitute α by

$$\alpha - \alpha(\mathrm{Id} \otimes (1+x_1)\partial_1)\mu \in \mathbb{F}_p\alpha + \mathbb{F}_p\beta,$$

so that

$$\alpha(h \otimes 1) \in \mathbb{F}_p^*, \quad \alpha(\mathrm{Id} \otimes (1+x_1)\partial_1) = 0.$$

Then α still is a root contained in Ω. Note that $L[\alpha,\beta](\alpha) \supset \mathfrak{s} \otimes F$ is non-solvable.

By Lemma 18.1.5 there exists $\kappa \in \Gamma(L, T)$ such that $L[\alpha, \kappa] \cong \mathcal{M}(1, 1)$ and $\alpha([L_{i\alpha}, L_{-i\alpha}], [L_\kappa, L_{-\kappa}]) \neq 0$ for some $i \in \mathbb{F}_p^*$.

(b) We now consider the T-semisimple 3-section $L[\alpha, \kappa, \mu]$. Set according to Equation (17.4.1)

$$\Psi_{\alpha,\kappa,\mu} : L(\alpha, \kappa, \mu)_{[p]} + T \to L[\alpha, \kappa, \mu]_{[p]} + \tilde{T} \subset \operatorname{Der} \tilde{S},$$

where $\tilde{T} := \Psi_{\alpha,\kappa,\mu}(T)$, and $\tilde{S}$ is the $\tilde{T}$-socle of $L[\alpha, \kappa, \mu]$. Given a Lie subalgebra M of $L[\alpha, \kappa, \mu]$ we denote by $M_{[p]}$ the p-envelope of M in $\operatorname{Der} \tilde{S}$. Note that the restricted Lie algebra $L[\alpha, \kappa, \mu]_{[p]} + \tilde{T} \subset \operatorname{Der} \tilde{S}$ is centerless. As T is a torus of maximal dimension in $L(\alpha, \kappa, \mu)_{[p]} + T$, it follows from Lemma 1.2.6(3) that $\tilde{T}$ is a torus of maximal dimension in $L[\alpha, \kappa, \mu]_{[p]} + \tilde{T}$. Let J be a minimal $\tilde{T}$-invariant ideal of $L[\alpha, \kappa, \mu]$. Then $TR(J) \leq TR(L[\alpha, \kappa, \mu]) \leq 3$ (see Theorems 1.2.7(1) and 1.3.11(3)).

(c) Suppose $TR(J) = 3$. Then it follows from Lemma 1.2.6(3) that $\tilde{T} \subset J_{[p]}$. From this it is immediate that $L[\alpha, \kappa, \mu] = J + \Psi_{\alpha,\kappa,\mu}(H)$. This in turn gives $J = \tilde{S}$. Since $\tilde{S}$ is now derivation simple, $\tilde{S} = \tilde{\mathfrak{s}} \otimes \mathcal{O}(m; \underline{1})$, where $\tilde{\mathfrak{s}}$ is a simple Lie algebra. In the present situation, Theorem 3.6.1 implies that the torus $\tilde{T} \subset \tilde{\mathfrak{s}}_{[p]} \otimes \mathcal{O}(m; \underline{1})$ is conjugate under $\operatorname{Aut}(\tilde{\mathfrak{s}} \otimes \mathcal{O}(m; \underline{1}))$ to $T_0 \otimes F$ for some torus T_0 in $\tilde{\mathfrak{s}}_{[p]}$. Hence we can choose $\Psi_{\alpha,\kappa,\mu}$ such that $\tilde{T} = T_0 \otimes F$ and identify the roots on $\tilde{T}$ and T_0. Then

$$L[\alpha, \kappa, \mu](\alpha) = \tilde{\mathfrak{s}}(\alpha) \otimes \mathcal{O}(m; \underline{1}) + \Psi_{\alpha,\kappa,\mu}(H).$$

Since α is non-solvable, one has $\alpha(C_{\tilde{\mathfrak{s}}}(T_0)) \neq 0$. Theorem 10.6.4(3) in combination with Theorem 17.1.7 now yields $m = 0$.

As a consequence, $\Psi_{\alpha,\kappa,\mu}(L_\gamma) \subset \tilde{\mathfrak{s}}$ for all nonzero $\gamma \in \Gamma(L[\alpha, \kappa, \mu], \tilde{T})$. The choice of κ in (a) now implies that the assumption $\alpha\big((\Psi_{\alpha,\kappa,\mu}(H) \cap \tilde{\mathfrak{s}})^{(1)}\big) = 0$ would imply $\Psi_{\alpha,\kappa,\mu}(L(\alpha)) \subset \Psi_{\alpha,\kappa,\mu}(H)$. But $L(\alpha)$ is not solvable by (a). Hence $\Psi_{\alpha,\kappa,\mu}(H) \cap \tilde{\mathfrak{s}}$ is a non-trigonalizable subalgebra of $\tilde{\mathfrak{s}}$. Set $\mathfrak{t} := \Psi_{\alpha,\kappa,\mu}(T \cap (\ker \mu))$ and $\mathfrak{h} := \tilde{\mathfrak{s}}(\mu)$. Since $L(\mu)$ is nilpotent by (a), then $\mathfrak{h}$ is a nilpotent subalgebra of $\tilde{\mathfrak{s}}$. Moreover, the inclusion $\Psi_{\alpha,\kappa,\mu}(H) \cap \tilde{\mathfrak{s}} \subset \mathfrak{h}$ shows that $\mathfrak{h}$ is a *non-trigonalizable* nilpotent subalgebra of $\tilde{\mathfrak{s}}$. As $\tilde{\mathfrak{s}}$ is simple and $\mathfrak{t}$ is a torus of dimension at most 2 in $\tilde{\mathfrak{s}}_{[p]}$, Corollary 12.5.9 yields $\tilde{\mathfrak{s}} \cong \mathcal{M}(1, 1)$. As $TR(\mathcal{M}(1, 1)) = 2 < TR(J)$, we reach a contradiction.

(d) As a consequence of (c), $TR(J) \leq 2$. We now put $\tilde{T}' := \tilde{T} \cap J_{[p]}$. Lemma 1.2.6(3) yields that $\tilde{T}'$ is a torus of maximal dimension in $J_{[p]}$. As J is not nilpotent, one has $\tilde{T}' \neq \{0\}$.

Suppose $\mu(\tilde{T}') \neq 0$. Then $\Psi_{\alpha,\kappa,\mu}(L_{i\mu}) \subset J$ for all $i \in \mathbb{F}_p^*$ and hence $\Psi_{\alpha,\kappa,\mu}(L_\alpha) \subset J$ as $\alpha({L_\mu}^{[p]}) \neq 0$ (cf. (a)). Since $L[\alpha, \kappa] \cong \mathcal{M}(1, 1)$ is simple, the T-equivariant homomorphism

$$L[\alpha, \kappa, \mu](\alpha, \kappa) \to L[\alpha, \kappa]$$

maps $J(\alpha, \kappa)$ onto a nonzero ideal, whence onto $L[\alpha, \kappa]$. As $\Psi_{\alpha,\kappa}(T)$ is a 2-dimensional torus in $L[\alpha, \kappa] \cong \mathcal{M}(1,1) \cong \operatorname{Der}\mathcal{M}(1,1)$ (cf. Equation (17.4.2)), the p-envelope of $\Psi_{\alpha,\kappa,\mu}(H) \cap J$ in $J_{[p]}$ contains a torus of dimension at least 2. This torus must be smaller than $\tilde{T}'$, because μ vanishes on H. This gives the contradiction $TR(J) > 2$.

Thus, $\mu(\tilde{T}') = 0$. Then $\alpha(\tilde{T}') \neq 0$ or $\kappa(\tilde{T}') \neq 0$. The choice of κ implies that $\sum_{i \in \mathbb{F}_p^*} \Psi_{\alpha,\kappa,\mu}(L_{i\alpha}) \subset J$. Therefore the homomorphism $L[\alpha, \kappa, \mu](\alpha, \kappa) \to L[\alpha, \kappa]$ maps $J(\alpha, \kappa)$ onto $\mathcal{M}(1,1)$ as before. It follows that

$$2 = TR(\mathcal{M}(1,1)) \leq TR(J) = \dim \tilde{T}' \leq 2.$$

Since this equation holds for every minimal $\tilde{T}$-invariant ideal of $L[\alpha, \kappa, \mu]$ and $TR(L[\alpha, \kappa, \mu]) \leq 3$, we may conclude at this point that the $\tilde{T}$-socle $\tilde{S}$ is the unique minimal ideal of $L[\alpha, \kappa, \mu]$ and $TR(\tilde{S}) = 2$ holds.

Since μ vanishes on $\tilde{T}'$ and $\tilde{T}'$ is 2-dimensional, α and κ are independent on $\tilde{T}'$. Therefore $\Psi_{\alpha,\kappa,\mu}(L_{i\alpha+j\kappa}) \subset \tilde{S}$ for all nonzero $(i, j) \in \mathbb{F}_p \times \mathbb{F}_p$. By the choice of κ, we then have $\alpha\big([\tilde{S}_{i\alpha}, \tilde{S}_{-i\alpha}], [\tilde{S}_\kappa, \tilde{S}_{-\kappa}]\big) \neq 0$ for some $i \in \mathbb{F}_p^*$. Since $\tilde{S}$ is a minimal ideal, it is of the form $\tilde{S} \cong \tilde{\mathfrak{s}} \otimes \mathcal{O}(m; \underline{1})$, where $\tilde{\mathfrak{s}}$ is simple. Moreover, $L[\alpha, \kappa, \mu]_{[p]} + \tilde{T}$ is mapped into $\big((\operatorname{Der}\tilde{\mathfrak{s}}) \otimes \mathcal{O}(m; \underline{1})\big) \rtimes \big(\mathrm{Id} \otimes W(m; \underline{1})\big)$ such that the projection into the second summand is a transitive subalgebra of $W(m; \underline{1})$. Arguing as before, we may normalize $\tilde{T}' = T_0 \otimes 1$, where T_0 is a (2-dimensional) torus of maximal dimension in $\tilde{\mathfrak{s}}_{[p]}$. The preceding observation means that T_0 is a *non-standard* torus in $\tilde{\mathfrak{s}}$. This gives $\tilde{\mathfrak{s}} = \mathcal{M}(1,1)$. In this case all derivations of $\tilde{\mathfrak{s}}$ are inner. Then

$$L[\alpha, \kappa, \mu] = \big(\mathcal{M}(1,1) \otimes \mathcal{O}(m; \underline{1})\big) \rtimes \big(\mathrm{Id} \otimes \mathcal{D}\big)$$

and

$$L[\alpha, \kappa, \mu]_{[p]} = \big(\mathcal{M}(1,1) \otimes \mathcal{O}(m; \underline{1})\big) \rtimes \big(\mathrm{Id} \otimes \mathcal{D}_{[p]}\big)$$

for some Lie subalgebra $\mathcal{D}$ of $W(m; \underline{1})$. Furthermore, Theorem 3.6.1 shows that $\Psi_{\alpha,\kappa,\mu}$ can be chosen such that

$$\tilde{T} = (T_0 \otimes 1) \oplus F(\mathrm{Id} \otimes t_0),$$

where $t_0 \in W(m; \underline{1})$ is toral (at this point it might be 0).

Suppose $\mathcal{D} \not\subset W(m; \underline{1})_{(0)}$. Adjusting indices we may assume that there is

$$d \in \mathcal{D}, \quad d \equiv \partial_1 \pmod{\sum_{l>1} F\partial_l + W(m; \underline{1})_{(0)}}.$$

Note that $\mathcal{D} \subset \pi_2(L[\alpha, \kappa, \mu](\mu))$ and

$$T_0 \otimes 1 \subset C_{\mathcal{M}(1,1)}(T_0) \otimes \mathcal{O}(m; \underline{1}) \subset L[\alpha, \kappa, \mu](\mu).$$

Then $\gamma_4(L[\alpha, \kappa, \mu](\mu))$ contains

$$(\operatorname{ad} d)^3(T_0 \otimes x_1^3) \equiv T_0 \otimes 1 \pmod{C_{\mathcal{M}(1,1)}(T_0) \otimes \mathcal{O}(m; \underline{1})_{(1)}}.$$

Therefore $\gamma_4(L[\alpha,\kappa,\mu](\mu))$ is not $[p]$-nilpotent. As $L(\mu)$ is nilpotent (see (a)), this result contradicts Theorem 17.1.2. Consequently, $\mathcal{D} \subset W(m;\underline{1})_{(0)}$. But $\mathcal{D}$ is invariant under the transitive subalgebra $\mathcal{D}_{[p]} + Ft_0$. As a result, $\mathcal{D} = \{0\}$. Therefore Ft_0 is a transitive subalgebra.

If $t_0 = 0$, then $\mu(\tilde{T}) = 0$ and this gives $L_\mu \subset \operatorname{rad}_T L(\alpha,\kappa,\mu)$. But then $\sum_{j\in\mathbb{F}_p^*} L_{j\alpha} \subset \operatorname{rad}_T L(\alpha,\kappa,\mu)$ by (a), a contradiction.

Thus we may assume further that $t_0 = (1+x_1)\partial_1$ (Theorem 3.6.1) and $m = 1$. Choose $h, h' \in C_{\tilde{\mathfrak{s}}}(T_0)$ such that $[h,h']$ acts non-nilpotently on $\tilde{\mathfrak{s}}$. Recall that $\mu(T_0 \otimes F) = 0$. Then $\mu(\mathrm{Id} \otimes t_0) \neq 0$ and hence there exists $r \in \mathbb{F}_p^*$ such that $h \otimes (1+x_1) \in \tilde{S}_{r\mu}$ and $h' \otimes (1+x_1)^{p-1} \in \tilde{S}_{-r\mu}$. Clearly, the element

$$[h,h'] \otimes 1 = [h \otimes (1+x_1), h' \otimes (1+x_1)^{p-1}] \in [\tilde{S}_{r\mu}, \tilde{S}_{-r\mu}]$$

acts non-nilpotently on $\tilde{S}$. But then $[L_{r\mu}, L_{-r\mu}]$ acts non-nilpotently on L. This contradicts Theorem 17.3.2. □

Lemma 18.1.7.

(1) *If $\kappa \in \Omega$, then all elements in $\bigcup_{i\in\mathbb{F}_p^*} ((\operatorname{rad} L(\kappa))_{i\kappa} \cup [(\operatorname{rad} L(\kappa))_{i\kappa}, L_{-i\kappa}])$ are $[p]$-nilpotent.*

(2) *Every non-solvable root of $\Gamma(L,T)$ is contained in Ω.*

Proof. (1) Set

$$\Omega' := \Big\{\alpha \in \Gamma(L,T) \mid \alpha\Big(\bigcup_{i\in\mathbb{F}_p^*} \big((\operatorname{rad} L(\kappa))_{i\kappa} \cup [(\operatorname{rad} L(\kappa))_{i\kappa}, L_{-i\kappa}]\big)\Big) \neq 0\Big\}.$$

If $\kappa \in \Omega'$, then $L(\kappa) = H + \operatorname{rad} L(\kappa)$ holds. Then $L(\kappa)$ were solvable, and the assumption would imply $\sum_{i\in\mathbb{F}_p^*} L_{i\kappa} = \{0\}$, a contradiction. Hence $\kappa \notin \Omega'$ is true.

Assume for a contradiction that $\Omega' \neq \emptyset$. As $\kappa(H) \neq 0$, there exists $\lambda \in \Omega'$ such that $\kappa([L_\lambda, L_{-\lambda}]) \neq 0$. Since $\kappa \notin \Omega'$, Proposition 17.1.1(a) or (b) is satisfied with

$$h_1 \in \bigcup_{i\in\mathbb{F}_p^*} \big((\operatorname{rad} L(\kappa))_{i\kappa} \cup [(\operatorname{rad} L(\kappa))_{i\kappa}, L_{-i\kappa}]\big), \quad h_2 \in [L_\lambda, L_{-\lambda}].$$

Put $\mathfrak{g} := \sum_{\mu\in(\mathbb{F}_p\kappa+\mathbb{F}_p\lambda)\setminus\{0\}} (L_\mu + [L_\mu, L_{-\mu}])$ and $\bar{\mathfrak{g}} := \mathfrak{g}/\operatorname{rad}\mathfrak{g}$. The proposition yields that $\bar{\mathfrak{g}}$ is simple of absolute toral rank 2 and $\operatorname{rad}\mathfrak{g}$ is invariant under H. It is easy to see that $T = Fh_{1,s} \oplus Fh_{2,s} \oplus T \cap (\ker\kappa) \cap (\ker\lambda)$, where $h_{1,s}, h_{2,s}$ are the semisimple parts of h_1, h_2, respectively. Therefore $\operatorname{rad}\mathfrak{g}$ is a T-invariant solvable ideal of $L(\kappa,\lambda)$, i.e., $\operatorname{rad}\mathfrak{g} = \mathfrak{g} \cap \operatorname{rad}_T L(\kappa,\lambda)$. As a result, $L[\kappa,\lambda]^{(\infty)} \cong \bar{\mathfrak{g}}$. Therefore $L[\kappa,\lambda]$ is ruled by Lemma 18.1.3. This lemma states that $\Psi_{\kappa,\lambda}((\operatorname{rad} L(\mu)_\mu)) = \{0\}$ for all nonzero roots $\mu \in (\mathbb{F}_p\kappa + \mathbb{F}_p\lambda)$. But then $L[\kappa,\lambda] \cong L[\kappa]$ by our choices and $TR(\bar{\mathfrak{g}}) \leq TR(L[\kappa,\lambda]) \leq 1$, a contradiction.

(2) Let $\beta \in \Gamma(L,T)$ be non-solvable. Then $\beta(H) \neq 0$, and hence there is $\alpha \in \Omega$ such that $\beta([L_{-\alpha}, L_\alpha]) \neq 0$. We check the possibilities for $L[\alpha,\beta]$. Clearly, $L[\alpha,\beta] \neq \{0\}$ as β is non-solvable.

(a) Suppose $r = 2$. Then $L[\alpha, \beta] \subset L[\mu_1] \oplus L[\mu_2]$ by Theorem 17.4.1. In particular, there are exactly 2 non-solvable 1-sections. One may take $\mu_1 = \beta$. Observe that $\mu_1([L_{-\gamma}, L_\gamma]) = 0$ for all $\gamma \in (\mathbb{F}_p\alpha + \mathbb{F}_p\beta) \setminus \mathbb{F}_p\mu_1$. This gives $\alpha \in \mathbb{F}_p\mu_1 = \mathbb{F}_p\beta$. As $\alpha \neq 0$, we obtain $\beta \in \mathbb{F}_p^*\alpha \subset \Omega$.

(b) The case $r = 1$ and $TR(\tilde{S}) = 2$ is ruled by Lemma 18.1.3. This lemma yields $\beta \in \Omega$.

(c) Finally, suppose $r = 1$ and $TR(\tilde{S}) = 1$, which case is dealt with in Theorem 17.4.4. In case (1) of that theorem, we argue as in (a). Lemma 18.1.4 excludes case (2), and Lemma 18.1.6 excludes case (3).

We have to show that cases (4)–(6) of Theorem 17.4.4 do not occur. So assume that

$$\tilde{S} = \mathfrak{s} \otimes \mathcal{O}(m; \underline{1}), \ L[\alpha, \beta] \subset \big((\operatorname{Der} \mathfrak{s}) \otimes \mathcal{O}(m; \underline{1})\big) \rtimes \big(\operatorname{Id} \otimes W(m; \underline{1})\big),$$

where $\mathfrak{s} \in \{\mathfrak{sl}(2), W(1; \underline{1}), H(2; \underline{1})^{(2)}\}$, and for some root κ

(i) $\pi_2(L[\alpha, \beta](\kappa))$ is classical and $m > 0$, or

(ii) $\pi_2(L[\alpha, \beta](\kappa)) \cong W(1; \underline{1})$ and $m = 1$, or

(iii) $H(2; \underline{1})^{(2)} \subset \pi_2(L[\alpha, \beta](\kappa)) \subset H(2; \underline{1})$ and $m = 2$.

According to Corollary 10.6.6 we can choose $\Psi_{\alpha,\beta}$ such that

$$\overline{T} := \Psi_{\alpha,\beta}(T) = F(h \otimes 1) \oplus F(d \otimes 1 + \operatorname{Id} \otimes t_0),$$

where Fh is a maximal torus of $\mathfrak{s}$, $d \in \operatorname{Der} \mathfrak{s}$ and $t_0 \in W(m; \underline{1})$ are toral, and $d = 0$ and $t_0 = (1 + x_1)\partial_1$, or $t_0 \in \sum_{i=1}^m Fx_i\partial_i$ is nonzero. Note that $\kappa(h \otimes 1) = 0$.

In case (i) one has $(\pi_2(\overline{H}))^{(1)} \subset \operatorname{rad} \pi_2(L[\alpha, \beta](\kappa))$ and in case (ii) one has $(\pi_2(\overline{H}))^{(1)} = \{0\}$.

Suppose in case (iii) that $t_0 \in W(2; \underline{1})_{(0)}$. Then Ft_0 is conjugate in $H(2; \underline{1})$ to $FD_H(x_1x_2)$ and this gives $\pi_2(\overline{H}) \subset FD_H(x_1x_2) + H(2; \underline{1})_{(1)}$. Therefore $(\pi_2(\overline{H}))^{(1)} \subset H(2; \underline{1})_{(1)}$. Finally consider the case that $t_0 \notin W(2; \underline{1})_{(0)}$. In this case we may assume $d = 0$ and $t_0 = (1 + x_1)\partial_1$. Then $h \otimes (1 + x_1) \in \Psi_{\alpha,\beta}(\operatorname{rad} L_{i\kappa})$ for some $i \in \mathbb{F}_p^*$. Lemma 18.1.7(1) yields that $\kappa \notin \Omega$, whence $\kappa(\overline{H}^{(1)}{}_{[p]} \cap \overline{T}) = 0$.

So we have in all cases that $\pi_2(\overline{H})^{(1)}$ acts nilpotently on $\pi_2\big(L[\kappa, \lambda](\kappa)\big)$. This gives

$$\overline{H}^{(1)}{}_{[p]} \cap \overline{T} \subset \overline{T} \cap (\ker \kappa) = Fh \otimes 1.$$

We also derive from Lemma 18.1.2, that for any $\lambda \in \mathbb{F}_p\alpha + \mathbb{F}_p\beta$ one has $\lambda \in \Omega \Leftrightarrow \lambda(\overline{H}^{(1)}{}_{[p]} \cap \overline{T}) \neq 0$. As $\alpha \in \Omega$, one obtains $\alpha(\overline{H}^{(1)}{}_{[p]} \cap \overline{T}) \neq 0$. Consequently,

$$\overline{H}^{(1)}{}_{[p]} \cap \overline{T} = Fh \otimes 1, \quad \alpha(h \otimes 1) \neq 0.$$

This gives $\Psi_{\alpha,\beta}(L_\alpha) \subset \tilde{S}$ and hence $\Psi_{\alpha,\beta}([L_{-\alpha}, L_\alpha])_{[p]} \cap \overline{T} \subset \tilde{S}_{[p]} \cap \overline{T} \subset Fh \otimes 1$. As β is non-solvable, $\Psi_{\alpha,\beta}(\sum_{i \in \mathbb{F}_p^*} L_{i\beta}) \neq \{0\}$. Therefore $\mathbb{F}_p^*\beta$ contains a $\overline{T}$-root on

$L[\alpha, \beta]$. Recall that $\beta([L_{-\alpha}, L_\alpha]) \neq 0$, whence even more $\beta(\Psi_{\alpha,\beta}([L_{-\alpha}, L_\alpha])_{[p]} \cap \overline{T}) \neq 0$. This in turn gives $\beta(h \otimes 1) \neq 0$ and $\beta \in \Omega$. □

Lemma 18.1.8. *Suppose $\alpha \in \Omega$ and $\beta \in \Gamma(L, T)$. Cases* (4)–(6) *of Theorem 17.4.4 do not occur.*

Proof. We proceed as in part (2)(c) of the preceding proof and show that

$$\overline{H}^{(1)}{}_{[p]} \cap \overline{T} = \overline{T} \cap (\ker \kappa) = Fh \otimes 1 \text{ as well as } \lambda \in \Omega \Leftrightarrow \lambda\big(\overline{H}^{(1)}{}_{[p]} \cap \overline{T}\big) \neq 0$$

for every $\lambda \in \mathbb{F}_p\alpha + \mathbb{F}_p\beta$. Note that κ is non-solvable and $\kappa(h \otimes 1) = 0$ holds. Then $\kappa \notin \Omega$ contradicting Lemma 18.1.7(2). □

We are now in a position to describe some items of the global structure of L.

Proposition 18.1.9.

(1) $\operatorname{rad} L(\mu) \subset C(H)$ *for all* $\mu \in \Gamma(L, T) \setminus \{0\}$.
(2) $\Gamma(L, T) = \Omega \cup \{0\}$.
(3) *Every nonzero root α is Hamiltonian, and* $\dim L_\alpha = 5$ *holds.*

Proof. (1) Let μ be any nonzero root and $\alpha \in \Omega$, and consider the 2-section $L(\alpha, \mu)$. We intend to show that $(\operatorname{rad} L(\mu))_\mu \subset \operatorname{rad}_T L(\alpha, \mu)$. This clearly holds if α and μ are dependent (as according to Theorem 17.1.8 the radical of $L(\mu)$ is T-stable) or if $L(\alpha, \mu)$ is solvable. So we may assume that α, μ are independent roots and $L[\alpha, \mu] \neq \{0\}$.

The case $r = 2$ is ruled by Theorem 17.4.1. Then

$$L[\mu_1]^{(1)} \oplus L[\mu_2]^{(1)} \subset L[\alpha, \mu] + \Psi_{\alpha,\mu}(T) \subset L[\mu_1] \oplus L[\mu_2],$$

whence all 1-sections in $L[\alpha, \mu]$ are semisimple. In case $r = 1$ and $TR(\tilde{S}) = 2$, Lemma 18.1.3 applies and yields the claim. If $r = 1$ and $TR(\tilde{S}) = 1$, then Lemmas 18.1.4, 18.1.6, and 18.1.8 show that $L[\alpha, \mu] = L[\kappa]$, and again all 1-sections in $L[\alpha, \mu]$ are semisimple. These results prove that

$$[L_\alpha, (\operatorname{rad} L(\mu))_\mu] \subset (\operatorname{rad} L(\alpha + \mu))_{\alpha+\mu} \text{ if } \alpha + \mu \neq 0,$$
$$[L_\alpha, [L_{-\mu}, (\operatorname{rad} L(\mu))_\mu]] \subset (\operatorname{rad} L(\alpha))_\alpha$$

for any nonzero root μ and any $\alpha \in \Omega$. Set

$$I := \sum_{\mu \in \Gamma(L,T) \setminus \{0\}} (\operatorname{rad} L(\mu))_\mu + [L_{-\mu}, (\operatorname{rad} L(\mu))_\mu].$$

Since $L = \sum_{\alpha,\beta \in \Omega} (L_\alpha + [L_\alpha, L_\beta])$ by Proposition 1.3.6, we obtain that I is an ideal of L. If $I \neq \{0\}$, then the simplicity of L implies that all 1-sections are solvable.

But Corollary 12.5.9 shows that there is a 2-section of Melikian type while, due to Theorem 10.5.13, all 1-sections in this 2-section are non-solvable. This contradiction proves $I = \{0\}$. We conclude that $\operatorname{rad} L(\mu) \subset H$.

Repeating the above arguments, one obtains that $[H, \operatorname{rad} L(\mu)] \subset \operatorname{rad}_T L(\alpha, \mu)$ for all $\alpha \in \Omega$ and all nonzero $\mu \in \Gamma(L, T)$, whence $[[H, \operatorname{rad} L(\mu)], L_\alpha] \subset (\operatorname{rad} L(\alpha))_\alpha = \{0\}$. As before, we conclude that $[H, \operatorname{rad} L(\mu)] \subset C(L) = \{0\}$. Then $\operatorname{rad} L(\mu) \subset C(H)$.

(2) By (1), no nonzero root is solvable. Apply Lemma 18.1.7(2).

(3) The preceding deliberations show that every nonzero root α is non-solvable and is contained in Ω. Then Lemma 18.1.1 gives that α is Hamiltonian. Moreover, since $\operatorname{rad} L(\alpha) \subset H$, one has $\dim L_\alpha = \dim L[\alpha]_\alpha = p = 5$. □

Proposition 18.1.10. $\dim \gamma_2(H) = 3$, $\dim \gamma_3(H) = 2$, $\gamma_4(H) = \{0\}$ *and* $\gamma_3(H) \subset C(H) = H \cap T$ *hold.*

Proof. Let $\alpha \in \Gamma(L, T)$ be nonzero. Then $\alpha \in \Omega$ by Proposition 18.1.9. In the realization of Lemma 18.1.1 one has

$$\Psi_\alpha(H) = \sum_{i=1}^{3} F D_H((1 + x_1)^i x_2^i) + F(1 + x_1)^4 \partial_2.$$

(a) It is immediate from the above realization that $\gamma_4(H) \subset \operatorname{rad} L(\alpha)$. Then $[\gamma_4(H), L_\alpha] \subset (\operatorname{rad} L(\alpha))_\alpha = \{0\}$ by Proposition 18.1.9. Since this holds for every nonzero root α and L is simple, we derive $\gamma_4(H) = \{0\}$.

(b) In the above realization one has $\gamma_2(\Psi_\alpha(H)) = F D_H((1 + x_1)^2 x_2^2) \oplus F D_H((1 + x_1)x_2)$. This shows that $\gamma_2(H) + \operatorname{rad} L(\alpha)$ has codimension 2 in H for every nonzero $\alpha \in \Gamma(L, T)$. Since $\operatorname{rad} L(\alpha) \subset C(H)$ by Proposition 18.1.9, there exist $x, y \in H$ such that $H = Fx + Fy + (\gamma_2(H) + C(H))$. As a consequence, $\gamma_2(H) = F[x, y] + \gamma_3(H)$. Since $\gamma_4(H) = \{0\}$, we have that $\gamma_3(H) \subset C(H)$ and $\gamma_3(H) = F[x, [x, y]] + F[y, [x, y]] + \gamma_4(H) = F[x, [x, y]] + F[y, [x, y]]$.

(c) Let $\alpha, \beta \in \Gamma(L, T)$ be such that $L[\alpha, \beta] \cong \mathcal{M}(1, 1)$ (Lemma 18.1.5). Due to Lemma 18.1.2, $\Psi_{\alpha,\beta}(T)$ is non-standard. Corollary 12.5.10 shows that we may take $\Psi_{\alpha,\beta}(T) = F(1 + x_1)\partial_1 \oplus F(1 + x_2)\partial_2$. Since

$$[(1 + x_1)^4 (1 + x_2)^3 \tilde{\partial}_1, (1 + x_1)^3 (1 + x_2)^4 \tilde{\partial}_2] = (1 + x_1)^2 (1 + x_2)^2,$$

the calculations in Volume 1, p. 209 imply that $\dim \Psi_{\alpha,\beta}(\gamma_3(H)) = 2$. Hence $\dim \gamma_3(H) \geq 2$. On the other hand, $\gamma_3(H) = F[x, [x, y]] + F[y, [x, y]]$. This gives $\dim \gamma_3(H) = 2$. It is clear that $\gamma_2(H) \neq \gamma_3(H)$. Then $\dim \gamma_2(H) = 3$.

(d) Let $h \in C(H)$ and write $h = h_s + h_n$ with $h_s \in T$ and $h_n \in C(H)_{[p]}$ being $[p]$-nilpotent. In the above realization $\Psi_\alpha(C(H)) \subset F D_H((1 + x_1)x_2)$ holds. But then

$\Psi_\alpha(h_n) = 0$, and this gives $[h_n, L(\alpha)] \subset \operatorname{rad} L(\alpha)$. In view of Proposition 18.1.9, this entails that $[h_n, L_\alpha] = 0$ for all nonzero $\alpha \in \Gamma(L,T)$. As a consequence, $h_n = 0$, forcing $C(H) = H \cap T$. □

Proposition 18.1.11. *Let $\alpha, \beta \in \Gamma(L,T)$ be independent roots. Then either $L[\alpha,\beta] \cong \mathcal{M}(1,1)$ or*

$$H(2;(1,2))^{(2)} \subset L[\alpha,\beta] \subset H(2;(1,2))$$

and $\Psi_{\alpha,\beta}(T) \cap H(2;(1,2))^{(2)}$ is not a torus.

Proof. We go through our list of 2-sections. Since α is not solvable, $L[\alpha,\beta] \neq \{0\}$.

Suppose $L[\alpha,\beta]$ is ruled by Theorem 17.4.1. Then $L[\alpha,\beta] + \Psi_{\alpha,\beta}(T) \subset L[\mu_1] \oplus L[\mu_2]$. As μ_i both are contained in Ω, we have $\dim \gamma_2(\Psi_{\mu_i}(H)) = 2$ (cf. the preceding proof). But this gives $\dim \gamma_2(H) \geq 4$, which contradicts Proposition 18.1.10.

Suppose $L[\alpha,\beta]$ is ruled by Theorem 17.4.4. Due to Lemmas 18.1.4, 18.1.6, and 18.1.8, only the case $L[\alpha,\beta] = L[\mu]$ for some μ is possible. Since α and β are independent and non-solvable, this case cannot occur either.

Then $L[\alpha,\beta]$ is ruled by Theorem 17.4.3. Due to Lemma 18.1.3, this is the claim. □

Proposition 18.1.12.

(1) *$\Gamma(L,T)$ is an $\mathbb{F}_p$-subspace of T^*, and* $\dim_F T = \dim_{\mathbb{F}_p} \Gamma(L,T)$ *holds.*

(2) *The p-envelope of $\gamma_3(H)$ in $L_{[p]}$ coincides with T.*

Proof. (1) Let $\alpha, \beta \in \Gamma(L,T)$ be arbitrary nonzero. If these are $\mathbb{F}_p$-dependent, then $\alpha - \beta$ is a root (since these are Hamiltonian). If they are independent, then $L[\alpha,\beta]$ is described in Proposition 18.1.11. Again $\alpha - \beta$ is a root.

The second statement in (1) is given by Theorem 1.3.11(1).

(2) Recall that $\gamma_3(H) \subset T$ (Proposition 18.1.10). Denote by T_0 the $[p]$-envelope of $\gamma_3(H)$ in T and suppose that $T_0 \neq T$. Then T_0 is a proper subtorus of T. Let according to (1) $\alpha_1, \dots, \alpha_{\dim T}$ be an $\mathbb{F}_p$-basis of $\Gamma(L,T)$, and $t_1, \dots, t_s$ an F-basis of T_0 consisting of toral elements. By assumption $s < \dim T$. Then $\alpha_1, \dots, \alpha_{\dim T}$ are F-dependent on T_0, which means that $\operatorname{rank}_F \left((\alpha_i(t_j))_{\substack{1\leq i\leq \dim T\\ 1\leq j\leq s}}\right) < \dim T$. Since all t_j are toral, one has $\alpha_i(t_j) \in \mathbb{F}_p$. But then even more $\operatorname{rank}_{\mathbb{F}_p} \left((\alpha_i(t_j))_{\substack{1\leq i\leq \dim T\\ 1\leq j\leq s}}\right) < \dim T$ holds. Hence there exists $\mu \in \Gamma(L,T)$ such that $\mu(T_0) = 0$. This implies $\mu(\gamma_3(H)) = 0$ contrary to Proposition 18.1.9(2) incombination with Lemma 18.1.1. □

18.2 Melikian pairs

According to Proposition 18.1.11, if $\alpha, \beta \in \Gamma(L, T)$ are $\mathbb{F}_p$-independent, then either $L[\alpha, \beta] \cong \mathcal{M}(1,1)$ or $H(2;(1,2))^{(2)} \subset L[\alpha, \beta] \subset H(2;(1,2))$. If $L[\alpha, \beta] \cong \mathcal{M}(1,1)$ we say that (α, β) is a *Melikian pair*. Recall from Proposition 18.1.10 that $\gamma_3(H)$ is a 2-dimensional subspace of T.

Lemma 18.2.1. *A pair (α, β) is Melikian if and only if $\gamma_3(H) \cap (\ker \alpha) \neq \gamma_3(H) \cap (\ker \beta)$, i.e., if and only if $\alpha|_{\gamma_3(H)}$ and $\beta|_{\gamma_3(H)}$ are linearly independent over F.*

Proof. Suppose that $L[\alpha, \beta] \cong \mathcal{M}(1,1)$. In this case, both $\gamma_3(H)$ and $\Psi_{\alpha,\beta}(\gamma_3(H))$ are 2-dimensional, whence $\gamma_3(H) \cap (\ker \alpha) \cap (\ker \beta) = \{0\}$. So α and β must be linearly independent on $\gamma_3(H)$.

Suppose $H(2;(1,2))^{(2)} \subset L[\alpha, \beta] \subset H(2;(1,2))$. In this case $\Psi_{\alpha,\beta}(\gamma_3(H))$ is 1-dimensional, whence $\gamma_3(H)/\gamma_3(H) \cap (\ker \alpha) \cap (\ker \beta)$ has dimension 1. It follows that α and β are linearly dependent as linear functions on $\gamma_3(H)$. □

Let in the following (α, β) be a Melikian pair.

Lemma 18.2.2. *One has $L_{[p]}(\alpha, \beta) = L(\alpha, \beta)^{(1)} \oplus \big(T \cap (\ker \alpha) \cap (\ker \beta)\big)$ and $L_{[p]}(\alpha, \beta)^{(1)} = L(\alpha, \beta)^{(1)} \cong \mathcal{M}(1,1)$.*

Proof. (a) Recall that $\operatorname{rad}_T L(\alpha, \beta) \subset \sum_{\mu \in (\mathbb{F}_p\alpha + \mathbb{F}_p\beta)\setminus\{0\}} \operatorname{rad} L(\mu) \subset C(H) = H \cap T$ by Propositions 18.1.9 and 18.1.10, and therefore

$$\operatorname{rad}_T L(\alpha, \beta) = H \cap T \cap (\ker \alpha) \cap (\ker \beta) = C(L(\alpha, \beta)).$$

Due to Theorem 16.4.1, Theorem 11.4.3 applies. As all derivations of $\mathcal{M}(1,1)$ are inner, this theorem gives that the central extension

$$0 \to C(L(\alpha, \beta)) \to L(\alpha, \beta) \to L[\alpha, \beta] \cong \mathcal{M}(1,1) \to 0$$

splits. This means that there is a subalgebra $\mathcal{P} \subset L(\alpha, \beta)$ such that $L(\alpha, \beta) = \mathcal{P} \oplus C(L(\alpha, \beta))$ and $\mathcal{P} \cong L[\alpha, \beta] \cong \mathcal{M}(1,1)$. Then $L(\alpha, \beta)^{(1)} = \mathcal{P}^{(1)} = \mathcal{P}$.

(b) As $\widetilde{H}$ acts on $L(\alpha, \beta)^{(1)}$ as derivations and all derivations of $\mathcal{M}(1,1)$ are inner, it must be that $\widetilde{H} = H \cap L(\alpha, \beta)^{(1)} + \widetilde{H}_0$, where $\widetilde{H}_0$ is the subalgebra of all elements of $\widetilde{H}$ which annihilate $L(\alpha, \beta)^{(1)}$. In particular, $[H \cap L(\alpha, \beta)^{(1)}, \widetilde{H}_0] = \{0\}$. From part (a) of this proof we conclude $H = H \cap L(\alpha, \beta)^{(1)} + H \cap T \cap (\ker \alpha) \cap (\ker \beta)$, and from this we deduce $[H, \widetilde{H}_0] = \{0\}$.

Let μ be any nonzero root. Recall that $\mu \in \Omega$ and $\dim L_\mu = 5$ (Proposition 18.1.9(2),(3)). Hence H acts non-trigonalizably on L_μ, and therefore the H-module L_μ is irreducible. But then $\widetilde{H}_0$ acts on L_μ as scalar operators. It follows that $\widetilde{H}_0$ acts semisimply on L, implying $\widetilde{H}_0 \subset T$. From this it is immediate that $\widetilde{H}_0 = T \cap (\ker\,\alpha) \cap (\ker\,\beta)$.

Finally, observe that $L_{[p]}(\alpha,\beta) = L(\alpha,\beta) + \widetilde{H} = L(\alpha,\beta)^{(1)} + \widetilde{H}_0$. Since $L(\alpha,\beta)^{(1)}$ is centerless, we have $L(\alpha,\beta)^{(1)} \cap \widetilde{H}_0 = \{0\}$. The claim follows. □

For any Melikian pair (α,β), one derives from the preceding lemma that $\tilde{H} = (\tilde{H} \cap L(\alpha,\beta)^{(1)}) + T = (H \cap L(\alpha,\beta)^{(1)}) + T$. This gives $H = (H \cap L(\alpha,\beta)^{(1)}) + (H \cap T)$. Then $\gamma_3(H \cap L(\alpha,\beta)^{(1)}) = \gamma_3(H)$ is a 2-dimensional subspace of $T \cap L(\alpha,\beta)^{(1)}$. This gives

$$T = \gamma_3(H) \oplus \big(T \cap (\ker\,\alpha) \cap (\ker\,\beta)\big).$$

Set $T_0 := T \cap (\ker \alpha) \cap (\ker \beta)$, which is a restricted ideal of $L_{[p]}(\alpha,\beta)$.

Recall from Volume 1, p. 200 the definitions of the natural grading and the natural filtration of the Melikian algebra $\mathcal{M}(1,1)$,

$$\mathcal{M}(1,1)_{(i)} := \bigoplus_{j \geq i} \mathcal{M}(1,1)_{[j]}, \quad i \geq -3.$$

This filtration is uniquely determined by the requirement that it is a standard filtration and $\mathcal{M}(1,1)_{(0)}$ is the (uniquely determined) maximal subalgebra of codimension 5 and depth 3 in $\mathcal{M}(1,1)$ (Theorem 4.3.3(2)). Therefore all components $\mathcal{M}(1,1)_{(i)}$ of this filtration are invariant under the automorphism group of $\mathcal{M}(1,1)$.

For any isomorphism $\Phi : L(\alpha,\beta)^{(1)} \xrightarrow{\sim} \mathcal{M}(1,1)$ set

$$L(\alpha,\beta)^{(1)}{}_{(i)} := \Phi^{-1}(\mathcal{M}(1,1)_{(i)}) \quad \forall i \geq -3.$$

Note that $L(\alpha,\beta)^{(1)}{}_{(0)}$ is a maximal subalgebra of $L(\alpha,\beta)^{(1)}$ of codimension 5. Since the natural filtration of $\mathcal{M}(1,1)$ is invariant under all automorphisms of $\mathcal{M}(1,1)$, the above definition of the subspaces $L(\alpha,\beta)^{(1)}{}_{(i)}$ is independent of a particular choice of Φ.

According to Corollary 12.5.10, any non-standard torus is conjugate under the automorphism group $\operatorname{Aut}\mathcal{M}(1,1)$ to the torus

$$\mathfrak{t} := F(1+x_1)\partial_1 \oplus F(1+x_2)\partial_2.$$

Note that $\mathfrak{t} \cap \mathcal{M}(1,1)_{(-2)} = \{0\}$.

Lemma 18.2.3. *There exists an isomorphism of Lie algebras*

$$\Phi_{\alpha;\beta}\colon L(\alpha,\beta)^{(1)} \xrightarrow{\sim} \mathcal{M}(1,1)$$

which satisfies

$$\Phi_{\alpha;\beta}(\gamma_3(H)) = \mathfrak{t},\ \Phi_{\alpha;\beta}\big(\gamma_3(H) \cap (\ker\,\alpha)\big) = F(1+x_1)\partial_1. \tag{18.2.1}$$

Proof. Note that $L_{[p]}(\alpha,\beta)/T_0 \cong L(\alpha,\beta)^{(1)} \cong \mathcal{M}(1,1)$ by our assumption. Both Lie algebras $L_{[p]}(\alpha,\beta)/T_0$ and $\mathcal{M}(1,1)$ are centerless and restricted, hence every isomorphism between $L_{[p]}(\alpha,\beta)/T_0$ and $\mathcal{M}(1,1)$ is an isomorphism of restricted Lie algebras. Any such isomorphism maps the torus T/T_0 of the restricted Lie algebra $L_{[p]}(\alpha,\beta)/T_0$ onto a 2-dimensional non-standard torus of $\mathcal{M}(1,1)$. Choose any isomorphism $\Phi : L(\alpha,\beta)^{(1)} \cong L_{[p]}(\alpha,\beta)/T_0 \cong \mathcal{M}(1,1)$, where the first isomorphism is the natural one. Then $\gamma_3(H)$ is mapped onto T/T_0. The preceding reasoning shows that we may take $\Phi(\gamma_3(H)) = \mathfrak{t}$. Due to Theorem 10.5.12(1), there is an automorphism of $\mathcal{M}(1,1)$ which maps $\mathfrak{t} \cap (\ker\alpha)$ onto $F(1+x_1)\partial_1$. □

We fix a mapping $\Phi_{\alpha;\beta}$ as in Lemma 18.2.3. In particular, it satisfies Equation (18.2.1).

Lemma 18.2.4. *The following holds.*

(1) $\Phi_{\alpha;\beta}\big(L(\alpha) \cap L(\alpha,\beta)^{(1)}\big)$ *is the span of*

$$\big\{x_2^i(1+x_1)\partial_1,\ x_2^i\partial_2,\ x_2^i(1+x_1)^2,\ x_2^i(1+x_1)^4\tilde{\partial}_1,\ x_2^i(1+x_1)^3\tilde{\partial}_2, \mid 0 \le i \le 4\big\}$$

and $\Phi_{\alpha;\beta}\big(H \cap L(\alpha,\beta)^{(1)}\big)$ *is the span of*

$$\mathfrak{t} \cup \{(1+x_2)^2(1+x_1)^2,\ (1+x_2)^3(1+x_1)^4\tilde{\partial}_1,\ (1+x_2)^4(1+x_1)^3\tilde{\partial}_2\};$$

(2) $L(\alpha,\beta)^{(1)} = L(\alpha,\beta)^{(1)}{}_{(0)} \oplus (H \cap L(\alpha,\beta)^{(1)})$;

(3) $L(\alpha,\beta)^{(1)}{}_{(-3)} = L(\alpha,\beta)^{(1)} = L(\alpha,\beta)^{(1)}{}_{(-2)} \oplus \gamma_3(H)$.

Proof. All claims are to be proved inside $\mathcal{M}(1,1)$. Observe that $\Phi_{\alpha;\beta}(L(\alpha) \cap L(\alpha,\beta)^{(1)})$ has been computed on the last line of Volume 2, p. 97 and accordingly is as claimed. As $\Phi_{\alpha;\beta}(\gamma_3(H)) = \mathfrak{t} = F(1+x_1)\partial_1 \oplus F(1+x_2)\partial_2$ and $H \cap L(\alpha,\beta)^{(1)}$ is 5-dimensional, it is not hard to see that claim (1) holds. From this it is now obvious that $\Phi_{\alpha;\beta}\big(H \cap L(\alpha,\beta)^{(1)}\big)$ intersects $\mathcal{M}(1,1)_{(0)}$ trivially and spans a complementary space. This is (2). As $\Phi_{\alpha;\beta}(\gamma_3(H)) = \mathfrak{t}$, one argues similarly to obtain claim (3). □

Define a p-mapping $u \mapsto u^p$ on $L(\alpha,\beta)^{(1)}$ by setting

$$u^p := {\Phi_{\alpha;\beta}}^{-1}\big(\Phi_{\alpha;\beta}(u)^p\big) \quad \forall u \in L(\alpha,\beta)^{(1)},$$

where $\Phi_{\alpha;\beta}(u)^p$ is the pth power in $\mathcal{M}(1,1)$. Next define a p-semilinear mapping $\Lambda\colon L(\alpha,\beta)^{(1)} \longrightarrow C_{L_{[p]}(\alpha,\beta)}(L(\alpha,\beta)) = T_0$ given by

$$\Lambda(u) := u^{[p]} - u^p \quad \forall u \in L(\alpha,\beta)^{(1)},$$

where $u \mapsto u^{[p]}$ is the pth power map in $L_{[p]}$.

Lemma 18.2.5. *The mapping* Λ *vanishes on the subspace* $L(\alpha,\beta)^{(1)}{}_{(-2)}$. *In particular,* $u^{[p]} \in L(\alpha,\beta)^{(1)}$ *holds for all* $u \in L(\alpha,\beta)^{(1)}{}_{(-2)}$.

Proof. Suppose $\Lambda(u) \neq 0$ for some $u \in L(\alpha,\beta)^{(1)}{}_{(-2)}$. Then there is $\gamma \in \Gamma(L,T)$ which does not vanish on $\Lambda(u) \in T_0$. Since $\Lambda(u) \subset T \cap (\ker\alpha) \cap (\ker\beta)$, the root γ is $\mathbb{F}_p$-independent of α and β. Set

$$M(\gamma;\alpha,\beta) := \oplus_{i,j\in\mathbb{F}_p} L_{\gamma+i\alpha+j\beta}.$$

By Propositions 18.1.12(1) and 18.1.9(3), $M(\gamma;\alpha,\beta)$ is a 125-dimensional submodule of the $L_{[p]}(\alpha,\beta)$-module L. Note that every element of $t \in T_0$ acts on this module by scalar multiplication $\gamma(t)$. The mapping $\mathrm{ad}_L \circ \Phi_{\alpha;\beta}{}^{-1}$ gives $M(\gamma;\alpha,\beta)$ an $\mathcal{M}(1,1)$-module structure. For $u \in L(\alpha,\beta)^{(1)}$ and $v := \Phi_{\alpha;\beta}(u) \in \mathcal{M}(1,1)$ one has

$$\begin{aligned}
&\big((\mathrm{ad}_L \circ \Phi_{\alpha;\beta}{}^{-1})(v)\big)^p - (\mathrm{ad}_L \circ \Phi_{\alpha;\beta}{}^{-1})(v^p)\\
&\quad = (\mathrm{ad}_L\, u)^p - \mathrm{ad}_L\,\big(\Phi_{\alpha;\beta}{}^{-1}(\Phi_{\alpha;\beta}(u)^p)\big)\\
&\quad = \mathrm{ad}_L\,\big(u^{[p]} - u^p\big) = \mathrm{ad}_L\,\Lambda(\Phi_{\alpha;\beta}{}^{-1}(v)).
\end{aligned}$$

This means that the $\mathcal{M}(1,1)$-module $M(\gamma;\alpha,\beta)$ has a p-character χ given by $\chi(v)^p = \gamma(\Lambda(u))$ for all $v \in \mathcal{M}(1,1)$. By our assumption, χ does not vanish on $\mathcal{M}(1,1)_{(-2)}$. This contradicts Theorem 16.4.6. □

Let $\mathcal{N}(H)$ denote the set of all $[p]$-nilpotent elements of H.

Proposition 18.2.6.

(1) $\mathrm{rad}\, L(\kappa) = \gamma_3(H) \cap (\ker\,\kappa)$ *for all* $\kappa \in \Gamma(L,T)\setminus\{0\}$*;*

(2) $L(\kappa,\lambda) = L(\kappa,\lambda)^{(1)}$ *whenever* (κ,λ) *is a Melikian pair;*

(3) $H = \mathcal{N}(H) \oplus \gamma_3(H)$ *and* $\dim\, H = 5$*;*

(4) $\Phi_{\alpha;\beta}(\mathcal{N}(H)) = F(1+x_2)^2(1+x_1)^2 \oplus F(1+x_2)^3(1+x_1)^4\tilde{\partial}_1 \oplus F(1+x_2)^4(1+x_1)^3\tilde{\partial}_2$.

Proof. (a) We claim that $(1+x_2)^2(1+x_1)^2, (1+x_2)^3(1+x_1)^4\tilde{\partial}_1, (1+x_2)^4(1+x_1)^3\tilde{\partial}_2 \in \Phi_{\alpha;\beta}\big(\mathcal{N}(H) \cap L(\alpha,\beta)^{(1)}\big)$.

These elements are contained in $\Phi_{\alpha;\beta}\big(H \cap L(\alpha,\beta)^{(1)}\big)$. Lemma 10.5.5(1) shows that the 5-th power of all these elements is 0. Since they are contained in $\mathcal{M}(1,1)_{(-2)}$, Lemma 18.2.5 yields

$$\begin{aligned}
\Phi_{\alpha;\beta}{}^{-1}\big((1+x_2)^2(1+x_1)^2\big)^{[p]} &= \Phi_{\alpha;\beta}{}^{-1}\big((1+x_2)^3(1+x_1)^4\tilde{\partial}_1\big)^{[p]}\\
&= \Phi_{\alpha;\beta}{}^{-1}\big((1+x_2)^4(1+x_1)^3\tilde{\partial}_2\big)^{[p]} = 0.
\end{aligned}$$

This is the claim.

(b) Since $\gamma_4(H) = \{0\}$ (Proposition 18.1.10), the mapping $[p]$ is semilinear on the $[p]$-envelope $H_{[p]}$ of H in $L_{[p]}$. Set $\mathcal{N}(H)'$ the 3-dimensional subspace of

$\mathcal{N}(H) \cap L(\alpha, \beta)^{(1)}$ spanned by the elements exposed in (a). We derive from Lemma 18.2.4(1) and (a) that

$$(*) \qquad H \cap L(\alpha, \beta)^{(1)} = \mathcal{N}(H)' + \Phi_{\alpha;\beta}{}^{-1}(\mathfrak{t}) = \mathcal{N}(H)' + \gamma_3(H).$$

Next recall that $H = H \cap L(\alpha, \beta)^{(1)} + H \cap T_0$. As $\gamma_3(H) \subset T$, this gives

$$H = \mathcal{N}(H)' + H \cap T.$$

Consequently, $\mathcal{N}(H) = \mathcal{N}(H)'$. In particular, (4) follows.

(c) Due to (b), $H \cap L(\alpha, \beta)^{(1)} = \mathcal{N}(H) + \gamma_3(H)$ holds. In particular, $[L_\alpha, L_{-\alpha}] \subset \mathcal{N}(H) + \gamma_3(H)$ is true.

Let κ be any nonzero root. Since every nonzero root is non-solvable and is contained in Ω (due to Proposition 18.1.9(1),(2)), Lemma 18.1.5 shows that κ fits into a Melikian pair. But then the preceding reasoning gives

$$H = \sum_{\kappa \in \Gamma(L,T) \setminus \{0\}} [L_\kappa, L_{-\kappa}] \subset \mathcal{N}(H) + \gamma_3(H).$$

Clearly, every element of $\mathcal{N}(H) \cap \gamma_3(H)$ is $[p]$-nilpotent and semisimple (as $\gamma_3(H) \subset T$). Then the sum is direct, whence $H = \mathcal{N}(H) \oplus \gamma_3(H)$. Then $\dim H = \dim \mathcal{N}(H) + \dim \gamma_3(H) = 3 + 2 = 5$. This is assertion (3).

(d) Let (κ, λ) be a Melikian pair. Since $H \subset \mathcal{N}(H) + \gamma_3(H) \subset L(\kappa, \lambda)^{(1)}$, we obtain $L(\kappa, \lambda)^{(1)} = L(\kappa, \lambda)$, which is (2).

(e) Let $\kappa \in \Omega$ be arbitrary, choose λ so that (κ, λ) is a Melikian pair and $\Phi_{\kappa;\lambda} : L(\kappa, \lambda)^{(1)} \to \mathcal{M}(1, 1)$ satisfies Equation (18.2.1). As $\Phi_{\kappa;\lambda}$ is an isomorphism and $L(\kappa, \lambda)^{(1)} = L(\kappa, \lambda)$ holds, one may do all the computations in $\mathcal{M}(1, 1)$ with the additional setting $\Phi_{\kappa;\lambda}(\gamma_3(H) \cap (\ker \kappa)) = F(1 + x_1)\partial_1$. It is not hard to see that $\Phi_{\kappa;\lambda}(\operatorname{rad} L(\kappa)) = F(1 + x_1)\partial_1$. This is claim (1). □

We explictely recall that whenever (α, β) is a Melikian pair

$$L_{[p]}(\alpha, \beta) = L(\alpha, \beta) \oplus T_0, \quad L(\alpha, \beta) = L(\alpha, \beta)^{(1)} \cong \mathcal{M}(1, 1), \quad H \cap T = \gamma_3(H),$$

and $\Phi_{\alpha;\beta}$ is defined on $L(\alpha, \beta)$.

18.3 Conclusion

If $TR(L) = 2$, then Proposition 18.1.9 in combination with Lemma 18.1.5 gives $L \cong \mathcal{M}(1, 1)$. Therefore we may assume in the following that $TR(L) > 2$.

The results obtained so far apply to all non-standard tori of maximal dimension in $L_{[p]}$. However, not all such tori are conjugate under the action of the automorphism group of L. In order to identify L with one of the Melikian algebras we will need a sufficiently generic non-standard torus of maximal dimension in $L_{[p]}$.

Proposition 18.3.1. *There exists a non-standard torus T' of maximal dimension in $L_{[p]}$ for which $\gamma_3(C_L(T'))$ contains no nonzero toral elements of $L_{[p]}$.*

Proof. Choose according to Lemmas 18.1.5 and 18.2.3 a Melikian pair $(\alpha,\beta) \in \Gamma(L,T) \times \Gamma(L,T)$ and an isomorphism $\Phi_{\alpha;\beta}\colon L(\alpha,\beta) \xrightarrow{\sim} \mathcal{M}(1,1)$ satisfying (18.2.1). Set $h_i := \Phi_{\alpha;\beta}{}^{-1}(x_i\partial_i)$, $q_i = \Phi_{\alpha;\beta}{}^{-1}(\partial_i)$ and $n_i := q_i{}^{[p]}$, where $i = 1,2$. As the elements $x_i\partial_i$ are toral in $\mathcal{M}(1,1)_{(0)}$, Lemma 18.2.5 says that both h_1 and h_2 are toral elements of $L_{[p]}$. Note that $T = F(q_1+h_1) \oplus F(q_2+h_2) \oplus T_0$, where $T_0 = T \cap (\ker\alpha) \cap (\ker\beta)$.

As $\Phi_{\alpha;\beta}$ is a Lie algebra isomorphism, it is straightforward to see that $[h_i,q_i] = -q_i$ and $[n_i, L(\alpha,\beta)] = (\operatorname{ad} q_i)^5(L(\alpha,\beta)) = \{0\}$. As due to Lemma 18.2.2 $L_{[p]}(\alpha,\beta) = L(\alpha,\beta) \oplus T_0$ holds, we conclude $n_i \in T_0$ for $i = 1,2$. So it follows from JACOBSON's formula that $(q_i+h_i)^{[p]^k} = q_i + h_i + \sum_{j=1}^{k-1} n_i{}^{[p]^j}$ for all $k \geq 1$. Since $\gamma_3(H)_{[p]} = T$ by Proposition 18.1.12(2) and $\gamma_3(H) = \Phi_{\alpha;\beta}{}^{-1}(\mathrm{t}) = F(q_1+h_1) \oplus F(q_2+h_2)$, it follows that the p-closure of $Fn_1 + Fn_2$ coincides with T_0.

Recall that $\dim T_0 \geq 1$. In particular, $n_1 \neq 0$ or $n_2 \neq 0$. Changing indices if necessary, one may take $n_1 \neq 0$. Observe that $[q_1,q_2] \in \ker \Phi_{\alpha;\beta} = \{0\}$. This gives $(q_1+q_2)^{[p]} = q_1^{[p]} + q_2^{[p]}$. Take $a \in F \setminus \mathbb{F}_p$ and choose according to Theorem 10.5.4 $\sigma \in \operatorname{Aut}_{[0]} \mathcal{M}(1,1)$, which maps ∂_1 onto itself and ∂_2 onto $a\partial_1 + \partial_2$. By definition, σ preserves the natural grading of $\mathcal{M}(1,1)$. Since $\mathcal{M}(1,1)$ is simple, σ respects the p-mapping. Then $\sigma(\mathrm{t})$ is a torus in $\mathcal{M}(1,1)$. Since σ is an automorphism, $\sigma(\mathrm{t})$ is non-standard. Recall that T_0 is a torus and

$$\left(\Phi_{\alpha;\beta}{}^{-1}(\sigma(\mathrm{t}))\right)^{[p]} \subset \Phi_{\alpha;\beta}{}^{-1}(\sigma(\mathrm{t})) + T_0$$

holds. Then

$$T' := \Phi_{\alpha;\beta}{}^{-1}(\sigma(\mathrm{t})) + T_0$$

is a non-standard torus of maximal dimension in $L_{[p]}$. Observe that

$$\Phi_{\alpha;\beta}{}^{-1}\big(\sigma((1+x_1)\partial_1)\big) = q_1' + h_1', \quad \Phi_{\alpha;\beta}{}^{-1}\big(\sigma((1+x_2)\partial_2)\big) = q_2' + h_2',$$

where

$$q_1' := \Phi_{\alpha;\beta}{}^{-1}(\sigma(\partial_1)) = q_1, \quad q_2' := \Phi_{\alpha;\beta}{}^{-1}(\sigma(\partial_2)) = aq_1 + q_2,$$

and $h_i' := \Phi_{\alpha;\beta}{}^{-1}(\sigma(x_i\partial_i))$ are toral. Moreover,

$$[h_i', q_j'] + \delta_{i,j}q_j',\ [q_1'+h_1', q_2'+h_2'] \in \ker(\sigma^{-1} \circ \Phi_{\alpha;\beta}) = \{0\}.$$

Note that $q_1'^{[p]} = n_1$, $q_2'^{[p]} = a^p n_1 + n_2$. Now suppose there is a toral element $u \in \gamma_3(C_L(T')) = \Phi_{\alpha;\beta}{}^{-1}(\sigma(\mathrm{t}))$. Set $u = r(q_1'+h_1') + s(q_2'+h_2')$ for some $(r,s) \neq (0,0)$. Then

$$\begin{aligned} r(q_1'+h_1') + s(q_2'+h_2') &= \big(r(q_1'+h_1') + s(q_2'+h_2')\big)^{[p]} \\ &= r^p(q_1'+n_1+h_1') + s^p(q_2'+a^p n_1 + n_2 + h_2'), \end{aligned}$$

implying $r = r^p$, $s = s^p$, $(r^p + s^p a^p)n_1 + s^p n_2 = 0$. As a first consequence, $r, s \in \mathbb{F}_p$. If $s^p n_2 = 0$, then $r^p + s^p a^p = 0$. This gives $s \neq 0$ and $a \in \mathbb{F}_p$ contrary to our chice of a. So $s^p n_2 \neq 0$ holds, and this gives $n_2 = -((r/s)^p + a^p)n_1$. In other words, if the claim is not true, then for any $a \in F \setminus \mathbb{F}_p$ there is $\mu_a \in \mathbb{F}_p$ such that

$$n_2 = -(\mu_a + a^p)n_1.$$

Now choose $a_1, a_2 \in F$ such that $1, a_1, a_2$ are $\mathbb{F}_p$-independent to obtain the contradiction

$$(\mu_{a_1} + a_1^p) = (\mu_{a_2} + a_2^p). \qquad \square$$

In the following we fix T as in Proposition 18.3.1 and some nonzero $\alpha \in \Gamma(L,T)$. Recall that (Proposition 18.2.6(1))

$$L(\alpha)/\gamma_3(H) \cap (\ker \alpha) \cong L[\alpha] \cong H(2;\underline{1})^{(2)} \oplus F x_1^4 \partial_2.$$

Let $H(2;\underline{1})_{(i)}$ $(i \geq -1)$ denote the natural filtration of this Cartan type Lie algebra and put

$$L(\alpha)_{(i)} := L(\alpha) \cap \Psi_\alpha^{-1}\big(H(2;\underline{1})_{(i)}\big).$$

We thus obtain a filtration $(L(\alpha)_{(i)})_{i \geq -1}$ of the Lie algebra $L(\alpha)$ with

$$\bigcap_{i \geq -1} L(\alpha)_{(i)} = \gamma_3(H) \cap (\ker \alpha), \quad \dim L(\alpha)/L(\alpha)_{(0)} = 2.$$

This filtration is independent of the chosen isomorphism, because $H(2;\underline{1})^{(2)}{}_{(0)}$ is the unique subalgebra of codimension 2 in the Cartan type Lie algebra $H(2;\underline{1})^{(2)}$.

Definition 18.3.2. *For $\alpha \in \Gamma(L,T) \setminus \{0\}$ define*

$$W(\alpha) := \{w \in L(\alpha)_{(0)} \mid w^{[p]} \in L(\alpha)\},$$
$$U(\alpha) := \{u \in W(\alpha) \mid [u, W(\alpha)] \subset W(\alpha)\}.$$

Because of the uniqueness of the filtration $(L(\alpha)_{(i)})_{i \geq -1}$ the definition of $W(\alpha)$ depends on T and α, but is independent of the isomorphism chosen for the construction of $L(\alpha)_{(0)}$. Note that $W(\alpha)$ is a vector space and therefore $U(\alpha)$ is a subalgebra.

Proposition 18.3.3. *If (α, β) is a Melikian pair, then*

$$U(\alpha) = L(\alpha) \cap L(\alpha,\beta)_{(0)}.$$

Proof. (a) Choose any Lie algebra isomorphism $\Phi_{\alpha;\beta}\colon L(\alpha,\beta) \xrightarrow{\sim} \mathcal{M}(1,1)$ satisfying (18.2.1). In Lemma 18.2.4 we described $\Phi_{\alpha;\beta}(L(\alpha))$ as the span of

$$\big\{x_2^i(1+x_1)\partial_1,\ x_2^i\partial_2,\ x_2^i(1+x_1)^2,\ x_2^i(1+x_1)^4\tilde{\partial}_1,\ x_2^i(1+x_1)^3\tilde{\partial}_2, \mid 0 \leq i \leq 4\big\}.$$

Using that result we get that $\Phi_{\alpha;\beta}\big(L(\alpha) \cap L(\alpha,\beta)_{(0)}\big)$ is spanned by

$$\big\{x_2^i(1+x_1)\partial_1,\ x_2^i\partial_2,\ x_2^i(1+x_1)^2,\ x_2^i(1+x_1)^4\tilde{\partial}_1,\ x_2^i(1+x_1)^3\tilde{\partial}_2, \mid 1 \le i \le 4\big\}.$$

(b) Next recall that $L(\alpha)_{(0)}$ is the uniquely determined subalgebra of codimension 2 containing $\gamma_3(H) \cap (\ker \alpha)$. So $\Phi_{\alpha;\beta}(L(\alpha)_{(0)})$ contains $(1+x_1)\partial_1$. The computations made in Volume 2, p. 98 show that $\Phi_{\alpha;\beta}(L(\alpha)_{(0)})$ is spanned by

$$\begin{aligned}&\big\{(1+x_1)\partial_1,\ (1+x_1)^2,\ (1+x_1)^4\tilde{\partial}_1\big\} \cup \\ &\quad \big\{x_2^i(1+x_1)\partial_1,\ x_2^i\partial_2,\ x_2^i(1+x_1)^2,\ x_2^i(1+x_1)^4\tilde{\partial}_1,\ x_2^i(1+x_1)^3\tilde{\partial}_2, \mid 1 \le i \le 4\big\}.\end{aligned}$$

(c) We now will determine $W(\alpha)$. Let $u \in L(\alpha)_{(0)}$ be such that $\Phi_{\alpha;\beta}(u)$ is contained in

$$\begin{aligned}\mathcal{S} := &\big\{(1+x_1)^2,\ (1+x_1)^4\tilde{\partial}_1\big\} \cup \\ &\quad \big\{x_2^i(1+x_1)\partial_1,\ x_2^i\partial_2,\ x_2^i(1+x_1)^2,\ x_2^i(1+x_1)^4\tilde{\partial}_1,\ x_2^i(1+x_1)^3\tilde{\partial}_2, \mid 1 \le i \le 4\big\}.\end{aligned}$$

The latter set is contained in $\mathcal{M}(1,1)_{(-2)} \cap C_{\mathcal{M}(1,1)}((1+x_1)\partial_1)$. Then Lemma 18.2.5 shows that $\Phi_{\alpha;\beta}(u^{[p]}) = \Phi_{\alpha;\beta}(u)^p \in C_{\mathcal{M}(1,1)}((1+x_1)\partial_1)$. Therefore $u^{[p]} \in L(\alpha)$, which gives $u \in W(\alpha)$. As a consequence, the space spanned by $\mathcal{S}$ is contained in $W(\alpha)$. Observe that this space has codimension 1 in $L(\alpha)_{(0)}$. On the other hand, suppose $\Phi_{\alpha;\beta}{}^{-1}((1+x_1)\partial_1) \in W(\alpha)$. By definition this yields

$$\begin{aligned}\big(\Phi_{\alpha;\beta}{}^{-1}((1+x_1)\partial_1)\big)^{[p]} - \Phi_{\alpha;\beta}{}^{-1}((1+x_1)\partial_1) &= \Lambda\big(\Phi_{\alpha;\beta}{}^{-1}((1+x_1)\partial_1)\big) \\ &\in L(\alpha) \cap (\operatorname{im}\Lambda) \subset L(\alpha,\beta) \cap \big(T \cap (\ker\alpha) \cap (\ker\beta)\big) = \{0\}.\end{aligned}$$

It follows that $F\Phi_{\alpha;\beta}{}^{-1}((1+x_1)\partial_1)$ is a 1-dimensional subtorus, contrary to our choice of T. We conclude that $W(\alpha)$ is the inverse image of the space spanned by the set $\mathcal{S}$.

(d) Combining (a) and (c) we conclude $L(\alpha) \cap L(\alpha,\beta)_{(0)} \subset W(\alpha)$. Even more, the multiplication table in Volume 2, p. 98 gives that $\Phi_{\alpha;\beta}\big(L(\alpha) \cap L(\alpha,\beta)_{(0)}\big)$ stabilizes $\Phi_{\alpha;\beta}(W(\alpha))$. Therefore $L(\alpha) \cap L(\alpha,\beta)_{(0)} \subset U(\alpha)$ holds.

Let $u = u_0 + r(1+x_1)^2 + s(1+x_1)^4\tilde{\partial}_1$ with $u_0 \in \Phi_{\alpha;\beta}(L(\alpha) \cap L(\alpha,\beta)_{(0)})$ be an element in $\Phi_{\alpha;\beta}(U(\alpha))$. Since $[(1+x_1)^2, (1+x_1)^4\tilde{\partial}_1] = (1+x_1)\partial_1 \notin \Phi_{\alpha;\beta}(W(\alpha))$, it follows that $r = s = 0$. This is the claim. □

Lemma 18.3.4.

(1) $L(\alpha) = H \oplus U(\alpha)$.

(2) *There exists a linear map* $l_{(\alpha)}\colon L(\alpha) \to H$ *such that* $x - l_{(\alpha)}(x) \in U(\alpha)$ *for all* $x \in L(\alpha)$. *Furthermore,*

$$U(\alpha) = \{x - l_{(\alpha)}(x) \mid x \in L(\alpha)\}.$$

Proof. Choose β such that (α, β) is a Melikian pair and a Lie algebra isomorphism $\Phi_{\alpha;\beta}\colon L(\alpha,\beta) \xrightarrow{\sim} \mathcal{M}(1,1)$ satisfying (18.2.1). Lemma 18.2.4 shows that $\Phi_{\alpha;\beta}(L(\alpha))$ and $\Phi_{\alpha;\beta}(H)$ are spanned by

$$\{x_2^i(1+x_1)\partial_1,\ x_2^i\partial_2,\ x_2^i(1+x_1)^2,\ x_2^i(1+x_1)^4\tilde{\partial}_1,\ x_2^i(1+x_1)^3\tilde{\partial}_2, \mid 0 \le i \le 4\}$$

and

$$\{(1+x_1)\partial_1,\ (1+x_2)\partial_2,\ (1+x_2)^2(1+x_1)^2,\ (1+x_2)^3(1+x_1)^4\tilde{\partial}_1,\ (1+x_2)^4(1+x_1)^3\tilde{\partial}_2\},$$

respectively. By Proposition 18.3.3,

$$\Phi_{\alpha;\beta}(U(\alpha)) = C_{\mathcal{M}(1,1)}((1+x_1)\partial_1) \cap \mathcal{M}(1,1)_{(0)}.$$

From this one concludes that $\Phi_{\alpha;\beta}(U(\alpha))$ is spanned by

$$\{x_2^i(1+x_1)\partial_1,\ x_2^i\partial_2,\ x_2^i(1+x_1)^2,\ x_2^i(1+x_1)^4\tilde{\partial}_1,\ x_2^i(1+x_1)^3\tilde{\partial}_2, \mid 1 \le i \le 4\}.$$

Therefore $\Phi_{\alpha;\beta}(L(\alpha)) = \Phi_{\alpha;\beta}(H) + \Phi_{\alpha;\beta}(U(\alpha))$ and $\Phi_{\alpha;\beta}(H) \cap \Phi_{\alpha;\beta}(U(\alpha)) = \{0\}$. This proves (1). Then claim (2) follows by setting $l_{(\alpha)}$ the projection into the first summand. □

Given $x \in L_\alpha$ we set $\widetilde{x} := x - l_{(\alpha)}(x) \in U(\alpha)$. Moreover, set

$$L_{(0)} := \sum_{\mu \in \Gamma(L,T)\setminus\{0\}} U(\mu).$$

Lemma 18.3.5. *Let (α, β) be a Melikian pair. Then $L(\alpha,\beta)_{(0)} = L_{(0)} \cap L(\alpha,\beta)$. In particular, $L_{(0)} \cap L(\alpha,\beta)$ is a subalgebra of $L(\alpha,\beta)$. Moreover, for $x \in L_\alpha$, $y \in L_\beta$ one has*

$$[\widetilde{x}, \widetilde{y}] \equiv \widetilde{[x,y]} \pmod{U(\alpha) + U(\beta)}$$

Proof. Recall that $L(\mu) = H \oplus U(\mu)$ for all nonzero roots μ (Lemma 18.3.4), and

$$\dim H = 5, \quad \dim L(\alpha,\beta)_{(0)} = \dim \mathcal{M}(1,1)_{(0)} = 120,$$

$$\dim U(\mu) = \dim L(\mu) - \dim H = \sum_{i=1}^{4} \dim L_{i\mu} = 20 \quad \forall \mu \neq 0.$$

Set $\Delta := \{\alpha\} \cup (\beta + \mathbb{F}_p\alpha)$. For any $\mu \in \Delta$ one has

$$U(\mu) \cap \Big(\sum_{\delta \in \Delta\setminus\{\mu\}} U(\delta)\Big) \subset U(\mu) \cap \Big(L(\mu) \cap \sum_{\delta \in \Delta\setminus\{\mu\}} L(\delta)\Big) = U(\mu) \cap H = \{0\}.$$

This shows that the sum $U(\alpha) + \sum_{j=0}^{4} U(\beta + j\alpha)$ is direct. Therefore

$$\dim \Big(U(\alpha) + \sum_{j=0}^{4} U(\beta + j\alpha)\Big) = 6 \cdot 20 = 120 = \dim L(\alpha, \beta)_{(0)}.$$

Since $U(\alpha) + \sum_{j\in\mathbb{F}_p} U(\beta + j\alpha) \subset L(\alpha, \beta)_{(0)}$ by Proposition 18.3.3, this gives

$$L(\alpha, \beta)_{(0)} = U(\alpha) + \sum_{j\in\mathbb{F}_p} U(\beta + j\alpha) \subset L_{(0)} \cap L(\alpha, \beta).$$

On the other hand, the same proposition yields

$$L_{(0)} \cap L(\alpha, \beta) = \sum_{\mu\in(\mathbb{F}_p\alpha+\mathbb{F}_p\beta)\setminus\{0\}} U(\mu) \subset L(\alpha, \beta)_{(0)}.$$

To derive the stated equivalence we compute

$$\begin{aligned}
[\widetilde{x}, \widetilde{y}] = \Big([x, y] &- l_{(\alpha+\beta)}([x, y]) - [x, l_{(\beta)}(y)] + l_{(\alpha)}([x, l_{(\beta)}(y)]) \\
&- [l_{(\alpha)}(x), y] + l_{(\beta)}([l_{(\alpha)}(x), y]) + [l_{(\alpha)}(x), l_{(\beta)}(y)]\Big) \\
&+ \Big(l_{(\alpha+\beta)}([x, y]) - l_{(\alpha)}([x, l_{(\beta)}(y)]) - l_{(\beta)}([l_{(\alpha)}(x), y])\Big) \\
&= \widetilde{[x, y]} - \widetilde{[x, l_{(\beta)}(y)]} - \widetilde{[l_{(\alpha)}(x), y]} + \widetilde{h},
\end{aligned}$$

where $\widetilde{h} = l_{(\alpha+\beta)}([x, y]) - l_{(\alpha)}([x, l_{(\beta)}(y)]) - l_{(\beta)}([l_{(\alpha)}(x), y]) + [l_{(\alpha)}(x), l_{(\beta)}(y)]$. Observe that $\widetilde{x}, \widetilde{y} \in L_{(0)} \cap L(\alpha, \beta) = L(\alpha, \beta)_{(0)}$, and therefore $[\widetilde{x}, \widetilde{y}] \in L(\alpha, \beta)_{(0)} = L_{(0)} \cap L(\alpha, \beta)$. As $\widetilde{[x, y]}, \widetilde{[x, l_{(\beta)}(y)]}, \widetilde{[l_{(\alpha)}(x), y]} \in L_{(0)}$, it must be that $\widetilde{h} \in H \cap L_{(0)} = H \cap (\sum_\gamma U(\gamma))$. Expressing $\widetilde{h} = \sum_\gamma (v_\gamma - l_{(\gamma)}(v_\gamma))$ with $v_\gamma \in L_\gamma$ we see that $v_\gamma = 0$ for all γ, whence $l_{(\gamma)}(v_\gamma) = 0$ and $\widetilde{h} = 0$. Consequently, $[\widetilde{x}, \widetilde{y}] - \widetilde{[x, y]} = \widetilde{[x, l_{(\beta)}(y)]} + \widetilde{[l_{(\alpha)}(x), y]} \in U(\alpha) + U(\beta)$. □

Proposition 18.3.6. *$L_{(0)}$ is a maximal subalgebra of L. Moreover, $L = H \oplus L_{(0)}$ holds.*

Proof. (a) It is clear from Lemma 18.3.4 that $L = H + L_{(0)}$. Let $h \in H \cap L_{(0)}$. Expressing $h = \sum_{\gamma\neq 0} (v_\gamma - l_{(\gamma)}(v_\gamma))$ with $v_\gamma \in L_\gamma$ we see that $v_\gamma = 0$ for all γ, whence $l_{(\gamma)}(v_\gamma) = 0$ and $h = 0$. Therefore $L = H \oplus L_{(0)}$ holds.

(b) Since $U(\alpha)$ is a subalgebra, we need to show that $[U(\alpha), U(\beta)] \subset L_{(0)}$ for all pairs of roots (α, β) such that α and β are $\mathbb{F}_p$-independent. If (α, β) is a Melikian pair, this follows from Lemma 18.3.5. Thus, assume in the following that (α, β) is not a Melikian pair.

For $\mu \in \Gamma(L, T) \setminus \{0\}$ set $\gamma_3(H) \cap (\ker \mu) =: Fh_\mu$ and

$$\Gamma(\mu) := \{\gamma \in \Gamma(L, T) \mid \gamma(h_\mu) \neq 0\}.$$

It is immediate from Lemma 18.2.1 that (λ, μ) is a Melikian pair if and only if $\mu \in \Gamma(\lambda)$. Note that $\Gamma(\lambda) = \Gamma(\mu)$ whenever (λ, μ) is not a Melikian pair. Since (α, β) is assumed to be a non-Melikian pair, one has $\gamma_3(H) \cap (\ker \alpha) = \gamma_3(H) \cap (\ker \beta)$. And so we may take $h_\alpha = h_\beta$.

If $\gamma \in \Gamma(\alpha)$, then $(\beta - \gamma)(h_\alpha) = -\gamma(h_\alpha) \neq 0$, and hence $(\alpha, \beta - \gamma)$ is a Melikian pair. Similarly, $\gamma(h_\beta) = \gamma(h_\alpha) \neq 0$, and hence (γ, β) is a Melikian pair. But then so is $(\gamma, \beta - \gamma)$.

Let $\lambda_1 \in \Gamma(\alpha)$ and λ_2 be a root independent of λ_1 such that $\lambda_1 + \lambda_2 \in \mathbb{F}_p\alpha + \mathbb{F}_p\beta$ and assume that (λ_1, λ_2) is not a Melikian pair. As $\dim \gamma_3(H) = 2$, the first assumption yields $\gamma_3(H) = Fh_\alpha \oplus Fh_{\lambda_1}$. Then $\lambda_1 + \lambda_2$ is a root (Proposition 18.1.12(1)) which vanishes on h_α (since $\lambda_1 + \lambda_2 \in \mathbb{F}_p\alpha + \mathbb{F}_p\beta$ and $\alpha(h_\alpha) = \beta(h_\alpha) = 0$) and on h_{λ_1} (as (λ_1, λ_2) is not a Melikian pair). But then $\lambda_1 + \lambda_2$ vanishes on $\gamma_3(H)$ and Proposition 18.1.12(2) gives the contradiction $\lambda_1 + \lambda_2 = 0$. As a consequence, for any $\gamma \in \Gamma(\alpha)$ one has that

$$(\alpha, \gamma),\ (\alpha, \beta - \gamma),\ (\alpha + \gamma, \beta - \gamma),\ (\gamma, \beta - \gamma),\ (\gamma, \alpha + \beta - \gamma)$$

are Melikian pairs.

We now take arbitrary $u \in L_\alpha$ and $v \in L_\beta$. Lemma 18.1.5 shows that $\Gamma(\alpha)$ is non-empty. Then $L_\beta = \sum_{\gamma \in \Gamma(\alpha)} [L_\gamma, L_{\beta-\gamma}]$. Thus there exist $\gamma_1, \dots, \gamma_N \in \Gamma(\alpha)$ such that $v = \sum_{i=1}^N [x_i, y_i]$ for some $x_i \in L_{\gamma_i}$ and $y_i \in L_{\beta-\gamma_i}$, where $1 \leq i \leq N$. Applying Lemma 18.3.5 and the preceding remarks we obtain

$$\begin{aligned}
[\widetilde{u}, \widetilde{v}] &\in \sum_{i=1}^N [\widetilde{u}, [\widetilde{x_i}, \widetilde{y_i}]] + \sum_{i=1}^N [U(\alpha), U(\gamma_i) + U(\beta - \gamma_i)] \\
&\subset \sum_{i=1}^N \big([[\widetilde{u}, \widetilde{x_i}], \widetilde{y_i}] + [\widetilde{x_i}, [\widetilde{u}, \widetilde{y_i}]]\big) + L_{(0)} \\
&\subset \sum_{i=1}^N \Big([U(\alpha + \gamma_i) + U(\alpha) + U(\gamma_i), U(\beta - \gamma_i)] \\
&\qquad\qquad + [U(\gamma_i), U(\alpha + \beta - \gamma_i) + U(\alpha) + U(\beta - \gamma_i)]\Big) + L_{(0)} \\
&\subset L_{(0)}.
\end{aligned}$$

Consequently, $[U(\alpha), U(\beta)] \subset L_{(0)}$ in all cases. This shows that $L_{(0)}$ is a subalgebra.

(c) Let $\mathcal{P} \supset L_{(0)}$ be a subalgebra containing $L_{(0)}$ properly. By (a), one has $\mathcal{P} \cap H \neq \{0\}$. Choose any Melikian pair (α, β) and a mapping $\Phi_{\alpha;\beta}$ satisfying (18.2.1). Lemma 18.3.5 yields

$$L_{(0)} \cap L(\alpha, \beta) = L(\alpha, \beta)_{(0)}.$$

Then $\mathcal{P} \cap L(\alpha,\beta) = L(\alpha,\beta)_{(0)} \oplus (\mathcal{P} \cap H)$ and $\Phi_{\alpha;\beta}\big(\mathcal{P} \cap L(\alpha,\beta)\big) = \mathcal{M}(1,1)_{(0)} \oplus \Phi_{\alpha;\beta}\big(\mathcal{P} \cap H\big)$. However, $\mathcal{M}(1,1)_{(0)}$ is a maximal subalgebra of $\mathcal{M}(1,1)$, whence $\Phi_{\alpha;\beta}\big(\mathcal{P} \cap L(\alpha,\beta)\big) = \mathcal{M}(1,1)$. Then $H \subset \mathcal{P}$ and $L = H + L_{(0)} \subset \mathcal{P}$. Consequently, $L_{(0)}$ is a maximal subalgebra. □

Recall that $\mathcal{N}(H)$ denotes the set of all $[p]$-nilpotent elements of H.

Lemma 18.3.7.

(1) *$\mathcal{N}(H)$ contains exactly one 2-dimensional subspace H_{-1} satisfying the condition $[H_{-1}, H_{-1}] \subset \mathcal{N}(H)$. Moreover, $[H_{-1}, [H_{-1}, H_{-1}]] = \gamma_3(H)$ and $H = H_{-1} \oplus [H_{-1}, H_{-1}] \oplus [H_{-1}, [H_{-1}, H_{-1}]]$ hold.*

(2) *The subspace $H_{-1} + L_{(0)}$ is stable under the adjoint action of $L_{(0)}$.*

Proof. (1) Choose an arbitrary Melikian pair (α,β) and a mapping $\Phi_{\alpha;\beta}$ satisfying (18.2.1). Proposition 18.2.6 shows that

$$\Phi_{\alpha;\beta}(\mathcal{N}(H)) = \\ F(1+x_2)^2(1+x_1)^2 \oplus F(1+x_2)^3(1+x_1)^4\tilde{\partial}_1 \oplus F(1+x_2)^4(1+x_1)^3\tilde{\partial}_2.$$

Recall from Volume 1, p. 199 that

$$\begin{aligned}
[(1+x_2)^3(1+x_1)^4\tilde{\partial}_1, (1+x_2)^4(1+x_1)^3\tilde{\partial}_2] &= (1+x_2)^2(1+x_1)^2,\\
[(1+x_2)^2(1+x_1)^2, (1+x_2)^3(1+x_1)^4\tilde{\partial}_1] &= (1+x_1)\partial_1,\\
[(1+x_1)^2(1+x_2)^2, (1+x_2)^4(1+x_1)^3\tilde{\partial}_2] &= (1+x_2)\partial_2.
\end{aligned}$$

Take any linearly independent elements

$$\begin{aligned}
u_1 &= a_1(1+x_2)^2(1+x_1)^2 + b_1(1+x_2)^3(1+x_1)^4\tilde{\partial}_1 + c_1(1+x_2)^4(1+x_1)^3\tilde{\partial}_2\\
u_2 &= a_2(1+x_2)^2(1+x_1)^2 + b_2(1+x_2)^3(1+x_1)^4\tilde{\partial}_1 + c_2(1+x_2)^4(1+x_1)^3\tilde{\partial}_2
\end{aligned}$$

in $\Phi_{\alpha;\beta}(\mathcal{N}(H))$ such that $[u_1,u_2] \in \Phi_{\alpha;\beta}(\mathcal{N}(H))$. Then

$$\begin{aligned}
\Phi_{\alpha;\beta}(\mathcal{N}(H)) \ni [u_1,u_2] = (a_1b_2 - a_2b_1)&(1+x_1)\partial_1 + (a_1c_2 - a_2c_1)(1+x_2)\partial_2\\
&+ (b_1c_2 - b_2c_1)(1+x_2)^2(1+x_1)^2,
\end{aligned}$$

forcing $a_1b_2 = a_2b_1$ and $a_1c_2 = a_2c_1$. If $a_2 \neq 0$, then $u_1 = \frac{a_1}{a_2}u_2$, which is false. Therefore, $a_2 = 0$. Arguing similarly, one obtains $a_1 = 0$. This shows that

$$H_{-1} := \Phi_{\alpha;\beta}{}^{-1}\big(F(1+x_2)^3(1+x_1)^4\tilde{\partial}_1 + F(1+x_2)^4(1+x_1)^3\tilde{\partial}_2\big)$$

is the only 2-dimensional subspace of $\mathcal{N}(H)$ with the property that $[H_{-1}, H_{-1}] \subset \mathcal{N}(H)$.

The above computations give

$$[H_{-1}, H_{-1}] = \Phi_{\alpha;\beta}{}^{-1}\big(F(1+x_2)^2(1+x_1^2)\big),$$
$$[H_{-1}, [H_{-1}, H_{-1}]] = \Phi_{\alpha;\beta}{}^{-1}\big(F(1+x_1)\partial_1 + F(1+x_2)\partial_2\big) = \gamma_3(H).$$

These deliberations prove (1).

(2) Using the spanning set

$$\{x_2^i(1+x_1)\partial_1,\ x_2^i\partial_2,\ x_2^i(1+x_1)^2,\ x_2^i(1+x_1)^4\tilde{\partial}_1,\ x_2^i(1+x_1)^3\tilde{\partial}_2, \mid 1 \le i \le 4\}$$

for $\Phi_{\alpha;\beta}(U(\alpha)) = \Phi_{\alpha;\beta}(L(\alpha) \cap L(\alpha,\beta)_{(0)})$ displayed in part (a) of the proof of Lemma 18.3.4, it is routine to check (using the computations in Volume 2, p. 98)

$$[\Phi_{\alpha;\beta}(U(\alpha)), F(1+x_2)^3(1+x_1)^4\tilde{\partial}_1 + F(1+x_2)^4(1+x_1)^3\tilde{\partial}_2]$$
$$\subset \Phi_{\alpha;\beta}(U(\alpha)) + F(1+x_1)^4(1+x_2)^3\tilde{\partial}_1 + F(1+x_1)^3(1+x_2)^4\tilde{\partial}_2.$$

This implies that $H_{-1} + U(\alpha)$ is invariant under the adjoint action of $U(\alpha)$.

We stress the fact that, although the definition of H_{-1} and all computations depend on the choice of the Melikian pair, H_{-1} is independently described by the requirement that $[H_{-1}, H_{-1}] \subset \mathcal{N}(H)$. Now let $\alpha \neq 0$ be arbitrary. As it fits into a Melikian pair, the above computations give

$$[U(\alpha), H_{-1}] \subset H_{-1} + U(\alpha) \subset H_{-1} + L_{(0)}.$$

Since this is true for all α, the definition of $L_{(0)}$ now yields claim (2). □

The result of all these deliberations is the following theorem.

Theorem 18.3.8. *Let L be a finite dimensional simple Lie algebra over an algebraically closed field of characteristic $p > 3$. If the minimal p-envelope of L contains a torus of maximal dimension which is not stardard, then $L \cong \mathcal{M}(n_1, n_2)$ is a Melikian algebra.*

Proof. Proposition 18.3.6 shows that $L_{(0)}$ is a maximal subalgebra. By the preceding lemma,

$$L_{(-1)} := H_{-1} + L_{(0)}$$

is stable under the adjoint action of the subalgebra $L_{(0)}$. We now consider the standard filtration of L associated with the pair $(L_{(0)}, L_{(-1)})$ (cf. Definition 3.5.1) and the associated graded algebra $\oplus G_i := \oplus L_{(i)}/L_{(i+1)}$. Since L is simple and $L_{(0)}$ is maximal, this filtration is exhaustive and separating. Since $L_{(-1)} = H_{-1} \oplus L_{(0)}$, we have that $L_{(-i)} = L_{(0)} \oplus \sum_{j=1}^{i} \gamma_j(H_{-1})$ for all $i > 0$. Lemma 18.3.7 shows that

$$G_{-1} \cong H_{-1}, \quad G_{-2} \cong [H_{-1}, H_{-1}], \quad G_{-3} \cong [H_{-1}, [H_{-1}, H_{-1}]], \quad G_{-4} = \{0\}.$$

As $\dim H_{-1} = 2$, we obtain by the same token that $\dim G_{-2} = 1$ and $\dim G_{-3} = 2$. As $\dim G_{-1} = 2$, and G_0 acts faithfully on G_{-1}, we have an embedding $G_0 \subset \mathfrak{gl}(2)$.

Let (α, β) be any Melikian pair. Recall that $L_{(0)} \cap L(\alpha, \beta) = L(\alpha, \beta)_{(0)}$ (Lemma 18.3.5). As the filtration under consideration has depth 3, it follows that the filtration of $L(\alpha, \beta) \cong \mathcal{M}(1, 1)$ induced by that of L is the natural filtration of $\mathcal{M}(1, 1)$. Then $L_{(0)} \cap L(\alpha, \beta)$ acts on $L_{(-1)} \cap L(\alpha, \beta)/L_{(0)} \cap L(\alpha, \beta)$ as $\mathfrak{gl}(2)$, and $L_{(4)} \cap L(\alpha, \beta) \neq \{0\}$ holds. As a consequence, $G_0 \cong \mathfrak{gl}(2)$ and $G_4 \neq \{0\}$. Applying the Weak Recognition Theorem Corollary 5.5.3 we now obtain the claim. □

Chapter 19

Solving the case when all T-roots are solvable

In this chapter we assume that T is a torus of maximal dimension in the semisimple p-envelope $L_{[p]}$ of L and **all elements of $\Gamma(L,T)$ are solvable roots**.

19.1 2-sections revisited

In the case under consideration, the 0-sections, 1-sections, and 2-sections have rather limited structures.

Proposition 19.1.1. *The following is true.*

(1) $H = \text{nil}\, H$.

(2) *Each* 1*-section of* L *relative to* T *is nilpotent and acts trigonalizably on* L.

(3) *Every extended root is linear on* $\tilde{H}$ *and on all root spaces* L_κ.

Proof. Since all roots in $\Gamma(L,T)$ are solvable, Corollary 17.3.3(2) shows that $[L_\kappa, L_{-\kappa}]$ consists of $[p]$-nilpotent elements for all nonzero roots κ. Consequently, $H = \sum_{\kappa\in\Gamma(L,T)\setminus\{0\}} [L_\kappa, L_{-\kappa}] = \text{nil}\, H$.

In particular, $\kappa(H) = 0$ holds for all roots κ, and therefore $L(\kappa)$ is nilpotent. Theorem 17.1.2 implies that $L(\kappa)$ acts trigonalizably on L.

Every T-root may be regarded an extended root (§1.3) and as such it is linear on every root space L_κ by Proposition 1.3.2(5). As $\tilde{H}^{(1)} \subset H = \text{nil}\, H$, Remark 1.3.3 shows that every T-root is linear on $\tilde{H}$. □

Next we turn to the determination of 2-sections.

Lemma 19.1.2. *Let* $\mathfrak{g}$ *be a simple Lie algebra with* $TR(\mathfrak{g}) \leq 2$ *and* $\mathfrak{t} \subset \text{Der}\,\mathfrak{g}$ *be a torus such that the centralizer* $C_{\mathfrak{g}}(\mathfrak{t})$ *acts nilpotently on* $\mathfrak{g}$ *and each* 1*-section of* $\mathfrak{g}$ *relative to* $\mathfrak{t}$ *is solvable. Then* $\mathfrak{g} \cong H(2;\underline{1};\Phi(\tau))^{(1)}$, $\dim\,\mathfrak{t} = 2$, *and* $\mathfrak{t} \subset \mathfrak{g}_{[p]} = \text{Der}\,\mathfrak{g}$.

Proof. (a) Suppose $\mathfrak{t}$ is a maximal torus of $\text{Der}\,\mathfrak{g}$. If every 2-section of $\mathfrak{g}$ relative to $\mathfrak{t}$ is solvable, then so is $\mathfrak{g}$ (Theorem 1.3.10). Since this is not the case, there are

$\kappa, \lambda \in \Gamma(\mathfrak{g}, \mathfrak{t})$ such that the 2-section $M := \mathfrak{g}(\kappa, \lambda)$ is non-solvable. For $i \in \mathbb{F}_p$, set

$$M_i := \sum_{j \in \mathbb{F}_p} M_{i\kappa + j\lambda}.$$

Clearly, the decomposition $M = \bigoplus_{i \in \mathbb{F}_p} M_i$ is an $\mathbb{F}_p$-grading of M. As M is non-solvable, M_0 does not act nilpotently on M (see Proposition 1.3.7). Since $C_{\mathfrak{g}}(\mathfrak{t})$ does act nilpotently on $\mathfrak{g}$ by our assumption, the ENGEL–JACOBSON Theorem shows that there is $x \in M_{r\lambda}$ for some $r \in \mathbb{F}_p^*$ such that $\mathrm{ad}_M x$ is not nilpotent. Since the torus $\mathfrak{t}$ is maximal, we have that $\lambda(x) = 0$ and $\kappa(x) \neq 0$. Interchanging the rôles of κ and λ in this argument, we find $y \in \mathfrak{g}_{s\kappa}$ for some $s \in \mathbb{F}_p^*$ such that $\kappa(y) = 0$ and $\lambda(y) \neq 0$. Since $\mathfrak{t}$ is a maximal torus, the semisimple parts x_s and y_s of x, y in $M_{[p]} \subset \mathrm{Der}\,\mathfrak{g}$ lie in the torus $\mathfrak{t}_0 := \mathfrak{t} \cap M_{[p]} \subset \mathfrak{g}_{[p]}$. By construction, x_s and y_s are linearly independent. Since $\dim \mathfrak{t}_0 \leq MT(\mathfrak{g}_{[p]}) = TR(\mathfrak{g})$, our assumption on $\mathfrak{g}$ implies that $\mathfrak{t}_0 = Fx_s + Fy_s$.

(b) Suppose that for some $\Lambda \in \Gamma(\mathfrak{g}, \mathfrak{t}_0)$ the 1-section $\mathfrak{g}(\Lambda)$ is non-solvable. Clearly, $\mathfrak{g}(\Lambda) = \mathfrak{g}(\alpha_1, \ldots, \alpha_l)$ for some $\alpha_i \in \Gamma(\mathfrak{g}, \mathfrak{t})$. So $\mathfrak{g}(\Lambda)$ is a section of $\mathfrak{g}$ relative to $\mathfrak{t}$. Since $\mathfrak{g}(\Lambda)$ is assumed to be non-solvable, it contains a non-solvable 2-section relative to $\mathfrak{t}$, say M' (again by Theorem 1.3.10). We now repeat the argument from part (a) with M replaced by M' (and with $\mathfrak{t}$ unchanged) to observe that the p-envelope $\mathfrak{g}(\Lambda)_{[p]} \subset \mathfrak{g}_{[p]}$ contains a 2-dimensional torus $\mathfrak{t}_1$ which acts faithfully on $\mathfrak{g}(\Lambda)$. But then $\mathfrak{t}_1 \oplus (\mathfrak{t}_0 \cap (\ker \Lambda))$ is a 3-dimensional torus in $\mathfrak{g}_{[p]}$ violating our assumption that $TR(\mathfrak{g}) = 2$. Thus all 1-sections of $\mathfrak{g}$ relative to the 2-dimensional torus $\mathfrak{t}_0$ in $\mathfrak{g}_{[p]}$ are solvable.

(c) By (a), $\mathfrak{t}_0$ is a torus of maximal dimension in $\mathfrak{g}_{[p]}$, and by (b), every $\mathfrak{t}_0$-root is solvable. Proposition 19.1.1(2) applied to $\mathfrak{g}$ and $\mathfrak{t}_0$ shows that every 1-section is nilpotent. Therefore any 1-section $\mathfrak{g}(\Lambda)$ with respect to $\mathfrak{t}_0$ is a CSA of $\mathfrak{g}$ of toral rank 1 (Proposition 11.2.6). Due to Corollary 9.2.12 $\mathfrak{g}$ is of types $\mathfrak{sl}(2)$, $W(1; \underline{n})$, $H(2; \underline{n}; \Phi)^{(2)}$. As $TR(\mathfrak{g}) = 2$, the Lie algebra $\mathfrak{g}$ is listed in Theorem 10.6.3. Then $\mathfrak{g}$ is one of $W(1; \underline{2})$, $H(2; \underline{1}; \Phi(\tau))^{(1)}$, $H(2; \underline{1}; \Phi(1))$, $H(2; (1,2))^{(2)}$. Theorem 7.2.2(1),(7),(5) shows that these algebras different from $H(2; \underline{1}; \Phi(\tau))^{(1)}$ have codimension 1 in their minimal p-envelopes. Therefore $\mathfrak{t}_0 \cap \mathfrak{g} \neq \{0\}$ holds in these cases, which contradicts Proposition 19.1.1(1). As a consequence, $\mathfrak{g} \cong H(2; \underline{1}; \Phi(\tau))^{(1)}$. Theorem 10.3.1 proves $\mathrm{Der}\,\mathfrak{g} = \mathfrak{g}_{[p]}$. In particular, $\mathfrak{t} \subset \mathfrak{g}_{[p]}$ holds.

(d) Now suppose that $\mathfrak{t}$ is not a maximal torus in $\mathrm{Der}\,\mathfrak{g}$, let $\mathfrak{t}' \supsetneq \mathfrak{t}$ be a maximal torus in $\mathrm{Der}\,\mathfrak{g}$. Clearly, the assumptions of the lemma hold for $\mathfrak{t}'$. So we get from (a) - (c) that $\dim \mathfrak{t} < \dim \mathfrak{t}' = 2$. Then $\mathfrak{t}$ is 1-dimensional, whence $\mathfrak{g}$ is a 1-section and hence is solvable. This is not true. □

The 2-sections are ruled by the following proposition.

Proposition 19.1.3. *If $L(\alpha, \beta)$ is non-solvable for some $\alpha, \beta \in \Gamma(L, T)$, then*

$$L[\alpha, \beta] \cong H(2; \underline{1}; \Phi(\tau))^{(1)}.$$

Proof. Suppose that $L[\alpha, \beta]$ is non-solvable and let $\{0\} \neq \tilde{S} = \bigoplus_{i=1}^{r} \tilde{S}_i$ be the sum of all minimal T-invariant ideals of the T-semisimple Lie algebra $L[\alpha, \beta]$. The structure of $\tilde{S}$ is described in Theorems 17.4.1, 17.4.3, and 17.4.4. The algebras described in Theorems 17.4.1, 17.4.4 do not occur in our case, since they all possess non-solvable 1-sections relative to $\overline{T}$. Thus $L[\alpha, \beta]$ is described in Theorem 17.4.3, so that $\tilde{S}$ is simple and $TR(\tilde{S}) = 2$. As all 1-sections of L relative to T are solvable, so are all 1-sections of $\tilde{S}$ relative to $\overline{T}$. Lemma 19.1.2 now says that only $L[\alpha, \beta] = \tilde{S} \cong H(2; \underline{1}; \Phi(\tau))^{(1)}$ does occur. □

Unfortunately, in the present case it is not sufficient to know the 2-sections, but one also needs strong information on 3-sections. This will be done next.

19.2 The case when $TR(L) = 3$

In a next step we consider the simple Lie algebras of absolute toral rank 3 and derive some structural features.

Proposition 19.2.1. *If $TR(L) = 3$ and L contains a solvable* 2*-section $L(\alpha, \beta)$ with $\mathbb{F}_p$-independent $\alpha, \beta \in \Gamma(L, T)$, then $L \cong H(2; (1, 2); \Phi(\tau))^{(1)}$.*

Proof. Let M be a maximal T-invariant subalgebra of L containing $L(\alpha, \beta)$.

(a) Suppose M is not solvable. Then M contains a non-solvable 2-section relative to T (Theorem 1.3.10). Since $M(\alpha, \beta) = L(\alpha, \beta)$ is solvable, there exist $\mathbb{F}_p$-independent $\gamma, \delta \in \Gamma(L, T)$ with $\gamma \notin \mathbb{F}_p\alpha + \mathbb{F}_p\beta$ such that $M(\gamma, \delta)$ is non-solvable. For $i \in \mathbb{F}_p$ set

$$M(\gamma, \delta)_i := \sum_{j \in \mathbb{F}_p} M_{i\delta + j\gamma}.$$

Then $M(\gamma, \delta) = \bigoplus_{i \in \mathbb{F}_p} M(\gamma, \delta)_i$ is $\mathbb{F}_p$-graded. By Proposition 1.3.7, the subalgebra $M(\gamma, \delta)_0$ does not act nilpotently on $M(\gamma, \delta)$. By Proposition 19.1.1(1), $H \subset M(\gamma, \delta)_0$ acts nilpotently on L. The ENGEL–JACOBSON Theorem now yields that there exists $x \in M_{j\gamma}$, for some $j \in \mathbb{F}_p^*$, with $\delta(x) \neq 0$. Let $L_1(x)$ denote the Fitting-1-component for $\operatorname{ad} x$. Since $\operatorname{ad} x$ is not nilpotent, $L_1(x) = (\operatorname{ad} x)^{p^e}(L) \neq \{0\}$ (where $e \gg 0$). By Proposition 1.3.5, the simplicity of L implies $L = L_1(x) + [L_1(x), L_1(x)]$. Hence $L_1(x) \not\subset M$. Note that $L_1(x) = (\operatorname{ad} x)^{p^e}(L)$ is T-invariant. Therefore the factor space $(L_1(x) + M)/M$ is nonzero, and invariant under T and $\operatorname{ad} x$, and $\operatorname{ad} x$ acts invertibly on it. From this it follows that there is a root μ such that $\mu + \mathbb{F}_p\gamma \subset \Gamma(L/M, T)$. Write $\mu = m\alpha + n\beta + r\gamma$ for some $m, n, r \in \mathbb{F}_p$. Then $m\alpha + n\beta$ is a T-weight of L/M. However $L(\alpha, \beta) \subset M$, a contradiction.

(b) As a consequence, M is solvable. Corollary 9.2.13 says that L is one of $\mathfrak{sl}(2)$, $W(1; \underline{n})$, $H(2; \underline{n}; \Phi)^{(2)}$ for some $\underline{n}$ and Φ.

As $TR(L) = 3$, we have $L \not\cong \mathfrak{sl}(2)$.

The semisimple p-envelope $W(1;\underline{n})_{[p]}$ is simply $W(1;\underline{n}) + \sum_{i=1}^{n-1} F\,\partial^{p^i}$. As $TR((W(1;\underline{n})) = n$ (Theorem 7.6.3(2)), then $W(1;\underline{n})$ intersects with any torus of maximal dimension in $W(1;\underline{n})_{[p]}$. As $H = \operatorname{nil} H$ by Proposition 19.1.1, L is not of Witt type.

Theorem 6.3.10 shows that there are 3 types of Hamiltonian algebras in 2 generators, namely $H(2;\underline{n})^{(2)}$, $H(2;\underline{n};\Phi(\tau))^{(1)}$, $H(2;\underline{n};\Phi(l))$ for some $n_1 \le n_2$. In the respective cases we have dimensions $p^{n_1+n_2}-2$, $p^{n_1+n_2}-1$, $p^{n_1+n_2}$ (Theorem 6.3.10) and absolute toral ranks $n_1+n_2-1, n_1+n_2, n_1+n_2$ (Theorem 10.1.4). Moreover, Theorem 7.2.2(5),(7) implies that dim $H(2;\underline{n})^{(2)}{}_{[p]} = (p^{n_1+n_2}-2) + (n_1-1+n_2-1)$ and dim $H(2;\underline{n};\Phi(l))_{[p]} = p^{n_1+n_2} + (1+n_1-1+n_2-1)$. We therefore have $H(2;\underline{n})^{(2)} \cap T \ne \{0\}$ and $H(2;\underline{n};\Phi(l)) \cap T \ne \{0\}$ for every torus T of maximal dimension in the respective p-envelopes. As in the Witt case, these cases are impossible. Then $L \cong H(2;\underline{n};\Phi(\tau))^{(1)}$ with $n_1 \le n_2$ and $n_1 + n_2 = 3$, completing the proof. □

Proposition 19.2.2. *Suppose $L \cong H(2;(1,2);\Phi(\tau))^{(1)}$ and T is a 3-dimensional torus in the minimal p-envelope $L_{[p]}$. Then*

$$0 \notin \Gamma(L,T) \text{ but } G := \Gamma(L,T) \cup \{0\} \text{ is an elementary } p\text{-group of order } p^3.$$

There is a skew-symmetric $\mathbb{F}_p$-bilinear form

$$f : G \times G \to F$$

and an eigenvector basis $(d_\mu)_{\mu\in\Gamma(L,T)}$ of L with respect to T such that

$$[d_\lambda, d_\mu] = f(\lambda,\mu) d_{\lambda+\mu}.$$

Moreover, $L_\mu = F d_\mu$ is 1-dimensional, $f(\lambda,\mu) = \mu(d_\lambda)$ and $d_\lambda{}^{[p]} \in T\setminus\{0\}$.

Proof. Set $\tau := (p-1, p^2-1)$. According to Theorems 6.3.10(1), 7.2.2(6), and 7.1.3(3), we have the following description of L and $L_{[p]}$:

$$L \cong H(2;(1,2);\Phi(\tau))^{(1)} = F(1-x^{(\tau)})\partial_1 + F(1-x^{(\tau)})\partial_2 + H(2;(1,2))^{(1)}{}_{(0)},$$
$$L \ne L_{[p]} \subset \operatorname{Der} L \cong H(2;(1,2);\Phi(\tau)) + F\big((1-x^{(\tau)})\partial_2\big)^p.$$

Note that

$$\big((1-x^{(\tau)})\partial_2\big)^p = \partial_2{}^p - (\operatorname{ad}\partial_2)^{p-1}(x^{(\tau)}\partial_2).$$

Consider $L_{[p]}$ as a subalgebra of $W(2;(1,2))_{[p]} = F\partial_1 \oplus F\partial_2 \oplus F{\partial_2}^p \oplus W(2;(1,2))_{(0)}$. We then have ${\partial_1}^p = 0$. Consequently,

$$L_{[p]} = H(2;(1,2);\Phi(\tau)) + F({\partial_2}^p - x_1^{(p-1)} x_2^{(p^2-p)}\partial_2).$$

Next, it is obvious that

$$L_{[p]} \cap W(2;(1,2))_{(0)}/L_{[p]} \cap W(2;(1,2))_{(1)} = L_{(0)}/L_{(1)} \cong \mathfrak{sl}(2).$$

Therefore

$$\dim T \cap W(2;(1,2))_{(0)} \leq 1.$$

Suppose that $T \cap W(2;(1,2))_{(0)} \neq \{0\}$. Then $T \cap W(2;(1,2))_{(0)}$ constitutes a CSA in $L_{[p]} \cap W(2;(1,2))_{(0)}/L_{[p]} \cap W(2;(1,2))_{(1)} \cong \mathfrak{sl}(2)$, and the 2-dimensional module $W(2;(1,2))/W(2;(1,2))_{(0)}$ for this $\mathfrak{sl}(2)$ is irreducible. Thus 0 is not a weight. This means that

$$T \cap W(2;(1,2)) \subset T \cap W(2;(1,2))_{(0)}.$$

Since $\dim L_{[p]}/L_{[p]} \cap W(2;(1,2)) = 1$, we obtain

$$\dim T \leq \dim L_{[p]}/L_{[p]} \cap W(2;(1,2)) + \dim T \cap W(2;(1,2))_{(0)} = 2,$$

a contradiction. Hence

$$T \cap W(2;(1,2))_{(0)} = \{0\}.$$

As a consequence, there are $E_1, E_2, E_3 \in W(2;(1,2))_{(0)}$ such that

$$T = F(\partial_1 + E_1) \oplus F(\partial_2 + E_2) \oplus F({\partial_2}^p + E_3).$$

We are now going to realize $H(2;(1,2);\Phi(\tau))^{(1)}$ as a Poisson Lie algebra (similar to what we have done in §10.3). Impose on the divided power algebra $\mathcal{O}(2;(1,2))$ a Poisson bracket

$$\{x^{(a)}, x^{(b)}\} := (x^{(a-\epsilon_1)}x^{(b-\epsilon_2)} - x^{(a-\epsilon_2)}x^{(b-\epsilon_1)})(1 - x^{(\tau)}).$$

It is easily checked that

$$\Big(\mathcal{O}(2;(1,2)), \{\ ,\ \}\Big)^{(1)} \cap F1 = \{0\}.$$

From this one derives that $\dim (\mathcal{O}(2;(1,2)), \{\ ,\ \})^{(1)} \leq p^3 - 1$. Define a linear mapping $\mathfrak{D} : \mathcal{O}(2;(1,2)) \to H(2;(1,2);\Phi(\tau))$ by

$$\begin{aligned} \mathfrak{D}(x^{(a)}) &= x^{(a-\epsilon_1)}\partial_2 - x^{(a-\epsilon_2)}\partial_1 \qquad \text{if } |a| \geq 2,\\ \mathfrak{D}(x_1) &= (1 - x^{(\tau)})\partial_2, \quad \mathfrak{D}(x_2) = -(1 - x^{(\tau)})\partial_1,\\ \mathfrak{D}(1) &= 0. \end{aligned}$$

This definition is as in Equation (6.5.4) in Volume 2, p. 338, where we now have $\omega = 2(1 + x^{(\tau)})dx_1 \wedge dx_2$ and $\omega_{1,2} = (1 + x^{(\tau)})$, $g_{1,2} = -(1 - x^{(\tau)})$. Equation (6.5.5) in Volume 2, p. 339 yields

$$[\mathfrak{D}(f), \mathfrak{D}(g)] = \mathfrak{D}(\mathfrak{D}(f)(g)) = \mathfrak{D}(\{f, g\}) \quad \forall\ f, g \in \mathcal{O}(2; (1,2)).$$

Thus $\mathfrak{D}$ is a homomorphism of algebras with $\ker \mathfrak{D} = F1$, and hence $\mathfrak{D}$ establishes an isomorphism of Lie algebras

$$\mathcal{O}(2; (1,2))^{(1)} \xrightarrow{\sim} H(2; (1,2); \Phi(\tau))^{(1)}.$$

The former equation implies that

$$\mathfrak{D}(\mathfrak{D}(f)^{p^j}(g)) = (\operatorname{ad} \mathfrak{D}(f))^{p^j}(\mathfrak{D}(g)) = [\mathfrak{D}(f)^{p^j}, \mathfrak{D}(g)].$$

Since $L_{[p]}$ may be considered a subalgebra of $W(2; (1,2))$ (which means $\mathfrak{D}(f)^{[p]^j} = \mathfrak{D}(f)^{p^j}$), this gives

$$\mathfrak{D}(E(g)) = [E, \mathfrak{D}(g)] \quad \forall E \in L_{[p]},\ \forall g \in \mathcal{O}(2; (1,2)),$$

whence

$$\begin{aligned}\mathfrak{D}(\{E(g), h\}) &= [\mathfrak{D}(E(g)), \mathfrak{D}(h)] = [[E, \mathfrak{D}(g)], \mathfrak{D}(h)] \\ &= [E, \mathfrak{D}(\{g, h\})] - [\mathfrak{D}(g), [E, \mathfrak{D}(h)]] \\ &= \mathfrak{D}(E(\{g, h\})) - \mathfrak{D}(\{g, E(h)\}),\end{aligned}$$

hence

$$\{E(g), h\} = E(\{g, h\}) - \{g, E(h)\}.$$

Applying this equation for $E = t \in T$ one obtains

$$\{t(g), h\} = t(\{g, h\}) - \{g, t(h)\} \quad \forall t \in T,\ \forall g, h \in \mathcal{O}(2; (1,2))^{(1)}.$$

Next we observe that there is a natural isomorphism $\mathcal{O}(2; (1,2)) \xrightarrow{\sim} \mathcal{O}(3; \underline{1})$ of commutative algebras via

$$x_1 \mapsto y_1, \quad x_2 \mapsto y_2, \quad x_2^{(p)} \mapsto y_3.$$

Note that according to Equations (2.1.3), (2.1.6) and Lemma 2.1.2

$$x_2^{(p^2-1)} = x_2^{(p-1)}(x_2^{(p)})^{(p-1)}$$

holds. Moreover, one has $\{x_1, x_2^{(p)}\} = x_2^{(p-1)}$. Thus $\mathcal{O}(3; \underline{1})$ carries a related Poisson bracket given by

$$\{y_1, y_2\} = (1 - y_1^{(p-1)} y_2^{(p-1)} y_3^{(p-1)}),\ \{y_1, y_3\} = y_2^{(p-1)},\ \{y_2, y_3\} = 0.$$

Moreover, T is mapped into $W(3;\underline{1})$, it acts on $(\mathcal{O}(3;\underline{1}), \{\ ,\ \})^{(1)}$ by derivations, and the above described form of T ensures that

$$T = F(\partial_1 + e_1) \oplus F(\partial_2 + e_2) \oplus F(\partial_3 + e_3), \quad \text{for some } e_1, e_2, e_3 \in W(3;\underline{1})_{(0)}.$$

Due to Corollary 7.5.2 there are generators u_1, u_2, u_3 of the truncated polynomial ring $\mathcal{O}(3;\underline{1})$ and toral elements $t_1, t_2, t_3 \in T$, such that

$$t_i = (1 + u_i)\partial_{u_i}, \ i = 1, 2, 3.$$

Set $z_i := 1 + u_i$, $z^a := \prod z_i^{a_i}$, and define linear forms (ϵ_i) dual to (t_i). Considering eigenvalues we obtain that

$$\{z_i, z_j\} = \alpha_{i,j} z_i z_j, \quad \alpha_{ij} \in F.$$

Then the Poisson bracket is given by

$$\{z^a, z^b\} = \sum_{i,j=1}^{3} a_i b_j z^{a+b-\epsilon_i-\epsilon_j} \{z_i, z_j\} = \sum_{i,j=1}^{3} a_i b_j \alpha_{i,j} z^{a+b}.$$

Applying these isomorphisms

$$L \cong H(2; (1,2); \Phi(\tau))^{(1)} \cong \mathcal{O}(2; (1,2))^{(1)} \cong \mathcal{O}(3;\underline{1})^{(1)}$$

one obtains (after an identification)

$$F z^a = L_{\sum a_i \epsilon_i}, \quad \Gamma(L,T) = \Big(\sum_{i=1}^{3} \mathbb{F}_p \epsilon_i\Big) \setminus \{0\}.$$

Identifying a and $\sum a_i \epsilon_i$ we finally set $d_a := z^a$, $f(a,b) := \sum_{i,j=1}^{3} a_i b_j \alpha_{i,j}$.

Since f is biadditive and skew-symmetric, one has $f(\lambda, i\lambda + \mu) = f(\lambda, \mu)$ and easily computes inductively

$$\mu({d_\lambda}^{[p]}) d_\mu = (\operatorname{ad} d_\lambda)^p (d_\mu) = f(\lambda, \mu)^p d_\mu.$$

In particular, ${d_\lambda}^{[p]}$ acts semisimply on L. As T is a maximal torus this gives ${d_\lambda}^{[p]} \in T$. Moreover, if ${d_\lambda}^{[p]} = 0$, then $f(\lambda, \mu) = 0$ for all roots μ, and this gives $[d_\lambda, L] = \{0\}$, a contradiction. Finally recall the definition $\mu(d_\lambda) = \mu({d_\lambda}^{[p]})^{1/p}$. □

Theorem 19.2.3. *Let $TR(L) = 3$. The following holds.*

(1) *$\Gamma(L,T) \cup \{0\}$ is an elementary p-group of order p^3.*

(2) *There is $k \in \mathbb{N}$ such that* $\dim L_\delta = k$ *for all $\delta \in \Gamma(L,T) \setminus \{0\}$.*

(3) *Every solvable 2-section of L relative to T is abelian.*

Proof. (a) Assume that L contains a solvable 2-section $L(\kappa, \lambda)$ relative to T. Then L is isomorphic to $H(2; (1,2); \Phi(\tau))^{(1)}$ by Proposition 19.2.1. The root space decomposition in $H(2; (1,2); \Phi(\tau))^{(1)}$ relative to a 3-dimensional torus in $L_{[p]}$ has been investigated in Proposition 19.2.2. As a result, all elements in $(\mathbb{F}_p\alpha \oplus \mathbb{F}_p\beta \oplus \mathbb{F}_p\gamma) \setminus \{0\}$ are roots, 0 is not a root, and all root spaces are 1-dimensional. This proves assertions (1) and (2) for this particular case.

In order to prove assertion (3), i.e., to show that $L(\kappa, \lambda)$ is abelian, we observe that there is a nonzero abelian ideal I of $L(\kappa, \lambda)$ which is invariant under T. As 0 is not a root, there is a nonzero root δ_1 for which $I_{\delta_1} \neq \{0\}$. We apply the notations and results of Proposition 19.2.2. Since root spaces are 1-dimensional, we get $Fd_{\delta_1} \subset I$. In particular, d_{δ_1} acts nilpotently on $L(\kappa, \lambda)$, whence

$$f(\delta_1, \delta_2) = \delta_2(d_{\delta_1}) = 0 \text{ for all } \delta_2 \in \mathbb{F}_p\kappa + \mathbb{F}_p\lambda.$$

Choose any δ_2 independent of δ_1. Then these roots span $\mathbb{F}_p\kappa + \mathbb{F}_p\lambda$. As f is biadditive and skew-symmetric, we get that f vanishes on $(\mathbb{F}_p\kappa + \mathbb{F}_p\lambda) \times (\mathbb{F}_p\kappa + \mathbb{F}_p\lambda)$. This means that the 2-section under consideration is abelian.

(b) Next we assume that no 2-section $L(\kappa, \lambda)$ is solvable. Proposition 19.1.3 then says that $L[\kappa, \lambda]$ is isomorphic to $H(2; \underline{1}; \Phi(\tau))^{(1)}$. By Theorem 10.2.3(5), there exist $x \in L_\kappa$ and $y \in L_\lambda$ with

$$\kappa(x) = 0, \;\; \lambda(x) \neq 0, \qquad \kappa(y) \neq 0, \;\; \lambda(y) = 0.$$

Thus $\dim L_{i\kappa} = \dim L_{i\kappa + j\lambda} = \dim L_{j\lambda}$ for all $\mathbb{F}_p$-independent $\kappa, \lambda \in \Gamma(L, T)$ and all $i, j \in \mathbb{F}_p^*$. This implies that all elements in $(\mathbb{F}_p\alpha \oplus \mathbb{F}_p\beta \oplus \mathbb{F}_p\gamma) \setminus \{0\}$ are roots and all root spaces are of the same dimension. □

Let $F[x_1, \ldots, x_m]$ denote the truncated polynomial ring in m variables. In this notation one has ${x_i}^p = 0$. Write $\mathfrak{m}$ for the maximal ideal of $F[x_1, \ldots, x_m]$.

Proposition 19.2.4. *Let $\mathfrak{g}$ be a transitive subalgebra of $W(m; \underline{1})$ and J be an ideal of $\mathfrak{g}$. Suppose J consists of nilpotent derivations only. Then there is an embedding of Lie algebras*

$$J \hookrightarrow \sum_{i=1}^{m} F[x_1, \ldots, x_{i-1}]\partial_i.$$

In particular, $\dim J \leq \frac{p^m - 1}{p-1}$.

Proof. (a) We may assume that $\mathfrak{g}$ is closed under p-th powers. Define inductively

$$V_0 := \{0\}, \quad B_0 := F, \quad V_i := \{v \in F[x_1, \ldots, x_m] \mid J(v) \subset B_{i-1}\}$$

and B_i the commutative and associative subalgebra with unit element of $F[x_1, \ldots, x_m]$ generated by V_i. It is not difficult to show by induction, that B_i is $\mathfrak{g}$-invariant. Since

J acts nilpotently on $F[x_1, \dots, x_m]$, one also has $B_{i-1} \subsetneq V_i \subset B_i$ whenever $B_{i-1} \neq F[x_1, \dots, x_m]$.

(b) Let $I \neq \{0\}$ be a $\mathfrak{g}$-invariant ideal of B_i. Since $\mathfrak{g}$ is a transitive subalgebra of $W(m; \underline{1})$, one has $I \not\subset \mathfrak{m}$. Hence I contains an invertible element, whence $I = B_i$. This shows that B_i is $\mathfrak{g}$-simple. Due to Corollary 3.3.3 $B_i \cong S_i \otimes \mathcal{O}(m_i; \underline{1})$, where S_i is a simple quotient of B_i. As B_i is commutative and associative, $S_i \cong F$. As a result,

$$B_i \cong F[x_1, \dots, x_{m_i}] \text{ and } m_{i-1} < m_i.$$

Set $\mathfrak{m}_i$ the unique maximal ideal of B_i. This ideal is the unique maximal nilpotent ideal. Therefore $\mathfrak{m}_i \cap B_{i-1}$ is a nilpotent ideal of B_{i-1} and $\mathfrak{m}_{i-1} B_i$ is a nilpotent ideal of B_i. This gives $\mathfrak{m}_i \cap B_{i-1} = \mathfrak{m}_{i-1}$.

Next we observe ${\mathfrak{m}_{i-1}}^2 \subset {\mathfrak{m}_i}^2 \cap B_{i-1}$. Suppose ${\mathfrak{m}_i}^2 \cap B_{i-1} \not\subset {\mathfrak{m}_{i-1}}^2$. Since B_{i-1} is $\mathfrak{g}$-simple, there is $u \in {\mathfrak{m}_i}^2 \cap B_{i-1}$ and $g \in \mathfrak{g}$ such that $g(u) \notin \mathfrak{m}_{i-1}$. But $g(u) \in (g(\mathfrak{m}_i)\mathfrak{m}_i) \cap B_{i-1} \subset \mathfrak{m}_{i-1}$. This contradiction shows that

$${\mathfrak{m}_i}^2 \cap B_{i-1} = {\mathfrak{m}_{i-1}}^2.$$

(c) Set $V_i' := V_i \cap \mathfrak{m}_i$, whence $V_i = V_i' + F$. Since V_i generates B_i, one has $V_i' + {\mathfrak{m}_i}^2 = \mathfrak{m}_i$. The canonical linear mapping

$$\varphi_i : V_i' \to \mathfrak{m}_i / {\mathfrak{m}_i}^2$$

is surjective. Put $U_0 = \{0\}$, and let inductively (for all $j < i$) U_j be a subspace for which

$$U_j \subset V_j', \quad U_j \cap {\mathfrak{m}_j}^2 = \{0\}, \quad U_j + {\mathfrak{m}_j}^2 = \mathfrak{m}_j, \quad U_0 \subset \cdots \subset U_{i-1}.$$

Note that $(\ker \varphi_i) \cap U_{i-1} \subset {\mathfrak{m}_i}^2 \cap U_{i-1} = {\mathfrak{m}_{i-1}}^2 \cap U_{i-1} = \{0\}$. Choose a preimage $U_i \subset V_i'$ such that $U_{i-1} \subset U_i$, $U_i + {\mathfrak{m}_i}^2 = \mathfrak{m}_i$ and $U_i \cap {\mathfrak{m}_i}^2 = \{0\}$.

By construction, $\dim U_i = m_i$. We thus constructed a flag $U_0 \subsetneq \cdots \subsetneq U_t$ such that $U_t \subset \mathfrak{m}$ generates $F[x_1, \dots, x_m]$ and $U_t \cap \mathfrak{m}^2 = \{0\}$ holds. Note that $J(U_i) \subset J(V_i) \subset B_{i-1}$ holds for all i. Choose generators $y_1, \dots, y_m \in U_t$ of $F[x_1, \dots, x_m]$ compatible with this flag. For $y_k \in U_i \setminus U_{i-1}$ one has $J(y_k) \subset B_{i-1}$. Hence $J \subset \sum_{k=1}^{m} F[y_1, \dots, y_{k-1}]\partial_{y_k}$. In particular, $\dim J \le \sum_{k=0}^{m-1} p^k = \frac{p^m - 1}{p-1}$ holds. □

Lemma 19.2.5. *Let* $\mathfrak{g} = \mathfrak{g}_{(-r)} \supset \cdots \supset \mathfrak{g}_{(0)} \supset \cdots \supset \mathfrak{g}_{(s)}$ *be a filtered Lie algebra and let* $\mathfrak{b}$ *be a trigonalizable subalgebra of* $\mathfrak{g}$. *Then* $\operatorname{gr} \mathfrak{b}$ *is a trigonalizable subalgebra of* $\operatorname{gr} \mathfrak{g}$.

Proof. Let $G = \operatorname{gr} \mathfrak{g}$, $B = \operatorname{gr} \mathfrak{b}$, and assume that $B^{(1)}$ does not act nilpotently on G. Clearly, $B^{(1)} = \sum_{i,j \in \mathbb{Z}} [B_i, B_j]$ where

$$B_i := \operatorname{gr}_i \mathfrak{b} = (\mathfrak{b} \cap \mathfrak{g}_{(i)} + \mathfrak{g}_{(i+1)})/\mathfrak{g}_{(i+1)} \subset \mathfrak{g}_{(i)}/\mathfrak{g}_{(i+1)} = \operatorname{gr}_i \mathfrak{g}.$$

Note that $\bigcup_{i,j\in\mathbb{Z}}[B_i, B_j]$ is a Lie set. So the ENGEL–JACOBSON Theorem implies that there are $k, l \in \mathbb{Z}$ such that $[B_k, B_l]$ contains an element which does not act nilpotently on G. Since for $k + l \neq 0$ the subspace $\mathrm{ad}_G\, B_{k+l}$ consists of nilpotent endomorphisms, one has $l = -k$. Then there exist $u_1, \dots, u_r \in \mathfrak{b} \cap \mathfrak{g}_{(k)}$ and $v_1, \dots, v_r \in \mathfrak{b} \cap \mathfrak{g}_{(-k)}$ such that the coset $\sum [u_i, v_i] + \mathfrak{g}_{(1)}$ contains an element which does not act nilpotently on $\mathfrak{g}$. Since $\sum [u_i, v_i] \in \mathfrak{g}_{(0)}$ and $[\mathfrak{g}_{(0)}, \mathfrak{g}_{(1)}] \subset \mathfrak{g}_{(1)}$, and $\mathfrak{g}_{(1)}$ acts nilpotently on $\mathfrak{g}$, it must be that $\sum [u_i, v_i]$ does not act nilpotently on $\mathfrak{g}$. But this is impossible as $\sum [u_i, v_i] \in \mathfrak{b}^{(1)}$ acts nilpotently on $\mathfrak{g}$. □

The proof of our next crucial result well illustrates some of the classification methods.

Theorem 19.2.6. *Suppose $TR(L) = 3$. Then $u_\mu{}^{[p]} \in T \setminus \{0\}$ for every nonzero $u_\mu \in L_\mu$, $\mu \in \Gamma(L, T)$.*

Proof. (a) Suppose the theorem is not true and let L be a counterexample of minimal dimension to it. Let $\mathcal{N}(L_{[p]})$ denote the set of all $[p]$-nilpotent elements in $L_{[p]}$.

By Proposition 19.1.1, all 1-sections $L(\delta)$ (where $\delta \neq 0$) relative to T are nilpotent and have the property that $L(\delta)^{(1)} \subset \mathcal{N}(L_{[p]})$. Let $n(\delta)$ denote the nilpotency class of $L(\delta)$ and let $\alpha \in \Gamma(L, T) \setminus \{0\}$ be such that

$$n(\alpha) = n := \max\{n(\delta) \mid \delta \in \Gamma(L, T)\}.$$

If $n \geq 3$, then $\{0\} \neq \gamma_{n-1}(L(\alpha)) \subset L(\alpha)^{(1)} \subset \mathcal{N}(L_{[p]})$. Since $\gamma_{n-1}(L(\alpha))$ is T-invariant, $\gamma_{n-1}(L(\alpha)) \cap L_{i\alpha} \neq \{0\}$ for some $i \in \mathbb{F}_p$. Let w be any nonzero element in the union $\bigcup_{i\in\mathbb{F}_p} \gamma_{n-1}(L(\alpha)) \cap L_{i\alpha}$.

If $n \leq 2$, then all 1-sections of L relative to T are abelian (by definition of n). In particular, $[H, L_\mu] = \{0\}$ for all roots μ. Then $H = \{0\}$ as L is centerless. Since L is a counterexample, there is a nonzero $v \in L_\alpha$ for some nonzero $\alpha \in \Gamma(L, T)$ such that $v^{[p]} = 0$ or $v^{[p]}$ is not $[p]$-semisimple. If $v^{[p]} = 0$, we set $w := v$. If $v^{[p]} \neq 0$ is not $[p]$-semisimple, we let w be the $[p]$-nilpotent part of $v^{[p]}$ in $\tilde{H}$. Clearly, $w = f(v)$ for some p-polynomial $f \in F[t]$ without a constant term.

Thus in all cases we find

$$\alpha \in \Gamma(L, T) \setminus \{0\} \quad \text{and} \quad w \in \tilde{H} \cup \bigcup_{i\in\mathbb{F}_p^*} L_{i\alpha},$$

such that

$$w \in \mathcal{N}(L_{[p]}) \setminus \{0\} \quad \text{and} \quad [w, L(\alpha)] = \{0\}.$$

From now we fix such α and w, set

$$L' := L + Fw + T$$

and denote by Q a maximal subalgebra of L' containing the normalizer of w in L'. Note that $w \in L$ or $[T, w] = 0$, whence $L'^{(1)} = L$. Since the $[p]$-subalgebra of $L_{[p]}$ generated by L' is $L_{[p]}$, and $L_{[p]}$ is centerless, one has $C(L') = \{0\}$.

(b) Let $\delta \in \Gamma(L,T)$. Since $\sum_{i\in\mathbb{F}_p} L_{\delta+i\alpha}$ is invariant under the nilpotent endomorphism $\operatorname{ad} w$, there exists $j = j(\delta)$ with $L_{\delta+j\alpha} \cap C_L(w) \neq \{0\}$. Since this holds for all roots δ we can find $\beta, \gamma \in \Gamma(L,T)$ such that α, β, γ are $\mathbb{F}_p$-independent and

(i) $L(\alpha)+Fw+T \subset Q, \quad Q_\beta \neq \{0\}, \quad Q_\gamma \neq \{0\}.$

Let $\delta \notin \mathbb{F}_p\alpha$. If $L(\alpha,\delta)$ is solvable, then $L \cong H(2;(1,2);\Phi(\tau))^{(1)}$ by Proposition 19.2.1. Proposition 19.2.2 now shows that L is not a counterexample. So this does not occur by assumption. Then Proposition 19.1.3 shows that $L[\alpha,\delta] \cong H(2;\underline{1};\Phi(\tau))^{(1)}$. Theorem 10.3.2(5) ensures that every nonzero root vector $v \in H(2;\underline{1};\Phi(\tau))^{(1)}{}_\alpha$ satisfies $Fv^{[p]} = \Psi_{\alpha,\delta}(T) \cap (\ker\alpha)$. Take an inverse image $v^{(\delta)} \in L_\alpha$ of v (such element depends on the choice of δ). Since T is a maximal torus and $\operatorname{rad} L(\alpha,\delta)$ is nilpotent and $L_\delta \neq \{0\}$, one obtains $(v^{(\delta)})^{[p]^e} \in T \setminus (\ker\delta)$ (for $e \gg 0$). As a result,

(ii) $\forall \delta \notin \mathbb{F}_p\alpha \quad \exists v^{(\delta)} \in L_\alpha : \quad v^{(\delta)}$ acts invertibly on $\sum_{i\in\mathbb{F}_p} L_{\delta+i\alpha}.$

(c) Suppose J is an ideal of L' contained in Q. Then $J \cap L$ is an ideal of L. So either $J \cap L = \{0\}$ or $J \cap L = L$. In the first case one has $[J, L'] \subset J \cap L = \{0\}$, and this gives $J \subset C(L') = \{0\}$ (see (a)). In the second case, $L \subset J \subset Q$, whence $Q = L + Fw + T$ contradicting the assumption $Q \neq L'$. Thus Q contains no non-trivial ideals of L'.

Choose a Q-invariant subspace $L'_{(-1)}$ in L' which contains Q and has the property that $L'_{(-1)}/Q$ is Q-irreducible. Let $(L'_{(i)})_{i\in\mathbb{Z}}$ denote the standard filtration associated with the pair $(L'_{(0)}, L'_{(-1)})$ where $L'_{(0)} = Q$. By the above, this filtration is exhaustive and separating, that is

$$L' = L'_{(-r)} \supset \cdots \supset L'_{(s+1)} = \{0\},$$

where $r > 0$ and $s \geq 0$ are finite. Let $G := \operatorname{gr} L'$ denote the associated graded Lie algebra and let $M(G)$ be the maximal ideal of G contained in $\sum_{i\leq -1} G_i$. Then $M(G)$ is a graded ideal of G and the quotient

$$\overline{G} := G/M(G)$$

inherits from G a graded Lie algebra structure (Proposition 3.5.3(3)).

If $G_1 \neq \{0\}$, then the graded Lie algebra $\overline{G} = \bigoplus_{i\in\mathbb{Z}} \overline{G}_i$ satisfies the conditions (g1)–(g4) of Notation 3.5.2 (by Proposition 3.5.3(4)). Then Theorem 3.5.6 says that $\overline{G}$ is semisimple and contains a unique minimal ideal $A(\overline{G})$ which is graded with grading induced by that of $\overline{G}$.

Suppose $G_1 = \{0\}$. Then $M(G) = \sum_{i<0} G_i$ which entails that

$$\overline{G} = \overline{G}_0 = L'_{(0)}/L'_{(1)} \cong L'_{(0)} = Q$$

as Lie algebras. Thanks to Equation (i) we have $\alpha, \beta, \gamma \in \Gamma(Q, T)$. Let A be an abelian ideal of Q. For every $\delta \in \Gamma(Q, T)$ any element $x \in A_\delta$ therefore has the property that $\alpha(x) = \beta(x) = \gamma(x) = 0$. Then any element in A_δ acts nilpotently on L', and hence on $G_{-1} = L'_{(-1)}/L'_{(0)}$. The ENGEL–JACOBSON Theorem now shows that A acts nilpotently on G_{-1}. The irreducibility of G_{-1} yields that A annihilates G_{-1} implying $A \subset L'_{(1)} = \{0\}$. Thus Q is semisimple.

Suppose Q has different minimal ideals I, J. Then

$$TR(I) + TR(J) = TR(I \oplus J) \le TR(L) = 3,$$

so we may assume $TR(I) = 1$. Note that I is T-invariant. As it is non-solvable, there is a non-solvable 2-section $I(\kappa, \lambda)$ (Theorem 1.3.10). Then $L(\kappa, \lambda)$ is non-solvable, whence $L[\kappa, \lambda] \cong H(2; \underline{1}; \Phi(\tau))^{(1)}$ (Proposition 19.1.3). Now $\Psi_{\kappa,\lambda}(I(\kappa, \lambda))$ is a non-solvable T-invariant subalgebra of $H(2; \underline{1}; \Phi(\tau))^{(1)}$ of absolute toral rank 1. Due to Theorem 10.3.2(5) the latter forces that $\Psi_{\kappa,\lambda}(I(\kappa, \lambda))$ carries only one T-root direction. But then it is solvable, a contradiction.

So in either case $\overline{G}$ is semisimple and has a unique minimal ideal $A(\overline{G})$ which naturally carries the grading induced by that of $\overline{G}$.

(d) By Corollary 3.3.3 there exist a simple algebra S, $m \in \mathbb{N}$, and an isomorphism

$$\tau : A(\overline{G}) \xrightarrow{\sim} S \otimes \mathcal{O}(m; \underline{1}).$$

We suppress the notion of τ and identify $A(\overline{G})$ with its image. The adjoint action of $\overline{G}_{[p]}$ on $A(\overline{G})$ gives rise to a restricted homomorphism

$$\Psi : \overline{G}_{[p]} \longrightarrow \operatorname{Der} A(\overline{G}) = \big((\operatorname{Der} S) \otimes \mathcal{O}(m; \underline{1})\big) \rtimes \big(\operatorname{Id} \otimes W(m; \underline{1})\big)$$

such that the subalgebra $(\pi_2 \circ \Psi)(\overline{G}_{[p]}) = (\pi_2 \circ \Psi)(\overline{G})_{[p]}$ is transitive in $W(m; \underline{1})$. As $A(\overline{G})$ is the unique minimal ideal, $\Psi|_{\overline{G}}$ is injective (otherwise $(\ker \Psi) \cap \overline{G}$ would contain a minimal ideal of $\overline{G}$ intersecting $A(\overline{G})$ trivially). We suppress the notion of Ψ when restricted to $\overline{G}$. According to this notation we then have that $\pi_2(\overline{G})_{[p]}$ is transitive in $W(m; \underline{1})$.

(e) Note that every nonzero element of $T \subset L'_{(0)}$ acts non-nilpotently on L. Therefore $T \cap L'_{(1)} = \{0\}$, whence $T \cong \operatorname{gr} T \subset G_0$. Let $\overline{T}$ denote the image of $\operatorname{gr} T$ in $\overline{G}$. Then $\overline{T}$ acts on $\overline{G}$ as a torus of derivations. Since by Equation (i) Q carries three $\mathbb{F}_p$-independent T-roots, and since $M(G) \cap \sum_{i \ge 0} G_i = \{0\}$, $\overline{G}$ carries three $\mathbb{F}_p$-independent $\overline{T}$-roots. As a consequence, $\overline{T}$ is a 3-dimensional torus. Theorem 1.4.6 yields

$$3 = \dim \overline{T} \le TR(\overline{G}) \le TR(G) \le TR(L) = 3,$$

forcing $\dim \overline{T} = TR(\overline{G}) = 3$.

Put $\mathfrak{t} := \Psi(\overline{T})$ and

$$\mathfrak{t}_0 := \mathfrak{t} \cap \Big(\big((\operatorname{Der}_0 S) \otimes \mathcal{O}(m; \underline{1})\big) \rtimes \big(\operatorname{Id} \otimes W(m; \underline{1})_{(0)}\big)\Big).$$

According to Theorem 3.6.1 we can choose τ such that

(iii) $\mathfrak{t} = \mathfrak{t}_0 \bigoplus \Big(\sum_{i=1}^{s} F\mathrm{Id}\otimes(1+x_i)\partial_i\Big)$ and $\mathfrak{t}_0 = \{\omega(t)\otimes 1+\mathrm{Id}\otimes\lambda(t) \mid t \in \mathfrak{t}_0\}$,

for restricted homomorphisms

$$\omega : \mathfrak{t}_0 \to \mathrm{Der}_0\, S \text{ and } \lambda : \mathfrak{t}_0 \to \sum_{i=s+1}^{m} F x_i \partial_i .$$

Lemma 1.2.6(3) shows that $\mathfrak{t} \cap (S \otimes \mathcal{O}(m;\underline{1}))_{[p]} \subset (\mathrm{Der}\, S) \otimes \mathcal{O}(m;\underline{1})$ is a torus of maximal dimension in $(S \otimes \mathcal{O}(m;\underline{1}))_{[p]}$. It is straightforward to see that

$$\mathfrak{t} \cap \big((\mathrm{Der}\, S) \otimes \mathcal{O}(m;\underline{1})\big) \subset \mathfrak{t}_0 \cap (\ker \lambda).$$

Set

$$\mathfrak{t} \cap \big(S \otimes \mathcal{O}(m;\underline{1})\big)_{[p]} =: \mathfrak{t}_0' \otimes F,$$

where $\mathfrak{t}_0'$ is a torus of maximal dimension in $S_{[p]} \subset \mathrm{Der}\, S$.

Since the image of T in $G = \mathrm{gr}\, L'$ lies in G_0, each 1-section $G(\delta)$ in G has the form $(\mathrm{gr}\, L')(\delta)$. Lemma 19.2.5 gives that $(\mathrm{gr}\, L')^{(1)}(\delta) + (\mathrm{gr}\, H) \subset (\mathrm{gr}\, L)(\delta)$ is trigonalizable. Therefore one has for every nonzero $\mathfrak{t}$-root δ that

(iv) $\overline{G}^{(1)}(\delta)+\overline{(\mathrm{gr}\, H)} \subset \overline{(\mathrm{gr}\, L)}(\delta)$, and $\overline{(\mathrm{gr}\, L)}(\delta)$ is solvable and trigonalizable.

(f) Next we intend to show that the case $TR(S) \le 2$ is impossible. So let us assume $TR(S) \le 2$ and derive a contradiction.

(f1) Note that $\mathfrak{t}$ acts on $S \otimes F \cong S$ as $\omega(\mathfrak{t}_0) \otimes F$. By (iv) and Lemma 19.1.2, $S \cong H(2;\underline{1};\Phi(\tau))^{(1)}$ and $\dim \omega(\mathfrak{t}_0) = 2$ hold, and $\omega(\mathfrak{t}_0) \subset S_{[p]} = \mathrm{Der}\, S$ is a torus of maximal dimension. By definition one has

$$\mathfrak{t}_0' = \omega(\mathfrak{t}_0' \otimes F) = \omega\Big(\mathfrak{t} \cap \big(S \otimes \mathcal{O}(m;\underline{1})\big)_{[p]}\Big) \subset \omega(\mathfrak{t}_0).$$

The maximality of $\mathfrak{t}_0'$ gives $\omega(\mathfrak{t}_0) = \mathfrak{t}_0'$. Then Equation (iii) simplifies to

(v) $\mathfrak{t} = (\mathfrak{t}_0' \otimes F) \oplus (F\,\mathrm{Id} \otimes d)$

for some toral derivation $d \in W(m;\underline{1})$ which is nonzero because $\dim \mathfrak{t} = 3$. As a consequence, $m > 0$.

According to Theorem 10.3.2, there exist $\mathbb{F}_p$-independent $\kappa', \nu' \in \Gamma(S, \mathfrak{t}_0')$ such that

$$S = \sum_{\delta \in (\mathbb{F}_p\kappa' + \mathbb{F}_p\nu')\setminus\{0\}} S_\delta$$

and each root space S_δ is 1-dimensional. Since $S \otimes F$ is $\mathfrak{t}$-invariant, we may extend these to $\mathfrak{t}$-roots κ and ν via

$$\kappa(t \otimes 1) := \kappa'(t),\ \nu(t \otimes 1) := \nu'(t)\ \forall t \in \mathfrak{t}_0',\ \kappa(\mathrm{Id} \otimes d) := \nu(\mathrm{Id} \otimes d) := 0.$$

Define $\mu \in \mathfrak{t}^*$ by setting

$$\mu(\mathfrak{t}_0' \otimes F) = 0, \ \mu(\mathrm{Id} \otimes d) = 1.$$

Since d is toral, every eigenvalue of $\mathrm{Id} \otimes d$ on $\overline{G}$ is contained in $\mathbb{F}_p$. Moreover, it is easy to see that the spectrum of the toral derivation d on $\mathcal{O}(m; \underline{1})$ equals $\mathbb{F}_p$. As a consequence,

(vi) $\Gamma(\overline{G}, \mathfrak{t}) \subset \mathbb{F}_p\kappa + \mathbb{F}_p\nu + \mathbb{F}_p\mu, \quad \Gamma(S \otimes \mathcal{O}(m; \underline{1}), \mathfrak{t}) = \big(\mathbb{F}_p\kappa + \mathbb{F}_p\nu + \mathbb{F}_p\mu\big) \backslash \mathbb{F}_p\mu.$

(f2) Recall the definition of α from (a). Suppose $\alpha(\mathfrak{t}_0' \otimes F) \neq 0$. Then $\alpha = k\kappa + n\nu + m\mu$ with $k\kappa + n\nu \neq 0$. In particular, $\mu \notin \mathbb{F}_p\alpha$ and $\mu - \alpha \in \Gamma(S \otimes \mathcal{O}(m; \underline{1}), \mathfrak{t})$. Due to Equation (ii) there is $v^{(\mu)} \in L_\alpha \subset Q$ which acts invertibly on $\sum L_{\mu+i\alpha}$. Then $\overline{\mathrm{gr}_0\, v^{(\mu)}}$ acts invertibly on $\sum \overline{G}_{\mu+i\alpha}$. Moreover, $\overline{\mathrm{gr}_0\, v^{(\mu)}} \in \overline{G}_\alpha \subset S \otimes \mathcal{O}(m; \underline{1})$ as $\alpha(\mathfrak{t}_0' \otimes F) \neq 0$. As $\mu - \alpha$ is a root, we obtain that $\overline{G}_\mu = [\overline{\mathrm{gr}_0\, v^{(\mu)}}, \overline{G}_{\mu-\alpha}]$ is nonzero and is contained in $S \otimes \mathcal{O}(m; \underline{1})$. This contradicts Equation (vi). Consequently, $\mathbb{F}_p\mu = \mathbb{F}_p\alpha$, and this gives

$$\overline{G} = \big(S \otimes \mathcal{O}(m; \underline{1})\big) \oplus \overline{G}(\alpha).$$

(f3) We will need some subtle estimates for the dimensions of root spaces.

We first observe that $\dim\, (S \otimes \mathcal{O}(m; \underline{1}))_\delta = p^{m-1}$ for every $\delta \in \Gamma(S \otimes \mathcal{O}(m; \underline{1}), \mathfrak{t})$. Next we recall that $L(\alpha) \subset Q$ and therefore $(\mathrm{gr}\, L(\alpha)) \cap M(G) = \{0\}$. This gives $\dim L_{i\alpha} = \dim \overline{G}_{i\alpha}$ for all $i \in \mathbb{F}_p$. Since all root spaces for nonzero roots have equal dimension (Theorem 19.2.3(2)), we obtain

(vii) $p^{m-1} \leq \dim \overline{G}_{i\alpha}, \quad i \in \mathbb{F}_p^*.$

Since $\mathrm{Der}\, S = \mathfrak{t}_0' \oplus S$, we have the inclusion

$$\overline{G}(\alpha) \subset \big(\mathfrak{t}_0' \otimes \mathcal{O}(m; \underline{1})\big) \rtimes \big(\mathrm{Id} \otimes W(m; \underline{1})\big).$$

Suppose $\bigcup_{i \in \mathbb{F}_p} \overline{G}^{(1)}{}_{i\alpha} \cap (\mathfrak{t}_0' \otimes \mathcal{O}(m; \underline{1})_{(1)}) \neq \{0\}$ and let $y = t_1 \otimes f + t_2 \otimes g$ be an element of this set, where $t_1, t_2 \in \mathfrak{t}_0'$ are linearly independent and $f, g \in \mathcal{O}(m; \underline{1})_{(1)}$. As $\overline{G}(\alpha) \subset (\mathfrak{t}_0' \otimes \mathcal{O}(m; \underline{1})) \rtimes (\mathrm{Id} \otimes \pi_2(\overline{G}(\alpha)))$, we have

$$[\overline{G}(\alpha), y] = t_1 \otimes \pi_2\big(\overline{G}(\alpha)\big)(f) + t_2 \otimes \pi_2\big(\overline{G}(\alpha)\big)(g).$$

Clearly $[\overline{G}_{j\alpha}, \bigcup_{i \in \mathbb{F}_p} \overline{G}^{(1)}{}_{i\alpha}] \subset \bigcup_{i \in \mathbb{F}_p} \overline{G}^{(1)}{}_{i\alpha}$ holds for all $j \in \mathbb{F}_p$. Note that $[\mathfrak{t}, y] \subset Fy$ since y is a root vector. Next observe that $y \in \overline{(\mathrm{gr}\, L)}(\alpha)$ (Equation (iv)). Therefore $[\overline{\mathrm{gr}\, w}, y] = 0$ holds by (a). Consequently,

$$[\overline{G}(\alpha), y] \equiv [\overline{(\mathrm{gr}\, L)}(\alpha), y] \pmod{\mathfrak{t}_0' \otimes \mathcal{O}(m; \underline{1})_{(1)}}.$$

Since $\overline{(\mathrm{gr}\, L)}(\alpha)$ is trigonalizable and t_1, t_2 are semisimple, it must be that both $\pi_2(\overline{G}(\alpha))(f)$, $\pi_2(\overline{G}(\alpha))(g)$ contain no invertible elements. As a result we obtain that

$$\Big[\bigcup_{j\in\mathbb{F}_p} \overline{G}_{j\alpha}, \bigcup_{i\in\mathbb{F}_p} \overline{G}^{(1)}{}_{i\alpha} \cap \big(\mathfrak{t}_0' \otimes \mathcal{O}(m;\underline{1})_{(1)}\big)\Big] \subset \bigcup_{i\in\mathbb{F}_p} \overline{G}^{(1)}{}_{i\alpha} \cap \big(\mathfrak{t}_0' \otimes \mathcal{O}(m;\underline{1})_{(1)}\big).$$

Since $\pi_2(\overline{G}(\alpha))_{[p]}$ is transitive in $W(m;\underline{1})$ (by (d)), we arrive at a contradiction. As

$$\overline{G}_{i\alpha} \cap (\ker \pi_2) = \overline{G}_{i\alpha} \cap \big(\mathfrak{t}_0' \otimes \mathcal{O}(m;\underline{1})\big) = \overline{G}_{i\alpha} \cap \Big((\mathfrak{t}_0' \otimes F) \oplus (\mathfrak{t}_0' \otimes \mathcal{O}(m;\underline{1})_{(1)})\Big)$$

and $\dim \mathfrak{t}_0' = 2$, we now have

(viii) $\quad \dim \overline{G}_{i\alpha} \cap (\ker \pi_2) \le 2 \quad \forall i \in \mathbb{F}_p^*$.

Set

$$K := \sum_{i\in\mathbb{F}_p^*} \pi_2(\overline{G}_{i\alpha}) + [\pi_2(\overline{G}_{i\alpha}), \pi_2(\overline{G}_{-i\alpha})]$$

and let K_p denote the p-envelope of K in $W(m;\underline{1})$. We claim that K_p consists of nilpotent elements of $W(m;\underline{1})$. So assume for a contradiction that it does not and consider the weight space decomposition $K = \sum_{i\in\mathbb{F}_p} K_i$ of K relative to d. By JACOBSON's formula, $K_p = K + \sum_{i\in\mathbb{F}_p} \sum_{j>0} K_i{}^{p^j}$. Since $L(\alpha) + Fw$ is nilpotent, so is K. Suppose one of the K_i's contains a non-nilpotent element of $W(m;\underline{1})$. Then $C(K_p)$ contains a non-zero toral element t commuting with d. By definition of K, we have $K \ne K_0$. Hence d does not centralize K. So $d \notin Ft$, which gives that $K_p + Fd$ contains a 2-dimensional torus. But then $\overline{G}$ contains a 4-dimensional torus, a contradiction. Thus all K_i's consist of nilpotent elements and our claim follows in view of the ENGEL–JACOBSON Theorem.

Note that K is an ideal in $\pi_2(\overline{G}(\alpha))_{[p]}$. Due to Proposition 19.2.4,

$$\sum_{i\in\mathbb{F}_p^*} \dim \pi_2(\overline{G}_{i\alpha}) \le \dim K \le \frac{p^m - 1}{p - 1}.$$

Combining this with Equations (vii) and (viii) we now get

$$(p-1)p^{m-1} \le \sum_{i\in\mathbb{F}_p^*} \dim \overline{G}_{i\alpha} \le \frac{p^m - 1}{p - 1} + 2(p-1),$$

which gives

$$m = 1, \qquad \sum_{i\in\mathbb{F}_p^*} \dim \pi_2(\overline{G}_{i\alpha}) \le 1, \qquad d = (1 + x_1)\partial_1 \text{ or } d \in \mathbb{F}_p^* x_1\partial_1.$$

(f4) Suppose $d = (1 + x_1)\partial_1$. Assume that there is a root vector $u \in \overline{G}_{l\alpha}$ and $l \ne 0$ satisfying $\pi_2(u) \ne 0$. Then $u = t\otimes(1+x_1)^k + r\mathrm{Id}\otimes(1+x_1)^{k+1}\partial_1$ with $t \in \mathfrak{t}_0', k \ne 0$,

$r \in F^*$. There exists a nonzero $v \in \overline{G}_{-l\alpha}$. Since $\sum_{i\in\mathbb{F}_p^*} \dim \pi_2(\overline{G}_{i\alpha}) \le 1$ and u and v have different weights for d, it must be that $\pi_2(v) = 0$. Then $v = t' \otimes (1+x_1)^{p-k}$ with $t' \in \mathfrak{t}_0' \setminus \{0\}$. Therefore,

$$[u, v] = -krt' \otimes 1 \neq 0.$$

Note that

$$[\overline{G}_{l\alpha}, \overline{G}_{-l\alpha}] \subset [\overline{(\operatorname{gr} L)_{l\alpha} + F \operatorname{gr} w}, \overline{(\operatorname{gr} L)_{-l\alpha} + F \operatorname{gr} w}] = [\overline{(\operatorname{gr} L)_{l\alpha}}, \overline{(\operatorname{gr} L)_{-l\alpha}}]$$

acts nilpotently. So this assumption cannot be true.

Next let $u \in \overline{G}_{l\alpha}$ and $l = 0$ be any root vector satisfying $\pi_2(u) \neq 0$. Then $u = t \otimes 1 + r\operatorname{Id} \otimes (1+x_1)\partial_1 = t \otimes 1 + r\operatorname{Id} \otimes d$ with $t \in \mathfrak{t}_0'$, $r \in F^*$. As a consequence,

$$\bigcup_{i\in\mathbb{F}_p} \overline{G}_{i\alpha} \subset \bigcup_{k\in\mathbb{F}_p} \mathfrak{t}_0' \otimes (1+x_1)^k \cup F\operatorname{Id} \otimes d.$$

No nonzero element in this set is $[p]$-nilpotent. Therefore there is no room for the distinguished element $\overline{\operatorname{gr} w}$. Therefore this case cannot occur.

(f5) As a consequence, $d \in \mathbb{F}_p^* x_1\partial_1$. Since $\pi_2(\overline{G}(\alpha))_{[p]}$ is transitive in $W(1;\underline{1})$, there is a root vector u satisfying $\pi_2(u) \notin W(1;\underline{1})_{(0)}$. Then $u = t \otimes x_1^{p-1} + r\operatorname{Id} \otimes \partial_1$, $t \in \mathfrak{t}_0'$, $r \in F^*$. Let k be such that $u \in \overline{G}_{k\alpha}$ and observe that $k \neq 0$.

There exists a nonzero $v \in \overline{G}_{-k\alpha}$. Since $\sum_{i\in\mathbb{F}_p^*} \dim \pi_2(\overline{G}_{i\alpha}) \le 1$ and u and v have different weights for d, it must be that $\pi_2(v) = 0$. Then $v = t' \otimes x_1$ with $t' \in \mathfrak{t}_0' \setminus \{0\}$. Therefore,

$$[u, v] = rt' \otimes 1 \neq 0.$$

As in the former case this cannot be true. We conclude that the case $TR(S) = 2$ is impossible.

(g) Suppose $TR(S) = 3$. Recall from (e) that $\mathfrak{t}_0'$ is a torus of maximal dimension in $S_{[p]}$. Therefore, $\dim \mathfrak{t}_0' = 3$, giving $\mathfrak{t} = \mathfrak{t}_0' \otimes F$. For $\delta' \in \Gamma(S, \mathfrak{t}_0')$ define $\delta \in \Gamma(\overline{G}, \mathfrak{t})$ by $(S \otimes F)_\delta := S_{\delta'} \otimes F$. Clearly, all 1-sections $S(\delta') \otimes F \subset \overline{G}(\delta)$ are solvable. Consequently, S and $\mathfrak{t}_0'$ satisfy the general assumptions of this chapter.

The above mentioned form of $\mathfrak{t}$ implies that every root space for a nonzero root is of the form

$$\text{(ix)} \quad \overline{G}_\mu = S_{\mu'} \otimes \mathcal{O}(m;\underline{1}), \qquad \forall \mu \in \Gamma(\overline{G}, \mathfrak{t}) \setminus \{0\}.$$

(g1) Suppose S is not a counterexample to our theorem. Then no root vector of S relative to $\mathfrak{t}_0'$ acts nilpotently on S. Recall that S is a graded algebra. Since all graded components of S are $\mathfrak{t}$-invariant and no root vector in S is $[p]$-nilpotent, one has $S = S_0$.

Since $L_\alpha \subset Q$, one obtains $G_\alpha \cap M(G) = \{0\}$. Equation (ix) gives

$$\dim L_\alpha = \dim \overline{G}_\alpha = p^m(\dim S_{\alpha'}).$$

Suppose $G_1 \neq \{0\}$. Recall that $A(\overline{G})$ denotes the minimal ideal of $\overline{G}$. If $A(\overline{G})^+ \neq \{0\}$, then Theorem 3.5.7 applies and shows that $S_{-1} \neq \{0\}$, which is not true. So $A(\overline{G})^+ = \{0\}$ and Theorem 3.5.8 applies. Since $L_\alpha \subset Q$, one has $\overline{G}_\alpha \subset \sum_{l\geq 0} A(\overline{G})_l = A(\overline{G})_0$. Combining this with the above-mentioned equation, Theorem 3.5.8(2) gives that all indeterminates have degree 0, which means $\overline{G}_{-1} = \{0\}$. But this is not true. Consequently, $G_1 = \{0\}$. Our construction in (c) gives $M(G) = \sum_{l<0} G_l$.

Theorem 19.2.3(2) applied to S and $\mathfrak{t}_0'$ as well as to L and T gives for all $\mu \neq 0$

$$p^m(\dim S_{\mu'}) = p^m(\dim S_{\alpha'}) = \dim L_\alpha = \dim L_\mu.$$

Then $G_\mu \cap M(G) = \{0\}$, and this yields the contradiction

$$\sum_{l<0} G_l = M(G) \subset \operatorname{gr}(H + Fw + T) \subset G_0.$$

This contradiction shows that S is a counterexample to our theorem.

(g2) It follows from our choice of L that $\dim S = \dim L$. We get $(\operatorname{gr} L) \cap M(G) = \{0\}$ and $m = 0$. Since $L' = L + Fw + T$ and $Fw + T \subset Q$, it is now straightforward to see that $M(G) = \{0\}$. Then

$$G \cong \overline{G} = S + F\overline{w} + \overline{T}$$

and $\overline{T}$ is a 3-dimensional torus in $S_{[p]}$.

If S contains a solvable 2-section $S(\eta, \delta)$, for some $\mathbb{F}_p$-independent $\eta, \delta \in \Gamma(S, \overline{T})$, then $S \cong H(2; (1,2); \Phi(\tau))^{(1)}$ by Proposition 19.2.1. Proposition 19.2.2 shows that S is not a counterexmple, a contradiction.

Thus no 2-section $S(\eta, \delta)$ with $\mathbb{F}_p$-independent η and δ is solvable. By Proposition 19.1.3 we then have $S[\eta, \delta] \cong H(2; \underline{1}; \Phi(\tau))^{(1)}$ whenever η and δ are $\mathbb{F}_p$-independent. Recall that no root vector in $H(2; \underline{1}; \Phi(\tau))^{(1)}$ acts nilpotently on $H(2; \underline{1}; \Phi(\tau))^{(1)}$. From this it is immediate that $S_i \cap S(\eta, \delta) \subset \operatorname{rad} S(\eta, \delta)$ for all $i \neq 0$. This shows that

$$S_0[\eta, \delta] := S_0(\eta, \delta)/\operatorname{rad} S_0(\eta, \delta) \cong S[\eta, \delta] \cong H(2; \underline{1}; \Phi(\tau))^{(1)}$$

for all $\mathbb{F}_p$-independent $\eta, \delta \in \Gamma(S, \overline{T})$.

Recall from (i) that there are $\mathbb{F}_p$-independent roots α, β, γ and $S_{-1, i\alpha} = \{0\}$ for all $i \in \mathbb{F}_p$. Note that $S_{-1} = G_{-1} \neq \{0\}$. Let $\nu \in \Gamma(S, \overline{T})$ be such that $S_{-1,\nu} \neq \{0\}$. Take any root δ independent of ν. Then $S_0[\nu, \delta] \cong H(2; 1; \Phi(\tau))^{(1)}$, hence there is $x \in S_{0,\delta}$ with $\nu(x) \neq 0$. Then

$$S_{-1,\nu+j\delta} = (\operatorname{ad} x)^j (S_{-1,\nu}) \neq \{0\} \qquad \forall j \in \mathbb{F}_p.$$

As a first consequence, there are 2 independent roots on S_{-1}. We may therefore take ν as $\mathbb{F}_p$-independent of γ and apply the preceding reasoning with $\delta := \gamma$. Consequently, $\Gamma(S_{-1}, \overline{T})$ contains $r\alpha + s\beta$ for some $r \in \mathbb{F}_p$ and $s \in \mathbb{F}_p^*$. Since α and $r\alpha + s\beta$ are $\mathbb{F}_p$-independent, we may take $\nu := r\alpha + s\beta$ and $\delta := (r-1)\alpha + s\beta$ and obtain $\alpha \in \Gamma(S_{-1}, \overline{T})$. This contradiction finally completes the proof of the theorem. □

Corollary 19.2.7. *Suppose $TR(L) = 3$. Then*

(1) $H = \{0\}$ *and* $\tilde{H} = T$;

(2) ${L_\delta}^{[p]} \subset T \cap (\ker \delta)$ *for all* $\delta \in \Gamma(L,T)$;

(3) *no T-root vector is $[p]$-nilpotent;*

(4) $\dim L_\delta \leq 2$ *for all* $\delta \in \Gamma(L,T)$.

Proof. Proposition 19.1.1 shows that every element of H is $[p]$-nilpotent. Due to the preceding theorem ($\delta = 0$ is allowed there) we have $H = \{0\}$. Next observe that

$$L_{[p]} = L + \sum_{\delta\in\Gamma(L,T)} \sum_{i>0} {L_\delta}^{[p]^i} = L + T.$$

Therefore $\tilde{H} = C_L(T) + T = H + T = T$. This is (1).

Let δ be any root. Then $\delta \neq 0$. As $\delta({L_\delta}^{[p]}) = 0$, assertions (2) and (3) are statements of the preceding theorem. As $H = \{0\}$, we have for $x, y \in L_\delta$

$$(x+y)^{[p]} - (x^{[p]} + y^{[p]}) \in \gamma_p(L_\delta) \subset H = \{0\}.$$

This shows that the mapping ${}^{[\,]} : L_\delta \to T \cap (\ker \delta)$ is p-semilinear. Due to the preceding theorem this mapping is injective. This gives $(\dim L_\delta) \leq (\dim T) - 1 = 2$. □

Theorem 19.2.8. *Suppose $TR(L) = 3$. Then*

$$[u,v]^{[p]} = \beta(u^{[p]})v^{[p]} - \alpha(v^{[p]})u^{[p]} \qquad \forall u \in L_\alpha, \forall v \in L_\beta.$$

Proof. (a) Put

$$\Delta(u,v) := [u,v]^{[p]} - \beta(u^{[p]})v^{[p]} + \alpha(v^{[p]})u^{[p]}.$$

According to Theorem 19.2.6 one has $\Delta \in T$.

If $L(\alpha,\beta)$ is solvable, then Proposition 19.2.1 shows that L is isomorphic to $H(2;(1,2);\Phi(\tau))^{(1)}$. Due to Proposition 19.2.2 there is a basis $(d_\delta)_{\delta\in\Gamma(L,T)}$ consisting of root vectors for T such that

$$[d_\lambda, d_\mu] = f(\lambda,\mu)d_{\lambda+\mu} \qquad \forall \lambda,\mu \in \Gamma(L,T),$$

where $f(\lambda,\mu)^p = \mu({d_\lambda}^{[p]})$ holds. Due to Theorem 19.2.3(3) $L(\alpha,\beta)$ is abelian. Then we have $[u,v] = 0$ and $\beta({d_\alpha}^{[p]}) = -\alpha({d_\beta}^{[p]}) = f(\beta,\alpha) = 0$. This is the claim.

(b) Thus we may assume that $L(\alpha,\beta)$ is non-solvable. We then have $L[\alpha,\beta] \cong H(2;\underline{1};\Phi(\tau))^{(1)}$ by Proposition 19.1.3.

Let $L(\alpha,\beta)_{[p]}$ denote the $[p]$-envelope of $L(\alpha,\beta)$ in $L_{[p]}$ and $\mathfrak{g}$ the minimal p-envelope of $L(\alpha,\beta)$. Theorem 1.1.6(1) yields an isomorphism of algebras $L(\alpha,\beta)_{[p]} \cong \mathfrak{g} \oplus C$ where $L(\alpha,\beta)$ is mapped into $\mathfrak{g}$ and C is central. We may identify (see Theorem 1.1.6(2) and Corollary 19.2.7(1))

$$L(\alpha,\beta) \hookrightarrow \mathfrak{g} \hookrightarrow L(\alpha,\beta)_{[p]} \subset L_{[p]}(\alpha,\beta) = L(\alpha,\beta) + T$$

where the composite mapping is the identity map on $L(\alpha,\beta)$. We conclude

$$\mathfrak{g} = L(\alpha,\beta) + \mathfrak{g} \cap T, \quad L(\alpha,\beta) + T = \mathfrak{g} \oplus \big(T \cap (\ker\,\alpha) \cap (\ker\,\beta)\big).$$

The last equation gives

$$\begin{aligned} &\mathfrak{g}/\operatorname{rad}\mathfrak{g} \\ &\cong \big(L(\alpha,\beta)+T\big)/\operatorname{rad}\big(L(\alpha,\beta)+T\big) \cong L[\alpha,\beta] + \big(T/T \cap (\ker\,\alpha) \cap (\ker\,\beta)\big) \\ &\cong \operatorname{Der} H(2;\underline{1};\Phi(\tau))^{(1)}. \end{aligned}$$

(c) Theorem 19.2.3(1),(2) shows that the root lattice $\Gamma(L,T)$ contains at least p^3-1 elements and all root spaces have equal dimension k. Corollary 19.2.7 proves that 0 is the missing element in $\Gamma(L,T)$, and $k \le 2$. Let γ be any root independent of α and β and set

$$M := \sum_{i,j \in \mathbb{F}_p} L_{\gamma+i\alpha+j\beta}.$$

This is a module for $\mathfrak{g}$ of dimension $\dim M = kp^2$. There is a composition factor V of the $\mathfrak{g}$-module M which is not annihilated by $L(\alpha,\beta)$, let $\rho : \mathfrak{g} \to \mathfrak{gl}(V)$ denote the representation. Note that $T \cap (\ker\,\alpha) \cap (\ker\,\beta)$ acts as $F\mathrm{Id}$ on M. Hence V is invariant under T. Therefore $\ker\,\rho$ is invariant under T. Suppose $w \in \ker\,\rho$ is a nonzero root vector. Theorem 19.2.6 shows that $w^{[p]}$ is a nonzero element of T. If $\alpha(w^{[p]}) \ne 0$ or $\beta(w^{[p]}) \ne 0$, then all of $L(\alpha,\beta)$ is contained in $\ker\,\rho$. We have excluded this. But then $w^{[p]}$ acts as $\gamma(w^{[p]})\mathrm{Id}$ on M. This gives the contradiction $\gamma(w^{[p]}) = 0$ and $w^{[p]} = 0$ as α,β,γ span $\Gamma(L,T)$. So ρ is faithful. Then V is a faithful irreducible $\mathfrak{g}$-module of dimension $\le 2p^2$.

If $k = 1$, then there is no room for a radical in $L(\alpha,\beta)$, whence $L(\alpha,\beta) \cong H(2;\underline{1};\Phi(\tau))^{(1)}$ and $\mathfrak{g} = \operatorname{Der} H(2;\underline{1};\Phi(\tau))^{(1)}$. Theorem 16.3.10 applies and gives that the character χ associated with the representation ρ vanishes on $H(2;\underline{1};\Phi(\tau))^{(1)}{}_{(0)}$.

Suppose $k = 2$. We check the assumptions of Theorem 16.3.14. Clearly, $TR(\mathfrak{g}) = TR(L(\alpha,\beta)) = 2$. Since 0 is not a T-root on L, one has $C(L(\alpha,\beta)) = \{0\}$ and this gives (by definition of a minimal p-envelope) $C(\mathfrak{g}) = \{0\}$. Therefore, $(\mathfrak{g} \cap T) \cap (\operatorname{rad}\mathfrak{g}) = \{0\}$. We conclude $\operatorname{rad}\mathfrak{g} = \operatorname{rad} L(\alpha,\beta)$ and hence

$$\dim\,\operatorname{rad}\mathfrak{g} = 2(p^2-1) - \dim\,H(2;\underline{1};\Phi(\tau))^{(1)} = p^2 - 1.$$

Theorem 16.3.14(2) gives that the character χ associated with the representation ρ vanishes on ${\mathfrak g_{(0)}}^{(1)}$.

We want to describe ${\mathfrak g_{(0)}}^{(1)}$, where $\mathfrak g_{(0)}$ is the unique subalgebra of maximal dimension containing $\operatorname{rad}\mathfrak g$. Theorem 16.3.14(3) provides a decomposition

$$\mathfrak g = \mathcal M \oplus \operatorname{rad}\mathfrak g, \;\; \mathcal M \cong \operatorname{Der} H(2;\underline{1};\Phi(\tau))^{(1)}$$

such that the image $T/T \cap (\ker\alpha) \cap (\ker\beta)$ is contained in $\mathcal M$. Proposition 16.3.12 shows that $\operatorname{rad}\mathfrak g$ is abelian, then Theorem 16.3.10(1) (with $V = \operatorname{rad}\mathfrak g = \operatorname{rad} L(\alpha,\beta)$) yields that

$$\operatorname{rad}\mathfrak g \cong H(2;\underline{1};\Phi(\tau))^{(1)}$$

as $(\mathfrak g/\operatorname{rad}\mathfrak g)$-module. The subalgebra of maximal dimension containing $\operatorname{rad}\mathfrak g$ is $\mathcal M_{(0)} + \operatorname{rad}\mathfrak g$, where

$$\mathcal M_{(0)} \cong \big(\operatorname{Der} H(2;\underline{1};\Phi(\tau))^{(1)}\big) \cap W(2;\underline{1})_{(0)}.$$

Then ${\mathcal M_{(0)}}^{(1)} \subset H(2;\underline{1};\Phi(\tau))^{(1)} \cap W(2;\underline{1})_{(0)} = {H(2;\underline{1};\Phi(\tau))^{(1)}}_{(0)}$. As $Fx_1^{(p-1)}\partial_2 + Fx_2^{(p-1)}\partial_1 \subset (\operatorname{Der} H(2;\underline{1};\Phi(\tau))^{(1)}) \cap W(2;\underline{1})_{(0)}$ (see §16.3), it is not hard to see by using Proposition 16.3.1 that

$${\mathcal M_{(0)}}^{(1)} = {H(2;\underline{1};\Phi(\tau))^{(1)}}_{(0)}.$$

(d) Due to Equation (6.5.5) in Volume 1, p. 339 there is a surjective Lie algebra homomorphism $D_{H,\omega} : \mathcal O(2;\underline{1}) \to H(2;\underline{1};\Phi(\tau))^{(1)}$ with kernel $F1$, and Theorem 10.3.2(1) implies that we may do the following calculations in $\mathcal O(2;\underline{1})$ with generators y_1, y_2, Poisson bracket

$$\{y_1,y_2\}_\omega = q(1+y_1)(1+y_2), \quad q \in F^*,$$

and the image of T in $\mathcal M$ is mapped onto $F(1+y_1)\partial_{y_1} \oplus F(1+y_2)\partial_{y_2}$. In this realization the maximal subalgebra of codimension 2 is $\mathcal O(2;\underline{1})_{(2)} + F$.

If $\dim L_\mu = 1$ for all μ, then $L(\alpha,\beta)$ has a basis $((1+y_1)^i(1+y_2)^j+F)_{\substack{0\le i,j\le p-1\\ i+j\ne 0}}$. If $\dim L_\mu = 2$ for all μ, then $\mathcal M$ has a basis $((1+y_1)^i(1+y_2)^j+F)_{\substack{0\le i,j\le p-1\\ i+j\ne 0}}$ and $\operatorname{rad} L(\alpha,\beta)$ has a basis $(\eta((1+y_1)^i(1+y_2)^j+F))_{\substack{0\le i,j\le p-1\\ i+j\ne 0}}$ where η denotes the module isomorphism $H(2;\underline{1};\Phi(\tau))^{(1)} \to \operatorname{rad} L(\alpha,\beta)$.

Set

$$\delta(u,v) := [u,v] - \beta(u^{[p]})^{1/p}v + \alpha(v^{[p]})^{1/p}u$$

for T-root vectors $u \in L_\alpha$, $v \in L_\beta$. For $u = (1+y_1)^a(1+y_2)^b + F \in L_\alpha$, $v = (1+y_1)^c(1+y_2)^d + F \in L_\beta$ one has

$$[u,v] = \{u,v\}_\omega + F = q(ad-bc)(1+y_1)^{a+c}(1+y_2)^{b+d} + F,$$
$$\beta(u^{[p]})v = (\operatorname{ad} u)^p v = q^p(ad-bc)^p v, \;\text{ i.e., }\; \beta(u^{[p]})^{1/p} = q(ad-bc),$$

$$\begin{aligned}
\alpha(v^{[p]})u &= (\operatorname{ad} v)^p u = q^p(bc-ad)^p u, \quad \text{i.e., } \alpha(v^{[p]})^{1/p} = q(bc-ad),\\
\delta(u,v) &= q(ad-bc)\big((1+y_1)^{a+c}(1+y_2)^{b+d} - (1+y_1)^c(1+y_2)^d\\
&\quad - (1+y_1)^a(1+y_2)^b\big) + F\\
&\equiv q(ad-bc)\big((a+c)y_1+(b+d)y_2 - (cy_1+dy_2) - (ay_1+by_2)\big) + F\\
&\equiv 0 \pmod{\mathcal{O}(2;\underline{1})_{(2)} + F}.
\end{aligned}$$

In case $\dim L_\mu = 2$ for all μ we observe that $(1+y_i)^2 - 2(1+y_i) + F \in \mathcal{M}_{(0)}$ $(i = 1,2)$, and therefore the subspace $[\mathcal{M}_{(0)}, \operatorname{rad} L(\alpha,\beta)]$ contains

$$\begin{aligned}
&\eta\big(\{(1+y_1)^2 - 2(1+y_1), (1+y_1)^i(1+y_2)^j\}_\omega + F\big)\\
&\quad = q2j\eta\big((1+y_1)^{i+2}(1+y_2)^j + F\big) - 2qj\eta\big((1+y_1)^{i+1}(1+y_2)^j + F\big),\\
&\eta\big(\{(1+y_2)^2 - 2(1+y_2) + F, (1+y_1)^i(1+y_2)^j\}_\omega + F\big)\\
&\quad = -q2i\eta\big((1+y_1)^i(1+y_2)^{j+2} + F\big) + 2qi\eta\big((1+y_1)^i(1+y_2)^{j+1} + F\big).
\end{aligned}$$

From this it is easy to deduce that $\eta((1+y_1)^i(1+y_2)^j - (1+y_1)(1+y_2) + F) \in [\mathcal{M}_{(0)}, \operatorname{rad} L(\alpha,\beta)]$ whenever $(i,j) \neq (0,0)$. As a consequence,

$$\eta\big((1+y_1)^c(1+y_2)^d - (1+y_1)^{a+c}(1+y_2)^{b+d} + F\big) \in [\mathcal{M}_{(0)}, \operatorname{rad} L(\alpha,\beta)]$$

if $(c,d) \neq (0,0), (-a,-b)$. We deduce for $u = (1+y_1)^a(1+y_2)^b + F \in L_\alpha$, $v = (1+y_1)^c(1+y_2)^d + F \in L_\beta$ (recall that $\beta(u^{[p]})^{1/p} = q(ad-bc)$ and $\eta(v)^{[p]} = 0$)

$$\begin{aligned}
\delta(u,\eta(v)) &= q(ad-bc)\eta\big((1+y_1)^{a+c}(1+y_2)^{b+d} - (1+y_1)^c(1+y_2)^d + F\big)\\
&\equiv 0 \pmod{[\mathcal{M}_{(0)}, \operatorname{rad} L(\alpha,\beta)]}
\end{aligned}$$

if $(c,d) \neq (0,0)$ and $(c,d) \neq (-a,-b)$. In these two exceptional cases one trivially has $\delta(u,\eta(v)) = q(ad-bc)\eta((1+y_1)^{a+c}(1+y_2)^{b+d} - (1+y_1)^c(1+y_2)^d + F) = 0$.

It is clear that $\delta(\eta(u),\eta(v)) = 0$ as $\operatorname{rad} L(\alpha,\beta)$ is abelian. As a result, we obtain $\delta(u,v) \in \mathfrak{g}_{(0)}{}^{(1)}$ in all cases under consideration.

(e) We stated in (c) that the character χ vanishes on $\mathfrak{g}_{(0)}{}^{(1)}$. Let $g \mapsto g^p$ denote the p-mapping on $\mathfrak{g}$. The vanishing of χ implies

$$\rho(g^{[p]} - g^p) = \rho(g)^p - \rho(g^p) = \chi(g)^p \mathrm{Id} = 0 \quad \forall g \in \mathfrak{g}_{(0)}{}^{(1)}.$$

Moreover, $\mathcal{M}_{(0)}{}^{(1)} \cong H(2;\underline{1};\Phi(\tau))^{(1)}{}_{(0)}$ is closed under the p-mapping, and the p-mapping vanishes on $\operatorname{rad}\mathfrak{g}$. Hence in both cases under consideration $\mathfrak{g}_{(0)}{}^{(1)}$ is closed under the p-mapping. Therefore $\delta(u,v)^{[p]} = \delta(u,v)^p \in \mathfrak{g}_{(0)}{}^{(1)} \subset L(\alpha,\beta)$. On the other hand, by Lemma 1.1.1, $\Delta(u,v) - \delta(u,v)^{[p]} \in \mathfrak{g}^{(1)} \subset L(\alpha,\beta)$. Consequently, $\Delta(u,v) \in L(\alpha,\beta) \cap T = \{0\}$. □

19.3 Solvable sections

We now turn to the general case.

Lemma 19.3.1. *Let $\alpha, \beta \in \Gamma(L,T)$ and $V_\alpha \subset L_\alpha$ be a subspace such that*

$$\Omega := \{\mu \in \Gamma(L,T) \mid \mu([V_\alpha, L_\beta]) \neq 0\} \neq \emptyset.$$

Then $L(\alpha,\beta)$ is non-solvable and there is a root $\lambda \in \Gamma(L,T)$ such that $\lambda(V_\alpha) \neq 0$ holds.

Proof. We observe that every 1-section $L(\mu)$ acts trigonalizably. Therefore every root ν is linear on $L(\mu)$. In particular, $\mu(L(\mu)) = 0$ holds.

(a) Suppose $\Omega \cap (\mathbb{F}_p\alpha + \mathbb{F}_p\beta) \neq \emptyset$.

By our introductoury remark, $\alpha + \beta$ is not contained in Ω. The present assumption then implies that $\alpha, \beta \in \Omega$. From this and Corollary 1.3.4(1) one deduces by induction that $V_\alpha, L_\beta \subset L(\alpha,\beta)^{(\infty)}$. Therefore $L(\alpha,\beta)$ is non-solvable, whence $L[\alpha,\beta] \cong H(2;\underline{1};\Phi(\tau))^{(1)}$. Similarly, we then conclude that $\Psi_{\alpha,\beta}(V_\alpha) \neq \{0\}$. Then $\beta(V_\alpha) \neq 0$ holds (Theorem 10.3.2(5)).

(b) Now, suppose that $\Omega \cap (\mathbb{F}_p\alpha + \mathbb{F}_p\beta) = \emptyset$.

If $\delta(L_\gamma) = 0$ for all $\gamma, \delta \in \mathbb{F}_p\alpha + \mathbb{F}_p\beta$, then every L_γ acts nilpotently on $L(\alpha,\beta)$, and hence $L(\alpha,\beta)$ is a nilpotent section. Consequently it acts trigonalizably on L. This contradicts the assumption $\Omega \neq \emptyset$. Choose $\gamma, \delta \in \mathbb{F}_p\alpha + \mathbb{F}_p\beta$ such that $\delta(L_\gamma) \neq 0$. Note that γ, δ are independent, hence $\mathbb{F}_p\alpha + \mathbb{F}_p\beta = \mathbb{F}_p\gamma + \mathbb{F}_p\delta$ holds. In the present case we have $\gamma \notin \Omega$.

The simplicity of L implies

$$L_\gamma = \sum_{\gamma-\mu,\mu\in\Omega} [L_{\gamma-\mu}, L_\mu].$$

Thus there is $\kappa \in \Gamma(L,T) \setminus (\mathbb{F}_p\alpha + \mathbb{F}_p\beta)$ such that

$$(\gamma-\kappa)([V_\alpha, L_\beta]) \neq 0, \quad \kappa([V_\alpha, L_\beta]) \neq 0, \quad \delta([L_{\gamma-\kappa}, L_\kappa]) \neq 0.$$

Suppose inductively that

$$L_\mu \subset L(\alpha,\beta,\kappa)^{(n)} \quad \forall \mu \in \Gamma(L(\alpha,\beta,\kappa),T) \setminus \mathbb{F}_p\gamma.$$

Observe that

$$(i\gamma + j\delta + k\kappa)([V_\alpha, L_\beta]) = k\kappa([V_\alpha, L_\beta]) \neq 0, \ \forall k \in \mathbb{F}_p^*,$$
$$(i\gamma + j\delta)([L_{\gamma-\kappa}, L_\kappa]) = j\delta([L_{\gamma-\kappa}, L_\kappa]) \neq 0, \ \forall j \in \mathbb{F}_p^*.$$

Then we get $L_\mu \subset L(\alpha,\beta,\kappa)^{(n+1)}$ by Corollary 1.3.4(1). This gives

$$L_\mu \subset L(\alpha,\beta,\kappa)^{(\infty)} \quad \forall \mu \in \Gamma(L(\alpha,\beta,\kappa),T) \setminus \mathbb{F}_p\gamma.$$

In particular, $L(\alpha,\beta,\kappa)$ is non-solvable.

If $[V_\alpha, L_\beta]$ is contained in $\ker \Psi_{\alpha,\beta,\kappa}$, then we may conclude as above that $[V_\alpha, L_\beta] \subset (\ker \Psi_{\alpha,\beta,\kappa})^{(\infty)} = \{0\}$, a contradiction. If $L_\kappa \subset \ker \Psi_{\alpha,\beta,\kappa}$ or $L_{\gamma-\kappa} \subset \ker \Psi_{\alpha,\beta,\kappa}$, then $\sum_{j\neq 0} L_{i\gamma+j\delta} \subset \ker \Psi_{\alpha,\beta,\kappa}$. This gives $[V_\alpha, L_\beta] \subset \ker \Psi_{\alpha,\beta,\kappa}$, which is not true. We obtain

$$\Psi_{\alpha,\beta,\kappa}(V_\alpha) \neq \{0\},\ \ \Psi_{\alpha,\beta,\kappa}(L_\beta) \neq \{0\},\ \ \Psi_{\alpha,\beta,\kappa}(L_\kappa) \neq \{0\},\ \ \Psi_{\alpha,\beta,\kappa}(L_{\gamma-\kappa}) \neq \{0\}.$$

In particular, $\Psi_{\alpha,\beta,\kappa}(T)$ is a 3-dimensional torus.

(c) Suppose $L[\alpha,\beta,\kappa]$ has different minimal T-invariant ideals I, J. Then

$$TR(I) + TR(J) = TR(I \oplus J) \le TR(L) = 3,$$

so we may assume $TR(I) = 1$. As I is non-solvable, there is a non-solvable 2-section $I(\lambda,\nu)$ (Theorem 1.3.10). Then $L(\lambda,\nu)$ is non-solvable, whence $L[\lambda,\nu] \cong H(2;\underline{1};\Phi(\tau))^{(1)}$ (Proposition 19.1.3). There is a natural homomorphism

$$\varphi : L[\alpha,\beta,\kappa](\lambda,\nu) \to L[\lambda,\nu]$$

whose kernel is solvable and T-invariant. Therefore $\varphi(I(\lambda,\nu))$ is a non-solvable T-invariant subalgebra of $H(2;\underline{1};\Phi(\tau))^{(1)}$ of absolute toral rank 1. Due to Theorem 10.3.2(5) the latter forces that $\varphi(I(\lambda,\nu))$ carries only one T-root direction. But then it is solvable, a contradiction.

So $L[\alpha,\beta,\kappa]$ has a unique minimal T-invariant ideal $\tilde{S}$, and $TR(\tilde{S}) \ge 2$ holds. There exist a simple algebra S and $m \in \mathbb{N}$ such that $\tilde{S} \cong S \otimes \mathcal{O}(m;\underline{1})$ under an isomorphism τ. This isomorphism induces a restricted Lie algebra homomorphism

$$\Psi : L(\alpha,\beta,\kappa)_{[p]} + T \longrightarrow \big((\operatorname{Der} S) \otimes \mathcal{O}(m;\underline{1})\big) \rtimes \big(\operatorname{Id} \otimes W(m;\underline{1})\big).$$

Let $\overline{T} := \Psi(T)$ and identify $\tilde{S}$ with $S \otimes \mathcal{O}(m;\underline{1})$. By Theorem 3.6.1 we can choose τ such that

$$\overline{T} = \overline{T}_0 \oplus \Big(\sum_{i=1}^{s} F\operatorname{Id} \otimes (1+x_i)\partial_i\Big)$$

where $\overline{T}_0$ is the normalizer of $S \otimes \mathcal{O}(m;\underline{1})_{(1)}$ in $\overline{T}$ and $s = \dim \overline{T}/\overline{T}_0$. Moreover,

$\overline{T}_0 = \{\omega(t) \otimes 1 + \mathrm{Id} \otimes \lambda(t) \mid t \in \overline{T}_0\}$ where

$$\omega : \overline{T}_0 \to \mathrm{Der}\, S \text{ and } \lambda : \overline{T}_0 \to \sum_{i=s+1}^{m} F x_i \partial_i$$

are restricted homomorphisms.

Lemma 1.2.6(3) shows that $\overline{T} \cap \tilde{S}_{[p]}$ is a torus of maximal dimension in $\tilde{S}_{[p]}$. Since $\tilde{S}_{[p]} = \tilde{S} + S_{[p]} \otimes F \subset (\mathrm{Der}\, S) \otimes \mathcal{O}(m; \underline{1})$, where $S_{[p]}$ is the p-envelope of S in $\mathrm{Der}\, S$, it is straightforward to see that $\overline{T} \cap \tilde{S}_{[p]} \subset \overline{T}_0 \cap (\ker \lambda)$. Set accordingly

$$\overline{T} \cap \tilde{S}_{[p]} = \overline{T}'_0 \otimes F,$$

where $\overline{T}'_0$ is a torus of maximal dimension in $S_{[p]}$.

(d) Note that $\overline{T}$ acts on $S \otimes F$ as $\omega(\overline{T}_0) \otimes F$. Each 1-section in $L(\alpha, \beta, \kappa)$ is nilpotent, then all 1-sections of $S \otimes F$ relative to $\omega(\overline{T}_0) \otimes F$ are nilpotent as well. If $TR(S) \leq 2$, then Lemma 19.1.2 yields that $S \cong H(2; \underline{1}; \Phi(\tau))^{(1)}$ and $\omega(\overline{T}_0) \subset S_{[p]}$ is 2-dimensional.

In particular, $TR(S) \geq 2$ holds.

Suppose $TR(S) = 2$ and $m = 0$. Then $\tilde{S} \cong S \cong H(2; \underline{1}; \Phi(\tau))^{(1)}$ and $\overline{T} = \overline{T}_0 = \overline{T}'_0 \otimes F$, and $\overline{T}'_0 \subset S_{[p]}$ is 2-dimensional. Moreover, we may consider $L[\alpha, \beta, \kappa] + \overline{T}$ a subalgebra of $\mathrm{Der}\, S$. However, there is no room for a 3-dimensional torus in $\mathrm{Der}\, S \cong \mathrm{Der}\, H(2; \underline{1}; \Phi(\tau))^{(1)}$. So this case is impossible.

(e) Now suppose that $TR(S) = 2$ and $m > 0$. Then $S \cong H(2; \underline{1}; \Phi(\tau))^{(1)}$ and $\overline{T}'_0 \subset S_{[p]}$ is 2-dimensional. By definition one has $\overline{T}'_0 \subset \omega(\overline{T}_0) \subset \mathrm{Der}\, S$. Since every torus of maximal dimension in $\mathrm{Der}\, S$ is 2-dimensional, we get $\overline{T}'_0 = \omega(\overline{T}_0)$, and hence

$$\overline{T} = (\overline{T}'_0 \otimes F) \oplus (F \mathrm{Id} \otimes d)$$

for some toral derivation $d \in W(m; \underline{1})$ which is nonzero because $\dim \overline{T} = 3$.

Since d is toral, every eigenvalue of $\mathrm{Id} \otimes d$ is contained in $\mathbb{F}_p$. Define $\epsilon \in \overline{T}^*$ by setting

$$\epsilon(\overline{T}'_0 \otimes F) = 0, \ \epsilon(\mathrm{Id} \otimes d) = 1.$$

Then $L[\alpha, \beta, \kappa] = S \otimes \mathcal{O}(m; \underline{1}) + L[\alpha, \beta, \kappa](\epsilon)$. We conclude $L[\alpha, \beta, \kappa]^{(\infty)} = S \otimes \mathcal{O}(m; \underline{1})$.

Let $\mathfrak{t}_1, \mathfrak{t}_2$ denote the tori generated by the semisemiple parts of $\Psi_{\alpha,\beta,\kappa}([V_\alpha, L_\beta])$ and $\Psi_{\alpha,\beta,\kappa}([L_{\gamma-\kappa}, L_\kappa])$. We mentioned in (b) that

$$\Psi_{\alpha,\beta,\kappa}([V_\alpha, L_\beta]), \Psi_{\alpha,\beta,\kappa}(L_{\gamma-\kappa}), \Psi_{\alpha,\beta,\kappa}(L_\kappa) \subset L[\alpha, \beta, \kappa]^{(\infty)} = S \otimes \mathcal{O}(m; \underline{1}).$$

Then $\mathfrak{t}_1, \mathfrak{t}_2 \subset \overline{T}'_0 \otimes F$. Note that

$$\gamma(\mathfrak{t}_1) = \delta(\mathfrak{t}_1) = 0, \quad \kappa(\mathfrak{t}_1) \neq 0, \quad \gamma(\mathfrak{t}_2) = 0, \quad \delta(\mathfrak{t}_2) \neq 0.$$

As δ, κ are roots on $L[\alpha, \beta, \kappa]$, we conclude by dimension reasons $\overline{T}'_0 \otimes F = \mathfrak{t}_1 \oplus \mathfrak{t}_2 = \overline{T} \cap (\ker \gamma)$.

According to Theorem 10.3.2, 0 is not a root for $\overline{T}'_0$ on S. Hence 0 is not a root for $\overline{T}_0$ on $\tilde{S}$, which means that $\mathbb{F}_p \gamma \cap \Gamma(\tilde{S}, \overline{T}) = \emptyset$. But then $\mathfrak{t}_2 = \{0\}$, a contradiction.

(f) Finally, suppose $TR(S) = 3$. Then $TR(\tilde{S}) = 3$ and hence $\dim \overline{T} \cap \tilde{S}_{[p]} = 3$. Therefore, $\overline{T} \subset \tilde{S}_{[p]}$. As before, $\overline{T} = \overline{T}'_0 \otimes F$ for some 3-dimensional torus $\overline{T}'_0 \subset S_{[p]}$. One has

$$L[\alpha, \beta, \kappa]_\mu = S_{\mu'} \otimes \mathcal{O}(m; \underline{1}) \quad \forall \mu \neq 0, \quad \text{where} \quad \mu'(t) := \mu(t \otimes 1) \quad \forall t \in \overline{T}'_0.$$

Suppose $L(\alpha, \beta)$ is solvable. Then $S(\alpha', \beta')$ is also solvable. Theorem 19.2.3(3) shows that $S(\alpha', \beta')$ is abelian. Therefore $[V_\alpha, L_\beta] \subset \ker \Psi_{\alpha,\beta,\kappa}$ holds. But we have stated in (b) that this is not true.

If $\Psi_{\alpha,\beta,\kappa}(V_\alpha) \subset S_{\alpha'} \otimes \mathcal{O}(m; \underline{1})_{(1)}$, then $\Psi_{\alpha,\beta,\kappa}([V_\alpha, L_\beta]) \subset S_{\alpha'+\beta'} \otimes \mathcal{O}(m; \underline{1})_{(1)}$ holds as well. But then $\kappa([V_\alpha, L_\beta]) = 0$, which is not true. Hence there are $v \in V_\alpha$, $v' \in S_{\alpha'} \setminus \{0\}$ such that

$$\Psi_{\alpha,\beta,\kappa}(v) \equiv v' \otimes 1 \pmod{S_{\alpha'} \otimes \mathcal{O}(m; \underline{1})_{(1)}}.$$

Theorem 19.2.6 gives $v'^{[p]} \in \overline{T}'_0 \setminus \{0\}$. Hence there is a $\overline{T}'_0$-root $\lambda' \in \mathbb{F}_p\alpha' + \mathbb{F}_p\beta' + \mathbb{F}_p\kappa'$ such that $\lambda'(v') \neq 0$. Then

$$\lambda(v^{[p]}) = \lambda(\Psi_{\alpha,\beta,\kappa}(v)^{[p]}) = \lambda'(v'^{[p]}) \neq 0. \qquad \square$$

It turned out during the classification work that whenever a Lie type theorem is proved we are able to classify a further big family of simple Lie algebras. Accordingly, the following proposition is a key result.

Proposition 19.3.2. *Every nonzero root vector of L with respect to T acts non-nilpotently on L. In particular, $H = \{0\}$ holds and every solvable section is abelian.*

Proof. (a) Put

$$N_\mu := \{x \in L_\mu \mid \kappa(x) = 0 \ \forall \kappa \in \Gamma(L, T)\}, \quad \mu \in \Gamma(L, T).$$

Since all roots are linear on every L_μ (Proposition 19.1.1), all N_μ are vector spaces. The assumption $[N_\alpha, L_\beta] \not\subset N_{\alpha+\beta}$ means $\Omega := \{\mu \in \Gamma(L, T) \mid \mu([N_\alpha, L_\beta]) \neq 0\}$ is non-void. Thus Lemma 19.3.1 applies and shows that there is a root μ such that $\mu(N_\alpha) \neq 0$, contrary to the definition of N_α.

As a consequence, $N := \sum N_\mu$ is an ideal of L. Hence $N = \{0\}$ and therefore for any root vector $u \in L_\alpha$ there exists $\beta \in \Gamma(L, T)$ such that $\beta(u) \neq 0$. Thus u acts non-nilpotently on L. In particular $H \subset N = \{0\}$ holds (cf. Proposition 19.1.1).

(b) To show that every solvable section is abelian it is sufficient to consider any solvable 2-section $L(\alpha, \beta)$.

If $L(\alpha, \beta)^{(1)}$ does not act nilpotently on L, then there are $\gamma, \delta \in \mathbb{F}_p\alpha + \mathbb{F}_p\beta$ such that $[L_\gamma, L_\delta]$ acts non-nilpotently on L. Note that γ, δ are necessarily independent, whence $L(\gamma, \delta) = L(\alpha, \beta)$. Put

$$\Omega := \{\mu \in \Gamma(L, T) \mid \mu([L_\gamma, L_\delta]) \neq 0\}.$$

The present assumption yields $\Omega \neq \emptyset$. Lemma 19.3.1 implies that $L(\gamma, \delta) = L(\alpha, \beta)$ is non-solvable, a contradiction.

As a consequence, $[L_\alpha, L_\beta] \subset N_{\alpha+\beta} = \{0\}$. This is the claim. □

We are now able to determine 3-sections of L.

Theorem 19.3.3. *Let $\alpha, \beta, \gamma \in \Gamma(L, T)$. One of the following holds.*

(1) *$L(\alpha, \beta, \gamma)$ is abelian.*

(2) *There exist $\delta_1, \delta_2 \in \Gamma(L, T)$ such that*

$$L(\alpha, \beta, \gamma) = L(\delta_1, \delta_2) \oplus C,$$

where $L[\delta_1, \delta_2] \cong H(2; \underline{1}; \Phi(\tau))^{(1)}$ and C is a T-invariant central ideal.

(3) *$L[\alpha, \beta, \gamma] \cong H(2; \underline{1}; \Phi(\tau))^{(1)} \otimes \mathcal{O}(1; \underline{1})$; moreover,*

$$\Psi_{\alpha,\beta,\gamma}(T) = \overline{T}_0 \otimes F + F\mathrm{Id} \otimes (1 + x_1)\partial_1,$$

where $\overline{T}_0$ is a 2-dimensional torus in Der $H(2; \underline{1}; \Phi(\tau))^{(1)}$*, and*

$$\Gamma(L[\alpha, \beta, \gamma], \Psi_{\alpha,\beta,\gamma}(T)) = (\mathbb{F}_p\alpha + \mathbb{F}_p\beta + \mathbb{F}_p\gamma) \setminus \mathbb{F}_p\epsilon,$$

where $\epsilon \in \mathbb{F}_p\alpha + \mathbb{F}_p\beta + \mathbb{F}_p\gamma$ and $\epsilon(\overline{T}_0 \otimes F) = 0$.

(4) *$L[\alpha, \beta, \gamma]$ is a simple Lie algebra of absolute toral rank 3.*

In all cases, (ker $\Psi_{\alpha,\beta,\gamma}$) $\cap$ *$L(\alpha, \beta, \gamma)$ is nilpotent.*

Proof. (a) If $L(\alpha, \beta, \gamma)$ is solvable then it is abelian by Proposition 19.3.2. We are in case (1) of the theorem. So assume that $L(\alpha, \beta, \gamma)$ is not solvable. Then $L[\alpha, \beta, \gamma] = L(\alpha, \beta, \gamma)/\mathrm{rad}_T\, L(\alpha, \beta, \gamma)$ is a nonzero T-semisimple Lie algebra.

Suppose $L[\alpha, \beta, \gamma]$ has different minimal T-invariant ideals I, J. Then

$$TR(I) + TR(J) = TR(I \oplus J) \leq TR(L) = 3,$$

so we may assume $TR(I) = 1$. As it is non-solvable, there is a non-solvable 2-section $I(\lambda, \nu)$ (Theorem 1.3.10). Then $L(\lambda, \nu)$ is non-solvable, whence $L[\lambda, \nu] \cong H(2; \underline{1}; \Phi(\tau))^{(1)}$ (Proposition 19.1.3). There is a natural homomorphism

$$\varphi : L[\alpha, \beta, \gamma](\lambda, \nu) \to L[\lambda, \nu]$$

whose kernel is solvable and T-invariant. Therefore $\varphi(I(\lambda,\nu))$ is a non-solvable T-invariant subalgebra of $H(2;\underline{1};\Phi(\tau))^{(1)}$ of absolute toral rank 1. Due to Theorem 10.3.2(5) the latter forces that $\varphi(I(\lambda,\nu))$ carries only one T-root direction. But then it is solvable, a contradiction.

So $L[\alpha,\beta,\gamma]$ has a unique minimal T-invariant ideal $\tilde{S}$, and $TR(\tilde{S}) \geq 2$ holds. There exist a simple algebra S and $m \in \mathbb{N}$ such that $\tilde{S} \cong S \otimes \mathcal{O}(m;\underline{1})$ under an isomorphism τ. This isomorphism induces a restricted Lie algebra homomorphism

$$\Psi : L(\alpha,\beta,\gamma)_{[p]} + T \longrightarrow \big((\mathrm{Der}\, S) \otimes \mathcal{O}(m;\underline{1})\big) \rtimes \big(\mathrm{Id} \otimes W(m;\underline{1})\big).$$

Let $\overline{T} := \Psi(T)$ and identify $\tilde{S}$ with $S \otimes \mathcal{O}(m;\underline{1})$. By Theorem 3.6.1 we can choose τ such that

$$\overline{T} = \overline{T}_0 \oplus \Big(\sum_{i=1}^{s} F\mathrm{Id} \otimes (1+x_i)\partial_i\Big)$$

where $\overline{T}_0$ is the normalizer of $S \otimes \mathcal{O}(m;\underline{1})_{(1)}$ in $\overline{T}$ and $s = \dim \overline{T}/\overline{T}_0$. Moreover, $\overline{T}_0 = \{\omega(t) \otimes 1 + \mathrm{Id} \otimes \lambda(t) \mid t \in \overline{T}_0\}$ where

$$\omega : \overline{T}_0 \to \mathrm{Der}\, S \quad \text{and} \quad \lambda : \overline{T}_0 \to \sum_{i=s+1}^{m} F x_i \partial_i$$

are restricted homomorphisms.

Lemma 1.2.6(3) shows that $\overline{T} \cap \tilde{S}_{[p]}$ is a torus of maximal dimension in $\tilde{S}_{[p]}$. Since $\tilde{S}_{[p]} = \tilde{S} + S_{[p]} \otimes F \subset (\mathrm{Der}\, S) \otimes \mathcal{O}(m;\underline{1})$, where $S_{[p]}$ is the p-envelope of S in $\mathrm{Der}\, S$, it is straightforward to see that $\overline{T} \cap \tilde{S}_{[p]} \subset \overline{T}_0 \cap (\ker \lambda)$. Set accordingly

$$\overline{T} \cap \tilde{S}_{[p]} = \overline{T}'_0 \otimes F,$$

where $\overline{T}'_0$ is a torus of maximal dimension in $S_{[p]}$.

Note that $\overline{T}$ acts on $S \otimes F$ as $\omega(\overline{T}_0) \otimes F$. Each 1-section in $L(\alpha,\beta,\gamma)$ is solvable, then all 1-sections of $S \otimes F$ relative to $\omega(\overline{T}_0) \otimes F$ are solvable as well. If $TR(S) \leq 2$, then Lemma 19.1.2 yields that $S \cong H(2;\underline{1};\Phi(\tau))^{(1)}$ and $\omega(\overline{T}_0) \subset S_{[p]}$ is 2-dimensional.

In particular, $TR(S) \geq 2$ holds.

(b) Suppose $TR(S) = 2$ and $m = 0$. Then $\tilde{S} \cong S \cong H(2;\underline{1};\Phi(\tau))^{(1)}$ and $\overline{T} = \overline{T}_0 = \overline{T}'_0 \otimes F$, and $\overline{T}'_0 \subset S_{[p]}$ is 2-dimensional. Moreover,

$$\Psi \text{ maps } L(\alpha,\beta,\gamma) \text{ into } \mathrm{Der}\, S = H(2;\underline{1};\Phi(\tau))^{(1)} + \overline{T}'_0.$$

So there exist $\mathbb{F}_p$-independent T-roots $\delta_1,\delta_2 \in \mathbb{F}_p\alpha + \mathbb{F}_p\beta + \mathbb{F}_p\gamma$ such that $\Psi(L(\alpha,\beta,\gamma)) = \Psi(L(\delta_1,\delta_2))$. Then the right-hand side algebra is T-semisimple, and one obtains $L[\alpha,\beta,\gamma] \cong L[\delta_1,\delta_2]$. Proposition 19.1.3 gives $L[\delta_1,\delta_2] \cong H(2;\underline{1};\Phi(\tau))^{(1)}$.

Put

$$C := \sum_{\mu\in(\mathbb{F}_p\alpha+\mathbb{F}_p\beta+\mathbb{F}_p\gamma)\setminus(\mathbb{F}_p\delta_1+\mathbb{F}_p\delta_2)} L_\mu \subset \operatorname{rad}_T L(\alpha,\beta,\gamma).$$

Take $\mu \in (\mathbb{F}_p\alpha + \mathbb{F}_p\beta + \mathbb{F}_p\gamma) \setminus (\mathbb{F}_p\delta_1 + \mathbb{F}_p\delta_2)$ and any $\kappa \in \mathbb{F}_p\alpha + \mathbb{F}_p\beta + \mathbb{F}_p\gamma$. As

$$\big(\operatorname{rad}_T L(\alpha,\beta,\gamma)\big) \cap L(\kappa,\mu) \subset \operatorname{rad}_T L(\kappa,\mu),$$

then $L[\kappa,\mu]$ misses all roots $\mathbb{F}_p^*\mu$ and therefore cannot be isomorphic to $H(2;\underline{1};\Phi(\tau))^{(1)}$. Then the 2-section $L(\kappa,\mu)$ is solvable, whence abelian. This shows that C is central in $L(\alpha,\beta,\gamma)$. So this is case (2) of the theorem.

(c) Now suppose that $TR(S) = 2$ and $m > 0$. Then $S \cong H(2;\underline{1};\Phi(\tau))^{(1)}$ and $\overline{T}_0' \subset S_{[p]} = \operatorname{Der} S$ is 2-dimensional. By definition one has $\overline{T}_0' \subset \omega(\overline{T}_0) \subset \operatorname{Der} S$. Since every torus of maximal dimension in $S_{[p]}$ is 2-dimensional, we get $\overline{T}_0' = \omega(\overline{T}_0)$, and hence

$$\overline{T} = (\overline{T}_0' \otimes F) \oplus (F\operatorname{Id} \otimes d)$$

for some derivation $d \in W(m;\underline{1})$. At the beginning $d = 0$ might be possible, however, if $d = 0$ then T acts as a 2-dimensional torus and this means that the root lattice on $L[\alpha,\beta,\gamma]$ is spanned by roots δ_1,δ_2. In this case $L[\alpha,\beta,\gamma] = L[\delta_1,\delta_2]$ holds. Due to Proposition 19.1.3 we conclude $m = 0$, which is contrary to our assumption.

So $d \neq 0$ is toral, hence every eigenvalue of $\operatorname{Id} \otimes d$ is contained in $\mathbb{F}_p$. Define $\epsilon \in \mathbb{F}_p\alpha + \mathbb{F}_p\beta + \mathbb{F}_p\gamma$ by setting

$$\epsilon(\overline{T}_0' \otimes F) = 0, \ \epsilon(\operatorname{Id} \otimes d) = 1.$$

Then $L[\alpha,\beta,\gamma] = S \otimes \mathcal{O}(m;\underline{1}) + L[\alpha,\beta,\gamma](\epsilon)$.

According to Theorem 10.3.2, there exist $\mathbb{F}_p$-independent $\kappa', \nu' \in \Gamma(S,\overline{T}_0')$ such that

$$S = \sum_{\delta'\in(\mathbb{F}_p\kappa'+\mathbb{F}_p\nu')\setminus\{0\}} S_{\delta'}.$$

Since $S \otimes F$ is $\overline{T}$-invariant, we may extend every $\overline{T}_0'$-root δ' to $\delta \in \overline{T}^*$ via

$$\delta(t \otimes 1) := \delta'(t)\ \forall t \in \overline{T}_0', \quad \delta(\operatorname{Id} \otimes d) := 0.$$

Note that

$$L[\alpha,\beta,\gamma](\delta,\epsilon) = S(\delta') \otimes \mathcal{O}(m;\underline{1}) + L[\alpha,\beta,\gamma](\epsilon).$$

Since $S(\delta')$ is abelian, $S(\delta') \otimes \mathcal{O}(m;\underline{1})$ is an abelian ideal of $L[\alpha,\beta,\gamma](\delta,\epsilon)$, and the quotient is an image of $L[\alpha,\beta,\gamma](\epsilon)$ which is solvable. Hence $L(\delta,\epsilon)$ is solvable, whence abelian. As a result, $L[\alpha,\beta,\gamma](\epsilon)$ annihilates the unique minimal ideal $S \otimes \mathcal{O}(m;\underline{1})$. We conclude $L[\alpha,\beta,\gamma](\epsilon) = \{0\}$.

This result proves that $S \otimes \mathcal{O}(m;\underline{1})$ is d-simple. Therefore $m = 1$ and $d = (1 + x_1)\partial_1$ holds (after an adjustment of d). This is case (3) of the theorem.

(d) Finally, suppose $TR(S) = 3$. Then $TR(\tilde{S}) = 3$ and hence $\dim \overline{T} \cap \tilde{S}_{[p]} = 3$. Therefore, $\overline{T} \subset \tilde{S}_{[p]}$. Then $L[\alpha, \beta, \gamma] = \tilde{S} + \Psi_{\alpha,\beta,\gamma}(H) = \tilde{S}$ holds (cf. Proposition 19.3.2). Since $L[\alpha, \beta, \gamma]$ is $\overline{T}$-semisimple, we have $m = 0$. Thus we are in case (4) of the theorem.

(e) It remains to show that $(\ker \Psi_{\alpha,\beta,\gamma}) \cap L(\alpha, \beta, \gamma)$ is nilpotent. This statement is obvious in case (1). In case (2) we have

$$\left(\ker \Psi_{\alpha,\beta,\gamma}\right) \cap L(\alpha, \beta, \gamma) = \left(\ker \Psi_{\delta_1,\delta_2}\right) \cap L(\delta_1, \delta_2) + C.$$

Set $T_1 := T \cap \left((\ker \Psi_{\delta_1,\delta_2}) \cap L(\delta_1, \delta_2)\right)_{[p]}$. Since δ_1, δ_2 are roots on $L[\delta_1, \delta_2]$, we conclude $\delta_1(T_1) = \delta_2(T_1) = 0$. But then $(\ker \Psi_{\delta_1,\delta_2}) \cap L(\delta_1, \delta_2)$ acts nilpotently on $L(\delta_1, \delta_2)$. This gives the claim.

One argues similarly in cases (3) and (4) of the theorem. □

Corollary 19.3.4.

(1) $0 \notin \Gamma(L, T)$.

(2) $\Gamma(L, T) \cup \{0\}$ *is an elementary* p*-group of rank* $TR(L)$.

(3) *There exists* $k > 0$ *such that*

$$\dim L_\mu = k \qquad \forall \mu \in \Gamma(L, T).$$

Proof. (1) 0 is not a root according to Proposition 19.3.2.

(2) (a) Let $\alpha, \beta \in \Gamma(L, T)$, $\alpha + \beta \neq 0$. In order to prove that $\alpha + \beta \in \Gamma(L, T)$ we make the following introductory remark. If $L(\kappa, \lambda)$ is solvable then it is abelian (Theorem 19.3.2). Hence we have $\kappa(L_\lambda) = 0$. Otherwise Proposition 19.1.3 shows $L[\kappa, \lambda] \cong H(2; \underline{1}; \Phi(\tau))^{(1)}$. Then all $i\kappa + j\lambda \neq 0$ are roots and $\mu(L_\lambda) \neq 0$ for all $\mu \in (\mathbb{F}_p\kappa + \mathbb{F}_p\lambda) \setminus \mathbb{F}_p\lambda$. Thus the following implications are true:

$$L(\kappa, \lambda) \text{ non-abelian} \Leftrightarrow \kappa(L_\lambda) \neq 0 \Leftrightarrow \lambda(L_\kappa) \neq 0 \Rightarrow \mathbb{F}_p\kappa + \mathbb{F}_p\lambda \subset \Gamma(L, T) \cup \{0\}.$$

If $\beta(L_\alpha) \neq 0$ the preceding remark shows $\alpha + \beta \in \Gamma(L, T)$. Thus suppose $\beta(L_\alpha) = 0, \alpha(L_\beta) = 0$. This implies that $L(\alpha, \beta)$ is abelian.

Suppose α, β are dependent. Since L_α does not annihilate L, there is a non-abelian 2-section $L(\alpha, \gamma)$. The introductory remark yields that $\mathbb{F}_p^*\alpha \subset \Gamma(L, T)$. In particular, $\alpha + \beta \in \Gamma(L, T)$ holds.

So assume that α, β are independent. The subalgebra $\sum_{\mu \in \Gamma(L,T)} \sum_{i \geq 0} {L_\mu}^{[p]^i}$ of $L_{[p]}$ is closed under $[p]$-th powers and hence coincides with $L_{[p]}$. In particular, this subalgebra contains the torus T. Since $\alpha + \beta$ is a nonzero linear form on T, there is $\gamma \in \Gamma(L, T)$ such that $(\alpha + \beta)({L_\gamma}^{[p]^i}) \neq 0$ for some $i \geq 0$. By symmetry we may assume $\alpha(L_\gamma{}^{[p]^i}) \neq 0$, which by definition means $\alpha(L_\gamma) \neq 0$. Then

$$\alpha(L_\gamma) \neq 0, \quad \gamma(L_\alpha) \neq 0, \quad (\alpha+\beta)(L_\gamma) \neq 0.$$

Note that α, β, γ are linearly independent.

Clearly, $L(\alpha, \beta, \gamma)$ is not abelian. So it is not ruled by case (1) of Theorem 19.3.3.

(b) Suppose $L(\alpha, \beta, \gamma)$ satisfies case (2) of Theorem 19.3.3. Since $L(\alpha, \gamma)$ is non-abelian, it is non-solvable and hence

$$L[\alpha, \beta, \gamma] = L[\alpha, \gamma].$$

Then β is not a root on $L[\alpha, \beta, \gamma]$, whence $L_\beta \in \operatorname{rad}_T L(\alpha, \beta, \gamma)$. As $L_\beta \neq \{0\}$ and γ is a root on $L[\alpha, \beta, \gamma] = L[\alpha, \gamma]$, this gives $\gamma(L_\beta) = 0$, and therefore $\beta(L_\gamma) = 0$ by our introductory remark.

Suppose some $\mu = i\alpha + j\beta, i, j \in \mathbb{F}_p^*$ is a root on L. Then $(i\alpha + j\beta)(L_\gamma) = i\alpha(L_\gamma) \neq 0$ holds, whence $L(\mu, \gamma)$ is non-abelian. Then (observe that we are in case (2) of Theorem 19.3.3) $L[\mu, \gamma] = L[\alpha, \gamma]$. But in the present case this is impossible, as $j \neq 0$. Thus no linear form $i\alpha + j\beta, i, j \in \mathbb{F}_p^*$ is a root.

As L_α does not centralize L, the set $\Omega := \{\lambda \in \Gamma(L, T) \mid \lambda(L_\alpha) \neq 0\}$ is non-void. Note that $\beta \notin \Omega$. The simplicity of L ensures the existence of $\delta, \beta - \delta \in \Omega$ such that $[L_\delta, L_{\beta-\delta}] \neq \{0\}$. The introductory remark yields $\alpha(L_\delta) \neq 0$. By definition we have $(\beta - \delta)(L_\alpha) \neq 0$, showing (again by the introductory remark)

$$\alpha + (\beta - \delta), \alpha + 2(\beta - \delta) \in \Gamma(L, T).$$

If $(\alpha + \beta - \delta)(L_\delta) \neq 0$, then we obtain $\alpha + \beta = (\alpha + \beta - \delta) + \delta \in \Gamma(L, T)$ contrary to the above.

If $(\alpha + \beta - \delta)(L_\delta) = 0$, then $(\alpha + 2(\beta - \delta))(L_\delta) = -\alpha(L_\delta) \neq 0$ holds. As above we obtain the contradiction $\alpha + 2\beta = (\alpha + 2(\beta - \delta)) + 2\delta \in \Gamma(L, T)$. Thus $L(\alpha, \beta, \gamma)$ cannot satisfy case (2) of Theorem 19.3.3.

(c) Suppose $L(\alpha, \beta, \gamma)$ satisfies case (3) of Theorem 19.3.3. In particular,

$$\Gamma(L[\alpha, \beta, \gamma], \Psi_{\alpha,\beta,\gamma}(T)) = (\mathbb{F}_p\alpha + \mathbb{F}_p\beta + \mathbb{F}_p\gamma) \setminus \mathbb{F}_p\epsilon$$

holds for some $\epsilon \in \mathbb{F}_p\alpha + \mathbb{F}_p\beta + \mathbb{F}_p\gamma$ which satisfies $\epsilon(\overline{T}_0' \otimes F) = 0$. Note that $L[\alpha, \gamma] \cong H(2; \underline{1}; \Phi(\tau))^{(1)}$. Therefore $L[\alpha, \beta, \gamma](\alpha, \gamma) \subset L[\alpha, \beta, \gamma]^{(\infty)} = S \otimes \mathcal{O}(1; \underline{1})$. This gives $\epsilon(L_\alpha) = \epsilon(L_\gamma) = 0$.

Since $\alpha(L_\gamma) \neq 0, \gamma(L_\alpha) \neq 0$, the linear forms α, γ, ϵ are $\mathbb{F}_p$-independent. Therefore they span $\mathbb{F}_p\alpha + \mathbb{F}_p\beta + \mathbb{F}_p\gamma$. Write $\alpha + \beta = r\epsilon + s\alpha + t\gamma$. One has

$$0 \neq (\alpha + \beta)(L_\gamma) = s\alpha(L_\gamma).$$

Then $s \neq 0$, and this gives $\alpha + \beta \in \Gamma(L, T)$.

If $L(\alpha, \beta, \gamma)$ satisfies case (4) of Theorem 19.3.3, then Theorem 19.2.3(1) gives the claim.

(3) Put $\Delta := \{\mu \in \Gamma(L,T) \mid \dim L_\mu \text{ is minimal}\}$ and $I := \sum_{\mu\in\Delta} L_\mu$. We prove that I is an ideal of L. Take any $\mu \in \Delta$ and $\alpha \in \Gamma(L,T)$. If $\mu(L_\alpha) \neq 0$, then there is $x \in L_\alpha$ with $\mu(x) \neq 0$, and this element acts invertibly on $\sum_{i\in\mathbb{F}_p} L_{\mu+i\alpha}$. Consequently, $\dim L_{\mu+\alpha} = \dim L_\mu$, whence $\mu + \alpha \in \Delta$. If $\mu(L_\alpha) = 0$, then $L(\alpha, \mu)$ is abelian, whence $[L_\mu, L_\alpha] = \{0\}$. □

Corollary 19.3.5.

(1) $\tilde{H} = T$.

(2) $[u,v]^{[p]} = \beta(u^{[p]})v^{[p]} - \alpha(v^{[p]})u^{[p]}$ *holds for all root vectors* $u \in L_\alpha$, $v \in L_\beta$.

Proof. (1) Let α, β, γ be any roots and take $x \in {L_\alpha}^{[p]}$. Let $x_n \in \tilde{H}$ denote the $[p]$-nilpotent part of x. In case (1) of Theorem 19.3.3, one has $\Psi_{\alpha,\beta,\gamma}(L_\alpha) = \{0\}$. In cases (2) and (3) of Theorem 19.3.3, one easily computes that $\Psi_{\alpha,\beta,\gamma}({L_\alpha}^{[p]}) = \Psi_{\alpha,\beta,\gamma}(L_\alpha)^{[p]} \subset \Psi_{\alpha,\beta,\gamma}(T)$. In case (4) of that theorem, Theorem 19.2.6 applies and again yields $\Psi_{\alpha,\beta,\gamma}({L_\alpha}^{[p]}) \subset \Psi_{\alpha,\beta,\gamma}(T)$. Therefore $\Psi_{\alpha,\beta,\gamma}(x_n) = 0$ holds. This shows that every root vector $v \in [x_n, L_\beta]$ is contained in $\ker\, \Psi_{\alpha,\beta,\gamma}$. As this is a nilpotent ideal of $L(\alpha, \beta, \gamma)$, then $v^{[p]}$ acts nilpotently on all L_γ. Proposition 19.3.2 yields $[x_n, L_\beta] = \{0\}$ for all β. But then $[x_n, L] = \{0\}$ and this gives $x_n = 0$. Consequently, $x \in T$.

The subalgebra $\sum_{\mu\in\Gamma(L,T)} \sum_{i\geq 0} {L_\mu}^{[p]^i}$ of $L_{[p]}$ is closed under $[p]$-th powers and hence coincides with $L_{[p]}$. We conclude

$$\tilde{H} = H + \sum_{\mu\in\Gamma(L,T)} \sum_{i>0} {L_\mu}^{[p]^i} = T.$$

(2) Take any $\gamma \in \Gamma(L,T)$. We intend to show that

$$[u,v]^{[p]} - \beta(u^{[p]})v^{[p]} + \alpha(v^{[p]})u^{[p]} \in \ker \gamma,$$

which result would imply the claim. As $(\ker\, \Psi_{\alpha,\beta,\gamma}) \cap L(\alpha, \beta, \gamma)$ is nilpotent, it is sufficient to prove the claim in $L[\alpha, \beta, \gamma]$. We also may assume that $\Psi_{\alpha,\beta,\gamma}(L(\alpha, \beta))$ is non-solvable, because otherwise $L(\alpha, \beta)$ would be abelian, in which case the claim is trivial.

By assumption, $L(\alpha, \beta, \gamma)$ is not solvable. Suppose that $L(\alpha, \beta, \gamma)$ satisfies case (2) of Theorem 19.3.3. Then $L[\alpha, \beta, \gamma] \cong H(2; \underline{1}, \Phi(\tau))^{(1)}$. We have shown in Theorem 10.3.2(4) that there is a basis of root vectors $d_\mu \in H(2; \underline{1}, \Phi(\tau))^{(1)}{}_\mu$ and a biadditive skew-symmetric form f such that

$$[d_\lambda, d_\mu] = f(\lambda, \mu) d_{\lambda+\mu}.$$

Consequently,

$$[d_\lambda{}^{[p]}, d_\mu] = (\operatorname{ad} d_\lambda)^p(d_\mu) = f(\lambda,\mu)^p d_\mu, \ i.e., \ \mu(d_\lambda) = f(\lambda,\mu).$$

We conclude

$$\begin{aligned}\gamma\big([d_\alpha, d_\beta]\big) &= f(\alpha,\beta)\gamma(d_{\alpha+\beta}) = f(\alpha,\beta) f(\alpha+\beta,\gamma)\\ &= -f(\beta,\alpha)f(\alpha,\gamma) + f(\alpha,\beta)f(\beta,\gamma) = -\alpha(d_\beta)\gamma(d_\alpha) + \beta(d_\alpha)\gamma(d_\beta).\end{aligned}$$

Therefore

$$\begin{aligned}\gamma([d_\alpha, d_\beta]^{[p]}) &= \gamma([d_\alpha, d_\beta])^p = -\alpha(d_\beta)^p\gamma(d_\alpha)^p + \beta(d_\alpha)^p\gamma(d_\beta)^p\\ &= -\alpha(d_\beta{}^{[p]})\gamma(d_\alpha{}^{[p]}) + \beta(d_\alpha{}^{[p]})\gamma(d_\beta{}^{[p]}),\end{aligned}$$

which implies $[u,v]^{[p]} - \beta(u^{[p]})v^{[p]} + \alpha(v^{[p]})u^{[p]} \in \ker\gamma$.

Suppose that $L(\alpha,\beta,\gamma)$ satisfies case (3) of Theorem 19.3.3. Then there is $\epsilon \in (\mathbb{F}_p\alpha + \mathbb{F}_p\beta + \mathbb{F}_p\gamma) \setminus \{0\}$ which vanishes on $\overline{T}_0' \otimes F$. Then α, β, ϵ are independent, and hence span $\mathbb{F}_p\alpha + \mathbb{F}_p\beta + \mathbb{F}_p\gamma$. Put $\gamma = i\alpha + j\beta + k\epsilon$. As $\epsilon(\overline{T}_0' \otimes F) = 0$, it is enough to prove our claim for $i\alpha + j\beta$ instead of γ. This was done in the former case.

Suppose that $L(\alpha,\beta,\gamma)$ satisfies case (4) of Theorem 19.3.3. As $L[\alpha,\beta,\gamma]$ is simple of absolute toral rank 3, Theorem 19.2.8 applies and proves the claim. □

19.4 Conclusion

As a result of the deliberations in this chapter, we may embed the Lie algebras under consideration into more convenient algebras. To that end we consider a finite dimensional vector space M and a finite subgroup G in M^*, the dual space of M. Put $M_\alpha := \ker\alpha = \{m \in M \mid \alpha(m) = 0\}$ for all $\alpha \in G$ and let $V(M,G)$ be the graded space

$$V(M,G) := \bigoplus_{\mu\in G} M_\mu.$$

For the sake of clarity we write the μ-part of $V(M,G)$ as $\{(m,\mu) \mid m \in M_\mu\} = (M_\mu, \mu)$. Together with the product

$$[(m,\alpha),(m',\beta)] := \big(\beta(m)m' - \alpha(m')m, \alpha+\beta\big)$$

$V(M,G)$ becomes a Lie algebra.

Note that $\Gamma(L,T) \cup \{0\}$ is an $\mathbb{F}_p$-vector space (Corollary 19.3.4(2)). For any root vector $u \in L_\mu$ let $u_s \in T$ denote the semisimple part of u in $L_{[p]}$ and u_n the $[p]$-nilpotent part. Since $u^{[p]} \in T$ due to Corollary 19.3.5(1), we have $u_n{}^{[p]} = 0, u_s{}^{[p]} = u^{[p]}$.

Proposition 19.4.1. *Let $G := \Gamma(L, T) \cup \{0\}$. The mapping*

$$\Psi : \begin{cases} L & \longrightarrow & V(T, G) \\ x & \longmapsto & (x_s, \mu) \quad \forall\, x \in L_\mu \end{cases}$$

constitutes an injective Lie algebra homomorphism.

Proof. (a) Every root space L_μ is abelian (Proposition 19.3.2). We therefore have for arbitrary $x, y \in L_\mu$ that $x_s + y_s$ is semisimple ([S-F, Proposition 2.3.3(3)]). As $x_n + y_n$ is $[p]$-nilpotent, we get

$$(x + y)_s = x_s + y_s.$$

Thus Ψ is linear on every root space and hence extends to a linear mapping on L.

(b) In order to prove that Ψ is a Lie homomorphism, we take $u \in L_\alpha$, $v \in L_\beta$. As $^{[p]}$ is semilinear on T, Corollary 19.3.5(2) gives

$$\begin{aligned} \big([u, v]_s - \beta(u_s)v_s + \alpha(v_s)u_s\big)^{[p]} &= [u, v]_s{}^{[p]} - \beta(u_s)^p v_s{}^{[p]} + \alpha(v_s)^p u_s{}^{[p]} \\ &= [u, v]^{[p]} - \beta(u^{[p]})v^{[p]} + \alpha(v^{[p]})u^{[p]} = 0. \end{aligned}$$

Since $^{[p]}$ is injective on T we get $[u, v]_s - \beta(u_s)v_s + \alpha(v_s)u_s = 0$, which is the claim.

(c) To prove injectivity of Ψ take $\sum_{\mu \in \Gamma(L,T)} u_\mu \in \ker \Psi$. By definition of Ψ we have $u_\mu \in \ker \Psi$ for all μ. Therefore u_μ is $[p]$-nilpotent and consequently acts nilpotently on L. Theorem 19.3.2 yields $u_\mu = 0$ for all μ and therefore $u = 0$. □

We now put

$$M_\mu := \Psi(L_\mu) \subset T, \quad M := \sum_{\mu \in \Gamma(L,T)} M_\mu \subset T,$$

and consider $L = \bigoplus_{\mu \in \Gamma(L,T)} (M_\mu, \mu)$ as a homogeneous subalgebra of $V(T, G)$.

Lemma 19.4.2. *Put $L_{(0)} := \{\sum (u_\mu, \mu) \mid \sum u_\mu = 0\}$. Then*

(1) *$L_{(0)}$ is a subalgebra of L;*
(2) *$L/L_{(0)} \cong M$ via $\varphi\big(\sum (u_\mu, \mu)\big) := \sum u_\mu$;*
(3) *$L_{(0)}$ acts on M via*

$$\Big(\sum_\lambda (u_\lambda, \lambda)\Big) \cdot v := -\sum_\lambda \lambda(v) u_\lambda.$$

Proof. (1) Suppose $\sum u_\mu = \sum v_\mu = 0$. Then

$$\Big[\sum_\lambda (u_\lambda, \lambda), \sum_\mu (v_\mu, \mu)\Big] = \sum_{\lambda,\mu} \big(\mu(u_\lambda)v_\mu - \lambda(v_\mu)u_\lambda, \lambda + \mu\big)$$

holds. The result follows from the observation that

$$\sum_{\lambda,\mu} \big(\mu(u_\lambda)v_\mu - \lambda(v_\mu)u_\lambda\big) = \sum_\mu \mu\Big(\sum_\lambda u_\lambda\Big)v_\mu - \sum_\lambda \lambda\Big(\sum_\mu v_\mu\Big)u_\lambda = 0.$$

(2) Obvious.

(3) The multiplication in L yields for $\sum_\lambda (u_\lambda,\lambda) \in L_{(0)}$, $(v_\mu,\mu) \in L$

$$\begin{aligned}\varphi\Big(\Big[\sum_\lambda (u_\lambda,\lambda),(v_\mu,\mu)\Big]\Big) &= \varphi\Big(\sum_\lambda \big(\mu(u_\lambda)v_\mu - \lambda(v_\mu)u_\lambda, \lambda+\mu\big)\Big)\\ &= \sum_\lambda \mu(u_\lambda)v_\mu - \lambda(v_\mu)u_\lambda = -\sum_\lambda \lambda(v_\mu)u_\lambda\\ &= \sum_\lambda (u_\lambda,\lambda)\cdot v_\mu = \sum_\lambda (u_\lambda,\lambda)\cdot\varphi((v_\mu,\mu)).\end{aligned}$$ □

Lemma 19.4.3. *Let $\kappa,\lambda \in \Gamma(L,T)$ satisfy $\lambda(M_\kappa) \neq 0$. The following holds.*

(1) $M_\kappa \subset \ker \kappa$;

(2) $M_\lambda \neq M_\kappa$;

(3) $\kappa(M_\mu) \neq 0\ \forall \mu \in \mathbb{F}_p\kappa + \mathbb{F}_p^*\lambda$, *and* $(\mathbb{F}_p\kappa + \mathbb{F}_p\lambda)\setminus\{0\} \subset \Gamma(L,T)$;

(4) $M_\kappa \cap (\ker \lambda) = M_\kappa \cap M_\lambda = M_\lambda \cap (\ker \kappa)$;

(5) $M_\mu \subset M_\kappa + M_\lambda$ *for all* $\mu \in (\mathbb{F}_p\kappa + \mathbb{F}_p\lambda)\setminus\{0\}$;

(6) *If* $\mu \in \Gamma(L,T)$ *and* κ,λ,μ *are F-independent on* $M_\kappa + M_\lambda$, *then* $M_\mu \subset M_\kappa + M_\lambda$.

Proof. (1) follows from the fact that $\kappa(L_\kappa) = 0\ \forall\ \kappa \in \Gamma(L,T)$.

(2) follows as $\lambda(M_\lambda) = 0,\ \lambda(M_\kappa) \neq 0$.

(3) By assumption there is $u \in L_\kappa$ satisfying $\lambda(u) \neq 0$. Then $L(\kappa,\lambda)$ is non-abelian (and therefore non-solvable), hence $L[\kappa,\lambda] \cong H(2;\underline{1};\Phi(\tau))^{(1)}$ (Proposition 19.1.3). Due to Theorem 10.3.2(5) there is $v \in L_\mu$ with $\kappa(v) \neq 0$. Thus $\kappa(M_\mu) \neq 0$. Theorem 10.3.2(2) proves the second statement.

(4) According to (3) we have $\kappa(M_{\lambda-\kappa}) \neq 0$. Choose $v \in M_{\lambda-\kappa}$ with $\kappa(v) = 1$. Then for every $w \in M_\kappa \cap (\ker \lambda)$ one has (as $(\lambda-\kappa)(w) = 0$)

$$(w,\lambda) = [(v,\lambda-\kappa),(w,\kappa)] \in (M_\lambda,\lambda).$$

Thus we have $M_\kappa \cap (\ker \lambda) \subset M_\kappa \cap M_\lambda$. Note that according to (3) $\kappa(M_\lambda) \neq 0$. Therefore $M_\kappa \cap M_\lambda \neq M_\kappa$. A dimension argument proves $M_\kappa \cap (\ker \lambda) = M_\kappa \cap M_\lambda$. Interchanging κ and λ we obtain the remaining assertion.

(5) Since $\kappa(L_\lambda) \neq 0$ and $\lambda(L_\kappa) \neq 0$ (see part (3) of this proof), it is not hard to see that $L(\kappa, \lambda)$ is generated as an algebra by $L_\kappa \cup L_\lambda$. The definition of the multiplication in $V(T, G)$ then means that M_μ is contained in $M_\kappa + M_\lambda$.

(6) Suppose $\mu(M_\kappa \cap M_\lambda) = 0$. According to (2) and (4) one has $\dim M_\kappa/(M_\kappa \cap M_\lambda) = \dim M_\lambda/(M_\kappa \cap M_\lambda) = 1$. Then $\dim (M_\kappa + M_\lambda)/(M_\kappa \cap M_\lambda) \leq 2$ holds. As κ, λ, μ vanish on $M_\kappa \cap M_\lambda$, these linear forms are F-dependent on $M_\kappa + M_\lambda$. This contradiction shows that $\mu(M_\kappa \cap M_\lambda) \neq 0$, and thereby gives $\mu(M_\kappa) \neq 0$, $\mu(M_\lambda) \neq 0$.

We now may apply (4) for various cases to obtain

$$M_\mu \cap (\ker \kappa) = M_\kappa \cap M_\mu = M_\kappa \cap (\ker \mu),$$
$$M_\mu \cap (\ker \lambda) = M_\lambda \cap M_\mu = M_\lambda \cap (\ker \mu).$$

Next suppose that $M_\mu \cap (\ker \kappa) = M_\mu \cap (\ker \lambda)$. This assumption yields

$$M_\kappa \cap M_\mu = M_\lambda \cap M_\mu \subset (\ker \kappa) \cap (\ker \lambda) \cap (\ker \mu).$$

Then $M_\kappa \cap M_\mu \subset M_\lambda$ and $\dim (M_\kappa + M_\lambda)/M_\kappa \cap M_\mu \leq 2$. As before, the linear forms κ, λ, μ are F-linearly dependent on $M_\kappa + M_\lambda$, a contradiction. Hence

$$M_\mu \cap (\ker \kappa) \neq M_\mu \cap (\ker \lambda),$$

and thus

$$M_\mu \subset M_\mu \cap (\ker \kappa) + M_\mu \cap (\ker \lambda) = M_\kappa \cap (\ker \mu) + M_\lambda \cap (\ker \mu) \subset M_\kappa + M_\lambda.$$

□

Lemma 19.4.4. *There is k such that* $\dim M_\mu = k$ *for all roots μ. If $k > 1$, then*

$$M_\mu = M \cap (\ker \mu) \quad \forall \mu \in \Gamma(L, T).$$

Proof. Corollary 19.3.4(3) in combination with Proposition 19.4.1 yields that $\dim M_\mu = k$ is independent of μ.

Suppose $k > 1$. Let $\alpha \in \Gamma(L, T)$ be arbitrary. Due to Proposition 19.3.2 one has that $\alpha \neq 0$, and there is $\beta \in \Gamma(L, T)$ such that $\beta(L_\alpha) \neq 0$, and $L(\alpha, \beta)$ is not solvable. Therefore $L[\alpha, \beta] \cong H(2; \underline{1}; \Phi(\tau))^{(1)}$ (Proposition 19.1.3). In this case

$$\alpha(M_\beta) \neq 0, \quad \beta(M_\alpha) \neq 0.$$

Set

$$\Omega := \{\mu \in \Gamma(L, T) \mid \alpha, \beta, \mu \text{ are } F\text{-independent on } M_\alpha + M_\beta\}.$$

(a) Suppose $\Omega = \emptyset$. Then

$$(M_\alpha + M_\beta) \cap (\ker \alpha) \cap (\ker \beta) = (M_\alpha + M_\beta) \cap (\ker \alpha) \cap (\ker \beta) \cap (\ker \mu)$$

for all $\mu \in \Gamma(L, T)$. Consequently,

$$M_\alpha \cap (\ker \beta) \subset (M_\alpha + M_\beta) \cap (\ker \alpha) \cap (\ker \beta)$$
$$\subset \bigcap_{\mu \in \Gamma(L,T)} (M_\alpha + M_\beta) \cap (\ker \mu) \subset \bigcap_{\mu \in \Gamma(L,T)} T \cap (\ker \mu) = \{0\}.$$

This implies $\dim M_\alpha = 1$, contrary to our assumption.

(b) Thus $\Omega \neq \emptyset$. If $\gamma \in \Omega$, then Lemma 19.4.3(6) gives $M_\gamma \subset M_\alpha + M_\beta$. Moreover, the simplicity of L implies

$$L = \sum_{\mu \in \Omega} L_\mu + \sum_{\lambda, \mu \in \Omega} [L_\lambda, L_\mu].$$

Thus for $\gamma \notin \Omega$ we have $L_\gamma = \sum_{\lambda, \gamma - \lambda \in \Omega} [L_\lambda, L_{\gamma-\lambda}]$. Then

$$M_\gamma \subset \sum_{\lambda, \gamma - \lambda \in \Omega} (M_\lambda + M_{\gamma - \lambda}) \subset M_\alpha + M_\beta$$

by Lemma 19.4.3(5).

(c) As a consequence of the above

$$M = \sum_{\gamma \in \Gamma(L,T)} M_\gamma = M_\alpha + M_\beta.$$

Thus Lemma 19.4.3(4) yields

$$M \cap (\ker \alpha) = (M_\alpha + M_\beta) \cap (\ker \alpha) = M_\alpha + M_\beta \cap (\ker \alpha) = M_\alpha. \qquad \square$$

Our first classification result in this chapter is the following

Theorem 19.4.5. *Let L be a finite dimensional simple Lie algebra over an algebraically closed field of characteristic $p > 3$. If the minimal p-envelope of L contains a torus T of maximal dimension such that all T-roots are solvable and that* $\dim L_\mu \geq 2$ *for some $\mu \in \Gamma(L, T)$, then*

$$L \cong S(m; \underline{n}; \Phi(\tau))^{(1)} \quad \textit{for some } m \geq 3.$$

Proof. Note that $\dim L_\mu = \dim M_\mu = (\dim M) - 1$ (by Lemma 19.4.4). Set $m := \dim M \geq 3$. Since L is centerless, there are F-linearly independent roots $\alpha_1, \dots, \alpha_m$ such that

$$\bigcap_{i=1}^{m} M \cap (\ker \alpha_i) = \{0\}.$$

Choose a basis $q_1, \ldots, q_m \in M$ dual to $\alpha_1, \ldots, \alpha_m$. The action of $L_{(0)}$ on M has been described in Lemma 19.4.2. With respect to the basis $(q_1, \ldots, q_m)$ the following are elements of $L_{(0)}$ and represent the respective matrices (for $i \neq j$) :

$$(q_i, 2\alpha_j) - (q_i, \alpha_j) : -E_{ij},$$
$$(q_i - q_j, \alpha_i + \alpha_j) - (q_i, \alpha_j) + (q_i, \alpha_i) : E_{jj} - E_{ii}.$$

Next we compute traces. For $x = \sum_\mu (u_\mu, \mu) \in L_{(0)}$ we have $\sum_\mu u_\mu = 0$. Express u_μ in terms of the α_j, q_j,

$$u_\mu = \sum_{j=1}^{m} \alpha_j(u_\mu) q_j.$$

As $u_\mu \in M_\mu \subset \ker \mu$, one obtains

$$(*) \qquad 0 = \mu(u_\mu) = \sum_{j=1}^{m} \alpha_j(u_\mu)\mu(q_j).$$

On the other hand,

$$x \cdot q_i = -\sum_{\mu} \mu(q_i) u_\mu = -\sum_{\mu, j} \mu(q_i)\alpha_j(u_\mu) q_j,$$

and hence $(*)$ yields

$$\operatorname{trace}(\operatorname{ad}_{L/L_{(0)}} x) = -\sum_{i=1}^{m} \sum_{\mu} \mu(q_i)\alpha_i(u_\mu) = 0.$$

Let $(L_{(i)})_{i \geq -1}$ denote the standard filtration defined by $L_{(0)}$ and L:

$$L_{(-1)} := L, \quad L_{(i+1)} := \{x \in L_{(i)} \mid [L, x] \subset L_{(i)}\} \quad (i \geq 0).$$

Thus it is proved that $L_{(0)}/L_{(1)} \cong \operatorname{ad}_{L/L_{(0)}} L_{(0)} \cong \mathfrak{sl}(m)$. We state as a consequence that $L/L_{(0)}$ is the standard irreducible m-dimensional $\mathfrak{sl}(m)$-module.

Set $t := \dim T \geq m \geq 3$. Note that there are exactly $p^t - 1$ roots each of dimension $m - 1$ (Corollary 19.3.4). Thus

$$\dim L = (m-1)(p^t - 1).$$

By definition $L_{(1)}/L_{(2)}$ embeds into $\operatorname{Hom}(L/L_{(0)}, L_{(0)}/L_{(1)})$ and hence

$$\begin{aligned}\dim L_{(2)} &= \dim L - (\dim L/L_{(0)} + \dim L_{(0)}/L_{(1)} + \dim L_{(1)}/L_{(2)}) \\ &\geq (m-1)(p^t - 1) - \big(m + (m^2 - 1) + m(m^2 - 1)\big) \\ &\geq 2(p^m - 1) - (m^3 + m^2 - 1) > 0.\end{aligned}$$

Therefore Theorem 5.2.7 applies and shows that there are embeddings

$$S(m;\underline{n})^{(3)} \subset \operatorname{gr} L \subset S(m;\underline{n}).$$

The Weak Recognition Theorem (Corollary 5.5.3) now gives $L \cong S(m;\underline{n};\Psi)^{(1)}$ for some $\underline{n}$ and Ψ.

According to Theorem 6.3.8, there are exactly the following three isomorphism types of Special algebras:

- $L \cong S(m;\underline{n})^{(1)}$: then (cf. [S-F, Theorem 4.3.7])
$$(m-1)(p^t-1) = \dim L = (m-1)\big(p^{\sum_{j=1}^m n_j} - 1\big),$$
whence $\dim T = t = \sum_{j=1}^m n_j$. The p-envelope is given by
$$S(m;\underline{n})^{(1)} + \sum_{i=1}^{m}\sum_{j_i=1}^{n_i-1} F\partial_i^{p^{j_i}}$$
(Theorem 7.2.2(2)). Therefore
$$\dim L_{[p]}/L = \sum_{i=1}^{m}(n_i-1) < t,$$
and hence $T \cap L \neq \{0\}$. This contradicts Theorem 19.3.2.
- $S(m;\underline{n};\Phi(\tau))^{(1)}$: this is the claim.
- $L \cong S(m;\underline{n};\Phi(\ell))$: from Theorem 6.3.7(2) we conclude that
$$\dim L = (m-1)(p^t-1) \neq (m-1)p^{\sum n_i} = \dim S(m;\underline{n};\Phi(\ell)).$$

Thus the only possible case is the one claimed. □

Next we investigate the remaining case that all root spaces are 1-dimensional. We set $M_\mu = Fm_\mu$ for all $\mu \in \Gamma(L,T) \cup \{0\}$. In particular, $m_\mu \neq 0$ if $\mu \neq 0$ and $m_\mu = 0$ if $\mu = 0$.

Lemma 19.4.6. *Let $\alpha, \beta, \gamma \in \Gamma(L,T)$. Suppose*

$$\alpha(m_\beta) = -\beta(m_\alpha) \neq 0, \quad \alpha(m_\gamma) = -\gamma(m_\alpha) \neq 0.$$

Then

(1) $M_{i\alpha+j\beta} = F(im_\alpha + jm_\beta)$, *for all* $i, j \in \mathbb{F}_p, (i,j) \neq (0,0)$;
(2) $\beta(m_\gamma) = -\gamma(m_\beta)$;
(3) *if* $\beta(m_\gamma) \neq 0$, *then* $m_\beta + m_\gamma \in F^* m_{\beta+\gamma}$.

Proof. (1) By definition of the Lie bracket and the present assumption on $\alpha(m_\beta)$ one has

$$[(m_\alpha,\alpha),(m_\beta,\beta)] = \big(\beta(m_\alpha)(m_\alpha+m_\beta),\alpha+\beta\big).$$

As $\alpha(m_\beta) = -\beta(m_\alpha) \neq 0$ and $\alpha(m_\alpha) = \beta(m_\beta) = 0$, we obtain by induction $(im_\alpha + jm_\beta) \in M_{i\alpha+j\beta}$ for all $i,j \in \mathbb{F}_p$, $(i,j) \neq (0,0)$.

Since $\alpha(im_\alpha + jm_\beta) = j\alpha(m_\beta)$ and $\beta(im_\alpha + jm_\beta) = i\beta(m_\alpha)$ and $(i,j) \neq (0,0)$, one has $im_\alpha + jm_\beta \neq 0$. The statement is now a direct consequence from the previous result and the fact that $M_{i\alpha+j\beta}$ is 1-dimensional.

(2)(i) Assume that $m_\alpha, m_\beta, m_\gamma$ are F-dependent. Then there are $r,s,t \in F$ not all vanishing such that $rm_\alpha + sm_\beta + tm_\gamma = 0$. If $t = 0$, then $0 = \beta(rm_\alpha + sm_\beta) = r\beta(m_\alpha)$. Hence $r = 0$. By symmetry we get $s = 0$, whence $r = s = t = 0$. As this is impossible, we may take $t = -1$, i.e., $m_\gamma = rm_\alpha + sm_\beta$. We obtain

$$s\alpha(m_\beta) = \alpha(m_\gamma) \neq 0, \text{ hence } s \neq 0, \text{ and } 0 = \gamma(m_\gamma) = r\gamma(m_\alpha) + s\gamma(m_\beta).$$

Consequently,

$$\begin{aligned}\gamma(m_\beta) &= -\frac{r}{s}\gamma(m_\alpha) = \frac{r}{s}\alpha(m_\gamma) = r\alpha(m_\beta) = -r\beta(m_\alpha)\\ &= -\beta(rm_\alpha + sm_\beta) = -\beta(m_\gamma).\end{aligned}$$

(ii) Assume that $m_\alpha, m_\beta, m_\gamma$ are F-independent. We compute

$$\begin{aligned}&[(m_\gamma,\gamma),[(m_\alpha,\alpha),(m_\beta,\beta)]] = [(m_\gamma,\gamma),\big(\beta(m_\alpha)(m_\alpha+m_\beta),\alpha+\beta\big)]\\ &\quad= \big(\beta(m_\alpha)(\alpha+\beta)(m_\gamma)(m_\alpha+m_\beta),\alpha+\beta+\gamma\big)\\ &\quad\quad- \big(\beta(m_\alpha)\gamma(m_\alpha+m_\beta)m_\gamma,\alpha+\beta+\gamma\big).\end{aligned}$$

The first component of this expression is

$$(*) \qquad \begin{aligned}\beta(m_\alpha)\big(\alpha(m_\gamma)+\beta(m_\gamma)\big)m_\alpha + \beta(m_\alpha)\big(\alpha(m_\gamma)+\beta(m_\gamma)\big)m_\beta&\\ - \beta(m_\alpha)\big(\gamma(m_\alpha)+\gamma(m_\beta)\big)m_\gamma&.\end{aligned}$$

If this term vanishes, then one has (as $\beta(m_\alpha) \neq 0$ and $m_\alpha, m_\beta, m_\gamma$ are F-linear independent)

$$\alpha(m_\gamma)+\beta(m_\gamma) = 0,\ \gamma(m_\alpha)+\gamma(m_\beta) = 0.$$

This gives $\beta(m_\gamma) = -\gamma(m_\beta)$.

Otherwise the vector $(*)$ spans $M_{\alpha+\beta+\gamma}$. Let us assume that this takes place. Next

$$\begin{aligned}&[(m_\alpha,\alpha),[(m_\beta,\beta),(m_\gamma,\gamma)]]\\ &\quad= [(m_\alpha,\alpha),\big(\gamma(m_\beta)m_\gamma - \beta(m_\gamma)m_\beta,\beta+\gamma\big)]\\ &\quad= \Big(\big(\beta(m_\alpha)+\gamma(m_\alpha)\big)\big(\gamma(m_\beta)m_\gamma - \beta(m_\gamma)m_\beta\big),\alpha+\beta+\gamma\Big)\\ &\quad\quad- \Big(\big(\gamma(m_\beta)\alpha(m_\gamma) - \beta(m_\gamma)\alpha(m_\beta)\big)m_\alpha,\alpha+\beta+\gamma\Big).\end{aligned}$$

The first component of this expression

$$(**) \quad \big(\beta(m_\alpha)+\gamma(m_\alpha)\big)\big(\gamma(m_\beta)m_\gamma-\beta(m_\gamma)m_\beta\big)-\big(\gamma(m_\beta)\alpha(m_\gamma)-\beta(m_\gamma)\alpha(m_\beta)\big)m_\alpha$$

is a vector in $M_{\alpha+\beta+\gamma}$. Therefore it has to be a multiple of $(*)$. Since the coefficients of m_α and m_β in $(*)$ are the same, this happens to hold also in $(**)$. We obtain

$$(\beta(m_\alpha) + \gamma(m_\alpha))\beta(m_\gamma) = \gamma(m_\beta)\alpha(m_\gamma) - \beta(m_\gamma)\alpha(m_\beta).$$

Since $\alpha(m_\beta) = -\beta(m_\alpha)$ and $\alpha(m_\gamma) = -\gamma(m_\alpha) \neq 0$, we get

$$\beta(m_\alpha)\beta(m_\gamma) + \gamma(m_\alpha)\beta(m_\gamma) = -\gamma(m_\beta)\gamma(m_\alpha) + \beta(m_\gamma)\beta(m_\alpha).$$

Then $\beta(m_\gamma) = -\gamma(m_\beta)$ holds.

(3) As before we compute

$$[(m_\beta, \beta), (m_\gamma, \gamma)] = \big(\gamma(m_\beta)(m_\beta + m_\gamma), \beta + \gamma\big)$$

and conclude that $m_\beta + m_\gamma = t m_{\beta+\gamma}$ for some $t \in F^*$ (as $\gamma(m_\beta) = -\beta(m_\gamma) \neq 0$). This is claim (3). □

Lemma 19.4.7. *There are* $(m_\mu)_{\mu\in\Gamma(L,T)}$ *satisfying* $M_\mu = F m_\mu$,

$$\lambda(m_\mu) = -\mu(m_\lambda) \quad \forall \lambda, \mu \in \Gamma(L, T),$$

and

$$m_\lambda + m_\mu = m_{\lambda+\mu} \quad \textit{if}\ \lambda + \mu \neq 0.$$

Proof. (a) Fix a root $\alpha \in \Gamma(L, T)$ and set

$$\Omega := \{\mu \in \Gamma(L, T) \mid \mu(M_\alpha) \neq 0\} \neq \emptyset.$$

Take any $m_\alpha \in M_\alpha \backslash \{0\}$. For $\mu \in \Omega$ we choose according to Lemma 19.4.3(3) a nonzero root vector $m_\mu \in M_\mu$ with

$$\alpha(m_\mu) = -\mu(m_\alpha) \neq 0.$$

Next let $\mu \notin \Omega$ be arbitrary, and let κ, λ be roots such that $\kappa, \lambda, \mu - \kappa, \mu - \lambda \in \Omega$ and

$$[L_\kappa, L_{\mu-\kappa}] = L_\mu = [L_\lambda, L_{\mu-\lambda}].$$

Lemma 19.4.6(2) with $\beta = \lambda$ and $\gamma = \kappa, \mu - \kappa, \mu - \lambda$ yields

$$\lambda(m_\kappa) = -\kappa(m_\lambda), \quad \lambda(m_{\mu-\kappa}) = -(\mu - \kappa)(m_\lambda), \quad \lambda(m_{\mu-\lambda}) = -(\mu - \lambda)(m_\lambda).$$

Then

$$\lambda(m_\kappa + m_{\mu-\kappa}) = -\kappa(m_\lambda) - (\mu - \kappa)(m_\lambda) = -\mu(m_\lambda),$$
$$\lambda(m_\lambda + m_{\mu-\lambda}) = \lambda(m_{\mu-\lambda}) = -(\mu - \lambda)(m_\lambda) = -\mu(m_\lambda).$$

Since $L(\mu-\lambda,\lambda)$ is not abelian, one has $(\mu-\lambda)(m_\lambda)\neq 0$. Lemma 19.4.6(3) yields $m_\lambda + m_{\mu-\lambda}\in F^* m_\mu$. Moreover, one has $\mu(m_\lambda)\neq 0$, and this gives $\lambda(m_\mu)\neq 0$, whence $m_\mu\neq 0$. Since $L(\mu-\kappa,\kappa)$ is not abelian, one obtains similarly $m_\kappa+m_{\mu-\kappa}\in F^* m_\mu$. Put $m_\lambda+m_{\mu-\lambda}=rm_\mu$, $m_\kappa+m_{\mu-\kappa}=sm_\mu$. The preceding computation yields $(r-s)\lambda(m_\mu)=0$, whence $r=s$. Consequently, $m_\kappa+m_{\mu-\kappa}=m_\lambda+m_{\mu-\lambda}$. As L is simple and $\Omega\neq\emptyset$, we have

$$L_\mu=\sum_{\kappa,\mu-\kappa\in\Omega}[L_\kappa,L_{\mu-\kappa}],\quad \mu\notin\Omega.$$

For $\mu\notin\Omega$ set

$$m_\mu:=m_\kappa+m_{\mu-\kappa},\qquad \text{for some } \kappa,\mu-\kappa\in\Omega,$$

which by the preceding reasoning is independent of the choice of κ. Thus m_μ is defined for all roots μ.

(b) Let $\rho\in\Omega$, $\mu\in\Gamma(L,T)$. If $\mu\in\Omega$, then Lemma 19.4.6(2) gives $\rho(m_\mu)=-\mu(m_\rho)$. If $\mu\notin\Omega$, then choose $\kappa\in\Omega$ such that $[L_\kappa,L_{\mu-\kappa}]=L_\mu$. The preceding definition gives $m_\mu=m_\kappa+m_{\mu-\kappa}$. We compute (as $\kappa,\mu-\kappa\in\Omega$)

$$\rho(m_\mu)=\rho(m_\kappa)+\rho(m_{\mu-\kappa})=-\kappa(m_\rho)-(\mu-\kappa)(m_\rho)=-\mu(m_\rho).$$

So we have

$$\rho(m_\mu)=-\mu(m_\rho)\quad \forall\rho\in\Omega,\ \forall\mu\in\Gamma(L,T).$$

Let $\rho\notin\Omega$, $\mu\in\Gamma(L,T)$. Choose $\kappa\in\Omega$ such that $[L_\kappa,L_{\rho-\kappa}]=L_\rho$. The definition gives $m_\rho=m_\kappa+m_{\rho-\kappa}$. A computation as before gives

$$\mu(m_\rho)=\mu(m_\kappa)+\mu(m_{\rho-\kappa})=-\kappa(m_\mu)-(\rho-\kappa)(m_\mu)=-\rho(m_\mu).$$

We conclude

$$\rho(m_\mu)=-\mu(m_\rho)\quad \forall\rho,\mu\in\Gamma(L,T).$$

(c) Let $\lambda,\mu\in\Gamma(L,T)$ be such that $\lambda+\mu\neq 0$. This condition means that $\lambda+\mu$ is a root. Applying (b) we get for an arbitrary root ρ

$$\rho(m_\lambda+m_\mu-m_{\lambda+\mu})=-\lambda(m_\rho)-\mu(m_\rho)+(\lambda+\mu)(m_\rho)=0.$$

Consequently, $m_\lambda+m_\mu-m_{\lambda+\mu}=0$. □

We now have a detailed description of the algebras under consideration.

Theorem 19.4.8. *Let L be a finite dimensional simple Lie algebra over an algebraically closed field of characteristic $p>3$. If the minimal p-envelope of L contains a torus T of maximal dimension such that all T-roots are solvable and that*

$\dim L_\mu = 1$ *for all* $\mu \in \Gamma(L,T)$*, then there is a non-degenerate skew-symmetric biadditive form*

$$f : (\Gamma(L,T) \cup \{0\}) \times (\Gamma(L,T) \cup \{0\}) \to F$$

and an eigenvector basis $(u_\mu)_{\mu\in\Gamma(L,T)}$ *satisfying* $L_\mu = Fu_\mu$ *and (with* $u_0 = 0$*)*

$$[u_\lambda, u_\mu] = f(\lambda,\mu)u_{\lambda+\mu} \quad \forall\, \lambda,\mu \in \Gamma(L,T).$$

Proof. With the notation of Lemma 19.4.7 we have

$$\lambda(m_\mu) = -\mu(m_\lambda),\ m_{\lambda+\mu} = m_\lambda + m_\mu$$

and

$$[(m_\lambda,\lambda),(m_\mu,\mu)] = \big(\mu(m_\lambda)m_{\lambda+\mu}, \lambda+\mu\big) \quad \text{if } \lambda+\mu \neq 0.$$

For $\lambda,\mu \in \Gamma(L,T)$ set

$$f(\lambda,\mu) := \mu(m_\lambda), \quad f(0,\mu) := f(\lambda,0) := f(0,0) := 0.$$

Recall the notion of the isomorphism $\Psi : L \to V(\Gamma(L,T) \cup \{0\}, T)$. Set $u_\mu := \Psi^{-1}(m_\mu,\mu)$.

Then f is biadditive and skew-symmetric and $[u_\lambda,u_\mu] = f(\lambda,\mu)u_{\lambda+\mu}$ holds for all $\lambda,\mu \in \Gamma(L,T)$. If $f(\alpha,\mu) = 0$ for all μ, then u_α centralizes L, which is absurd. So f is non-degenerate. □

The algebras described in this theorem are algebras of R. E. Block [Blo 58] of type $L(G,\delta,f)$ with $G = \Gamma(L,T) \cup \{0\}$ and $\delta = 0$. R. D. Schafer has connected these algebras with Hamiltonian type Lie algebras via the nodal non-commutative Jordan algebras in [Sch 60]. We will now explicitely show that the algebras under consideration are of Hamiltonian type.

Theorem 19.4.9. *Let L be a finite dimensional simple Lie algebra over an algebraically closed field of characteristic $p > 3$. If the minimal p-envelope of L contains a torus T of maximal dimension such that all T-roots are solvable and that* $\dim L_\mu = 1$ *for all* $\mu \in \Gamma(L,T)$*, then L is of Hamiltonian type.*

Proof. (a) With the notation of Lemma 19.4.7 we have for all $\lambda,\mu \in \Gamma(L,T)$ satisfying $\lambda+\mu \neq 0$

$$[(m_\lambda,\lambda),(m_\mu,\mu)] = \big(\mu(m_\lambda)m_{\lambda+\mu}, \lambda+\mu\big).$$

Define a skew-symmetric F-bilinear form $g : L \times L \to F$ by

$$g\Big(\sum_{\alpha\in\Gamma(L,T)} (r_\alpha m_\alpha,\alpha),\ \sum_{\beta\in\Gamma(L,T)} (s_\beta m_\beta,\beta)\Big) := \sum_{\alpha,\beta\in\Gamma(L,T)} r_\alpha s_\beta \beta(m_\alpha),\ r_\alpha, s_\beta \in F.$$

Next we compute

$$\begin{aligned}\operatorname{rad} g &= \{u \in L \mid g(u, L) = 0\}\\ &= \Big\{\sum (r_\alpha m_\alpha, \alpha) \mid g\Big(\sum (r_\alpha m_\alpha, \alpha), (m_\beta, \beta)\Big) = 0\ \forall \beta \in \Gamma(L, T)\Big\}.\end{aligned}$$

Note that $\sum (r_\alpha m_\alpha, \alpha) \in \operatorname{rad} g$ if and only if $\sum r_\alpha \beta(m_\alpha) = 0$ for all β, which means $\sum r_\alpha m_\alpha \in \bigcap_{\beta \in \Gamma(L,T)} T \cap (\ker \beta) = \{0\}$. As a result, $\operatorname{rad} g = L_{(0)}$ (cf. Lemma 19.4.2). Then g induces a non-degenerate skew-symmetric F-bilinear form

$$\bar{g} : L/L_{(0)} \times L/L_{(0)} \to F.$$

As $L/L_{(0)} \cong M$ as $L_{(0)}$-modules (Lemma 19.4.2), this gives a non-degenerate skew-symmetric form $\bar{g} : M \times M \to F$.

(b) Let $u = \sum_\alpha (r_\alpha m_\alpha, \alpha) \in L_{(0)}$. This means $\sum r_\alpha m_\alpha = 0$. We compute

$$\begin{aligned}&g\big([u, (m_\gamma, \gamma)], (m_\beta, \beta)\big) = g\big((\sum_\alpha r_\alpha \gamma(m_\alpha)(m_\alpha + m_\gamma), \alpha + \gamma), (m_\beta, \beta)\big)\\ &= \sum_\alpha r_\alpha \gamma(m_\alpha)(\beta(m_\alpha) + \beta(m_\gamma)) = \sum_\alpha r_\alpha \gamma(m_\alpha)\beta(m_\alpha);\end{aligned}$$

$$\begin{aligned}&g\big((m_\gamma, \gamma), [u, (m_\beta, \beta)]\big) = g\big((m_\gamma, \gamma), (\sum_\alpha r_\alpha \beta(m_\alpha)(m_\alpha + m_\beta), \alpha + \beta)\big)\\ &= \sum_\alpha r_\alpha \beta(m_\alpha)(\alpha(m_\gamma) + \beta(m_\gamma)) = \sum_\alpha r_\alpha \beta(m_\alpha)\alpha(m_\gamma)\\ &= -\sum_\alpha r_\alpha \gamma(m_\alpha)\beta(m_\alpha).\end{aligned}$$

This gives

$$g([u, v], w) = -g(v, [u, w])\ \forall u \in L_{(0)},\ \forall v, w \in L,$$

and therefore

$$\bar{g}(u \cdot m, m') = -\bar{g}(m, u \cdot m') \quad \forall u \in L_{(0)},\ \forall m, m' \in M.$$

We obtain a homomorphism $L_{(0)} \to \mathfrak{sp}(M)$. Denote the kernel of this homomorphism by $L_{(1)}$, i. e.,

$$L_{(1)} := \{u \in L_{(0)} \mid u \cdot M = \{0\}\} = \{u \in L_{(0)} \mid [u, L] \subset L_{(0)}\}.$$

(c) Let $q := \dim M$ and choose roots β_i such that $(m_{\beta_i})_{1 \le i \le q}$ is a basis of M. Then the family $(m_{\beta_i}, \beta_i) \in L$ $(i = 1, \ldots, q)$ establishes a basis of $L/L_{(0)}$.

Suppose the family $(\beta_i)_{1 \le i \le q}$ is F-dependent on M. Then

$$0 = \Big(\sum_{i=1}^{q} r_i \beta_i\Big)(m_{\beta_j}) = -\beta_j\Big(\sum_{i=1}^{q} r_i m_{\beta_i}\Big), \text{ for some } r_i \in F, \quad \forall j = 1, \ldots, q.$$

As a consequence, $g(\sum_{i=1}^q (r_i m_{\beta_i}, \beta_i), (m_{\beta_j}, \beta_j)) = \sum_{i=1}^q \beta_j(r_i m_{\beta_i}) = 0$. Since $(m_{\beta_j}, \beta_j)_{1\le i\le q}$ establishes a basis of $L/L_{(0)}$ and $L_{(0)} = \operatorname{rad} g$, we obtain $\sum_{i=1}^q (r_i m_{\beta_i}, \beta_i) \in \operatorname{rad} g = L_{(0)}$. But then $r_1 = \dots = r_q = 0$. This shows that the family $(\beta_i)_{1\le i\le q}$ is F-independent on M.

Consider the elements

$$v_{i,j} := (m_{\beta_i+\beta_j}, \beta_i + \beta_j) - (m_{\beta_i}, \beta_i) - (m_{\beta_j}, \beta_j),\ 1 \le i, j \le q.$$

As $m_{\beta_i+\beta_j} = m_{\beta_i} + m_{\beta_j}$, one has $v_{i,j} \in L_{(0)}$. Suppose

$$v := \sum_{1\le i,j\le q} r_{i,j} v_{i,j} \in L_{(1)}, \quad r_{i,j} \in F.$$

Then for any $w \in M$ one has

$$\begin{aligned} 0 &= -v \cdot w \\ &= \sum_{i,j} r_{i,j}(\beta_i(w) + \beta_j(w))(m_{\beta_i} + m_{\beta_j}) - \sum_{i,j} r_{i,j}\beta_i(w)m_{\beta_i} - \sum_{i,j} r_{i,j}\beta_j(w)m_{\beta_j} \\ &= \sum_{i,j} r_{i,j}\beta_i(w)m_{\beta_j} + \sum_{i,j} r_{i,j}\beta_j(w)m_{\beta_i}. \end{aligned}$$

Since $(m_{\beta_i})_{i=1,\dots,q}$ is linearly independent, this implies

$$\sum_i r_{i,k}\beta_i(w) + r_{k,i}\beta_i(w) = 0 \quad \forall w \in M, \quad \forall k = 1, \dots, q.$$

As a consquence, $r_{i,k} = -r_{k,i}$ for all i, k. Since $v_{i,k} = v_{k,i}$, this gives $2v = 0$ whence $v = 0$. As a result (recall that $q = \dim M$),

$$\dim L_{(0)}/L_{(1)} \ge \frac{q^2+q}{2} = \dim \mathfrak{sp}(M).$$

On the other hand, by definition there is an embedding $L_{(0)}/L_{(1)} \hookrightarrow \mathfrak{sp}(M)$ and this gives an isomorphism $L_{(0)}/L_{(1)} \cong \mathfrak{sp}(M)$ obtained by the natural action on $M \cong L/L_{(0)}$. In particular, $L/L_{(0)}$ is an irreducible $L_{(0)}$-module.

Let $(L_{(i)})_{i\ge -1}$ denote the standard filtration defined by $L_{(0)}$ and L:

$$L_{(-1)} := L, \quad L_{(i+1)} := \{x \in L_{(i)} \mid [L, x] \subset L_{(i)}\} \quad (i \ge 0).$$

Set $t := \dim T \ge q$. Note that there are exactly $p^t - 1$ roots each of dimension 1 (Corollary 19.3.4). Thus $\dim L = p^t - 1$. By definition $L_{(1)}/L_{(2)}$ embeds into $\operatorname{Hom}(L/L_{(0)}, L_{(0)}/L_{(1)})$ and hence

$$\begin{aligned} \dim L_{(2)} &= \dim L - (\dim L/L_{(0)} + \dim L_{(0)}/L_{(1)} + \dim L_{(1)}/L_{(2)}) \\ &\ge (p^t - 1) - \left(q + \frac{q^2+q}{2} + q\frac{q^2+q}{2}\right) \\ &\ge (p^q - 1) - \frac{1}{2}(q^3 + 2q^2 + 3q) \ge p^q - 3q^3 - \delta_{q,1} > 0. \end{aligned}$$

Therefore Theorem 5.2.7 applies and shows that there are embeddings

$$H(q;\underline{n})^{(3)} \subset \operatorname{gr} L \subset H(q;\underline{n}).$$

The Weak Recognition Theorem (Corollary 5.5.3) now gives $L \cong H(q;\underline{n};\Psi)^{(2)}$ for some $\underline{n}$ and Ψ. □

Remark 19.4.10. Let us combine these realizations of $L(G,0,f)$ with the one obtained by Equation (16.1.1). Choose accordingly an $\mathbb{F}_p$-basis $(\alpha_1,\dots,\alpha_m)$ of G, for $\alpha = \sum r_k\alpha_k \in G$ put $z^\alpha := \prod_{k=1}^m (1+x_k)^{r_k}$, and define

$$\Theta(u_\alpha) := \sum_{k=1}^m z^\alpha f(\alpha,\alpha_k) z_k\partial_k.$$

Put $M := \sum_{i=1}^m F z_k\partial_k$, which according to Theorem 16.1.5 is a torus of maximal toral rank in the minimal p-envelope of $\Theta(L(G,0,f))$. Observe that $\big(\sum_{k=1}^m f(\alpha,\alpha_k)z_k\partial_k\big)(z^\alpha) = \sum_{k=1}^m z^\alpha f(\alpha,\alpha_k) r_k = z^\alpha f(\alpha,\alpha) = 0$. One has (for $e \gg 0$)

$$\begin{aligned}\Theta(u_\alpha)^{[p]^e} - \big(\sum_{k=1}^m f(\alpha,\alpha_k)z_k\partial_k\big)^{[p]^e} &= \big(\Theta(u_\alpha) - (\sum_{k=1}^m f(\alpha,\alpha_k)z_k\partial_k)\big)^{[p]^e}\\ = \big(\sum_{k=1}^m (z^\alpha - 1) f(\alpha,\alpha_k)z_k\partial_k)\big)^{[p]^e} &= \sum_{k=1}^m (z^\alpha-1)^{p^e} f(\alpha,\alpha_k)^{p^e} z_k\partial_k = 0.\end{aligned}$$

Consequently, the semisimple part of $\Theta(u_\alpha)$ is $\tilde{u}_\alpha := \sum_{k=1}^m f(\alpha,\alpha_k)z_k\partial_k$. Thus the realization of $\Theta(u_\alpha)$ in $V(M,G)$ is

$$(\tilde{u}_\alpha,\alpha) = \Big(\sum_{k=1}^m f(\alpha,\alpha_k)z_k\partial_k, \alpha\Big).$$

The definition in Lemma 19.4.2 gives $L_{(0)} = \{\sum (m_\mu,\mu) \mid \sum m_\mu = 0\}$. In the above realization this condition means (setting $m_\mu = s_\mu \tilde{u}_\mu$ for $s_\mu \in F$)

$$0 = \sum_\mu m_\mu = \sum_\mu \sum_{k=1}^m s_\mu f(\mu,\alpha_k) z_k\partial_k,$$

i.e., $\sum_\mu s_\mu f(\mu,\alpha_k) = 0$ for all $k = 1,\dots,m$. As a consequence, the image of $L_{(0)}$ in $\Theta(L(G,0,f))$ is $\Theta(L(G,0,f)) \cap W(m;\underline{1})_{(0)}$. By our proof of Theorem 19.4.9 $L_{(0)}$ is a subalgebra of codimension q, where $L \cong H(q;\underline{n};\Psi)^{(2)}$. This means that $\Theta(L(G,0,f)) \cap W(m;\underline{1})_{(0)}$ is the unique subalgebra of maximal dimension in $\Theta(L(G,0,f))$.

We conclude the chapter by proving that the exposed algebras do indeed satisfy the assumption of Chapter 19 for every torus of maximal dimension.

Theorem 19.4.11. *Let L be one of*

$$S(m;\underline{n};\Phi(\tau))^{(1)},\ L(G,0,f).$$

For every torus T of maximal dimension in $L_{[p]}$ one has

$$L_{[p]} = L \oplus T,\ C_L(T) = \{0\}.$$

In particular, every T-1-section is nilpotent.

Proof. (1) As to $S(m;\underline{n};\Phi(\tau))^{(1)}$, Theorem 16.1.4 states

$$TR\big(S(m;\underline{n};\Phi(\tau))^{(1)}\big) = \sum n_i.$$

Theorem 7.1.3(1) gives

$$\dim \operatorname{Der} S(m;\underline{n};\Phi(\tau))^{(1)}/S(m;\underline{n};\Phi(\tau))^{(1)} = \sum n_i,$$

while Theorem 7.2.2(3) states

$$\operatorname{Der} S(m;\underline{n};\Phi(\tau))^{(1)} = S(m;\underline{n};\Phi(\tau))^{(1)}{}_{[p]}.$$

Let $T \subset S(m;\underline{n};\Phi(\tau))^{(1)}{}_{[p]}$ be a torus of maximal dimension, i.e., $\dim T = \sum n_i$. There is a natural embedding of restricted algebras

$$\varphi : S(m;\underline{n};\Phi(\tau))^{(1)}{}_{[p]} \hookrightarrow W(\textstyle\sum n_i;\underline{1})$$

induced by the isomorphism $\mathcal{O}(m;\underline{n}) \cong \mathcal{O}(\sum n_i;\underline{1})$. Note that $\varphi(T)$ is a torus of maximal dimension in $W(\sum n_i;\underline{1})$. Therefore $C_{W(\sum n_i;\underline{1})}(\varphi(T)) = \varphi(T)$ holds. We conclude

$$C_{S(m;\underline{n};\Phi(\tau))^{(1)}{}_{[p]}}(T) = T.$$

Since

$$S(m;\underline{n};\Phi(\tau))^{(1)}{}_{[p]} = S(m;\underline{n};\Phi(\tau))^{(1)} + C_{S(m;\underline{n};\Phi(\tau))^{(1)}{}_{[p]}}(T),$$

a dimension argument gives $C_{S(m;\underline{n};\Phi(\tau))^{(1)}}(T) = \{0\}$. Then every 1-section with respect to T is nilpotent.

(2) In order to treat the algebra $L = L(G,0,f)$, we refer to Theorem 16.1.5. That theorem states that there is a torus T' of maximal toral rank such that $L_{[p]} = L \oplus T'$ and $C_L(T') = \{0\}$. Now let T be any torus of maximal dimension in $L_{[p]}$. Theorem 1.5.5 shows that T' and T are Winter conjugate. Due to Theorem 1.5.1 then $T = C_{L_{[p]}}(T)$ holds. Then we argue as in the former case to prove the claim. □

Chapter 20

Attacking the general case

In Chapter 18 we solved the case where a torus of maximal dimension exists which is non-standard. A suitable 2-section of such a torus is of Melikian type (Lemma 18.1.5). Conversely, if T is a torus of maximal dimension and a 2-section $L(\alpha, \beta)$ is of Melikian type, then one can find a torus T_0 in $L(\alpha, \beta)_{[p]}$ which is non-standard on $L(\alpha, \beta)$, hence $T_0 + T \cap (\ker \alpha) \cap (\ker \beta)$ is a non-standard torus of maximal dimension in $L_{[p]}$. Theorem 18.3.8 then shows that L is Melikian. We have three equivalent descriptions:

- L is Melikian,
- $L_{[p]}$ contains a torus of maximal dimension which is non-standard,
- $L_{[p]}$ contains a torus of maximal dimension for which a 2-section is Melikian.

In Chapter 19 we treated the case when there is a torus of maximal dimension for which all roots are solvable. The algebras occurring in this case are exactly the Special algebras of the form $S(m; \underline{n}; \Phi(\tau))^{(1)}$ and the BLOCK algebras of type $L(G, 0, f)$.

We are now going to attack the remaining case that **every torus of maximal dimension is standard and has a non-solvable root**. Let me roughly indicate how to proceed. It turns out that not all tori are good for our purposes. As an example, the Witt algebra $W(1; \underline{1})$ has two essentially different tori of maximal dimension, namely $Fx\partial$ and $F(1 + x)\partial$. The first torus respects the natural maximal subalgebra $\sum_{i>0} Fx^i \partial$, while the second does not. Using Definition 11.2.2 and Theorem 11.2.5, the first is a proper torus. In a first step we intend to show that one can choose a torus T of maximal dimension which respects the maximal subalgebras $Q(\alpha)$ for all T-1-sections $L(\alpha)$ simultaneously. Then we will show that for such a torus the sum of all subalgebras $Q(\alpha)$ is a maximal T-invariant subalgebra $Q(L, T)$ of L. All these items can be proved inside 2-sections $L(\alpha, \beta)$, and even better inside the T-semisimple quotients $L[\alpha, \beta]$. The subalgebra $Q(L, T)$ defines an exhausting and separating T-invariant filtration of L. It turns out that the associated graded algebra contains a unique minimal T-invariant ideal $\mathfrak{A}(L, T)$ and the semisimple quotient $\mathfrak{S}(L, T) = \mathfrak{A}(L, T)/\operatorname{rad}\mathfrak{A}(L, T)$ is a simple algebra. As a main step we will show that $\mathfrak{S}(L, T)$ is restricted of Cartan type with natural grading. From this we eventually construct a filtration of L closely connected with the natural filtration of $\mathfrak{S}(L, T)$, and thereby prove that L is Cartan type as well. In fact, it will turn out that L and $\mathfrak{S}(L, T)$ are of the same type W, S, H, or K.

20.1 Optimal tori

Since no 2-section is of Melikian type in the present case, Theorems 17.4.1, 17.4.3, and 17.4.4 (and Theorem 17.1.8(1) for the final case of the next theorem) show that the 2-sections are ruled by the following theorem.

Theorem 20.1.1. *One of the following cases occurs.*

(1) $L[\alpha,\beta] = \{0\}$.

(2) $L[\alpha,\beta] = L[\kappa]$ *for some* $\kappa \in (\mathbb{F}_p\alpha + \mathbb{F}_p\beta)\setminus\{0\}$. *Moreover,* $\dim \Psi_{\alpha,\beta}(T) = 1$.

(3) *There are* $\mu_1, \mu_2 \in \Gamma(L,T)$ *such that*

$$L[\mu_1]^{(1)} \oplus L[\mu_2]^{(1)} \subset L[\alpha,\beta] + \Psi_{\alpha,\beta}(T) \subset L[\mu_1] \oplus L[\mu_2].$$

Moreover, $\Psi_{\alpha,\beta}(T) = Fh_1 \oplus Fh_2$ *with* $h_i \in L[\mu_i]^{(1)}$.

(4) $L[\alpha,\beta]$ *is one of the simple classical algebras* A_2, B_2, G_2.

(5) $L[\alpha,\beta]$ *is one of the restricted Cartan type algebras* $W(2;\underline{1})$, $S(3;\underline{1})^{(1)}$, $H(4;\underline{1})^{(1)}$, $H(4;\underline{1})^{(1)} + FD_H(x_1^{p-1}\cdots x_4^{p-1})$, $K(3;\underline{1})$.

(6) $L[\alpha,\beta]$ *is one of the simple non-restricted Cartan type algebras* $W(1;\underline{2})$, $H(2;\underline{1};\Phi(\tau))^{(1)}$, $H(2;\underline{1};\Phi(1))$.

(7) $H(2;(1,2))^{(2)} \subset L[\alpha,\beta] \subset H(2;(1,2))$.

(8) $H(2;\underline{1})^{(2)} \subset L[\alpha,\beta] = H(2;\underline{1})^{(1)}$*; moreover,*

$$\Psi_{\alpha,\beta}(T) = F(\delta_1 + x_1)\partial_1 \oplus F(\delta_2 + x_2)\partial_2,\ \delta_i \in \{0,1\}.$$

(9) $\mathfrak{s} \otimes \mathcal{O}(1;\underline{1}) \subset L[\alpha,\beta] \subset (\operatorname{Der}\mathfrak{s})^{(1)} \otimes \mathcal{O}(1;\underline{1})$, *where* $\mathfrak{s}$ *is one of* $\mathfrak{sl}(2)$, $W(1;\underline{1})$, $H(2;\underline{1})^{(2)}$*; moreover,*

$$\Psi_{\alpha,\beta}(T) = (Fh \otimes 1) \oplus (F\operatorname{Id} \otimes (1+x_1)\partial_1)$$

where h *is a nonzero toral element in* $\mathfrak{s}$.

(10) $\mathfrak{s} \otimes \mathcal{O}(1;\underline{1}) \subset L[\alpha,\beta] \subset ((\operatorname{Der}\mathfrak{s})^{(1)} \otimes \mathcal{O}(1;\underline{1})) \rtimes (\operatorname{Id} \otimes W(1;\underline{1}))$, *where* $\mathfrak{s}$ *is one of* $\mathfrak{sl}(2)$, $W(1;\underline{1})$, $H(2;\underline{1})^{(2)}$*; there is a Witt root* $\kappa \in \mathbb{F}_p\alpha + \mathbb{F}_p\beta$ *such that*

$$\pi_2(L[\alpha,\beta]) \cong L[\kappa] = W(1;\underline{1});$$

moreover,

$$\Psi_{\alpha,\beta}(T) = (Fh \otimes 1) \oplus F(d \otimes 1 + \operatorname{Id} \otimes t_0),$$

where $h \in \mathfrak{s}$, $d \in \operatorname{Der}\mathfrak{s}$ *are toral; in addition,* $d = 0$ *and* $t_0 = (1+x_1)\partial_1$, *or* $t_0 = x_1\partial_1$*; the root* κ *satisfies* $\kappa(h \otimes 1) = 0$.

(11) $\mathfrak{s} \otimes \mathcal{O}(2;\underline{1}) \subset L[\alpha,\beta] \subset ((\operatorname{Der}\mathfrak{s})^{(1)} \otimes \mathcal{O}(2;\underline{1})) \rtimes (\operatorname{Id} \otimes W(2;\underline{1}))$, *where* $\mathfrak{s}$ *is one of* $\mathfrak{sl}(2)$, $W(1;\underline{1})$, $H(2;\underline{1})^{(2)}$*; there is a Hamiltonian root* $\kappa \in \mathbb{F}_p\alpha + \mathbb{F}_p\beta$ *such that*

$$H(2;1)^{(2)} \subset \pi_2(L[\alpha,\beta]) \cong L[\kappa] \subset H(2;\underline{1})^{(1)};$$

moreover,

$$\Psi_{\alpha,\beta}(T) = (Fh \otimes 1) \oplus F(d \otimes 1 + \mathrm{Id} \otimes t_0),$$

where $h \in \mathfrak{s}$, $d \in \mathrm{Der}\,\mathfrak{s}$ *are toral; in addition,* $d = 0$ *and* $t_0 = D_H((1+x_1)x_2)$, *or* $t_0 = D_H(x_1x_2)$*; the root* κ *satisfies* $\kappa(h \otimes 1) = 0$. □

Let $\alpha, \beta \in \Gamma(L, T)$ be arbitrary nonzero. Since $(\mathrm{rad}_T\, L(\alpha, \beta)) \cap L(\alpha) \subset \mathrm{rad}\, L(\alpha)$, there are surjective homomorphisms

$$L(\alpha) \overset{\Psi_{\alpha,\beta}}{\twoheadrightarrow} L[\alpha, \beta](\alpha) \overset{\sigma_{\alpha,\beta}}{\twoheadrightarrow} L[\alpha], \quad \sigma_{\alpha,\beta} \circ \Psi_{\alpha,\beta}|_{L(\alpha)} = \Psi_\alpha. \tag{20.1.1}$$

Recall the definition of the maximum compositionally classical subalgebra $Q(\alpha)$ introduced in Definition 17.1.9. In order to simplify notation we set

$$\bar{Q}(\alpha) := \Psi_{\alpha,\beta}(Q(\alpha))$$

(keeping in mind that this depends on the choice of β). It is clear from the definition and (20.1.1) that $\bar{Q}(\alpha)$ is the maximum compositionally classical subalgebra $Q(L[\alpha, \beta](\alpha))$ of $L[\alpha, \beta](\alpha)$.

Recall that α is proper if and only if $[T, Q(\alpha)] \subset Q(\alpha)$. Looking at $\Psi_\alpha(L(\alpha))$ it is easy to see that this implies $H \subset Q(\alpha)$. Since $(\ker\, \Psi_{\alpha,\beta}) \cap L(\alpha) \subset \mathrm{rad}\, L(\alpha) \subset Q(\alpha)$, then

$$\alpha \in \Gamma(L, T) \text{ is proper if and only if } [\Psi_{\alpha,\beta}(T), \bar{Q}(\alpha)] \subset \bar{Q}(\alpha).$$

To unify notation we consider the root 0 a proper root. Keeping this in mind we may identify proper roots on $L[\alpha, \beta]$ with the respective roots on $L(\alpha, \beta)$. However, a proper root on $L(\alpha, \beta)$ might not be a root on $L[\alpha, \beta]$ at all. Set

$$\Gamma_{prop}(\cdot, \cdot) \text{ the set of proper roots and } \Gamma_{imp}(\cdot, \cdot) := \Gamma(\cdot, \cdot) \setminus \Gamma_{prop}(\cdot, \cdot).$$

Recall that the 1-sections are ruled by Theorem 17.1.8.

Lemma 20.1.2. *The mapping*

$$\begin{aligned} {}^{-} : \Gamma_{imp}(L(\alpha, \beta), T) &\to \Gamma_{imp}\big(L[\alpha, \beta], \Psi_{\alpha,\beta}(T)\big), \\ \bar{\mu}(\Psi_{\alpha,\beta}(t)) &:= \mu(t) \quad \forall t \in T \end{aligned}$$

is well defined and bijective.

Proof. Let $\mu \in \Gamma_{imp}(L(\alpha, \beta), T)$. Then $L[\mu] \cong W(1; \underline{1})$ or $H(2; \underline{1})^{(2)} \subset L[\mu] \subset H(2; \underline{1})$. In both cases one has $L[\mu]_\mu \neq \{0\}$. This implies that $\Psi_{\alpha,\beta}(L_\mu) \neq \{0\}$. Then $\mu(T \cap (\ker\, \Psi_{\alpha,\beta})) = 0$ holds, so $\bar{\mu}$ is well-defined and $\bar{\mu} \in \Gamma(L[\alpha, \beta], \Psi_{\alpha,\beta}(T))$ is a nonzero root. We mentioned above that $\bar{\mu}$ is not proper, hence it is improper.

If $\mu^\times \in \Gamma_{imp}(L[\alpha, \beta], \Psi_{\alpha,\beta}(T))$ is any root, then there is a nonzero root $\mu \in \Gamma(L(\alpha, \beta), T)$ satisfying $\mu(T \cap (\ker\, \Psi_{\alpha,\beta})) = 0$, so that $\bar{\mu}$ is defined and coincides with $\mu^\times$. We mentioned above that μ is not proper, hence it is improper.

Finally, let $\kappa, \mu \in \Gamma_{imp}(L(\alpha,\beta),T)$. The above shows that $\bar{\kappa}$, $\bar{\mu}$ are roots. If $\bar{\kappa} = \bar{\mu}$, then $T = T \cap (\ker(\kappa - \mu)) + T \cap (\ker \Psi_{\alpha,\beta})$. As κ, μ vanish on $T \cap (\ker \Psi_{\alpha,\beta})$, this means $\kappa - \mu = 0$. Hence the mapping is injective. □

Next we want describe proper roots in the semisimple quotients $L[\alpha,\beta]$ of some 2-sections $L(\alpha,\beta)$ as described in Theorem 20.1.1.

Lemma 20.1.3. *Let $H(2;\underline{1})^{(2)} \subset \mathfrak{g} \subset H(2;\underline{1})^{(1)}$ and $\mathfrak{t} \subset \operatorname{Der}\mathfrak{g}$ a 2-dimensional torus.*

(1) *There are (p^2-1) $\mathfrak{t}$-root spaces $H(2;\underline{1})^{(1)}{}_{\mu}$, all these are 1-dimensional. In particular, $C_{\mathfrak{g}}(\mathfrak{t}) = \mathfrak{t} \cap \mathfrak{g}$ holds.*
(2) *For any nonzero root μ one has*

$$Q(\mathfrak{g}(\mu)) = \mathfrak{g}(\mu) \cap \big(W(2;\underline{1})_{(0)} + \mathfrak{t} \cap (\ker \mu)\big).$$

(3) *Suppose $\mathfrak{t} = Fx_1\partial_1 \oplus Fx_2\partial_2$. Define $\epsilon_i(x_j\partial_j) = \delta_{i,j}$, $(i,j=1,2)$. The roots*

$$\mathbb{F}_p^*\epsilon_1 \cup \mathbb{F}_p^*\epsilon_2 \text{ are Witt,}$$

the roots

$$\mathbb{F}_p^*(\epsilon_2 - \epsilon_1) \text{ are classical,}$$

and all other roots are solvable. Moreover, $\operatorname{rad}\mathfrak{g}(\mu) \subset \mathfrak{t} + \mathfrak{g}_{(1)}$ *holds for all nonzero roots.*

Proof. (1) Due to Theorem 7.5.5(2) we may normalize $\mathfrak{t} = F(\delta_1 + x_1)\partial_1 \oplus F(\delta_2 + x_2)\partial_2$ with $\delta_i \in \{0,1\}$. Set $z_i := \delta_i + x_i$. Let μ be any root, set $r := \mu(z_1\partial_1)$, $s := \mu(z_2\partial_2)$. One computes

$$H(2;\underline{1})^{(1)}{}_{\mu} = FD_H(z_1^a z_2^b) \text{ where } a \equiv r+1,\ b \equiv s+1 \pmod{(p)}.$$

So $H(2;\underline{1})^{(1)}{}_{\mu}$ is nonzero if and only if $(a,b) \neq (0,0)$, and if so, then this space is 1-dimensional. In particular, $C_{\mathfrak{g}}(\mathfrak{t}) = \mathfrak{t} \cap \mathfrak{g} = FD_H(z_1z_2)$ holds.

(2) Let μ be nonzero, set $r := \mu(z_1\partial_1)$, $s := \mu(z_2\partial_2)$, and let $\mathfrak{t} \cap (\ker \mu) = Ft_{\mu}$ with $t_{\mu} := rz_2\partial_2 - sz_1\partial_1$, where $r, s \in \mathbb{F}_p$. If $\delta_1 \neq 0$ or $\delta_2 \neq 0$, we may then normalize $r = 0$. As before,

$$\mathfrak{g}(\mu) = \sum_{i=0}^{p-1} FD_H(z_1^{a_i} z_2^{b_i}) + FD, \qquad a_i \equiv ir + 1, \quad b_i \equiv is + 1 \pmod{(p)},$$

where $D = 0$ if $\mathfrak{g} = H(2;\underline{1})^{(2)}$ or $r \neq s$, and $D = D_H(z_1^{p-1} z_2^{p-1})$ otherwise.

If $r = 0, s \neq 0$, then $\mathfrak{g}(\mu) = \sum_{i=0}^{p-1} FD_H(z_1 z_2^i)$. In this case $\mathfrak{g}(\mu) \cong W(1;\underline{1})$ under the isomorphism $D_H(z_1 z_2^i) \mapsto (\delta_2 + x)^i \partial$. Hence $Q(\mathfrak{g}(\mu)) = \sum_{i=1}^{p-1} FD_H(z_1 x_2^i) = H(2;\underline{1})^{(1)}{}_{(0)} \cap \mathfrak{g}(\mu) + FD_H(z_1 x_2)$. As $D_H(z_1 x_2) = x_2\partial_2 - z_1\partial_1 = x_2\partial_2 + s^{-1}t_\mu$, we conclude $Q(\mathfrak{g}(\mu)) \subset \mathfrak{g}(\mu) \cap (W(2;\underline{1})_{(0)} + \mathfrak{t} \cap (\ker \mu))$. In addition, the latter space does not contain $D_H(z_1) = \partial_2$. So we have equality. The case $s = 0, r \neq 0$ is similar.

Next consider the case $rs \neq 0$, whence we may assume in addition $\delta_1 = \delta_2 = 0$. It is not hard to prove that $\mathfrak{g}(\mu) = C_{\mathfrak{g}}(t_\mu) \subset W(2;\underline{1})_{(0)}$, that the case $r + s = 0$ gives a classical root, and that all other 1-sections are solvable. Then $Q(\mathfrak{g}(\mu)) = \mathfrak{g}(\mu) = \mathfrak{g}(\mu) \cap W(2;\underline{1})_{(0)}$ holds. We finally mention that in the present case $\mathfrak{t} \subset W(2;\underline{1})_{(0)}$.

(3) Suppose $\mathfrak{t} = Fx_1\partial_1 \oplus Fx_2\partial_2$ and set $\epsilon_i(x_j\partial_j) = \delta_{i,j}$, $(i, j = 1, 2)$. We have already shown in (2) that ϵ_1, ϵ_2 are Witt roots, $\epsilon_2 - \epsilon_1$ is classical, and all roots $s\epsilon_2 + r\epsilon_1$ with $rs \neq 0$, $r - s \neq 0$ are solvable.

It remains to prove the statement on the radical. Note that $\mathfrak{g}(\epsilon_i) \cong W(1;\underline{1})$ is p-dimensional. Hence the radical vanishes. So consider the case $rs \neq 0$. If $r + s = 0$, then $\mathfrak{g}(\mu)/\operatorname{rad}\mathfrak{g}(\mu) \cong \mathfrak{sl}(2) \cong H(2;\underline{1})^{(2)}{}_{(0)}/H(2;\underline{1})^{(2)}{}_{(1)}$. The claim follows. Finally consider the case $rs \neq 0$, $r + s \neq 0$. As in (2) we have $\mathfrak{g}(\mu) = \sum_{i=0}^{p-1} FD_H(x_1^{a_i} x_2^{b_i}) + FD$, where $a_i \equiv ir + 1\ (p)$, $b_i \equiv is + 1\ (p)$. One has

$$\mathfrak{g}(\mu) \subset W(2;\underline{1})_{(1)} + \sum_{a_i + b_i = 2} FD_H(x_1^{a_i} x_2^{b_i}).$$

So we are only interested in summands with $(a_i, b_i) \in \{(2,0), (1,1), (0,2)\}$. The cases $(a_i, b_i) \in \{(2,0), (0,2)\}$ contradict the assumption $r + s \neq 0$, the case $(a_i, b_i) = (1,1)$ gives an element in $\mathfrak{t}$. □

Lemma 20.1.4.

(1) *Let $\mathfrak{g}$ be one of the simple restricted Cartan type algebras $W(2;\underline{1})$, $S(3;\underline{1})^{(1)}$, $H(4;\underline{1})^{(1)}$, $K(3;\underline{1})$, and $\mathfrak{t} \subset \mathfrak{g}$ a 2-dimensional torus. A nonzero root $\mu \in \Gamma(\mathfrak{g}, \mathfrak{t})$ is proper if and only if $\mu(\mathfrak{t} \cap \mathfrak{g}_{(0)}) \neq 0$.*

(2) *Let $\mathfrak{g} = H(2;\underline{1};\Phi(1))$ and $\mathfrak{t} \subset \mathfrak{g}_{[p]}$ a 2-dimensional torus, or $\mathfrak{g} = L[\alpha,\beta]$ and $\mathfrak{t} = \Psi_{\alpha,\beta}(T)$ as in case (8) of Theorem 20.1.1. A nonzero root $\mu \in \Gamma(\mathfrak{g}, \mathfrak{t})$ is proper if and only if $\mu(\mathfrak{t} \cap \mathfrak{g}) = 0$ (in which case μ is solvable) or $\mu(\mathfrak{t} \cap W(2;\underline{1})_{(0)}) \neq 0$.*

Proof. (1) Suppose $\mu \neq 0$ is proper. We mentioned that this implies $C_{\mathfrak{g}}(\mathfrak{t}) \subset Q(\mathfrak{g}(\mu))$. As $\mathfrak{t} \subset \mathfrak{g}$, this gives $\mathfrak{t} \subset Q(\mathfrak{g}(\mu))$. Due to Corollary 17.2.10 one has $Q(\mathfrak{g}(\mu)) = \mathfrak{g}_{(0)} \cap \mathfrak{g}(\mu) + \mathfrak{t} \cap (\ker \mu)$. Thus $\mathfrak{t} = (\mathfrak{t} \cap \mathfrak{g}_{(0)}) + \mathfrak{t} \cap (\ker \mu)$, whence $\mu(\mathfrak{t} \cap \mathfrak{g}_{(0)}) \neq 0$.

Conversely, if $\mu(\mathfrak{t} \cap \mathfrak{g}_{(0)}) \neq 0$, then $\mathfrak{t} = \mathfrak{t} \cap \mathfrak{g}_{(0)} + \mathfrak{t} \cap (\ker \mu)$, and hence $\mathfrak{g}_{(0)} \cap \mathfrak{g}(\mu) + \mathfrak{t} \cap (\ker \mu)$ is $\mathfrak{t}$-invariant. This is the proof of claim (1).

(2) Note that $C_{\mathfrak{g}}(\mathfrak{t}) = \mathfrak{t} \cap \mathfrak{g}$ and $Q(\mathfrak{g}(\mu)) = \mathfrak{g}(\mu) \cap \big(W(2;\underline{1})_{(0)} + \mathfrak{t} \cap (\ker \mu)\big)$ holds in both cases by Theorem 10.4.6, Corollary 17.2.10, and Lemma 20.1.3. We argue as in (1) to obtain that $\mathfrak{t} \cap \mathfrak{g} \subset Q(\mathfrak{g}(\mu))$ if μ is proper. Therefore we then have

$$\mu(\mathfrak{t} \cap \mathfrak{g}) = 0 \text{ or } \mathfrak{t} = \mathfrak{t} \cap \mathfrak{g} + \mathfrak{t} \cap (\ker \mu) = \mathfrak{t} \cap W(2;\underline{1})_{(0)} + \mathfrak{t} \cap (\ker \mu).$$

Conversely, if $\mu\big(\mathfrak{t} \cap \mathfrak{g}\big) = 0$, then $\mu\big(C_{\mathfrak{g}}(\mathfrak{t})\big) = 0$ and therefore μ is solvable. In this case μ is proper. If $\mu(\mathfrak{t} \cap W(2;\underline{1})_{(0)}) \neq 0$, then we obtain $\mathfrak{t} = \mathfrak{t} \cap W(2;\underline{1})_{(0)} + \mathfrak{t} \cap (\ker \mu)$ and therefore

$$\begin{aligned}[\mathfrak{t}, Q(\mathfrak{g}(\mu))] &= [\mathfrak{t} \cap W(2;\underline{1})_{(0)} + \mathfrak{t} \cap (\ker \mu), \mathfrak{g}(\mu) \cap \big(W(2;\underline{1})_{(0)} + \mathfrak{t} \cap (\ker \mu)\big)] \\ &\subset \mathfrak{g}(\mu) \cap W(2;\underline{1})_{(0)} \subset Q(\mathfrak{g}(\mu)).\end{aligned}$$

This means that μ is proper. □

Lemma 20.1.5.

(1) *Let $\mathfrak{g}$ be one of*

$$W(2;\underline{1}),\ S(3;\underline{1})^{(1)},\ H(4;\underline{1})^{(1)},\ H(4;\underline{1})^{(1)} + FD_H(x_1^{p-1}\cdots x_4^{p-1}),\ K(3;\underline{1})$$

and $\mathfrak{t} \subset \mathfrak{g}$ a 2-dimensional torus. Put $\mathfrak{g}_{(0)}$ the natural subalgebra of maximal dimension. If $\mathfrak{t} \subset \mathfrak{g}_{(0)}$, then all roots are proper. If $\mathfrak{t} \cap \mathfrak{g}_{(0)}$ is 1-dimensional, then $\Gamma_{imp}(\mathfrak{g}, \mathfrak{t}) = \mathbb{F}_p^* \kappa$ for some nonzero root κ. This root line is given by $\kappa(\mathfrak{t} \cap \mathfrak{g}_{(0)}) = 0$. If $\mathfrak{t} \cap \mathfrak{g}_{(0)} = \{0\}$, then all nonzero roots are improper.

(2) *Suppose $\mathfrak{g} \cong W(1;\underline{2})$ and $\mathfrak{t}$ is a 2-dimensional torus in $\mathfrak{g}_{[p]}$. If $\mathfrak{t} \cap \mathfrak{g}_{(0)} \neq \{0\}$, then all roots are proper. Otherwise all nonzero roots are improper.*

(3) *If $\mathfrak{g} = H(2;\underline{1};\Phi(\tau))^{(1)}$ and $\mathfrak{t}$ is a 2-dimensional torus in $\mathfrak{g}_{[p]}$, then all roots are solvable and hence proper.*

(4) *Let $\mathfrak{g} = H(2;\underline{1};\Phi(1))$ and $\mathfrak{t} \subset \mathfrak{g}_{[p]}$ a 2-dimensional torus, or $\mathfrak{g} = L[\alpha,\beta]$ and $\mathfrak{t} = \Psi_{\alpha,\beta}(T)$ as in case (8) of Theorem 20.1.1. If $\mathfrak{t} \subset W(2;\underline{1})_{(0)}$, then all roots are proper. If $\dim \mathfrak{t} \cap W(2;\underline{1})_{(0)} = 1$, then $\Gamma_{imp}(\mathfrak{g}, \mathfrak{t}) = \mathbb{F}_p^* \kappa$ where $\kappa \neq 0$ and $\kappa\big(\mathfrak{t} \cap W(2;\underline{1})_{(0)}\big) = 0$. If $\mathfrak{t} \cap W(2;\underline{1})_{(0)} = \{0\}$, then $\Gamma_{prop}(\mathfrak{g}, \mathfrak{t}) = \mathbb{F}_p \kappa$ where $\kappa\big(\mathfrak{t} \cap \mathfrak{g}\big) = 0$ and κ is solvable. Here we allow $\kappa = 0$.*

(5) *Suppose $H(2;(1,2))^{(2)} \subset \mathfrak{g} \subset H(2;(1,2))$ and $\mathfrak{t}$ is a 2-dimensional torus in $\mathfrak{g}_{[p]}$. If $\mathfrak{t} \cap H(2;(1,2))_{(0)} \neq \{0\}$, then all roots are proper. Otherwise $\Gamma_{prop}(\mathfrak{g}, \mathfrak{t}) = \mathbb{F}_p \kappa$ where $\kappa(\mathfrak{t} \cap \mathfrak{g}) = 0$ and κ is solvable. Here we allow $\kappa = 0$.*

Proof. (1) is a direct consequence of Lemma 20.1.4(1) (recall that $\mathfrak{t} \cap \mathfrak{g}_{(0)}$ is a torus and a root on $H(4;\underline{1})^{(1)} + FD_H(x_1^{p-1}\cdots x_4^{p-1})$ is proper if and only if it is on $H(4;\underline{1})^{(1)}$).

(2) Suppose $\mathfrak{g} \cong W(1;\underline{2})$. In case of Theorem 7.6.2 all roots are improper and in case of Theorem 7.6.5 all roots are proper.

(3) If $\mathfrak{g} = H(2;\underline{1};\Phi(\tau))^{(1)}$, then all roots are solvable and therefore proper (Theorem 10.3.2(3)).

(4) In this case Lemma 20.1.4(2) yields that a nonzero root μ is proper if and only if μ does vanish on $\mathfrak{t} \cap \mathfrak{g}$ or does not vanish on $\mathfrak{t} \cap W(2;\underline{1})_{(0)}$.

If $\mathfrak{t} \subset W(2;\underline{1})_{(0)}$, then no nonzero root does vanish on this space.

Suppose $\mathfrak{t}_0 := \mathfrak{t} \cap W(2;\underline{1})_{(0)}$ is 1-dimensional. This is a subtorus and there is a nonzero root κ which vanishes on $\mathfrak{t}_0$. This is the only root direction which is possibly improper. Note that we may normalize in both cases under consideration $\mathfrak{t} = F(\delta_1 + x_1)\partial_1 \oplus F(\delta_2 + x_2)\partial_2$ (applying an automorphism of $W(2;\underline{1})$ and $\operatorname{Der} H(2;\underline{1})^{(2)}$, respectively). In the present case we may also assume $\delta_1 = 0, \delta_2 = 1$. Then κ is given by $\kappa(x_1\partial_1) = 0$. In case $\mathfrak{g} = H(2;\underline{1};\Phi(1))$ Proposition 16.2.2(1) shows that $x_1\partial_1 \notin \mathfrak{g}$, whence $\kappa(\mathfrak{t} \cap \mathfrak{g}) \neq 0$. In case $\mathfrak{g} = H(2;\underline{1})^{(2)}$ one has $\mathfrak{t} \cap \mathfrak{g} = F(x_1\partial_1 - (1 + x_2)\partial_2)$ and again $\kappa(\mathfrak{g} \cap \mathfrak{t}) \neq 0$. So, indeed, κ is an improper root.

If $\mathfrak{t} \cap W(2;\underline{1})_{(0)} = \{0\}$, then a root κ is proper if and only if $\kappa(\mathfrak{t} \cap \mathfrak{g}) = 0$. If $\mathfrak{t} \cap \mathfrak{g}$ is a torus, then there is at most 1 such root direction, and if $\mathfrak{t} \cap \mathfrak{g}$ is not a torus, then only $\kappa = 0$ is possible.

(5) Suppose $H(2;(1,2))^{(2)} \subset \mathfrak{g} \subset H(2;(1,2))$. Note that

$$H(2;(1,2))_{[p]}/H(2;(1,2))^{(2)}{}_{[p]}$$

is $[p]$-nilpotent. Therefore $\mathfrak{t} \subset H(2;(1,2))^{(2)}{}_{[p]}$ holds. Lemma 10.2.3 shows $\dim \mathfrak{t} \cap \mathfrak{g} = 1$.

If $\mathfrak{t} \cap H(2;(1,2))_{(0)} \neq \{0\}$, then all roots are proper (Theorem 10.2.16). If $\mathfrak{t} \cap H(2;(1,2))$ is a torus not contained in $H(2;(1,2))_{(0)}$, then Theorem 10.2.14 applies and shows that all roots except the direction $\mathbb{F}_p\kappa$ given by $\kappa(\mathfrak{t} \cap H(2;(1,2))) = 0$ are improper, and κ is solvable. If $\mathfrak{t} \cap H(2;(1,2))$ is not a torus, then all nonzero roots are improper (Corollary 10.2.9). Note that $\mu(\mathfrak{t} \cap H(2;(1,2))) \neq 0$ holds for all nonzero μ and $\mathfrak{t} \cap H(2;(1,2))_{(0)} = \{0\}$ in this case. These 3 cases cover all possilities. □

Lemma 20.1.6. *Suppose $\mathfrak{g} = L[\alpha,\beta]$ and $\mathfrak{t} = \Psi_{\alpha,\beta}(T)$ are as in case (9) of Theorem 20.1.1. If Fh is a proper torus of $\mathfrak{s}$, then all roots are proper. If Fh is an improper torus of $\mathfrak{s}$, then $\Gamma_{prop}(\mathfrak{g},\mathfrak{t}) = \mathbb{F}_p\kappa$ where $\kappa(h \otimes 1) = 0$ and κ is solvable.*

Proof. Let μ be any nonzero root and set $\bar{\mu} := \mu(h \otimes 1), \tilde{\mu} := \mu(\mathrm{Id} \otimes (1 + x_1)\partial_1) \in \mathbb{F}_p$. Then $\mathfrak{g}_\mu = \mathfrak{s}_{\bar{\mu}} \otimes (1 + x_1)^{\tilde{\mu}}$ and $\mathfrak{g}(\mu) = \sum_{i=1}^{p-1} \mathfrak{s}_{i\bar{\mu}} \otimes (1 + x_1)^{i\tilde{\mu}} + C_{\mathfrak{g}}(\mathfrak{t})$. From this it is clear that $\mathfrak{g}(\mu)^{(\infty)} = \sum_{i=0}^{p-1} \mathfrak{s}_{i\bar{\mu}} \otimes (1 + x_1)^{i\tilde{\mu}} \cong \sum_{i=0}^{p-1} \mathfrak{s}_{i\bar{\mu}}$ if $\mu(h \otimes 1) \neq 0$, and $\mathfrak{g}(\mu)^{(\infty)} = \{0\}$ if $\mu(h \otimes 1) = 0$. So if Fh is a proper torus of $\mathfrak{s}$, then all roots are proper, and if Fh is an improper torus of $\mathfrak{s}$, then all roots satisfying $\mu(h \otimes 1) \neq 0$ are improper. The root κ satisfying $\kappa(h \otimes 1) = 0$ is solvable, and thus proper. □

Finally we consider cases (10) and (11) of Theorem 20.1.1 and $\mathfrak{t} := \Psi_{\alpha,\beta}(T)$. With the notation there we have

$$\mathfrak{s} \in \{\mathfrak{sl}(2), W(1;\underline{1}), H(2;\underline{1})^{(2)}\}, \ \tilde{S} = \mathfrak{s} \otimes \mathcal{O}(m;\underline{1}), \ m = 1, 2,$$

and $\mathfrak{t} = Fh \otimes 1 + F(d \otimes 1 + \mathrm{Id} \otimes t_0)$ with $t_0 = (\delta + x_1)\partial_1$ in case (10) and $t_0 = D_H((\delta + x_1)x_2)$ in case (11), where $\delta \in \{0, 1\}$ and $d \in \mathrm{Der}\,\mathfrak{s}$ is toral and $d = 0$ if $\delta = 1$. If $\mathfrak{s} \not\cong H(2;\underline{1})^{(2)}$ then $\mathrm{Der}\,\mathfrak{s} = \mathfrak{s}$ and again we may take $d = 0$.

The root direction $\mathbb{F}_p\kappa$ is defined by $\kappa(\mathfrak{t} \cap \tilde{S}) = 0$. Observe that $\delta = 0$ if and only if $[\mathfrak{t}, \mathrm{rad}\,\tilde{S}] \subset \mathrm{rad}\,\tilde{S}$.

Note that $\tilde{S}/\mathrm{rad}\,\tilde{S} \cong \mathfrak{s}$, and $\mathfrak{t} \cap \tilde{S} = Fh \otimes 1$ is a maximal torus of $\tilde{S}$. Therefore we speak of $\mathfrak{t} \cap \tilde{S}$ being a proper or improper torus of $\tilde{S}$ (cf. Definition 11.2.2).

Lemma 20.1.7. *Let $L[\alpha, \beta]$ be as in cases* (10) *and* (11) *of Theorem 20.1.1 and* $\mathfrak{t} = \Psi_{\alpha,\beta}(T)$.

(1) *Suppose* $[\mathfrak{t}, \mathrm{rad}\,\tilde{S}] \not\subset \mathrm{rad}\,\tilde{S}$. *Then κ is improper. If $\mathfrak{t} \cap \tilde{S}$ is an improper torus of $\tilde{S}$, then all nonzero roots are improper. Otherwise* $\Gamma_{imp}(L[\alpha, \beta], \mathfrak{t}) = \mathbb{F}_p^*\kappa$.

(2) *Suppose* $[\mathfrak{t}, \mathrm{rad}\,\tilde{S}] \subset \mathrm{rad}\,\tilde{S}$. *Then κ is proper. The $\mathfrak{t}$-invariant homomorphism $\tilde{S} \to \tilde{S}/\mathrm{rad}\,\tilde{S} \cong \mathfrak{s}$ induces a bijection $\Gamma_{imp}(\mathfrak{g}, \mathfrak{t}) \xrightarrow{\sim} \Gamma_{imp}(\mathfrak{s}, \mathfrak{t})$. In particular, Fh is a proper torus of $\mathfrak{s}$ if and only if* $\Gamma_{imp}(L[\alpha, \beta], \mathfrak{t}) = \emptyset$.

Proof. (1) Suppose $[\mathfrak{t}, \mathrm{rad}\,\tilde{S}] \not\subset \mathrm{rad}\,\tilde{S}$, i.e., $\delta = 1$. Recall that by assumption $d = 0$. In case (10) $L[\kappa] \cong W(1;\underline{1})$ holds and the torus acts on this as $Ft_0 = F(1 + x_1)\partial_1$. So κ is improper. One argues similarly for case (11). Let μ be any root satisfying $\mu(h \otimes 1) \neq 0$. Then $L[\alpha, \beta](\mu) = \tilde{S}(\mu) + C_{L[\alpha,\beta]}(\mathfrak{t})$, and $\mathfrak{t} = Fh \otimes 1 + \mathfrak{t} \cap (\ker \mu)$. This shows that μ is proper if and only if μ is proper when restricted to $\tilde{S}$ and $Fh \otimes 1$. This is the claim.

(2) Suppose $[\mathfrak{t}, \mathrm{rad}\,\tilde{S}] \subset \mathrm{rad}\,\tilde{S}$, i.e., $\delta = 0$. Arguing as in (1) we see that $L[\kappa]$ is isomorphic to $W(1;\underline{1})$ or $H(2;\underline{1})^{(2)} \subset L[\kappa] \subset H(2;\underline{1})^{(1)}$ and the torus acts as $Fx_1\partial_1$ and $FD_H(x_1x_2)$, respectively. This shows that κ is proper. Therefore there is a natural bijection $\Gamma_{imp}(\mathfrak{g}, \mathfrak{t}) \to \Gamma_{imp}(\tilde{S}, \mathfrak{t})$. Since $\delta = 0$, the radical $\mathrm{rad}\,\tilde{S} = \mathfrak{s} \otimes \mathcal{O}(m;\underline{1})_{(1)}$ is $\mathfrak{t}$-invariant. By definition, $\mathfrak{t}$ acts on $\tilde{S}(\mu)$ as a proper torus if and only if it does on the quotient $\tilde{S}(\mu)/\tilde{S}(\mu) \cap \mathrm{rad}\,\tilde{S} \cong \mathfrak{s}$. This gives the required bijection.

If $\mathfrak{t}$ acts on $\mathfrak{s}$ as a 1-dimensional torus, then $\Gamma_{imp}(\mathfrak{s}, \mathfrak{t}) = \Gamma_{imp}(\mathfrak{s}, Fh)$. So Fh acts as a proper torus on $\mathfrak{s}$ if and only if all $\mathfrak{t}$-roots on $L[\alpha, \beta]$ are proper.

Now suppose that $\mathfrak{t}$ acts as a 2-dimensional torus. Then $\mathfrak{s} \cong H(2;\underline{1})^{(2)}$ and $\mathfrak{t} \hookrightarrow \mathrm{Der}\,H(2;\underline{1})^{(2)} = CH(2;\underline{1})$ is 2-dimensional. Theorem 7.5.5(2) tells us that one may normalize the image $\iota(\mathfrak{t}) \subset CH(2;\underline{1})$ to $\iota(\mathfrak{t}) = F(\delta_1 + x_1)\partial_1 \oplus F(\delta_2 + x_2)\partial_2$ with $\delta_1, \delta_2 \in \{0, 1\}$. Observe that $h \in \mathfrak{s}$, whence $\iota(h) \in \mathbb{F}_p^*((\delta_1 + x_1)\partial_1 - (\delta_2 + x_2)\partial_2)$. Therefore Fh is a proper torus if and only if $\delta_1 = \delta_2 = 0$ by Theorem 11.2.5(3). Observe that we are now investigating an algebra with torus as in case (8) of

Theorem 20.1.1. Lemma 20.1.5(4) yields that this condition is equivalent with the equation $\Gamma_{imp}(\mathfrak{s},\mathfrak{t}) = \emptyset$. □

Next we refer to the concept of elementary switching described in §1.5. Let A be a restricted Lie algebra and $\mathfrak{t}$ a torus in A. Let β be a nonzero root and $u \in A_\beta$. With the notations in that section we have (see Equations (1.5.5) and (1.5.6))

$$\begin{aligned} t_u &:= t - \beta(t)(u + q(u)) \text{ for } t \in \mathfrak{t}, \\ \mathfrak{t}_u &:= \{t_u \mid t \in \mathfrak{t}\}, \\ \gamma_{(u,\lambda)}(t_u) &:= \gamma(t) - \gamma(\xi(u,\lambda))\beta(t) \quad \text{for } \gamma \in \Gamma(A,\mathfrak{t}), \\ A(\beta) &= A(\beta_{(u,\lambda)}). \end{aligned}$$

Here (u,λ) and $q(u), \xi(u,\lambda) \in \sum_{l>0} Fu^{[p]^l}$ are defined in Volume 1, pp. 47/48. Let us specialize $A = L(\alpha,\beta)_{[p]} + T$ for any T-roots α,β and $\mathfrak{t} = T$. As a consequence,

$$\Psi_{\alpha,\beta}(t_u) = \Psi_{\alpha,\beta}(t) - \beta(t)\Psi_{\alpha,\beta}(u+q(u)),\ \Psi_{\alpha,\beta}(q(u)) \in \sum_{l>0} F\Psi_{\alpha,\beta}(u)^{[p]^l}$$

for all $t \in T$.

Lemma 20.1.8. *Let $\mathfrak{g} = L[\alpha,\beta]$ and $\mathfrak{t} = \Psi_{\alpha,\beta}(T)$ be as in Theorem 20.1.1. Suppose $\beta \in \Gamma_{imp}(L,T)$ and $u \in L_\beta$ is such that β_u is proper. Set $\tilde{\mathfrak{t}} := \Psi_{\alpha,\beta}(T_u)$. Then*

$$|\Gamma_{imp}(\mathfrak{g},\tilde{\mathfrak{t}})| \le |\Gamma_{imp}(\mathfrak{g},\mathfrak{t}) \setminus \mathbb{F}_p^*\beta|.$$

Proof. (a) In case (1) of Theorem 20.1.1 there exists no nonzero root at all. In case (2) one has $\mathfrak{g} = L[\beta]$. Then $\tilde{\mathfrak{t}}$ is a proper torus. In case (3) we may take $\mathfrak{g} = L[\alpha] \oplus L[\beta]$. Choose $t \in T$ with $\Psi_{\alpha,\beta}(t) = h_1 \in L[\alpha]^{(1)}$. As $\beta(h_1) = 0$, one has $\Psi_{\alpha,\beta}(t_u) = h_1$. Then $\alpha_{(u,\lambda)}$ is improper if and only if α is so. In case (4) all roots are proper, so this does not occur here.

(b) Let $\mathfrak{g}$ be one of $W(2;\underline{1})$, $S(3;\underline{1})^{(1)}$, $H(4;\underline{1})^{(1)}$, $H(4;\underline{1})^{(1)} + FD_H(x_1^{p-1}\cdots x_4^{p-1})$, $K(3;\underline{1})$. Due to Lemma 20.1.5 for both tori $\mathfrak{t}' \in \{\mathfrak{t},\tilde{\mathfrak{t}}\}$ there occur 3 cases:

- $\Gamma_{imp}(\mathfrak{g},\mathfrak{t}') = \emptyset$;
- $\Gamma_{imp}(\mathfrak{g},\mathfrak{t}') = \mathbb{F}_p^*\kappa$, $\dim \mathfrak{t}' \cap \mathfrak{g}_{(0)} = 1$, $\kappa(\mathfrak{t}' \cap \mathfrak{g}_{(0)}) = 0$;
- $\Gamma_{imp}(\mathfrak{g},\mathfrak{t}') = \Gamma(\mathfrak{g},\mathfrak{t}') \setminus \{0\}$.

Note that $\mathfrak{t}$ is not of the first type because there is an improper root, and if it is of the last type then the claim follows (as $|\Gamma(\mathfrak{g},\mathfrak{t})| = |\Gamma(\mathfrak{g},\tilde{\mathfrak{t}})|$). So we may assume that $\Gamma_{imp}(\mathfrak{g},\mathfrak{t}) = \mathbb{F}_p^*\beta$, $\dim \mathfrak{t} \cap \mathfrak{g}_{(0)} = 1$, and $\beta(\mathfrak{t} \cap \mathfrak{g}_{(0)}) = 0$. Put $\mathfrak{t} \cap \mathfrak{g}_{(0)} = Ft_0$, a 1-dimensional torus. Similarly, $\tilde{\mathfrak{t}}$ is not of the last type because there exists the proper nonzero root $\beta_{(u,\lambda)}$, and if it is of the first type then the claim follows. So we may assume that $\Gamma_{imp}(\mathfrak{g},\tilde{\mathfrak{t}}) = \mathbb{F}_p^*\kappa$, $\dim \tilde{\mathfrak{t}} \cap \mathfrak{g}_{(0)} = 1$, and $\kappa(\tilde{\mathfrak{t}} \cap \mathfrak{g}_{(0)}) = 0$. As $\beta(t_0) = 0$, the definitions concerning elementary switchings yield $(t_0)_u = t_0$ and $\beta_{(u,\lambda)}(t_0) = 0$. So $\tilde{\mathfrak{t}} \cap \mathfrak{g}_{(0)} = Ft_0$ and $\mathbb{F}_p\beta_{(u,\lambda)}$ is the only root line which vanishes on $\tilde{\mathfrak{t}} \cap \mathfrak{g}_{(0)}$. Since $\beta_{(u,\lambda)}$ is proper and κ is not, this case is impossible.

(c) Let $\mathfrak{g}$ be one of $W(1;\underline{2})$, $H(2;\underline{1};\Phi(\tau))^{(1)}$. Due to Lemma 20.1.5, all nonzero roots are improper or all roots are proper. The claim follows.

(d) Let $\mathfrak{g}$ be as in part (4) of Lemma 20.1.5. Then $\mathfrak{g}$ is the only missing part in case (6) or is as in case (8) of Theorem 20.1.1. For both tori $\mathfrak{t}' \in \{\mathfrak{t}, \tilde{\mathfrak{t}}\}$ there occur 3 cases:

- $\Gamma_{imp}(\mathfrak{g}, \mathfrak{t}') = \emptyset$;
- $\Gamma_{imp}(\mathfrak{g}, \mathfrak{t}') = \mathbb{F}_p^*\kappa$, $\dim \mathfrak{t}' \cap W(2;\underline{1})_{(0)} = 1$, $\kappa(\mathfrak{t}' \cap W(2;\underline{1})_{(0)}) = 0$;
- $\Gamma_{imp}(\mathfrak{g}, \mathfrak{t}') = \Gamma(\mathfrak{g}, \mathfrak{t}') \setminus \mathbb{F}_p\kappa$, κ solvable.

If $\tilde{\mathfrak{t}}$ is of the first type, the claim follows. If $\tilde{\mathfrak{t}}$ is of the third type, then $\mathbb{F}_p\kappa = \mathbb{F}_p\beta_{(u,\lambda)}$. Since β is improper, it is Witt or Hamiltonian. Then so is $\beta_{(u,\lambda)}$. However, κ is solvable. So we may assume that $\tilde{\mathfrak{t}}$ is of the second type, whence $\dim \tilde{\mathfrak{t}} \cap W(2;\underline{1})_{(0)} = 1$ and $\kappa(\tilde{\mathfrak{t}} \cap W(2;\underline{1})_{(0)}) = 0$. Next, $\mathfrak{t}$ is not of the first type because there is an improper root, and if it is of the last type then the claim follows (because $|\Gamma_{imp}(\mathfrak{g}, \tilde{\mathfrak{t}})| = p-1 < |\Gamma_{imp}(\mathfrak{g}, \mathfrak{t})| - p$). So we may assume that $\mathfrak{t}$ is of the second type as well. Arguing as in part (b) of this proof, we arrive at a contradiction.

(e) Let $\mathfrak{g}$ be as in cases (7) or (9) of Theorem 20.1.1. Due to Lemmas 20.1.5(5) and 20.1.6, all roots are proper or all proper roots are solvable. Since β is improper, it is Witt or Hamiltonian. Then so is $\beta_{(u,\lambda)}$. Hence $\tilde{\mathfrak{t}}$ cannot be of the second type. Therefore $\Gamma_{imp}(\mathfrak{g}, \tilde{\mathfrak{t}}) = \emptyset$ and the claim follows.

(f) It remains to consider the cases treated in Lemma 20.1.7.

Suppose $[\mathfrak{t}, \operatorname{rad} \tilde{S}] \not\subset \operatorname{rad} \tilde{S}$. If $\mathfrak{t} \cap \tilde{S}$ is an improper torus of $\tilde{S}$, then all nonzero $\mathfrak{t}$-roots are improper. The claim follows. So we may assume that $\mathfrak{t} \cap \tilde{S} = Fh \otimes 1$ is a proper torus of $\tilde{S}$ (i.e., Fh is a proper torus of $\mathfrak{s}$). The lemma gives $\mathbb{F}_p\beta = \mathbb{F}_p\kappa$, whence $\beta(h \otimes 1) = 0$. As in former cases, this implies $(h \otimes 1)_u = h \otimes 1$ and $\beta_{(u,\lambda)}((h \otimes 1)_u) = \beta(h \otimes 1) = 0$. The lemma applied for $\tilde{\mathfrak{t}}$ now says (as $\beta_{(u,\lambda)}$ is proper and $L[\beta_{(u,\lambda)}] = L[\beta] = L[\kappa]$) that $[\tilde{\mathfrak{t}}, \operatorname{rad} \tilde{S}] \subset \operatorname{rad} \tilde{S}$. Since $\tilde{\mathfrak{t}} \cap \tilde{S} = \mathfrak{t} \cap \tilde{S} = Fh \otimes 1$ and Fh is a proper torus of $\mathfrak{s}$, we obtain in this case that all $\tilde{\mathfrak{t}}$-roots are proper. The claim follows.

Suppose $[\mathfrak{t}, \operatorname{rad} \tilde{S}] \subset \operatorname{rad} \tilde{S}$. Then κ is proper. As β is improper, this gives $\beta(h \otimes 1) \neq 0$. In particular, $\Psi_{\alpha,\beta}(u) \in \tilde{S}$ holds. This implies $\pi_2(\tilde{\mathfrak{t}}) = \pi_2(\mathfrak{t})$. Therefore $[\tilde{\mathfrak{t}}, \operatorname{rad} \tilde{S}] \subset \operatorname{rad} \tilde{S}$ holds as well. Part (2) of Lemma 20.1.7 reduces the proof to the algebra $\mathfrak{s}$,

$$|\Gamma_{imp}(\mathfrak{g}, \mathfrak{t}) \setminus \mathbb{F}_p^*\beta| = |\Gamma_{imp}(\mathfrak{s}, \mathfrak{t}) \setminus \mathbb{F}_p^*\beta|, \qquad |\Gamma_{imp}(\mathfrak{s}, \tilde{\mathfrak{t}})| = |\Gamma_{imp}(\mathfrak{g}, \tilde{\mathfrak{t}})|.$$

If $\mathfrak{t}$ acts on $\mathfrak{s}$ as a 1-dimensional torus, then we argue as for case (2) of Theorem 20.1.1, and if $\mathfrak{t}$ acts on $\mathfrak{s}$ as a 2-dimensional torus, then we argue as for case (8) of Theorem 20.1.1. The claim follows. □

Applying elementary switchings, one can transform improper roots into proper ones. Namely, let $A = W(1;\underline{1})$ and $\mathfrak{t} = F(1+x)\partial$. We take $u = -\partial$. Observe that $\partial^p = 0$, whence $q(u) = \xi(u,\lambda) = 0$. For $t = (1+x)\partial$ one obtains $\beta(t)u = [(1+x)\partial, -\partial] = \partial$, and hence $t_u = t - \partial = x\partial$. Then $\mathfrak{t}_u$ is a proper torus. The second interesting case is when $H(2;\underline{1})^{(2)} \subset A \subset H(2;\underline{1})$ and $\mathfrak{t} = FD_H((1+x_1)x_2)$. In this case we put $u := D_H(x_2) = -\partial_1$ and obtain as above $D_H((1+x_1)x_2)_u = D_H(x_1x_2)$. Finally, Theorem 17.1.8 shows that $L_{[p]}[\beta]$ is as in the preceding cases if β is Witt or Hamiltonian. Then we choose a root vector $\tilde{u} \in L_{[p]}(\beta)$ which is mapped under Ψ_β onto u. It is clear that $\Psi_\beta(T_{\tilde{u}})$ is a proper torus in $L_{[p]}[\beta]$.

Proposition 20.1.9. *There exists a standard torus in $L_{[p]}$ of maximal dimension for which all roots are proper.*

Proof. As a general assumption in this chapter we only deal with the case that all tori of maximal dimension in $L_{[p]}$ are standard. Choose a torus $T \in L_{[p]}$ of maximal dimension which has a minimal number of improper roots. Suppose there is an improper root β. We just mentioned that there is a root vector $\tilde{u} \in \cup_{i\in\mathbb{F}_p^*} L_{i\beta}$ such that after a suitable switching the root $\beta_{(\tilde{u},\lambda)}$ is proper. Let $\sim$ denote the equivalence relation on $\Gamma_{imp}(L,T)$ defined by $\kappa \sim \mu$ if and only if $\mathbb{F}_p\kappa + \mathbb{F}_p\beta = \mathbb{F}_p\mu + \mathbb{F}_p\beta$. The equivalence classes are of the form $\mathbb{F}_p^*\beta$ or $\Gamma_{imp}(L(\alpha,\beta),T) \setminus \mathbb{F}_p^*\beta$ with α,β being $\mathbb{F}_p$-independent. Let Y be a complete set of representatives of the equivalence classes containing β. Then Lemma 20.1.2 shows that

$$\begin{aligned}|\Gamma_{imp}(L,T)| &= (p-1) + \sum_{\alpha\in Y\setminus\{\beta\}} |\Gamma_{imp}(L(\alpha,\beta),T) \setminus \mathbb{F}_p^*\beta| \\ &= (p-1) + \sum_{\alpha\in Y\setminus\{\beta\}} |\Gamma_{imp}(L[\alpha,\beta],\Psi_{\alpha,\beta}(T)) \setminus \mathbb{F}_p^*\beta|.\end{aligned}$$

We also have the corresponding expression for $\Gamma_{imp}(L,T_{\tilde{u}})$, which, as $\beta_{(\tilde{u},\lambda)}$ is proper by hypothesis, reads as

$$|\Gamma_{imp}(L,T_{\tilde{u}})| = \sum_{Y\setminus\{\beta\}} |\Gamma_{imp}(L[\alpha_{(\tilde{u},\lambda)},\beta_{(\tilde{u},\lambda)}],\Psi_{\alpha,\beta}(T_{\tilde{u}}))|.$$

Furthermore,

$$|\Gamma_{imp}(L[\alpha_{(\tilde{u},\lambda)},\beta_{(\tilde{u},\lambda)}],\Psi_{\alpha,\beta}(T_{\tilde{u}}))| \leq |\Gamma_{imp}(L[\alpha,\beta],\Psi_{\alpha,\beta}(T)) \setminus \mathbb{F}_p^*\beta|$$

by Lemma 20.1.8. Then $T_{\tilde{u}}$ has less improper roots, contrary to our choice of T. Hence T has no improper roots at all, and so all roots are proper. □

Definition 20.1.10. *Every torus $T \subset L_{[p]}$ of maximal dimension which is standard and has only proper roots is called* optimal.

From here on we fix an optimal torus T and abbreviate $\Gamma := \Gamma(L,T)$.

20.2 Root spaces in 2-sections

In this section we develop some structural insight into the root space structure of the 2-sections mentioned in Theorem 20.1.1 with respect to the image of T, now under the assumption that all T-roots are proper. We are particularly interested in the problem how the subalgebras $Q(\alpha)$ sit in 2-sections.

To simplify notation we set (cf. §20.1)

$$\bar{Q}_\alpha := \Psi_{\alpha,\beta}(Q_\alpha), \quad \bar{Q}(\alpha) = \Psi_{\alpha,\beta}(Q(\alpha)), \quad \bar{Q}(\alpha,\beta) := \sum_{\mu\in(\mathbb{F}_p\alpha+\mathbb{F}_p\beta)\setminus\{0\}} \bar{Q}(\mu).$$

Recall that $Q(L[\alpha,\beta](\alpha)) = \bar{Q}(\alpha)$. Set $\mathfrak{t} := \Psi_{\alpha,\beta}(T)$. Since T is optimal, all $\mathfrak{t}$-roots on $\mathfrak{g} := L[\alpha,\beta]$ are proper by Lemma 20.1.2 (which means by definition that $Q(L[\alpha])$ is $\mathfrak{t}$-invariant or, equivalently, that $\bar{Q}(\alpha)$ is $\mathfrak{t}$-invariant).

It will often be the case that $\mathfrak{g} \subset \operatorname{Der}\tilde{S} \subset \operatorname{ad} W(m;\underline{n}) \simeq W(m;\underline{n})$ for some m and n, and when that happens we will define $\mathfrak{g}_{(i)} = \mathfrak{g} \cap W(m;\underline{n})_{(i)}$ and $(\operatorname{Der}\tilde{S})_{(i)} = (\operatorname{Der}\tilde{S}) \cap W(m;\underline{n})_{(i)}$.

Put

$$\Delta := \Delta(L,T) := \{\mu \in \Gamma \mid \mu(H) = 0\}.$$

Proposition 20.2.1. *Let $\mathfrak{g} := L[\alpha,\beta]$ be one of the restricted Cartan type algebras $W(2;\underline{1})$, $S(3;\underline{1})^{(1)}$, $H(4;\underline{1})^{(1)}$, $H(4;\underline{1})^{(1)} + FD_H(x_1^{p-1}\cdots x_4^{p-1})$, $K(3;\underline{1})$, and $\mathfrak{t} = \Psi_{\alpha,\beta}(T)$. Then*

$$\mathfrak{t} \subset \mathfrak{g}_{(0)}, \quad \bar{Q}(\alpha,\beta) = \mathfrak{g}_{(0)}, \quad \big(\operatorname{rad}\bar{Q}(\mu)\big)_\mu \subset \mathfrak{g}_{(1)} \text{ for all nonzero roots } \mu.$$

The 1-sections can be described as follows.

(1) $\mathfrak{g} = W(2;\underline{1})$*: exactly two roots ζ,η stick out of $W(2;\underline{1})_{(0)}$; these roots are of Witt type, while $\zeta-\eta$ is classical. All other root directions are spanned by solvable roots.*

(2) $\mathfrak{g} = S(3;\underline{1})^{(1)}$*: exactly three roots $\zeta,\eta,-(\zeta+\eta)$ stick out of $S(3;\underline{1})^{(1)}{}_{(0)}$; these roots are of Witt type, while $\zeta-\eta$, $2\zeta+\eta$, $\zeta+2\eta$ are classical. All other root directions are spanned by solvable roots.*

(3) $\mathfrak{g} = H(4;\underline{1})^{(1)}$, $H(4;\underline{1})^{(1)} + FD_H(x_1^{p-1}\cdots x_4^{p-1})$*: exactly four roots $\pm\zeta,\pm\eta$ stick out of $H(4;\underline{1})_{(0)}$; these roots are of Hamiltonian type, while $\zeta\pm\eta$ are classical. All other root directions are spanned by solvable roots.*

(4) $\mathfrak{g} = K(3;\underline{1})$*: exactly three roots $\zeta,\eta,\zeta+\eta$ stick out of $K(3;\underline{1})_{(0)}$; these roots are of Witt type, while $\zeta-\eta$ is classical. All other root directions are spanned by solvable roots. Moreover, ζ,η are the roots on $K(3;\underline{1})_{(-1)}/K(3;\underline{1})_{(0)}$ and $\zeta+\eta$ is the root on $K(3;\underline{1})/K(3;\underline{1})_{(-1)}$.*

Proof. (a) Since all $\mathfrak{t}$-roots are proper, Lemma 20.1.5(1) proves $\mathfrak{t} \subset \mathfrak{g}_{(0)}$. Then we apply Corollary 17.2.10(1) to obtain $\bar{Q}(\mu) = \mathfrak{g}_{(0)} \cap \mathfrak{g}(\mu)$ for all nonzero roots μ. Since $\mathfrak{g}_{(0)}$ is $\mathfrak{t}$-invariant, this gives $\bar{Q}(\alpha,\beta) = \mathfrak{g}_{(0)}$.

Since every $\mathfrak{t}$-section of $\mathfrak{g}_{(0)}/\mathfrak{g}_{(1)}$ is classical reductive in all cases under consideration, one has that $\mathfrak{t}$ acts trivially on the image of rad $\bar{Q}(\mu)$ for all nonzero roots μ. This implies $(\operatorname{rad} \bar{Q}(\mu))_\mu \subset \mathfrak{g}_{(1)}$.

(b) Suppose $\mathfrak{g} = W(2;\underline{1})$. Recall that one may normalize $\psi(\mathfrak{t}) = Fx_1\partial_1 \oplus Fx_2\partial_2$ under an automorphism ψ of $W(2;\underline{1})$. Note that $\psi(W(2;\underline{1})_{(i)}) = W(2;\underline{1})_{(i)}$ for all i. We suppress the notion of this automorphism. Define roots $\epsilon_i(x_j\partial_j) = \delta_{i,j}$ $(i, j = 1, 2)$. Let μ be an arbitrary nonzero root and t_μ a toral element which spans $\mathfrak{t} \cap (\ker \mu)$. Write $t_\mu = ax_1\partial_1 + bx_2\partial_2$. Theorem 17.2.2 shows the following. The case $b = 0$ yields that the root ϵ_2 is of Witt type. By symmetry the same result is true for the case $a = 0$, i.e., for ϵ_1. The roots $-\epsilon_1$ and $-\epsilon_2$ are related to ∂_1, ∂_2, resp., and these are the only roots sticking out of $W(2;\underline{1})_{(0)}$. The case $a = b \neq 0$ yields that the root $\epsilon_2 - \epsilon_1$ is of classical type, and all other root directions are spanned by solvable roots.

(c) Suppose $\mathfrak{g} = S(3;\underline{1})^{(1)}$. Recall that $\mathfrak{t}$ is conjugate to $F(x_1\partial_1 - x_3\partial_3) \oplus F(x_2\partial_2 - x_3\partial_3)$ under an automorphism of $S(3;\underline{1})^{(1)}$ by Theorem 7.5.5. As before we suppress the notion of this automorphims. Define $\hat{\epsilon}_i(x_j\partial_j - x_3\partial_3) = \delta_{i,j}$ for $i, j = 1, 2$. Let μ be an arbitrary nonzero root and t_μ a toral element which spans $\mathfrak{t} \cap (\ker \mu)$. Write $t_\mu = ax_1\partial_1 + bx_2\partial_2 + cx_3\partial_3$ with $a, b, c \in \mathbb{F}_p$. As $t_\mu \in S(3;\underline{1})^{(1)}$, one has $a + b + c = 0$. If $b = 0$, Theorem 17.2.4(1) applies with the present assumption $z_1 = x_1$, and shows that the root $\hat{\epsilon}_2$ is Witt. The proof shows that $F\partial_2$ determines the root in $\mathbb{F}_p\hat{\epsilon}$ sticking out of ${S(3;\underline{1})^{(1)}}_{(0)}$. The cases $a = 0$ and $c = 0$ give the symmetric results (by interchanging indices). So there are 3 Witt roots and the respective root spaces sticking out of ${S(3;\underline{1})^{(1)}}_{(0)}$ are $F\partial_1$, $F\partial_2$, $F\partial_3$. The respective roots are $-\hat{\epsilon}_1$, $-\hat{\epsilon}_2$, $\hat{\epsilon}_1 + \hat{\epsilon}_2$.

The homogeneous part ${S(3;\underline{1})^{(1)}}_0 \cong \mathfrak{sl}(3)$ acts naturally on the space $F\partial_1 \oplus F\partial_2 \oplus F\partial_3$. Hence there are non-solvable roots $\pm(\hat{\epsilon}_1 - \hat{\epsilon}_2)$, $\pm(2\hat{\epsilon}_1 + \hat{\epsilon}_2)$, $\pm(\hat{\epsilon}_1 + 2\hat{\epsilon}_2)$ on ${S(3;\underline{1})^{(1)}}_0$. These roots relate to the cases $a = b$, $b = c$, $a = c$. Theorem 17.2.4(2),(3) shows that these roots are classical, and that all other 1-sections are contained in ${S(3;\underline{1})^{(1)}}_{(1)} + \mathfrak{t}$, hence are solvable.

(d) Suppose $\mathfrak{g} = H(4;\underline{1})^{(1)}$, $H(4;\underline{1})^{(1)} + FD_H(x_1^{p-1} \cdots x_4^{p-1})$. Note that $\mathfrak{t}$ is conjugate to $FD_H(x_1x_3) \oplus FD_H(x_2x_4)$ under an automorphism of $H(4;\underline{1})^{(1)}$ (Theorem 7.5.8). Define $\tilde{\epsilon}_i(D_H(x_jx_{2+j})) = \delta_{i,j}$ for $i, j = 1, 2$. Let μ be an arbitrary nonzero root and t_μ a toral element which spans $\mathfrak{t}\cap(\ker \mu)$. Write $t_\mu = aD_H(x_1x_3) + bD_H(x_2x_4)$ with $a, b \in \mathbb{F}_p$. Theorem 17.2.6(1) shows that the root $\tilde{\epsilon}_2$ (the case $b = 0$) is Hamiltonian. By symmetry we obtain that also $\tilde{\epsilon}_1$ is Hamiltonian.

Part (2) of Theorem 17.2.6 shows that the classical 1-sections are obtained if $t_\mu = aD_H(x_1x_3) + bD_H(x_2x_4)$, where $a, b \neq 0$ and $a \pm b = 0$. The respective root directions are $\tilde{\epsilon}_1 \mp \tilde{\epsilon}_2$.

By part (3) of Theorem 17.2.6, all other 1-sections are solvable.

(e) Suppose $\mathfrak{g} = K(3;\underline{1})$. Note that we may normalize $\mathfrak{t} = FD_K(x_1x_2) \oplus FD_K(x_3)$ (Theorem 7.5.13). Define $(\epsilon_1^*, \epsilon_2^*)$ dual to $(D_K(x_1x_2), D_K(x_3))$. Let μ be any nonzero root and let $t_\mu = aD_K(x_1x_2) + bD_K(x_3)$ be toral and span $\mathfrak{t} \cap (\ker \mu)$. In the following cases Theorem 17.2.8(3) applies:

$$b = 0, \text{ which means } \mu \in \mathbb{F}_p^* \epsilon_2^*, \qquad a = b, \text{ which means } \mu \in \mathbb{F}_p^*(\epsilon_2^* - \epsilon_1^*),$$

in which cases μ is of Witt type;

$$a = 0, \text{ which means } \mu \in \mathbb{F}_p^* \epsilon_1^*,$$

in which case μ is classical;

$$a \neq 0, \quad b \neq 0, \quad a \pm b \neq 0,$$

in which case μ is solvable. There is only the case $a = -b$ still missing. The automorphism $x_1 \mapsto x_2$, $x_2 \mapsto -x_1$, $x_3 \mapsto x_3$ brings this case to the case $a = b$ and changes ϵ_1^* into $-\epsilon_1^*$ (Theorem 7.3.8).

In summary: the Witt roots are given by $\mathbb{F}_p^*\epsilon_2^*$, $\mathbb{F}_p^*(\epsilon_2^* - \epsilon_1^*)$, $\mathbb{F}_p^*(\epsilon_2^* + \epsilon_1^*)$, the respective root vectors sticking out of $K(3; \underline{1})_{(0)}$ are $D_K(1)$, $D_K(x_2)$, $D_K(x_1)$, and the respective roots are $\zeta + \eta = -2\epsilon_2^*$, $\zeta = \epsilon_1^* - \epsilon_2^*$, $\eta = -\epsilon_1^* - \epsilon_2^*$. □

Proposition 20.2.2. *Let* $\mathfrak{g} := L[\alpha, \beta] \cong W(1; \underline{2})$ *and* $\mathfrak{t} := \Psi_{\alpha,\beta}(T)$, *or* $\mathfrak{g} := L[\alpha, \beta]$ *be as in case* (7) *of Theorem 20.1.1. The following holds.*

(1) $\mathfrak{t} \cap \mathfrak{g} \subset \mathfrak{g}_{(0)}$ *and* $\bar{Q}(\mu) = \mathfrak{g}(\mu) \cap \mathfrak{g}_{(0)}$ *for all nonzero roots* μ.

(2) $(\operatorname{rad} \bar{Q}(\mu))_\mu \subset \mathfrak{g}_{(1)}$ *for all roots* $\mu \notin \Delta$.

There exist independent roots $\zeta \in \Gamma$, $\eta \in \Delta$ *such that* $\zeta + i\eta$ *is Witt (respectively Hamiltonian) for all* $i \in \mathbb{F}_p$ *and* η *is solvable. The roots sticking out of* $\mathfrak{g}_{(0)}$ *are* $\zeta + i\eta$ *(respectively* $\pm(\zeta + i\eta)$*) for all* $i \in \mathbb{F}_p$. *Every root space* $Q_{i\eta}$ $(i \in \mathbb{F}_p)$ *contains an element* u_i *for which* $\zeta(u_i) \neq 0$.

Proof. (1) If $\mathfrak{g} = W(1; \underline{2})$, then due to Lemma 20.1.5(2) one has $\mathfrak{t} \cap \mathfrak{g}_{(0)} \neq \{0\}$. Theorem 7.6.5 applies and gives all claims.

(2) Suppose $H(2; (1, 2))^{(2)} \subset \mathfrak{g} \subset H(2; (1, 2))$. Lemma 20.1.5(5) gives $\mathfrak{t} \cap \mathfrak{g}_{(0)} \neq \{0\}$. Since $\dim \mathfrak{t} \cap \mathfrak{g} = 1$ (see Volume 2, p. 37) this implies $\mathfrak{t} \cap \mathfrak{g} = \mathfrak{t} \cap \mathfrak{g}_{(0)} \neq \{0\}$. Since $H(2; (1, 2))_{[p]}/H(2; (1, 2))^{(2)}{}_{[p]}$ is spanned by $[p]$-nilpotent elements, one has $\mathfrak{t} \subset H(2; (1, 2))^{(2)}{}_{[p]}$. Theorem 10.2.16 applies and proves the following.

If $\mu(\mathfrak{t} \cap \mathfrak{g}_{(0)}) \neq 0$, then μ is Hamiltonian. Corollary 17.2.10 proves $\bar{Q}(\mathfrak{g}(\mu)) = \mathfrak{g}_{(0)} \cap \mathfrak{g}(\mu)$ for every such root. Since $\mathfrak{g}_{(0)}/\mathfrak{g}_{(1)} \cong \mathfrak{sl}(2) \cong H(2; \underline{1})^{(2)}{}_{(0)}/H(2; \underline{1})^{(2)}{}_{(1)}$ is classical, it can only be that $\mathfrak{g}_{(0)} \subset \mathfrak{g}_{(1)} + \bar{Q}(\mathfrak{g}(\mu))$. Then $\operatorname{rad} \bar{Q}(\mathfrak{g}(\mu)) \subset \mathfrak{g}_{(1)}$. Clearly, $\mu(H) \neq 0$, whence $\mu \notin \Delta$.

If $\mu(\mathfrak{t} \cap \mathfrak{g}_{(0)}) = 0$, it is not hard to see (looking at the associated graded algebra) that $\mathfrak{g}(\mu) \subset \mathfrak{g}_{(0)}$. Therefore $\bar{Q}(\mathfrak{g}(\mu)) = \mathfrak{g}(\mu) \cap \mathfrak{g}_{(0)}$ holds. Theorem 10.2.16(2) shows that every $L_{i\mu}$ contains an element u_i acting non-nilpotently on $\mathfrak{g}$. Corollary 17.3.3 shows that $\mu \in \Delta$. □

Proposition 20.2.3. *Let* $\mathfrak{g} := L[\alpha, \beta] \cong H(2; \underline{1}; \Phi(1))$ *and* $\mathfrak{t} := \Psi_{\alpha,\beta}(T)$, *or* $\mathfrak{g} := L[\alpha, \beta]$ *be as in case (8) of Theorem 20.1.1. The following holds.*

(1) $\mathfrak{t} \subset W(2; \underline{1})_{(0)}$ *and* $\bar{Q}(\alpha, \beta) = \mathfrak{g}_{(0)}$.

(2) $(\mathrm{rad}\, \bar{Q}(\mu))_\mu \subset \mathfrak{g}_{(1)}$ *for all nonzero roots* μ.

(3) *Exactly two roots* ζ, η *stick out of* $W(2; \underline{1})_{(0)}$*; these roots are of Witt type, while* $\zeta - \eta$ *is classical. In particular,* $\zeta - \eta$ *does not vanish on* $\mathfrak{t} \cap \mathfrak{g}$. *All other root directions are spanned by solvable roots.*

Proof. (1) Let $\mathfrak{g} \cong H(2; \underline{1}; \Phi(1))$. Lemma 20.1.5(4) proves $\mathfrak{t} \subset W(2; \underline{1})_{(0)}$. Corollary 17.2.10(2) then gives $\bar{Q}(\mu) = \mathfrak{g}(\mu) \cap W(2; \underline{1})_{(0)}$ for all nonzero μ, and this implies (as $W(2; \underline{1})_{(0)}$ is invariant under $\mathfrak{t}$) $\bar{Q}(\alpha, \beta) = \mathfrak{g}_{(0)}$.

One may normalize $\mathfrak{t} = F x_1 \partial_1 \oplus F x_2 \partial_2$ under an automorphism of $W(2; \underline{1})$. Let μ be any nonzero root, and choose a toral element t_μ which spans $\mathfrak{t} \cap (\ker \mu)$. Write $t_\mu = a x_1 \partial_1 + b x_2 \partial_2$. If $b = 0$, then Proposition 16.2.2(2) applies and shows that μ is Witt. By symmetry we get the same if $a = 0$. Next suppose $a \neq 0, b \neq 0$. Then $\mathfrak{g}(\mu) \subset \mathfrak{g}_{(0)}$. If $b = a$, then we get a classical root, and all other roots are solvable. As a result, there are two roots ζ, η sticking out of $\mathfrak{g}_{(0)}$; these are Witt, and $\zeta - \eta$ is classical. Since $\mathfrak{g}_{(0)}/\mathfrak{g}_{(1)} \cong \mathfrak{sl}(2)$ is $\mathfrak{t}$-invariant, it can only be that $\mathrm{rad}\, \bar{Q}(\mu) \subset \mathfrak{t} \cap \mathfrak{g} + \mathfrak{g}_{(1)}$ for all μ.

(2) If case (8) of Theorem 20.1.1 holds, then Lemmas 20.1.5(4) and 20.1.3 apply and yield a proof of all claims as in (1). □

Proposition 20.2.4. *Let* $\mathfrak{g}$ *and* $\mathfrak{t} := \Psi_{\alpha,\beta}(T)$ *be as in case* (9) *of Theorem 20.1.1. The following holds.*

(1) Fh *is a proper torus of* $\mathfrak{s}$ *and* $\bar{Q}(\alpha, \beta) = \mathfrak{g} \cap \big((\mathrm{Der}\, \mathfrak{s})^{(1)}{}_{(0)} \otimes \mathcal{O}(1; \underline{1})\big)$.

(2) $\mathrm{rad}\, \bar{Q}(\mu) = \mathfrak{g}(\mu) \cap \big((\mathrm{Der}\, \mathfrak{s})^{(1)}{}_{(1)} \otimes \mathcal{O}(1; \underline{1})\big)$ *for all roots* $\mu \notin \Delta$.

(3) *There exists a root* $\eta \in \Delta \cap (\mathbb{F}_p \alpha + \mathbb{F}_p \beta)$. *Every root space* $Q_{i\eta}$ *contains an element* u_i *which acts non-nilpotently on* $\mathfrak{g}$.

If $\mathfrak{s} \cong \mathfrak{sl}(2)$, *then* $\bar{Q}(\alpha, \beta) = \mathfrak{g}$. *If* $\mathfrak{s} \cong W(1; \underline{1})$, *then there is a Witt root* ζ *such that the set of roots sticking out of* $\bar{Q}(\alpha, \beta)$ *is exactly* $\zeta + \mathbb{F}_p \eta$. *If* $\mathfrak{s} \cong H(2; \underline{1})^{(2)}$, *then there is a Hamiltonian root* ζ *such that the set of roots sticking out of* $\bar{Q}(\alpha, \beta)$ *is exactly* $\pm\zeta + \mathbb{F}_p \eta$.

Proof. (a) Lemma 20.1.6 shows that Fh is a proper torus of $\mathfrak{s}$. Let μ be any root, and set $\bar{\mu} := \mu(h \otimes 1)$ and $\tilde{\mu} := \mu(\mathrm{Id} \otimes (1 + x_1)\partial_1)$. If $\bar{\mu} \neq 0$, then $\mathfrak{g}_\mu = \mathfrak{s}_{\bar{\mu}} \otimes (1 + x_1)^{\tilde{\mu}}$.

(b) Suppose $\bar{\mu} \neq 0$. The mapping $x_1 \mapsto 0$ establishes a $\mathfrak{t}$-equivariant isomorphism $\mathfrak{g}(\mu) \cong \mathfrak{s} + \mathfrak{h}$, where $\mathfrak{h} \subset C_{(\mathrm{Der}\, \mathfrak{s})^{(1)}}(Fh)$ is the image of $C_{\mathfrak{g}}(Fh \otimes 1)$. Observe that $Q(\mathfrak{s} + \mathfrak{h}) = \mathfrak{s}_{(0)} + \mathfrak{h}$ and $Q(\mathfrak{s} + \mathfrak{h})_{(1)} \subset \mathfrak{s}_{(1)} + C_{(\mathrm{Der}\, \mathfrak{s})^{(1)}{}_{(1)}}(Fh)$. As a consequence (as Fh is a proper torus),

$$\bar{Q}(\mathfrak{g}(\mu)) = \mathfrak{g}(\mu) \cap \big((\operatorname{Der}\mathfrak{s})^{(1)}{}_{(0)} \otimes \mathcal{O}(1;\underline{1})\big)$$

and

$$\operatorname{rad}\bar{Q}(\mathfrak{g}(\mu)) = \mathfrak{g}(\mu) \cap \big((\operatorname{Der}\mathfrak{s})^{(1)}{}_{(1)} \otimes \mathcal{O}(1;\underline{1})\big).$$

(c) Suppose $\bar{\mu} = 0$. Then

$$\mathfrak{g}(\mu) = \mathfrak{g}(\mu) \cap \big(C_{(\operatorname{Der}\mathfrak{s})^{(1)}}(Fh) \otimes \mathcal{O}(1;\underline{1})\big) \subset (\operatorname{Der}\mathfrak{s})^{(1)}{}_{(0)} \otimes \mathcal{O}(1;\underline{1})$$

(as Fh is a proper torus).

The root space $\bar{Q}_\mu$ contains an element of the form $h \otimes (1 + x_1)^i$ which acts non-nilpotently. Corollary 17.3.3 proves $\mu \in \Delta$.

Define η by $\bar{\eta} = 0$, $\tilde{\eta} = 1$. Then η is a root because $h \otimes (1 + x_1)$ is a root vector for it. We conclude $\eta \in \Delta$.

(d) Since $(\operatorname{Der}\mathfrak{s})^{(1)}{}_{(0)} \otimes \mathcal{O}(1;\underline{1})$ is invariant under $\mathfrak{t}$, we obtain $\bar{Q}(\alpha,\beta) = \mathfrak{g} \cap \big((\operatorname{Der}\mathfrak{s})^{(1)}{}_{(0)} \otimes \mathcal{O}(1;\underline{1})\big)$. This completes the proof of (1). We mentioned that $\mu \notin \Delta$ holds if and only if $\bar{\mu} \neq 0$. For this case we have proved (2). We also have already shown that (3) holds.

If $\mathfrak{s} \cong \mathfrak{sl}(2)$, then the above deliberations show $\bar{Q}(\alpha,\beta) = \mathfrak{g}$.

There is a root ζ satisfying $\zeta(\operatorname{Id} \otimes (1 + x_1)\partial_1) = 0$. If $\mathfrak{s} \cong W(1;\underline{1})$, then we may assume that ζ sticks out of $\mathfrak{s}_{(0)} \otimes F$. This shows that the set of roots which stick out of $\mathfrak{s}_{(0)} \otimes \mathcal{O}(1;\underline{1})$ is exactly $\zeta + \mathbb{F}_p\eta$. Argue similarly if $\mathfrak{s} \cong H(2;\underline{1})^{(2)}$. □

Finally we deal with both cases (10) and (11) of Theorem 20.1.1. Set $r = 1, 2$ in the respective cases and $\mathcal{D} := \pi_2(L[\alpha,\beta])$. Here $r = 1$ if $\mathcal{D} \cong W(1;\underline{1})$ (case (10)) and $r = 2$ if $\mathcal{D}$ is Hamiltonian (case (11)). Recall that $\mathfrak{s}$ is one of $\mathfrak{sl}(2)$, $W(1;\underline{1})$, $H(2;\underline{1})^{(2)}$. In particular, $d = 0$ in the first and second cases.

Proposition 20.2.5. *Suppose* $\mathfrak{g} := L[\alpha,\beta]$ *and* $\mathfrak{t} := \Psi_{\alpha,\beta}(T)$ *are as in cases* (10) *or* (11) *of Theorem 20.1.1. Set* $\mathcal{D} := \pi_2(L[\alpha,\beta])$. *The following holds.*

(1) $\mathfrak{t} = Fh \otimes 1 + F(d \otimes 1 + \operatorname{Id} \otimes t_0)$, *where* Fh *is a proper torus in* $\mathfrak{s}$ *and* Ft_0 *is a proper torus in* $\mathcal{D}$. *In particular, one may take* $t_0 = x\partial$ *if* $r = 1$ *and* $t_0 = D_H(y_1y_2)$ *if* $r = 2$, *and if* $\mathfrak{s}$ *is Hamiltonian, then* $d = m(x_1\partial_1 + x_2\partial_2)$ *for some* $m \in \mathbb{F}_p$.

(2) $\bar{Q}(\alpha,\beta) = \mathfrak{g} \cap \Big(\big((\operatorname{Der}\mathfrak{s})^{(1)}{}_{(0)} \otimes F + (\operatorname{Der}\mathfrak{s}) \otimes \mathcal{O}(r;\underline{1})_{(1)}\big) \rtimes (\operatorname{Id} \otimes \mathcal{D}_{(0)})\Big)$.

(3) $(\operatorname{rad}\bar{Q}(\mu))_\mu \subset \big((\operatorname{Der}\mathfrak{s})^{(1)}{}_{(1)} \otimes F + (\operatorname{Der}\mathfrak{s}) \otimes \mathcal{O}(r;\underline{1})_{(1)}\big) \rtimes (\operatorname{Id} \otimes \mathcal{D}_{(1)})$ *for every nonzero root* μ.

(4) *The* $\mathfrak{t}$*-invariant space* $\mathfrak{g}/\bar{Q}(\alpha,\beta)$ *is represented by* $(\mathfrak{s}/\mathfrak{s}_{(0)}) \oplus (\mathcal{D}/\mathcal{D}_{(0)})$.
Suppose $d = 0$. *There is a root* ζ *such that* $\mathfrak{s} \otimes F \subset \mathfrak{g}(\zeta)$. *The set of roots sticking out of* $\bar{Q}(\alpha,\beta)$ *is (after an adjustment)*

$$\{-\kappa\},\ \{\pm\kappa\};\ \{\zeta,-\kappa\},\ \{\zeta,\pm\kappa\};\ \{\pm\zeta,-\kappa\},\ \{\pm\zeta,\pm\kappa\}$$

in the respective cases that $\mathfrak{s}$ is classical, Witt, or Hamiltonian, and κ is Witt or Hamiltonian. The only case where a classical root occurs is when $\mathfrak{s}$ is classical. In this case ζ is classical. All other root directions are solvable.

If $d \neq 0$, then $\mathfrak{s} \cong H(2;\underline{1})^{(2)}$ and $m \neq 0$, and the roots sticking out of $\bar{Q}(\alpha,\beta)$ are (after an adjustment)

$$\{\pm\zeta - m\kappa, -\kappa\},\ \{\pm\zeta - m\kappa, \pm\kappa\}$$

in the respective cases where κ is Witt or Hamiltonian. Here $\pm\zeta - m\kappa$ are Witt roots and 2ζ is a classical root. All other root directions are solvable.

(5) *If u_λ, u_μ are root vectors sticking out of $\bar{Q}(\alpha,\beta)$, then $[u_\lambda, u_\mu] \in \bar{Q}(\alpha,\beta)$.*

Proof. (1) Suppose $\mathfrak{g}$ is as in case (10) of Theorem 20.1.1. Lemma 20.1.7 shows that $\mathfrak{t} = (Fh \otimes 1) \oplus F(d \otimes 1 + \mathrm{Id} \otimes t_0)$, where Fh is a proper torus in $\mathfrak{s}$ and $t_0 = x\partial$. If $Fh + Fd$ is 1-dimensional, this deliberation proves (1) with $m = 0$ in this case (10). Now suppose $Fh + Fd$ is 2-dimensional, whence $\mathfrak{s} \cong H(2;\underline{1})^{(2)}$. Theorem 7.5.5(2) shows that we may normalize $Fh + Fd = F(\delta_1 + x_1)\partial_1 \oplus F(\delta_2 + x_2)\partial_2$ with $\delta_1, \delta_2 \in \{0,1\}$. Then $\mathfrak{s}$ and $Fh + Fd$ are as in case (8) of Theorem 20.1.1. Lemma 20.1.5(4) yields $\delta_1 = \delta_2 = 0$. Adjusting h by a nonzero scalar and adding a multiple of h to d one may take $h = D_H(x_1x_2)$ and $d \in F(x_1\partial_1 + x_2\partial_2)$. As d is assumed to be toral, one obtains $d = m(x_1\partial_1 + x_2\partial_2)$ with $m \in \mathbb{F}_p^*$. This proves claim (1) in case (10).

If $\mathfrak{g}$ is as in case (11) of Theorem 20.1.1, then one may argue as in the preceding case.

(2) We intend to show that

$$\bar{Q}(\mu) \subset \mathfrak{g}(\mu) \cap \Big(\big((\mathrm{Der}\,\mathfrak{s})^{(1)}{}_{(0)} \otimes F + (\mathrm{Der}\,\mathfrak{s}) \otimes \mathcal{O}(r;\underline{1})_{(1)}\big) \rtimes \big(\mathrm{Id} \otimes \mathcal{D}_{(0)}\big)\Big)$$

for all nonzero μ. Here we set $(\mathrm{Der}\,\mathfrak{s})^{(1)}{}_{(0)} = \mathfrak{s}$ if $\mathfrak{s} \cong \mathfrak{sl}(2)$.

- Consider the case that $\mathfrak{s} \cong \mathfrak{sl}(2)$. Then $\mathfrak{g} = (\mathfrak{s} \otimes \mathcal{O}(r;\underline{1})) \rtimes (\mathrm{Id} \otimes \mathcal{D})$. In particular, $d = 0$ and (recall that κ is defined by the requirement $\kappa(h \otimes 1) = 0$) $\mathfrak{g}(\kappa) = Fh \otimes \mathcal{O}(r;\underline{1}) + \mathrm{Id} \otimes \mathcal{D}$. Then all roots except those from the κ-direction are solvable or classical, whence $\bar{Q}(\mu) = \mathfrak{g}(\mu)$ for these roots. Since $Fh \otimes \mathcal{O}(r;\underline{1}) \subset \mathrm{rad}\,\mathfrak{g}(\kappa)$, it is not hard to see that $\bar{Q}(\kappa) = Fh \otimes \mathcal{O}(r;\underline{1}) + \mathrm{Id} \otimes \mathcal{D}_{(0)}$. The claim follows.

- Next suppose that $\mathfrak{s}$ is Witt or Hamiltonian. As before,

$$\bar{Q}(\kappa) \subset C_{(\mathrm{Der}\,\mathfrak{s})^{(1)}}(Fh) \otimes \mathcal{O}(r;\underline{1}) + \mathrm{Id} \otimes \mathcal{D}_{(0)}.$$

Since Fh is proper in $\mathfrak{s}$, one may normalize $Fh = Fx_1\partial_1$ if $\mathfrak{s}$ is Witt, and normalize $Fh = FD_H(x_1x_2)$ if $\mathfrak{s}$ is Hamiltonian. In either case $C_{(\mathrm{Der}\,\mathfrak{s})^{(1)}}(Fh) \otimes \mathcal{O}(r;\underline{1}) \subset (\mathrm{Der}\,\mathfrak{s})^{(1)}{}_{(0)} \otimes \mathcal{O}(r;\underline{1})$ holds. So the claim holds for $\mu = \kappa$.

Since $\pi_2(\Psi_{\alpha,\beta}(H)) = C_{\mathcal{D}}(t_0) \subset \mathcal{D}_{(0)} \subset W(r;\underline{1})_{(0)}$, one has that $(\operatorname{Der}\mathfrak{s}) \otimes \mathcal{O}(r;\underline{1})_{(1)}$ is a solvable $(\Psi_{\alpha,\beta}(H)+\mathfrak{t})$-invariant ideal of $(\operatorname{Der}\mathfrak{s}) \otimes \mathcal{O}(r;\underline{1})$. There is a homomorphism

$$\begin{aligned}\sigma :&\big((\operatorname{Der}\mathfrak{s}) \otimes \mathcal{O}(r;\underline{1})\big) \rtimes \big(\mathrm{Id} \otimes \mathcal{D}_{(0)}\big) \to \\ &\big((\operatorname{Der}\mathfrak{s}) \otimes \mathcal{O}(r;\underline{1})\big) \rtimes \big(\mathrm{Id} \otimes \mathcal{D}_{(0)}\big)/\big((\operatorname{Der}\mathfrak{s}) \otimes \mathcal{O}(r;\underline{1})_{(1)}\big) \rtimes \big(\mathrm{Id} \otimes \mathcal{D}_{(0)}\big) \\ &\cong \operatorname{Der}\mathfrak{s}.\end{aligned}$$

Now let μ be a root independent of κ, i.e., assume $\mu(h \otimes 1) \neq 0$. Then $\mathfrak{g}(\mu) \subset \mathfrak{s} \otimes \mathcal{O}(r;\underline{1}) + \Psi_{\alpha,\beta}(H)$. Since $\mathfrak{g}(\mu) \cap (\ker\sigma)$ is a solvable ideal, it is contained in $\bar{Q}(\mu)$. We conclude $\sigma(\bar{Q}(\mu)) = Q(\sigma(\mathfrak{g}(\mu))$. Moreover, by (1), $\mathfrak{t}$ acts as a proper torus on $\mathfrak{s}$. If $\sigma(\mathfrak{t}) = Fh$, then we normalize $Fh = x_1\partial_1$ or $Fh = FD_H(x_1x_2)$ as above, and if $\sigma(\mathfrak{t}) = Fh + Fd$ is 2-dimensional, we argue as in Lemma 20.1.3(2). As a result, in either case one has $\sigma(\bar{Q}(\mu)) \subset (\operatorname{Der}\mathfrak{s})^{(1)}{}_{(0)}$. We conclude

$$\bar{Q}(\mu) \subset \mathfrak{g}(\mu) \cap \Big(\big((\operatorname{Der}\mathfrak{s})^{(1)}{}_{(0)} \otimes F + (\operatorname{Der}\mathfrak{s}) \otimes \mathcal{O}(r;\underline{1})_{(1)}\big) \rtimes \big(\mathrm{Id} \otimes \mathcal{D}_{(0)}\big)\Big).$$

Thus the required inclusion holds for all μ. As a consequence, we have

$$\bar{Q}(\alpha,\beta) \subset \mathfrak{g} \cap \Big(\big((\operatorname{Der}\mathfrak{s})^{(1)}{}_{(0)} \otimes F + (\operatorname{Der}\mathfrak{s}) \otimes \mathcal{O}(r;\underline{1})_{(1)}\big) \rtimes F\mathrm{Id} \otimes \mathcal{D}_{(0)}\Big).$$

On the other hand, the right hand side algebra is invariant under $\mathfrak{t}$ and is a compositionally classical algebra. Claim (2) follows.

(3) We set $(\operatorname{Der}\mathfrak{s})^{(1)}{}_{(1)} = \{0\}$ if $\mathfrak{s} \cong \mathfrak{sl}(2)$. Note that $\big((\operatorname{Der}\mathfrak{s})^{(1)}{}_{(1)} \otimes F + (\operatorname{Der}\mathfrak{s}) \otimes \mathcal{O}(r;\underline{1})_{(1)}\big) \rtimes (\mathrm{Id}\otimes\mathcal{D}_{(1)})$ is a nilpotent ideal of $\big((\operatorname{Der}\mathfrak{s})^{(1)}{}_{(0)} \otimes F + (\operatorname{Der}\mathfrak{s})\otimes\mathcal{O}(r;\underline{1})_{(1)}\big) \rtimes (\mathrm{Id}\otimes\mathcal{D}_{(0)})$, and the quotient is isomorphic to $(\mathfrak{s}_{(0)}/\mathfrak{s}_{(1)}) \oplus (\mathcal{D}_{(0)}/\mathcal{D}_{(1)})$. This quotient is $\mathfrak{t}$-invariant and classical reductive. Therefore the image of $(\operatorname{rad}\bar{Q}(\mu))_\mu$ (for any nonzero root μ) meets it only trivially.

(4) By (2), it is obvious that the $\mathfrak{t}$-invariant space $\mathfrak{g}/\bar{Q}(\alpha,\beta)$ is represented by $(\mathfrak{s}/\mathfrak{s}_{(0)}) \oplus (\mathcal{D}/\mathcal{D}_{(0)})$.

Note that $((\operatorname{Der}\mathfrak{s})^{(1)}{}_{(1)} \otimes F + (\operatorname{Der}\mathfrak{s}) \otimes \mathcal{O}(r;\underline{1})_{(1)}) \rtimes (\mathrm{Id} \otimes \mathcal{D}_{(1)})$ is a $\mathfrak{t}$-invariant nilpotent subalgebra of $((\operatorname{Der}\mathfrak{s})^{(1)} \otimes F + (\operatorname{Der}\mathfrak{s}) \otimes \mathcal{O}(r;\underline{1})) \rtimes (\mathrm{Id}\otimes\mathcal{D})$ and the quotient space is isomorphic to $(\mathfrak{s}/\mathfrak{s}_{(1)}) \oplus (\mathcal{D}/\mathcal{D}_{(1)})$. In order to determine all non-solvable roots we have to determine the roots of $(\mathfrak{s}/\mathfrak{s}_{(1)}) \oplus (\mathcal{D}/\mathcal{D}_{(1)})$.

If $\mathfrak{s} \cong \mathfrak{sl}(2)$, the claim is obvious.

Next suppose that $\mathfrak{s}$ is Witt or Hamiltonian and $d = 0$. Since Fh is proper in $\mathfrak{s}$, one may normalize $Fh = Fx_1\partial_1$ if $\mathfrak{s}$ is Witt, and normalize $Fh = FD_H(x_1x_2)$ if $\mathfrak{s}$ is Hamiltonian. With this normalization we find roots

$$\zeta(h \otimes 1) = -1, \quad \zeta(\mathrm{Id} \otimes t_0) = 0, \quad \kappa(h \otimes 1) = 0, \quad \kappa(\mathrm{Id} \otimes t_0) = 1.$$

Statement (4) is quite obvious.

So assume $\mathfrak{s} \cong H(2;\underline{1})^{(2)}$ and $h = D_H(x_1x_2)$, $d = m(x_1\partial_1 + x_2\partial_2) \neq 0$. Define

$$\zeta(h \otimes 1) = -1,\ \zeta(d \otimes 1 + \mathrm{Id} \otimes t_0) = 0,\ \kappa(h \otimes 1) = 0,\ \kappa(d \otimes 1 + \mathrm{Id} \otimes t_0) = 1.$$

In this case ζ might not be a root but just a linear form on $\mathfrak{t}$. The roots sticking out of $\big(((\mathrm{Der}\,\mathfrak{s})^{(1)}{}_{(1)} \otimes F + (\mathrm{Der}\,\mathfrak{s}) \otimes \mathcal{O}(r;\underline{1})_{(1)}) \rtimes (\mathrm{Id} \otimes \mathcal{D}_{(1)})\big)$ are represented by

$$D_H(x_1) \otimes 1,\ D_H(x_2) \otimes 1,\ D_H(x_1^2) \otimes 1,\ D_H(x_1x_2) \otimes 1,\ D_H(x_2^2) \otimes 1,$$
$$\mathrm{Id} \otimes \partial,\ \mathrm{Id} \otimes x\partial$$

if $r = 1$ and by

$$D_H(x_1) \otimes 1,\ D_H(x_2) \otimes 1,\ D_H(x_1^2) \otimes 1,\ D_H(x_1x_2) \otimes 1,\ D_H(x_2^2) \otimes 1,$$
$$\mathrm{Id} \otimes D_H(y_1),\ \mathrm{Id} \otimes D_H(y_2),\ \mathrm{Id} \otimes D_H(y_1^2),\ \mathrm{Id} \otimes D_H(y_1y_2),\ \mathrm{Id} \otimes D_H(y_2^2)$$

if $r = 2$. We have taken $t_0 = x\partial$ if $r = 1$ and $t_0 = D_H(y_1y_2)$ if $r = 2$. One computes

$$\begin{aligned}
[h, D_H(x_1)] &= -D_H(x_1), & [h, D_H(x_2)] &= D_H(x_2),\\
[d, D_H(x_1)] &= -mD_H(x_1), & [d, D_H(x_2)] &= -mD_H(x_2),\\
[h, D_H(x_1^2)] &= -2D_H(x_1^2), & [h, D_H(x_2^2)] &= 2D_H(x_2^2),\\
[d, D_H(x_1^2)] &= 0, & [d, D_H(x_2^2)] &= 0,\\
[t_0, \partial] &= -\partial, \text{ if } r = 1, & &\\
[t_0, D_H(y_1)] &= -D_H(y_1), & [t_0, D_H(y_2)] &= D_H(y_2), \text{ if } r = 2.
\end{aligned}$$

This proves claim (4).

(5) Suppose there are $u_\lambda \in \mathfrak{g}_\lambda \setminus \bar{Q}(\alpha,\beta)$ and $u_\mu \in \mathfrak{g}_\mu \setminus \bar{Q}(\alpha,\beta)$ such that $[u_\lambda, u_\mu] \notin \bar{Q}(\alpha,\beta)$. So $\lambda, \mu, \lambda+\mu$ stick out of $\bar{Q}(\alpha,\beta)$. Looking at roots (see (4)) it is only possible that $d \neq 0$ and $\{\lambda,\mu\} = \{\pm\zeta - m\kappa\}$. Then $u_\lambda, u_\mu \in (\mathrm{Der}\,\mathfrak{s})^{(1)} \otimes \mathcal{O}(r;\underline{1})$ and hence $[u_\lambda, u_\mu] \in ((\mathrm{Der}\,\mathfrak{s})^{(1)} \otimes \mathcal{O}(r;\underline{1}))_{-2m\kappa} \subset \bar{Q}(\kappa) \subset \bar{Q}(\alpha,\beta)$, a contradiction. □

20.3 The distinguished subalgebra $Q(L,T)$

We consider the following subspace of L

$$Q := Q(L,T) := \sum_{\alpha \in \Gamma \setminus \{0\}} Q(\alpha).$$

Our assumption of T being optimal implies that $Q(L,T)$ is T-invariant.

Put $Q_\alpha := Q(L,T) \cap L_\alpha$. In establishing that $Q(L,T)$ is a subalgebra, we have to prove $[Q_\alpha, Q_\beta] \subset Q_{\alpha+\beta}$ for roots α, β. In order to establish this we argue in $L[\alpha,\beta]$. We refer to the notation of the preceding section:

$$\bar{Q}_\alpha := \Psi_{\alpha,\beta}(Q_\alpha),\quad \bar{Q}(\alpha) = \Psi_{\alpha,\beta}(Q(\alpha)),\quad \bar{Q}(\alpha,\beta) := \sum_{\mu \in (\mathbb{F}_p\alpha + \mathbb{F}_p\beta)\setminus\{0\}} \bar{Q}(\mu)$$

and $\mathfrak{t} := \Psi_{\alpha,\beta}(T)$.

Proposition 20.3.1. *Q is a T-invariant subalgebra of L containing H.*

Proof. (a) Recall that $H \subset Q(\mu)$ for all nonzero μ by definition. If α and β are $\mathbb{F}_p$-dependent, then $[Q_\alpha, Q_\beta] \subset Q(\alpha)$ because $Q(\alpha)$ is a subalgebra.

(b) Let α, β be $\mathbb{F}_p$-independent roots. It suffices to show that $[Q_\alpha, Q_\beta] \subset Q(\alpha+\beta)$ for arbitrary roots. We first prove the statement in $L[\alpha,\beta]$. These semisimple quotients of 2-sections are described in Theorem 20.1.1. In case (1) of that theorem we have $\bar{Q}_\alpha = \bar{Q}_\beta = \{0\}$. In case (2) we have $\bar{Q}_\alpha = \{0\}$ or $\bar{Q}_\beta = \{0\}$. In case (3) we have $\bar{Q}(\alpha,\beta) = \bar{Q}(\mu_1) \oplus \bar{Q}(\mu_2)$. So in all these cases one has $[\bar{Q}_\alpha, \bar{Q}_\beta] = \{0\}$.

In case (4) all roots are classical or solvable, whence $\bar{Q}(\alpha,\beta) = L[\alpha,\beta]$. If $L[\alpha,\beta] = H(2;\underline{1};\Phi(\tau))^{(1)}$ then all roots are solvable, whence $\bar{Q}(\alpha,\beta) = L[\alpha,\beta]$. Cases (5)–(11) of Theorem 20.1.1 (except the case $L[\alpha,\beta] = H(2;\underline{1};\Phi(\tau))^{(1)}$) are treated in Propositions 20.2.1–20.2.5. In all these cases there exists a subalgebra $\mathfrak{n}$ of $\mathfrak{g} = L[\alpha,\beta]$, such that $\bar{Q}(\mu) = \mathfrak{g}(\mu) \cap \mathfrak{n}$ for all nonzero roots μ. Note that $\bar{Q}(\mu)$ is $\Psi_{\alpha,\beta}(T)$-invariant. Hence $\bar{Q}_\mu \subset \bar{Q}(\mu) \subset \mathfrak{n}$. We obtain $[\bar{Q}_\alpha, \bar{Q}_\beta] \subset \mathfrak{g}_{\alpha+\beta} \cap \mathfrak{n} \subset \bar{Q}(\alpha+\beta)$.

(c) As a result,

$$[Q_\alpha, Q_\beta] \subset Q(\alpha+\beta) + L(\alpha+\beta) \cap \operatorname{rad}_T L(\alpha,\beta).$$

Now observe that $L(\alpha+\beta) \cap \operatorname{rad}_T L(\alpha,\beta) \subset \operatorname{rad} L(\alpha+\beta) \subset Q(\alpha+\beta)$. This is the claim. □

For further investigations we will collect some information about roots sticking out of Q. Set

$$\Gamma_- := \{\mu \in \Gamma \mid L_\mu \not\subset Q_\mu\}, \quad \Gamma_-(L(\alpha,\beta),T) := \Gamma(L(\alpha,\beta),T) \cap \Gamma_-.$$

Note that $\Gamma_- \cap \Delta = \emptyset$, since all solvable 1-sections are contained in Q.

Proposition 20.3.2.

(1) *If $\mu \in \Gamma_-$, then* $\dim L_\mu/Q_\mu = 1$.

(2) $\mathbb{F}_p\alpha \cap \Gamma_- = \emptyset$ *if α is solvable or classical,* $\mathbb{F}_p\alpha \cap \Gamma_- = \{i_0\alpha\}$ *if α is Witt,* $\mathbb{F}_p\alpha \cap \Gamma_- = \{\pm i_0\alpha\}$ *if α is Hamiltonian (for some $i_0 \in \mathbb{F}_p^*$).*

(3) *Suppose $L[\alpha,\beta]$ is not one of the following cases of Theorem 20.1.1: case* (6) *with* $L[\alpha,\beta] = W(1;\underline{2})$, *case* (7), *and case* (9) *with $\mathfrak{s}$ being non-classical in Theorem 20.1.1. Then* $|(\zeta + \mathbb{F}_p\eta) \cap \Gamma_-| \leq 2$ *for all $\zeta, \eta \in \mathbb{F}_p\alpha + \mathbb{F}_p\beta$. Moreover,* $|(\mathbb{F}_p\alpha + \mathbb{F}_p\beta) \cap \Gamma_-| \leq 4$ *and H distinguishes the elements of* $(\mathbb{F}_p\alpha + \mathbb{F}_p\beta) \cap \Gamma_-$.

(4) *Suppose $L[\alpha,\beta]$ is one of the following cases of Theorem 20.1.1: case* (6) *with* $L[\alpha,\beta] = W(1;\underline{2})$, *case* (7), *case* (9) *with $\mathfrak{s}$ being non-classical in Theorem 20.1.1. Then there exist*

$$\zeta \in \Gamma_-,\ \eta \in \Delta, \qquad \mathbb{F}_p\zeta + \mathbb{F}_p\eta = \mathbb{F}_p\alpha + \mathbb{F}_p\beta.$$

For every such choice there are $u_{i\eta} \in Q_{i\eta}$ *(for all* $i \in \mathbb{F}_p^*$*) such that* $\zeta(u_{i\eta}) \neq 0$. *In addition,*

$$(\mathbb{F}_p\zeta + \mathbb{F}_p\eta) \cap \Gamma_- = \zeta + \mathbb{F}_p\eta \quad \text{if } \zeta \text{ is Witt,}$$
$$(\mathbb{F}_p\zeta + \mathbb{F}_p\eta) \cap \Gamma_- = \pm\zeta + \mathbb{F}_p\eta \quad \text{if } \zeta \text{ is Hamiltonian.}$$

(5) *If* $\alpha, \beta, \alpha + \beta \in \Gamma_-$, *then* $L[\alpha, \beta]$ *is as in case* (5) *with* $K(3; \underline{1})$, *or in cases* (10) *and* (11) *with* $d \neq 0$.

Proof. (1), (2) These assertions are checked in $L[\alpha]$, where they are obviously true.

(3) – (5) The statements follow for cases (1) – (4) of Theorem 20.1.1 from assertion (2) of this theorem. If $L[\alpha, \beta] \cong H(2; \underline{1}; \Phi(\tau))^{(1)}$, then all roots are solvable and hence $L(\alpha, \beta) \subset Q$. For all other cases we refer to Propositions 20.2.1 – 20.2.5. □

Next we determine the T-semisimple quotients of $Q(\alpha, \beta)$.

Proposition 20.3.3. *Let* $\mathrm{rad}_T\, Q(\alpha, \beta)$ *denote the maximal* T*-invariant solvable ideal of* $Q(\alpha, \beta)$ *and set*

$$Q[\alpha, \beta] := Q(\alpha, \beta)/\mathrm{rad}_T\, Q(\alpha, \beta).$$

Then $Q[\alpha, \beta] \cong \bar{Q}(\alpha, \beta)/\mathrm{rad}_{\Psi_{\alpha,\beta}(T)}\, \bar{Q}(\alpha, \beta)$, *and* $Q[\alpha, \beta]$ *is one of the following:*

(A) $Q[\alpha, \beta] = \{0\}$;

(B) $Q[\alpha, \beta] = Q[\mu] \cong \mathfrak{sl}(2)$ *for some root* μ;

(C) $Q[\alpha, \beta] = Q[\mu_1] \oplus Q[\mu_2] \cong \mathfrak{sl}(2) \oplus \mathfrak{sl}(2)$ *for some roots* $\mu_1, \mu_2 \in \mathbb{F}_p\alpha + \mathbb{F}_p\beta$;

(D) $Q[\alpha, \beta] \cong \mathfrak{s} \otimes \mathcal{O}(1; \underline{1})$ *where* $\mathfrak{s} \cong \mathfrak{sl}(2)$, *and* $\Psi_{\alpha,\beta}(T) = (Fh \otimes 1) \oplus (F\mathrm{Id} \otimes (1 + x_1)\partial_1)$ *where* h *is a nonzero toral element in* $\mathfrak{s}$;

(E) $Q[\alpha, \beta] \cong H(2; \underline{1}; \Phi(\tau))^{(1)}$; *all 1-sections* $L(\mu)$ $(\mu \in \mathbb{F}_p\alpha + \mathbb{F}_p\beta)$ *are solvable;*

(F) $Q[\alpha, \beta]$ *is classical simple of type* A_2, B_2, G_2.

In particular, $Q_\alpha \cap \mathrm{rad}_T\, Q(\alpha, \beta) = (\mathrm{rad}\, Q(\alpha))_\alpha$ *holds for all* $\alpha \notin \Delta$ *and all* $\beta \in \Gamma(L, T)$.

Proof. (a) Note that $L(\mu) \cap \mathrm{rad}_T\, L(\alpha, \beta) \subset \mathrm{rad}\, L(\mu) \subset Q(\mu)$ for all $\mu \in (\mathbb{F}_p\alpha + \mathbb{F}_p\beta) \setminus \{0\}$, and therefore one has $\mathrm{rad}_T\, L(\alpha, \beta) \subset \mathrm{rad}_T\, Q(\alpha, \beta)$. Consequently,

$$Q[\alpha, \beta] \cong \bar{Q}(\alpha, \beta)/\mathrm{rad}_{\Psi_{\alpha,\beta}(T)}\, \bar{Q}(\alpha, \beta).$$

In case (1) of Theorem 20.1.1, $Q[\alpha, \beta] = \{0\}$ holds. In case (2) one has $Q[\alpha, \beta] = \bar{Q}(\kappa)/\mathrm{rad}\, \bar{Q}(\kappa) = \{0\}$ if κ is Witt, otherwise $Q[\alpha, \beta] = \bar{Q}(\kappa)/\mathrm{rad}\, \bar{Q}(\kappa) \cong \mathfrak{sl}(2)$. We argue similarly to obtain that in case (3) one of cases (A), (B), or (C) occur. In case (4) one has $\bar{Q}(\alpha, \beta) = L[\alpha, \beta]$. This is case (F).

For the remaining cases set $\mathfrak{g} := L[\alpha, \beta]$ and $\mathfrak{t} := \Psi_{\alpha,\beta}(T)$.

(b) Next suppose that $\mathfrak{g}$ is one of $W(2; \underline{1})$, $S(3; \underline{1})^{(1)}$, $H(4; \underline{1})^{(1)}$, $H(4; \underline{1})^{(1)} + FD_H(x_1^{p-1} \cdots x_4^{p-1})$, $K(3; \underline{1})$. Then $\mathfrak{t} \subset \mathfrak{g}_{(0)} = \bar{Q}(\alpha, \beta)$ by Proposition 20.2.1.

Therefore $\mathrm{rad}_{\mathfrak{t}}\, \bar{Q}(\alpha,\beta) = \mathrm{rad}\, \bar{Q}(\alpha,\beta)$, hence $Q[\alpha,\beta] \cong \bar{Q}(\alpha,\beta)/\mathrm{rad}\, \bar{Q}(\alpha,\beta)$. As a consequence, we are in case (B) if $\mathfrak{g}$ is $W(2;\underline{1})$ or $K(3;\underline{1})$ and in case (F) if $\mathfrak{g}$ is $S(3;\underline{1})^{(1)}$ or $H(4;\underline{1})^{(1)}$.

(c) Suppose $\mathfrak{g} \cong W(1;\underline{2})$. Due to Proposition 20.2.2, $\bar{Q}(\mu) = \mathfrak{g}(\mu) \cap \mathfrak{g}_{(0)}$ holds for all nonzero μ. As a consequence, $\bar{Q}(\alpha,\beta) \subset W(1;\underline{2})_{(0)}$ is solvable. This is case (A).

(d) Suppose $\mathfrak{g} \cong H(2;\underline{1};\Phi(\tau))^{(1)}$. Then all roots are solvable, whence $\bar{Q}(\alpha,\beta) = \mathfrak{g}$ is simple. This is case (E).

(e) Suppose $\mathfrak{g} \cong H(2;\underline{1};\Phi(1))$. By Proposition 20.2.3, $\mathfrak{t} \subset W(2;\underline{1})_{(0)}$ and $\bar{Q}(\alpha,\beta) = \mathfrak{g}_{(0)}$. Argue as before. We are in case (B).

These deliberations cover cases (5) and (6) of Theorem 20.1.1.

(f) Suppose $\mathfrak{g}$ is as in case (7) of Theorem 20.1.1. Proposition 20.2.2 shows that $\bar{Q}(\mu) = \mathfrak{g}(\mu) \cap \mathfrak{g}_{(0)}$ holds for all nonzero roots μ, and all roots $\zeta + i\eta$ ($i \in \mathbb{F}_p$) are Hamiltonian. As a consequence, $\bar{Q}(\alpha,\beta) \subset H(2;(1,2))_{(0)}$ and $\bar{Q}(\alpha,\beta)$ is non-solvable. Note that there is a homomorphism with solvable kernel J

$$\bar{Q}(\alpha,\beta) \to H(2;(1,2))_{(0)}/H(2;(1,2))_{(1)} \cong \mathfrak{sl}(2).$$

Since $\bar{Q}(\alpha,\beta)$ is non-solvable, this homomorphism is surjective and $J = \mathrm{rad}\, \bar{Q}(\alpha,\beta)$ holds. This gives

$$\mathfrak{sl}(2) \cong \bar{Q}(\alpha,\beta)/\mathrm{rad}\, \bar{Q}(\alpha,\beta) \cong Q[\alpha,\beta]/\mathrm{rad}\, Q[\alpha,\beta].$$

As $Q[\alpha,\beta] \neq \{0\}$ is $\mathfrak{t}$-semisimple, one has

$$\begin{aligned} \oplus_{j=1}^{k} \mathfrak{s}_j \otimes \mathcal{O}(m_j;\underline{1}) &\subset Q[\alpha,\beta]_{[p]} + \mathfrak{t} \\ &\subset \oplus_{j=1}^{k} \big((\mathrm{Der}\, \mathfrak{s}_j) \otimes \mathcal{O}(m_j;\underline{1})\big) \rtimes \big(\mathrm{Id} \otimes W(m_j;\underline{1})\big), \end{aligned}$$

where $\mathfrak{s}_j$ is simple and $\pi_2\big(Q[\alpha,\beta]_{[p]}+\mathfrak{t}\big)$ is a transitive subalgebra of $\oplus_{j=1}^{k} W(m_j;\underline{1})$. As a consequence, $k = 1$ and $\mathfrak{s}_1 \cong \mathfrak{sl}(2)$. From this we conclude

$$Q[\alpha,\beta]_{[p]} \cong \big(\mathfrak{sl}(2) \otimes \mathcal{O}(m;\underline{1})\big) \rtimes \big(\mathrm{Id} \otimes \mathcal{D}\big),$$

where $\mathcal{D} \subset W(m;\underline{1})$. In addition, as $Q[\alpha,\beta]/\mathrm{rad}\, Q[\alpha,\beta] \cong \mathfrak{sl}(2)$,

$$\mathrm{rad}\, Q[\alpha,\beta] \cong \big(\mathfrak{sl}(2) \otimes \mathcal{O}(m;\underline{1})_{(1)}\big) \rtimes \big(\mathrm{Id} \otimes \mathcal{D}\big).$$

But then $\mathcal{D} \subset W(m;\underline{1})_{(0)}$. On the other hand, $\mathcal{D}$ is invariant under $\mathcal{D} + \pi_2(\mathfrak{t})$, and the latter is a transitive subalgebra. Hence $\mathcal{D} = \{0\}$. Since $Q[\alpha,\beta]$ is $\mathfrak{t}$-semisimple, we now have $m = 0$ or $\pi_2(\mathfrak{t}) \not\subset W(m;\underline{1})_{(0)}$.

If $m = 0$, then $Q(\mu)$ is solvable for all but one root direction. However, in the present case there are several Hamiltonian root directions. So we have $m \neq 0$. Next we recall that $\dim\, \mathfrak{t} \cap \bar{Q}(\alpha,\beta) = 1$ (Theorem 1.2.9). Hence

$$\mathrm{ad}_{\,Q[\alpha,\beta]}\, \mathfrak{t} = F(\mathrm{ad}_{\,Q[\alpha,\beta]}\, h') \oplus Ft, \quad h' \in Q[\alpha,\beta],$$

where t is toral. Due to Theorem 3.6.1 one may normalize $t = \mathrm{Id} \otimes (1 + x_1)\partial_1$ and $h' = h \otimes 1$. Then $Q[\alpha,\beta]$ is Ft-simple, whence $m = 1$. This is case (D).

(g) Suppose $\mathfrak{g}$ is as in case (8) of Theorem 20.1.1. Proposition 20.2.3 shows that $\mathfrak{t} \subset W(2;\underline{1})_{(0)}$ and $H(2;\underline{1})^{(2)}{}_{(0)} \subset \bar{Q}(\alpha, \beta) \subset H(2;\underline{1})^{(1)}{}_{(0)}$. This is case (B).

(h) Suppose $\mathfrak{g}$ is as in case (9) of Theorem 20.1.1. Proposition 20.2.4 shows that $\bar{Q}(\alpha, \beta) = \mathfrak{g} \cap \big((\operatorname{Der} \mathfrak{s})^{(1)}{}_{(0)} \otimes \mathcal{O}(1;\underline{1})\big)$.

If $\mathfrak{s} = \mathfrak{sl}(2)$, then $\bar{Q}(\alpha, \beta) = L[\alpha, \beta]$. As $\operatorname{Der} \mathfrak{s} = \mathfrak{s}$ in this case, this is (D).

If $\mathfrak{s} \cong W(1;\underline{1})$, then $(\operatorname{Der} \mathfrak{s})_{(0)} \cong W(1;\underline{1})_{(0)}$. In this case $\bar{Q}(\alpha, \beta) \subset W(1;\underline{1})_{(0)} \otimes \mathcal{O}(1;\underline{1})$ is solvable. We are in case (A).

Suppose $\mathfrak{s} \cong H(2;\underline{1})^{(2)}$. Recall (see Theorem 20.1.1) that $\mathfrak{t} = F(h \otimes 1) \oplus F(\operatorname{Id} \otimes (1 + x_1)\partial_1)$, where Fh is a proper torus of $H(2;\underline{1})^{(2)}$. Then $H(2;\underline{1})_{(1)} \otimes \mathcal{O}(1;\underline{1})$ is a solvable $\mathfrak{t}$-invariant ideal of $H(2;\underline{1})_{(0)} \otimes \mathcal{O}(1;\underline{1})$ and the quotient is isomorphic to $\mathfrak{sl}(2) \otimes \mathcal{O}(1;\underline{1})$. Since $\bar{Q}(\alpha, \beta)$ is not solvable in the present situation, we are in case (B) or in case (D). However, in case (B) there is only one non-solvable 1-section of $Q[\alpha, \beta]$, which is not true in the present case.

(i) Suppose $\mathfrak{g}$ is as in cases (10) or (11) of Theorem 20.1.1. Proposition 20.2.5 shows that

$$\bar{Q}(\alpha, \beta) = \mathfrak{g} \cap \Big(\big((\operatorname{Der} \mathfrak{s})^{(1)}{}_{(0)} \otimes F + (\operatorname{Der} \mathfrak{s}) \otimes \mathcal{O}(r;\underline{1})_{(1)}\big) \rtimes (\operatorname{Id} \otimes W(r;\underline{1})_{(0)})\Big)$$

with $r = 1, 2$. Moreover, $\mathfrak{t} = Fh \otimes 1 + F(d \otimes 1 + \operatorname{Id} \otimes t_0)$ holds where $t_0 \in W(r;\underline{1})_{(0)}$, and $d = 0$, or $\mathfrak{s} \cong H(2;\underline{1})^{(2)}$ and $d = m(x_1\partial_1 + x_2\partial_2)$.

Note that $((\operatorname{Der} \mathfrak{s})^{(1)}{}_{(0)} \otimes F + (\operatorname{Der} \mathfrak{s}) \otimes \mathcal{O}(r;\underline{1})_{(1)}) \rtimes (\operatorname{Id} \otimes W(r;\underline{1})_{(0)})$ has the $\mathfrak{t}$-invariant solvable ideal $((\operatorname{Der} \mathfrak{s})^{(1)}{}_{(1)} \otimes F + (\operatorname{Der} \mathfrak{s}) \otimes \mathcal{O}(r;\underline{1})_{(1)}) \rtimes (\operatorname{Id} \otimes W(r;\underline{1})_{(1)})$ and the quotient is isomorphic to $(\mathfrak{s}_{(0)}/\mathfrak{s}_{(1)}) \oplus (W(r;\underline{1})_{(0)}/W(r;\underline{1})_{(1)})$. As a result, there is a homomorphism σ from $\bar{Q}(\alpha, \beta)$ into the $\mathfrak{t}$-semisimple quotient of $(\mathfrak{s}_{(0)}/\mathfrak{s}_{(1)}) \oplus (W(r;\underline{1})_{(0)}/W(r;\underline{1})_{(1)})$. Note that $Q[\alpha, \beta]$ is the homomorphic image of $\bar{Q}(\alpha, \beta)/\bar{Q}(\alpha, \beta) \cap (\ker \sigma)$ where the kernel of this homomorphism is solvable and $\mathfrak{t}$-invariant.

If $r = 1$, then $W(r;\underline{1})_{(0)}/W(r;\underline{1})_{(1)}$ is solvable. In this case

$$\sigma : \bar{Q}(\alpha, \beta) \to \mathfrak{s}_{(0)}/\operatorname{rad} \mathfrak{s}_{(0)}$$

is surjective. As a consequence, $\bar{Q}(\alpha, \beta) \cap (\ker \sigma) = \operatorname{rad}_{\mathfrak{t}} \bar{Q}(\alpha, \beta)$, whence $Q[\alpha, \beta] \cong \mathfrak{s}_{(0)}/\operatorname{rad} \mathfrak{s}_{(0)}$. If $\mathfrak{s}$ is classical or Hamiltonian, then we are in case (B). If $\mathfrak{s}$ is Witt, then we are in case (A).

If $r = 2$, we similarly have (see Theorem 20.1.1) an isomorphism from $Q[\alpha, \beta]$ onto the $\mathfrak{t}$-semisimple quotient of $(\mathfrak{s}_{(0)}/\operatorname{rad} \mathfrak{s}_{(0)}) \oplus (H(2;\underline{1})_{(0)}/H(2;\underline{1})_{(1)})$. If $\mathfrak{s}$ is classical or Hamiltonian, then we are in case (C). If $\mathfrak{s}$ is Witt, then we are in case (B). □

Next we will construct an ideal in Q.

Lemma 20.3.4. *Let $\alpha, \beta \in \Gamma(Q, T)$. The following holds.*

(1) *If $Q(\alpha)$ is non-solvable and β arbitrary, then $Q[\alpha, \beta](\alpha) \cong \mathfrak{sl}(2) \oplus \big(\Psi_{\alpha,\beta}(H) \cap (\ker \alpha)\big)$.*
(2) *If $Q(\alpha)$ is non-solvable and $\alpha - \beta \in \Delta$, then only cases* (B) *or* (D) *in Proposition 20.3.3 can occur.*
(3) *If α, β are $\mathbb{F}_p$-independent and $Q(\alpha)$ and $Q(\beta)$ are non-solvable and $\alpha - \beta \in \Delta$, then only case* (D) *in Proposition 20.3.3 can occur.*
(4) *Suppose $Q(\alpha)$ is non-solvable and $Q(\beta)$ is solvable. Then $\alpha(Q_\beta) \neq 0$ if and only if $Q[\alpha, \beta]$ is as in case* (D) *of Proposition 20.3.3.*

Proof. (1) Since $Q(\alpha)$ is non-solvable, cases (A) and (E) of Proposition 20.3.3 cannot occur. In case (B) one has $Q[\alpha, \beta](\alpha) \cong \mathfrak{sl}(2)$ and $\Psi_{\alpha,\beta}(H) \cap (\ker \alpha) = \{0\}$, and in case (C) $Q[\alpha, \beta](\alpha) \cong \mathfrak{sl}(2) \oplus \Psi_{\alpha,\beta}(H) \cap (\ker \alpha)$ holds. In case (D) choose an $\mathfrak{sl}(2)$-triple (e, h, f) in $\mathfrak{s}$ and observe that $Q[\alpha, \beta](\alpha) = Fe \otimes (1 + x_1)^k \oplus Fh \otimes 1 \oplus Ff \otimes (1 + x_1)^{-k} \cong \mathfrak{sl}(2)$ for some $k \in \mathbb{F}_p^*$. Finally, it is well known that the claim is true in classical algebras.

(2) In cases (A) and (E) all 1-sections are solvable, and so these cases do not occur. In cases (C) and (F) $\Psi_{\alpha,\beta}(H)$ distinguishes $\mathbb{F}_p$-independent roots. Therefore only cases (B) and (D) remain possible.

(3) In view of claim (2) only cases (B) and (D) are possible. However, there are two independent non-solvable 1-sections in $Q(\alpha, \beta)$. This excludes case (B).

(4) Since there is a non-solvable section $Q(\alpha)$ in $Q(\alpha, \beta)$, cases (A) and (E) cannot occur. In cases (B), (C), and (F) the solvable sections are contained in $H + \operatorname{rad}_T Q(\alpha, \beta)$ while some $i\alpha$ ($i \in \mathbb{F}_p^*$) is a root on $Q[\alpha, \beta]$. Therefore $\alpha(Q_\beta) = 0$.

Conversely, in case (D) choose an $\mathfrak{sl}(2)$-triple (e, h, f) in $\mathfrak{s}$ (where h is the element mentioned in Theorem 20.3.3(D)) and define

$$\epsilon_1\big(h \otimes 1\big) = 1,\ \epsilon_2\big(h \otimes 1\big) = 0,\ \epsilon_1\big(\mathrm{Id} \otimes (1 + x_1\partial_1)\big) = 0,\ \epsilon_2\big(\mathrm{Id} \otimes (1 + x_1\partial_1)\big) = 1.$$

Then $Q[\alpha, \beta]_{2\epsilon_1 + k\epsilon_2} = Fe \otimes (1 + x_1)^k$, $Q[\alpha, \beta]_{-2\epsilon_1 + k\epsilon_2} = Ff \otimes (1 + x_1)^k$, and $Q[\alpha, \beta]_{k\epsilon_2} = Fh \otimes (1 + x_1)^k$. This shows in particular, that $\beta = k\epsilon_2$ for some $k \in \mathbb{F}_p^*$ and $\alpha(h \otimes 1) \neq 0$. As $\big(h \otimes (1 + x_1)^k\big)^{[p]} = h \otimes 1$, the claim follows. □

Proposition 20.3.5. *Set*

$$\begin{aligned} J_\mu &:= Q_\mu \cap \operatorname{rad} Q(\mu) && \text{if } \mu \notin \Delta, \\ J_\mu &:= \big\{x \in Q_\mu \mid \kappa(x) = 0 \quad \text{for all } \kappa \in \Gamma(Q, T) \\ &\qquad\qquad \text{for which } Q(\kappa) \text{ is non-solvable}\big\}, && \text{if } \mu \in \Delta. \end{aligned}$$

Then $J := \sum_{\mu \in \Gamma(Q,T)} J_\mu$ is a T-invariant ideal of Q. One has $J \neq Q$ if and only if there is a root κ such that $Q(\kappa)$ is non-solvable.

Proof. Let α, β be arbitrary roots. We consider the various 2-sections $Q(\alpha, \beta)$ and observe that these are ruled by Proposition 20.3.3 and Lemma 20.3.4.

(a) Suppose $\alpha + \beta \notin \Delta$.

In case (E) of Proposition 20.3.3, all roots vanish on H (Theorem 10.3.2(2)), which is not true here as $(\alpha + \beta)(H) \neq 0$ by definition of Δ.

In case (A), $Q[\alpha, \beta]$ is solvable. Hence $J_\beta \subset \mathrm{rad}_T\, Q(\alpha, \beta)$.

In cases (B), (C), and (F), every non-vanishing 1-section $Q[\alpha, \beta](\mu)$ is isomorphic to $\mathfrak{sl}(2) \oplus \Psi_{\alpha,\mu}(H) \cap (\ker \mu)$ (Lemma 20.3.4(1)), and hence $J_\beta \subset \mathrm{rad}_T\, Q(\alpha, \beta)$.

In case (D) the 1-section $Q[\alpha, \beta](\beta)$ is isomorphic to $\mathfrak{sl}(2) \oplus \big(\Psi_{\alpha,\beta}(H) \cap (\ker \beta)\big)$ or else $Q[\alpha, \beta]_\beta$ contains an element which acts non-nilpotently on $Q[\alpha, \beta]$ (Lemma 20.3.4(1),(4)). Observe that in case (D) all root spaces are 1-dimensional. Thus also in this case $J_\beta \subset \mathrm{rad}_T\, Q(\alpha, \beta)$ holds. Consequently,

$$[Q_\alpha, J_\beta] \subset Q_{\alpha+\beta} \cap \mathrm{rad}_T\, Q(\alpha, \beta) \subset Q_{\alpha+\beta} \cap \mathrm{rad}\, Q(\alpha + \beta) = J_{\alpha+\beta}$$

holds in all cases.

(b) Suppose $\alpha + \beta \in \Delta$. If all 1-sections $Q(\mu)$ $(\mu \in \Gamma)$ are solvable, then $J_{\alpha+\beta} = Q_{\alpha+\beta}$ by definition. So assume that there is a non-solvable section $Q(\kappa)$. There is $i_0 \in \mathbb{F}_p^*$ such that $\pm i_0\kappa$ are roots on Q. Clearly, $i_0\kappa(H) \neq 0$. Choosing the right sign for i_0 we therefore may assume $(\beta + i_0\kappa)(H) \neq 0$. For $\rho_i := i_0\kappa + i(\alpha + \beta), i \in \mathbb{F}_p$, we have $(\beta + \rho_i)(H) = (\beta + i_0\kappa)(H) \neq 0$, $(\alpha + \beta + \rho_i)(H) = i_0\kappa(H) \neq 0$. The preceding case then yields

$$[[Q_\alpha, J_\beta], Q_{\rho_i}] \subset [J_\beta, Q_{\alpha+\rho_i}] + [Q_\alpha, [J_\beta, Q_{\rho_i}]] \subset J_{\alpha+\beta+\rho_i} = J_{\rho_{i+1}}.$$

Inductively we obtain $(\mathrm{ad}\,[Q_\alpha, J_\beta])^p(Q_{i_0\kappa}) \subset J_{i_0\kappa} \neq Q_{i_0\kappa}$. Then $\kappa([Q_\alpha, J_\beta]) = 0$, and this proves $[Q_\alpha, J_\beta] \subset J_{\alpha+\beta}$.

(c) By our arguments in (a) and (b) J is a T-invariant ideal of Q. By definition, $J = Q$ if all $Q(\mu)$ are solvable. On the other hand, if there is a non-solvable section $Q(\kappa)$, then $J_0 \neq H$ as $\kappa(H) \neq 0$. This proves the proposition. □

20.4 Pushing the classical case

In this section we will detect the classical algebras. Note that every 1-section in a classical algebra with respect to any torus of maximal dimension (i.e., with respect to a CSA) is classical or is contained in this CSA.

We assume in this section that all *T*-roots are solvable or classical and at least one root is non-solvable.

By definition this implies $Q = L$. Then Q is simple, whence $J = \{0\}$.

Lemma 20.4.1.

(1) *A root* $\alpha \in \Gamma$ *is non-solvable if and only if* $\alpha(H) \neq 0$.

(2) *If* α *is non-solvable, then* $L(\alpha) \cong \mathfrak{sl}(2) \oplus (H \cap (\ker \alpha))$.

(3) $\tilde{H} = T$.

Proof. (1) Clearly, if α is non-solvable, then $\alpha(H) \neq 0$. If α is solvable and $\alpha(H) \neq 0$, then by definition $L_\alpha = L_\alpha \cap \operatorname{rad} L(\alpha) \subset J = \{0\}$. So this is impossible.

(2) Suppose α is non-solvable. By definition, $L(\alpha)/\operatorname{rad} L(\alpha) \cong \mathfrak{sl}(2)$. As $L_{i\alpha} \cap \operatorname{rad} L(\alpha) \subset J = \{0\}$ for all $i \in \mathbb{F}_p^*$, we get the claim.

(3) Let $h \in \tilde{H}$ be $[p]$-nilpotent. Then h acts nilpotently on every L_α. If $\alpha(H) \neq 0$, then L_α is 1-dimensional and hence $[h, L_\alpha] = \{0\}$. In addition, $L = \sum_{\alpha(H) \neq 0} L_\alpha + [L_\alpha, L_{-\alpha}]$. Then $[h, L] = \{0\}$, and this gives $h = 0$. As a consequence, no element of $\tilde{H}$ is $[p]$-nilpotent. Then $\tilde{H} = T$. □

Given a non-solvable root $\alpha \in \Gamma$ we set

$$E(\alpha) := \{\mu \in \Delta \mid \alpha(L_\mu) \neq 0\},$$
$$\mathcal{B} := \{\alpha \in \Gamma \mid \alpha \text{ is a non-solvable root and } E(\alpha) \neq \{0\}\},$$
$$P(\alpha) := \sum_{\mu \in E(\alpha)} L_\mu, \qquad \text{for } \alpha \in \mathcal{B}.$$

Note that 0 is contained in $E(\alpha)$. If $\mu \in E(\alpha) \setminus \{0\}$, then $\mu(H) = 0$ while $\alpha(H) \neq 0$. Hence α and μ are $\mathbb{F}_p$-independent. Moreover, $L(\mu)$ is solvable. Lemma 20.3.4(4) applies and shows (as an equivalent condition) that $L[\alpha, \mu]$ is as in (D) of Proposition 20.3.3, i.e, $L[\alpha, \mu] \cong \mathfrak{sl}(2) \otimes \mathcal{O}(1; \underline{1})$.

Lemma 20.4.2. *If* $\alpha \in \mathcal{B}$, *then* $E(\alpha)$ *is an* $\mathbb{F}_p$*-vector space and* $P(\alpha)$ *is a subalgebra of* L.

Proof. Take $\kappa, \lambda \in E(\alpha) \setminus \{0\}$. Suppose κ and λ are $\mathbb{F}_p$-dependent. Note that $L[\alpha, \kappa](\kappa) = Fh \otimes \mathcal{O}(1; \underline{1})$. This shows that all of $\mathbb{F}_p\kappa$ are roots.

Next assume that κ and λ are $\mathbb{F}_p$-independent. Since $L[\alpha, \lambda]$ is as in case (D), all of $\alpha + \mathbb{F}_p\lambda$ are non-solvable roots. As $\alpha(L_\kappa) \neq 0$, one has $(\alpha + k\lambda)(L_\kappa) \neq 0$ for some $k \in \mathbb{F}_p^*$. Lemma 20.3.4(4) shows that the section $L[\alpha + k\lambda, \kappa]$ is as in case (D). In particular, $\alpha + k(\kappa + \lambda)$ is a non-solvable root. Lemma 20.3.4(2) implies that the section $L[\alpha, \alpha + k(\kappa + \lambda)]$ is as in cases (B) or (D). However, case (B) is impossible because there are 2 different non-solvable 1-sections $L(\alpha)$ and $L(\alpha + k(\kappa + \lambda))$. So only (D) is possible, whence $\alpha(L_{\kappa+\lambda}) \neq 0$ (Lemma 20.3.4(4)). This is the claim. □

Lemma 20.4.3. *If $\mathcal{B} \neq \emptyset$, then there are $\alpha \in \mathcal{B}$ and $\beta, \gamma \in E(\alpha)$ satisfying $L[\beta, \gamma] \cong H(2; \underline{1}; \Phi(\tau))^{(1)}$.*

Proof. (a) Choose a root $\alpha \in \mathcal{B}$ for which $E(\alpha)$ has maximal $\mathbb{F}_p$-dimension. Assume that $P(\alpha)$ is solvable. Choose in the derived series the member $P(\alpha)^{(m)}$ for which $\sum_{\mu \in E(\alpha) \setminus \{0\}} L_\mu$ is contained in $P(\alpha)^{(m)}$ but not in $P(\alpha)^{(m+1)}$. There is a root $\kappa \in E(\alpha) \setminus \{0\}$ such that $L_\kappa \not\subset P(\alpha)^{(m+1)}$. As a consequence, $\kappa(L_\mu) = 0$ for all $\mu \in E(\alpha) \setminus \{0\}$ (Proposition 1.3.2). Set

$$E(\alpha)' := \{\lambda \in E(\alpha) \setminus \{0\} \mid \lambda(L_\mu) = 0 \ \forall \mu \in E(\alpha)\}.$$

By the above (and the definition $\lambda(H) = 0$ for $\lambda \in \Delta$), $E(\alpha)'$ is non-empty.

(b) Choose any $\beta \in E(\alpha)'$. Since $\beta \neq 0$ and $\beta(H) = 0$ by definition, there is a non-zero root $\gamma \in \Gamma$ such that $\beta({L_\gamma}^{[p]}) \neq 0$ (which means $\beta(L_\gamma) \neq 0$). According to the definition of $E(\alpha)'$ we have $\gamma \notin E(\alpha)$. Since $\beta \in E(\alpha)$, $\alpha + i\beta$ is a non-solvable root for every $i \in \mathbb{F}_p$. Therefore there is $i_0 \in \mathbb{F}_p$ such that $\alpha' := \alpha + i_0\beta$ is a non-solvable root which in addition does not vanish on L_γ. Consider the section $L[\alpha', \gamma]$. It is not as in cases (A) or (E) of Proposition 20.3.3, because there is a non-solvable root. In cases (B), (C), and (F), $\Psi_{\alpha',\gamma}(L_\gamma)$ is either $\{0\}$ or else is spannend by an element which acts non-invertibly on $\Psi_{\alpha',\gamma}(L_{\alpha'})$. None of these are possible. Hence it is as in case (D). This shows that $\alpha' \in \mathcal{B}$ and $\gamma \in E(\alpha')$.

(c) Let $\mu \in E(\alpha)$ be arbitrary. Recall that $\alpha(L_\mu) \neq 0$, but $\beta(L_\mu) = 0$ by definition of $E(\alpha)'$. Then $\alpha'(L_\mu) \neq 0$ holds. Since $\mu \in \Delta$ (as $\mu \in E(\alpha)$), we get $\mu \in E(\alpha')$. Consequently,

$$E(\alpha') \supset E(\alpha) \cup \{\gamma\} \neq E(\alpha).$$

Since α was chosen such that $E(\alpha)$ has maximal dimension, this is the contradiction we were heading for. This contradiction shows that there is a root $\alpha \in \mathcal{B}$ for which $P(\alpha)$ is non-solvable.

(d) Take according to (c) $\alpha \in \mathcal{B}$ for which $P(\alpha)$ is non-solvable. Theorem 1.3.10 shows that there are roots $\beta, \gamma \in E(\alpha)$ such that $L(\beta, \gamma)$ is non-solvable. By definition $\beta(H) = \gamma(H) = 0$. Then Proposition 20.3.3 shows $L[\beta, \gamma] = Q[\beta, \gamma] \cong H(2; \underline{1}; \Phi(\tau))^{(1)}$. □

Lemma 20.4.4. $\mathcal{B} = \emptyset$.

Proof. (a) Assume the contrary and choose α, β, γ, as in the preceding lemma. We are going to construct a suitable non-solvable root λ. Note that $\gamma(L_\beta) \neq 0$ (Theorem 10.3.2(5)). The simplicity of L gives $L_\beta = \sum_{\mu(H) \neq 0} [L_\mu, L_{\beta-\mu}]$. Due to Theorem 17.3.2 all roots are linear on L_β. Hence there is $\delta \in \Gamma$ such that $\delta(H) \neq 0$ and $\gamma([L_\delta, L_{\beta-\delta}]) \neq 0$. In particular, $\beta - \delta$ is a root, and as $(\beta - \delta)(H) = \delta(H) \neq 0$, these roots are non-solvable (Lemma 20.4.1). Then $L(\delta, \beta - \delta)$ is as in case (D) of

Proposition 20.3.3 by Lemma 20.3.4(3). This implies

$$\delta, \beta - \delta \in \mathcal{B},\ \beta \in E(\delta) \cap E(\beta - \delta).$$

If $\gamma \in E(\delta)$, then set $\lambda := \delta$.

If $\gamma \notin E(\delta)$, then $\delta(L_\gamma) = 0$ by definition. In this case $(\beta-\delta)(L_\gamma) = \beta(L_\gamma) \neq 0$. Therefore $\gamma \in E(\beta - \delta)$. Put $\lambda := \beta - \delta$.

In both cases we have

$$\lambda \in \mathcal{B}, \quad \beta, \gamma \in E(\lambda), \quad \gamma([L_\lambda, L_{\beta-\lambda}]) \neq 0.$$

(b) Put $G := L(\lambda, \beta, \gamma)$. As $G(\beta, \gamma)$ is of type $H(2; \underline{1}; \Phi(\tau))^{(1)}$, one has (cf. Theorem 10.3.2, Corollary 17.3.3(4))

$$\beta(G_\gamma) \neq 0,\ \gamma(G_\beta) \neq 0,\ \beta(H) = \gamma(H) = 0.$$

Since $\lambda(H) \neq 0$, this shows that $T \cap G_{[p]}$ acts as a 3-dimensional torus on G. In particular, $TR(G) = 3$ and every ideal of G is T-invariant.

Let $\mu = i\lambda + j\beta + k\gamma$ with $i \in \mathbb{F}_p^*$ be a root. Then $j\beta + k\gamma \in E(\lambda)$ (Lemma 20.4.2) and therefore $L[\lambda, j\beta + k\gamma]$ of type (D) in Proposition 20.3.3 if $(j,k) \neq (0,0)$. This shows that $i\lambda$ also is a root, hence $i = \pm 1$ as λ is a root. We obtain $\Gamma(G,T) = \{\pm\lambda, 0\} + \mathbb{F}_p\beta + \mathbb{F}_p\gamma$.

Let κ be any root. If $\kappa = i\lambda + j\beta + k\gamma$ with $i \in \mathbb{F}_p^*$, then $\kappa(H) \neq 0$ and therefore $G(\kappa) \cong \mathfrak{sl}(2) \oplus (H \cap (\ker \kappa))$ (Lemma 20.4.1). As $H \cap (\ker \kappa) = H \cap (\ker \lambda)$, one has $[H \cap (\ker \lambda), L_\kappa] = \{0\}$. Next let $\kappa \in \mathbb{F}_p\beta + \mathbb{F}_p\gamma$. Then $L(\kappa)$ is nilpotent. By Theorem 17.1.2 $[H \cap (\ker \lambda), L_\kappa]$ acts nilpotently on L. The definition of J (Proposition 20.3.5) then gives $[H \cap (\ker \lambda), L_\kappa] \subset J \cap G$. We mentioned that $J = \{0\}$.

Consequently, $H \cap (\ker \lambda)$ is an ideal of G. It is proper because $H \neq H \cap (\ker \lambda)$.

(c) Let I be a maximal ideal of G containing $H \cap (\ker \lambda)$. Note that I is T-invariant by a previous argument. Consider any root $\mu \in \mathbb{F}_p\lambda + \mathbb{F}_p\beta + \mathbb{F}_p\gamma$ satisfying $\mu(H) \neq 0$. Then μ is non-solvable (Lemma 20.4.1), hence $G(\mu) \cong \mathfrak{sl}(2) \oplus (H \cap (\ker \mu))$. Recall that $H \cap (\ker \mu) = H \cap (\ker \lambda) \subset I$. If $G_\mu \subset I$, then $H \subset I$ and hence $H + \sum_{\nu(H)\neq 0} G_\nu \subset I$. Due to the construction (see (a)) there is $x \in [G_\lambda, G_{\beta-\lambda}] \subset I$ with $\gamma(x) \neq 0$. Then $\sum_{j\in\mathbb{F}_p} \sum_{i\in\mathbb{F}_p^*} G_{i\gamma+j\beta} \subset I$. Finally, as $\beta(G_\gamma) \neq 0$, then $\sum_{j\in\mathbb{F}_p^*} G_{j\beta} \subset I$. But then $G = I$, which is not true. Consequently, $G_\mu \cap I = \{0\}$ if $\mu(H) \neq 0$.

Consider any nonzero root satisfying $\mu(H) = 0$ which means $\mu \in \mathbb{F}_p\beta + \mathbb{F}_p\gamma$. Suppose $G_\mu \subset I$. Since $G(\beta, \gamma)$ is of type $H(2; \underline{1}; \Phi(\tau))^{(1)}$, there is $y \in G_\mu$ and a root $\kappa \in \mathbb{F}_p\beta + \mathbb{F}_p\gamma$ for which $\kappa(y) \neq 0$. Then $(\lambda + i_0\kappa)(y) \neq 0$ for some $i_0 \in \mathbb{F}_p^*$. But then $G_{\lambda+i_0\kappa} \subset I$ and $\lambda + i_0\kappa$ is non-solvable. The preceding case gives a contradiction.

As a result, $\overline{G} := G/I \neq \{0\}$ is a T-invariant and simple algebra and carries the same roots as G.

(d) Set $\overline{G}_{\pm 1} := \sum_{i,j\in\mathbb{F}_p} \overline{G}_{\pm\lambda+i\beta+j\gamma}$ and $\overline{G}_0 := \overline{G}(\beta,\gamma)$. Since $\Gamma(G,T) = \{\pm\lambda, 0\} + \mathbb{F}_p\beta + \mathbb{F}_p\gamma$, one has a $\mathbb{Z}$-grading

$$\overline{G} = \overline{G}_{-1} \oplus \overline{G}_0 \oplus \overline{G}_1.$$

Corollary 4.1.5 now yields that $\overline{G}$ is classical of rank 3, forcing $\beta(H) \neq 0$. This contradiction proves the lemma. □

Our classification result in this context is as follows.

Theorem 20.4.5. *Let L be a finite dimensional simple Lie algebra over an algebraically closed field F of characteristic $p > 3$, and assume that all tori of maximal dimension in the minimal p-envelope of L are standard. Suppose there is such a torus T for which all roots in $\Gamma(L,T)$ are solvable or classical and one root in $\Gamma(L,T)$ is classical. Then L is of classical type.*

Proof. Since there is a classical root, one has $J \neq L$. Then $J = \{0\}$ holds. Suppose there is a solvable root $\mu \neq 0$. Then $\mu(H) = 0$ (Lemma 20.4.1) and therefore, by definition of J, there is a non-solvable root α such that $\alpha(L_\mu) \neq 0$. But this means $\alpha \in \mathcal{B}$. This contradicts Lemma 20.4.4. As a consequence, no root is solvable.

Therefore all 1-sections are of the form $L(\mu) \cong \mathfrak{sl}(2) \oplus (H \cap (\ker \mu))$ and all 2-sections are classical of types (C) or (F) of Proposition 20.3.3. The MILLS–SELIGMAN Classification Theorem 4.1.2 applies and yields the claim. □

20.5 The filtration defined by $Q(L,T)$

From now on we assume that

T is standard, all T-roots are proper, and there is a non-solvable and non-classical T-root.

It will turn out that Q defines a suitable filtration of L. Our first result in this direction is the following proposition.

Proposition 20.5.1. *Q is a maximal T-invariant subalgebra of L.*

Proof. (a) By the present assumption there is a non-solvable and non-classical T-root α', which is then Witt or Hamiltonian. As a consequence of this assumption, $Q(\alpha') \neq L(\alpha')$, whence $Q \neq L$. Assume that G is a T-invariant subalgebra of L such that

$$Q \subsetneq G \subset L.$$

As every T-module is semisimple, there is a decomposition $G = Q + \sum_\mu G \cap L_\mu$. The construction of Q ensures the existence of a non-solvable and non-classical root

α with $L(\alpha) \cap G \neq L(\alpha) \cap Q$. As $L(\alpha) \cap Q = Q(\alpha)$ is a maximal subalgebra of $L(\alpha)$, this implies that $L(\alpha) \subset G$. As $L(\alpha) \not\subset Q$, one has $\alpha \notin \Delta$.

(b) Let β be any root. We now refer to Proposition 20.3.2(3),(4). If $L[\alpha, \beta]$ is as in (4) of that proposition, then there exist roots $\zeta \in \Gamma_-$ and $\eta \in \Delta$ which span $\mathbb{F}_p\alpha + \mathbb{F}_p\beta$, and there exists (for all $i \in \mathbb{F}_p^*$) $u_{i\eta} \in Q_{i\eta}$ such that $\zeta(u_{i\eta}) \neq 0$. Since $\alpha \notin \Delta$, one has $\alpha = j_0\zeta + i_0\eta$ with $j_0 \neq 0$. Then $\alpha(u_\eta) \neq 0$ holds and therefore the element u_η acts invertibly on $\sum_{i\in\mathbb{F}_p} L_{j\alpha+i\eta}$ for every nonzero j. As $L(\alpha)$ and u_η are contained in G, this implies that $L(\alpha, \beta) \subset G + L(\eta)$. Moreover, as $\eta \in \Delta$, $L(\eta)$ is solvable and so it is contained in $Q \subset G$. Thus we obtain $L_\beta \subset G$.

Otherwise case (3) of Proposition 20.3.2 holds. Put

$$W_j := \sum_{i\in\mathbb{F}_p} L_{i\alpha+j\beta}/G_{i\alpha+j\beta}, \text{ whence } L(\alpha,\beta)/G(\alpha,\beta) = \sum_{j\in\mathbb{F}_p^*} W_j.$$

Note that each W_j is a $L(\alpha)$-module. Let U denote any composition factor and ρ the respective representation. Due to the proposition we have

$$\dim U \leq \dim \sum_{i\in\mathbb{F}_p} L_{i\alpha+j\beta}/G_{i\alpha+j\beta} \leq |(j\beta + \mathbb{F}_p\alpha) \cap \Gamma_-| \leq 2$$

and hence $\dim \rho\big(L(\alpha)\big) \leq (\dim U)^2 \leq 4 < p$. As α is Witt or Hamiltonian, one has $L[\alpha]^{(1)} = L[\alpha]^{(2)}$ and $\dim L[\alpha]^{(1)} \geq p$ (Theorem 17.1.8). Therefore it can only be that the image under the mapping

$$L[\alpha] \to L(\alpha)/(\operatorname{rad} L(\alpha) + \ker \rho)$$

is abelian. This means that $\rho(L(\alpha))$ is solvable. Representation theory of solvable algebras implies that U is 1-dimensional and hence $L(\alpha)^{(1)}$ annihilates U.

(c) As a result of (b), it can only be that $L_\beta \subset G$ or $L(\alpha)^{(1)}$ acts nilpotently on $L(\alpha, \beta)/G(\alpha, \beta)$. Put $N := \bigcap_{n=0}^{\infty} L(\alpha)^{(n)}$. Then $N \neq \{0\}$, $N^{(1)} = N \subset G$, and $[N, L_\beta] \subset G$ holds in both cases for all β. As a consequence, $[N, L] \subset G$. Due to Proposition 1.3.5, for every $x \in N$, G contains the ideal $I_x := L_1(\operatorname{ad} x) + [L_1(\operatorname{ad} x), L_1(\operatorname{ad} x)]$ generated by the Fitting-1-component of $\operatorname{ad} x$. ENGEL's theorem proves that there is $x \in N$, such that $N \cap L_1(\operatorname{ad} x) \neq \{0\}$. For such x one has $G \supset I_x = L$. □

To define a filtration of L we set $L_{(0)} := Q$, choose a $(L_{(0)}+T)$-invariant subspace $L_{(-1)} \supset L_{(0)}$ which is $(L_{(0)} + T)$-irreducible and define for $n > 0$

$$L_{(n)} := \{x \in L_{(n-1)} \mid [x, L_{(-1)}] \subset L_{(n-1)}\}$$
$$L_{(-n-1)} := [L_{(-1)}, L_{(-n)}] + L_{(n-1)}.$$

There are r, s such that

$$L+T = (L_{(-r)}+T) \supsetneq \cdots \supsetneq (L_{(-1)}+T) \supsetneq (L_{(0)}+T) \supsetneq L_{(1)} \cdots \supsetneq L_{(s+1)} = \{0\}$$

is a standard filtration for $L+T$ in the sense of Definition 3.5.1. We have depth $r>0$ and height $s\ge 0$ for this filtration. Every member $L_{(i)}$ of this filtration is T-invariant.

Next we define sets of roots

$$\Gamma_n := \{\mu\in\Gamma \mid L_\mu\cap L_{(n)}\not\subset L_{(n+1)}\},$$
$$\Gamma_+ := \cup_{n>0}\Gamma_n,$$
$$\Delta_n := \Delta\cap\Gamma_n.$$

Note that $\Gamma_- = \cup_{n<0}\Gamma_n$.

Let us determine $L_{(1)}$.

Lemma 20.5.2. *Let $\alpha\in\Delta$, $\beta\in\Gamma$, and $x\in L_\alpha$. If $\beta(x)=0$, then $[x,L_\beta]\subset Q$.*

Proof. Observe that $x\in Q$. We only have to consider the case where $L_\beta\not\subset Q$. If $\alpha=0$, then $x\in H$ acts nilpotently L_β/Q_β, which is 1-dimensional. So assume $\alpha\ne 0$. Consider the 2-section $L(\alpha,\beta)$. Since H does not distinguish the roots β and $\alpha+\beta$, case (3) of Proposition 20.3.2 does not apply. So we are in case (4) of this proposition.

Suppose $L[\alpha,\beta]=W(1;\underline{2})$. Due to Theorem 7.6.5(4) $L[\alpha,\beta]_\alpha$ is 1-dimensional and spanned by a non-nilpotent element. Hence $x\in\ker\Psi_{\alpha,\beta}$, whence $[x,L_\beta]\subset\ker\Psi_{\alpha,\beta}\subset Q$.

In case (7) of Theorem 20.1.1 we have $L[\alpha,\beta]\subset H(2;(1,2))$. Then $\Psi_{\alpha,\beta}(L(\alpha))$ is a CSA contained in $\bar{Q}(\alpha,\beta)\subset H(2;(1,2))_{(0)}$ (Proposition 20.2.2(1)), and $\Psi_{\alpha,\beta}(x)$ is an element of this CSA which acts nilpotently on $H(2;(1,2))$. This can only be if $\Psi_{\alpha,\beta}(x)\in H(2;(1,2))_{(1)}$. As a consequence,

$$[\Psi_{\alpha,\beta}(x),H(2;(1,2))]\subset H(2;(1,2))_{(0)}.$$

This in combination with Proposition 20.2.2(1) gives $[x,L(\alpha,\beta)]\subset Q$.

Suppose $L[\alpha,\beta]$ is as in case (9) of Proposition 20.1.1 with $\mathfrak{s}$ non-classical. Then $L[\alpha,\beta]\subset(\mathrm{Der}\,\mathfrak{s})^{(1)}\otimes\mathcal{O}(1;\underline{1})$ and $\Psi_{\alpha,\beta}(T)=(Fh\otimes 1)\oplus(F\mathrm{Id}\otimes(1+x_1)\partial_1))$, Fh a proper torus in $\mathfrak{s}$ (see also Proposition 20.2.4). The assumption "$\alpha\in\Delta$" yields $\alpha(h\otimes 1)=\alpha(H)=0$, proving $\Psi_{\alpha,\beta}(x)\in C_{(\mathrm{Der}\,\mathfrak{s})^{(1)}}(h)\otimes\mathcal{O}(1;\underline{1})$. In both cases $\mathfrak{s}=W(1;\underline{1})$ and $\mathfrak{s}=H(2;\underline{1})^{(2)}$, $C_{(\mathrm{Der}\,\mathfrak{s})^{(1)}}(h)$ is a CSA contained in $(\mathrm{Der}\,\mathfrak{s})^{(1)}{}_{(0)}$ (as Fh is a proper torus). As $\beta(x)=0$, we find that x acts nilpotently on $L(\alpha,\beta)$. Then $\Psi_{\alpha,\beta}(x)\in C_{(\mathrm{Der}\,\mathfrak{s})^{(1)}{}_{(1)}}(h)\otimes\mathcal{O}(1;\underline{1})$. As before, this proves the claim. □

Lemma 20.5.3. *Let $\alpha\notin\Delta$ and $\beta\in\Gamma$. Then $[Q(\alpha,\beta),\mathrm{rad}_T\,Q(\alpha,\beta)]$ acts nilpotently on L/Q.*

Proof. (a) Let us assume that the statement is not true. Choose a T-invariant ideal $I\subset\mathrm{rad}_T\,Q(\alpha,\beta)$ of $Q(\alpha,\beta)$ minimal with the property such that $[Q(\alpha,\beta),I]$ acts non-nilpotently on L/Q. The set $\cup_{\mu\in\Gamma}[Q(\alpha,\beta),I]_\mu$ is closed under the Lie bracket.

By the ENGEL–JACOBSON Theorem there is a root μ such that $[Q(\alpha, \beta), I]_\mu$ acts non-nilpotently on L/Q. Corollary 17.3.3(4) shows that $[Q(\alpha, \beta), I]_\mu \subset Q_\mu$ would act nilpotently if $\mu \notin \Delta$. So we have $\mu \in \Delta$. Theorem 17.1.2 yields that $L(\mu)^{(1)}$ acts nilpotently on L. Now we consider the Lie set

$$\bigcup_{\zeta \in \mathbb{F}_p\alpha + \mathbb{F}_p\beta} \left(\bigcup_{x \in Q_\zeta,\, y \in I_{\mu-\zeta}} [x, y]\right) \bigcup L(\mu)^{(1)}$$

and conclude that there are $\zeta \in \mathbb{F}_p\alpha + \mathbb{F}_p\beta$, $x \in Q_\zeta$, $y \in I_{\mu-\zeta}$ such that $[x, y]$ does not act nilpotently on L/Q. If ζ is contained in Δ, then ζ, μ cannot span the full lattice $\mathbb{F}_p\alpha + \mathbb{F}_p\beta$, because $\alpha \notin \Delta$. This gives that ζ, μ are dependent, whence $[x, y] \in L(\mu)^{(1)}$, a contradiction.

Consequently, $\zeta \notin \Delta$. Let $\kappa \in \Gamma_-$ be any root. In the case that $L(\zeta, \kappa)$ is as in Proposition 20.3.2(3) not all of $\{\kappa, \kappa + \zeta, \kappa + 2\zeta\}$ is contained in Γ_-. In the case that $L(\zeta, \kappa)$ is as in Proposition 20.3.2(4), there is $\eta \in \Delta$ so that κ, η span $\mathbb{F}_p\zeta + \mathbb{F}_p\kappa$. Write $\zeta = i_0\kappa + j_0\eta$ with $i_0 \neq 0$. Then

$$\{\kappa, \kappa + \zeta, \kappa + 2\zeta\} = \{\kappa, (1 + i_0)\kappa + j_0\eta, (2 + i_0)\kappa + 2j_0\eta\},$$

and again not all of $\{\kappa, \kappa+\zeta, \kappa+2\zeta\}$ is contained in Γ_-. Consequently, $(\operatorname{ad} x)^2(L_\kappa) \subset Q$. One gets $(\operatorname{ad} y)^2(L_\kappa) \subset Q$ by the same reason.

(b) Put $W := \sum_{i,j \in \mathbb{F}_p} L_{\kappa+i\alpha+j\beta}$. This is a $Q(\alpha, \beta)$-module. By construction of I we have that $[Q(\alpha, \beta), I^{(1)}]$ is a T-invariant ideal of $Q(\alpha, \beta)$ which acts nilpotently on W. In particular, $\gamma_3(I)$ acts nilpotently on W.

Next regard W a module for the solvable algebra $T + Fx + \operatorname{rad}_T Q(\alpha, \beta)$. Let U denote any composition factor with representation ϱ_U and set G the p-envelope of $\varrho_U\big(T + Fx + \operatorname{rad}_T Q(\alpha, \beta)\big)$ in $\mathfrak{gl}(U)$. Representation theory of solvable Lie algebras shows that $U = u(G) \otimes_{u(K)} F$ is induced by a 1-dimensional representation of a restricted subalgebra K of G. This implies that every element $g \in G \setminus K$ has the property that $\varrho_U(g)^{p-1} \neq 0$. Combining this with our result in (a) gives $\varrho_U(x), \varrho_U(y) \in K$. Then $[x, y]$ annihilates the 1-dimensional K-submodule. As a consequence, $\varrho_U([x, y])$ has eigenvalue 0.

(c) Recall that $\gamma_3(I)$ acts nilpotently on W. ENGEL's Theorem yields $\gamma_3(\varrho_U(I)) = \{0\}$. Then U is an eigenvalue module for $\varrho_U(I)$ with eigenvalue function say λ_U (Lemma 3.7.2). Combining this with the result of (b), we get $\lambda_U([x, y]) = 0$. As this is true for every composition factor, one obtains that $[x, y]$ acts nilpotently on W. As this is true for all $\kappa \in \Gamma_-$, $[x, y]$ acts nilpotently on L/Q, a contradiction. □

Lemma 20.5.4. *One has* $L_{(1)} = \oplus_{\alpha \in \Gamma} L_{(1)} \cap L_\alpha$ *and*

$$L_{(1)} \cap L_\alpha = (\operatorname{rad} Q(\alpha))_\alpha \text{ if } \alpha \notin \Delta,$$

$$L_{(1)} \cap L_\alpha = \{x \in L_\alpha \mid x \text{ is } [p]-\text{nilpotent}\} \text{ if } \alpha \in \Delta.$$

Proof. (a) By definition $L_{(1)}$ is T-invariant.

Set $I_\alpha := (\text{rad}\, Q(\alpha))_\alpha$ if $\alpha \notin \Delta$, $I_\alpha := \{x \in L_\alpha \mid x \text{ is } [p] - \text{nilpotent}\}$ if $\alpha \in \Delta$ and $I := \sum_{\alpha \in \Gamma} I_\alpha$. This is a T-invariant subspace. Since $L_{(1)}(\alpha)$ is a solvable ideal of $L_{(0)}(\alpha) = Q(\alpha)$, one has $L_{(1)}(\alpha) \subset \text{rad}\, Q(\alpha)$. As $L_{(1)} \cap L_\alpha$ acts nilpotently on L, it contains only $[p]$-nilpotent elements. Hence we have $L_{(1)} \subset I$.

(b) Suppose $\alpha \in \Delta$. Lemma 20.5.2 shows that $[I_\alpha, L] \subset Q$. By definition, this gives $I_\alpha \subset L_{(1)}$.

(c) Suppose $\alpha \notin \Delta$. Recall the definition of the ideal J of Q given in Proposition 20.3.5 and observe that $I_\mu = J_\mu$ if $\mu \notin \Delta$. Therefore $[I_\alpha, Q_\beta] \subset I_{\alpha+\beta}$ if $\alpha + \beta \notin \Delta$. If $\alpha + \beta \in \Delta$, then the definition of J gives that $[I_\alpha, Q_\beta]$ acts nilpotently on L/Q. Lemma 20.5.2 shows $[[I_\alpha, Q_\beta], L] \subset Q$, whence $[I_\alpha, Q_\beta] \subset L_{(1)}$.

As a consequence, $\sum_{\alpha \in \Gamma \setminus \Delta} I_\alpha + L_{(1)}$ is a T-invariant ideal of Q which contains I. By Corollary 17.3.3(4) and the ENGEL–JACOBSON Theorem this ideal acts nilpotently on L, hence annihilates $L_{(-1)}/Q$. By definition, it therefore is contained in $L_{(1)}$. □

Lemma 20.5.5. *Let $L[\alpha, \beta] \cong K(3; \underline{1})$. Then*

$$K(3; \underline{1})_{(1)} \subset \Psi_{\alpha,\beta}\big(H + L_{(1)} \cap L(\alpha, \beta)\big)$$

and

$$L_{(-1)} \cap L(\alpha, \beta) = \Psi_{\alpha,\beta}{}^{-1}\big(K(3; \underline{1})_{(-1)}\big) \cap L.$$

Proof. (a) Proposition 20.2.1 shows that $K(3; \underline{1})_{(0)} = \Psi_{\alpha,\beta}(Q \cap L(\alpha, \beta))$. Note that every $K(3; \underline{1})_{(i)}$ is $\Psi_{\alpha,\beta}(T)$-invariant, and $K(3; \underline{1})_{(1)}$ is an ideal of $K(3; \underline{1})_{(0)}$ acting nilpotently on $K(3; \underline{1})$. Therefore $K(3; \underline{1})_{(1)}(\mu)$ is a solvable ideal of $K(3; \underline{1})_{(0)}(\mu)$, for every nonzero root $\mu \in \mathbb{F}_p\alpha + \mathbb{F}_p\beta$. We conclude

$$K(3; \underline{1})_{(1)}(\mu) \subset \text{rad}\, K(3; \underline{1})_{(0)}(\mu) = \text{rad}\,\big(\Psi_{\alpha,\beta}(Q(\mu))\big) = \Psi_{\alpha,\beta}\big(\text{rad}\, Q(\mu)\big)$$

and

$$K(3; \underline{1})_{(1)} \subset \Psi_{\alpha,\beta}(H) + \sum_{\mu \in (\mathbb{F}_p\alpha + \mathbb{F}_p\beta) \setminus \{0\}} \Psi_{\alpha,\beta}\big((\text{rad}\, Q(\mu))_\mu\big).$$

Note that $\Psi_{\alpha,\beta}(T) \subset L[\alpha, \beta]$ in the present case and therefore $\Delta \cap (\mathbb{F}_p\alpha + \mathbb{F}_p\beta) = \{0\}$ holds. Then Lemma 20.5.4 shows that $(\text{rad}\, Q(\mu))_\mu \subset L_{(1)}$. This proves the first inclusion.

(b) The inclusion just proved implies that every root vector of $K(3; \underline{1})_{(1)}$ for a nonzero root maps $\Psi_{\alpha,\beta}\big(L_{(-1)} \cap L(\alpha, \beta)\big)$ into $\Psi_{\alpha,\beta}\big(Q \cap L(\alpha, \beta)\big) = K(3; \underline{1})_{(0)}$. Normalize $\Psi_{\alpha,\beta}(T) = FD_K(x_1x_2) + FD_K(x_3)$. Note that $K(3; \underline{1}) = FD_K(1) \oplus K(3; \underline{1})_{(-1)}$ and $D_K(x_1x_3) \in K(3; \underline{1})_{(1)}$. Therefore the above property can only happen if $\Psi_{\alpha,\beta}\big(L_{(-1)} \cap L(\alpha, \beta)\big) \subset K(3; \underline{1})_{(-1)}$. As $K(3; \underline{1})_{(-1)}/K(3; \underline{1})_{(0)}$ is an irreducible $K(3; \underline{1})_{(0)}$-module, we have equality.

As a consequence, ${\Psi_{\alpha,\beta}}^{-1}(K(3;\underline{1})_{(-1)}) \cap L = L_{(-1)} \cap L(\alpha,\beta) + (\ker \Psi_{\alpha,\beta}) \cap L = L_{(-1)} \cap L(\alpha,\beta)$. □

Recall the definition $L(\alpha)_{(n)} = {\Psi_\alpha}^{-1}(L[\alpha]_{(n)})$.

Lemma 20.5.6.

(1) *If α is Hamiltonian with $\pm\alpha \in \Gamma_-$, then*

$$L(\alpha)_{(n)} \subset L_{(n)} + H + \operatorname{rad} L(\alpha)\ \forall n \geq 0, \quad Q_{i\alpha} \subset L_{(1)}\ \forall i \in \mathbb{F}_p \setminus \{0, \pm 2\}.$$

(2) *If α is Witt with $-\alpha \in \Gamma_{-1}$, then*

$$L(\alpha)_{(n)} \subset L_{(n)} + \operatorname{rad} L(\alpha)\ \forall n \geq 0, \quad Q_{i\alpha} \subset L_{(1)}\ \forall i \in \mathbb{F}_p^*.$$

(3) *If $-\alpha \in \Gamma_{-2}$, then α is Witt and*

$$L(\alpha)_{(n)} \subset L_{(2n)} + \operatorname{rad} L(\alpha)\ \forall n \geq 0, \quad Q_{i\alpha} \subset L_{(2)}\ \forall i \in \mathbb{F}_p^*.$$

(4) *$r \leq 2$ and $s \geq r + 2$ hold.*

Proof. (1) Recall that $\Psi_\alpha(L(\alpha)) \subset H(2;\underline{1})^{(2)} + \Psi_\alpha(H)$. One may normalize $\psi_\alpha(T) = FD_H(x_1x_2)$. Lemma 20.5.4 yields

$$\begin{aligned} &D_H(x_1^3), D_H(x_1^2x_2), D_H(x_1x_2^2), D_H(x_2^3) \\ &\quad \in H(2;\underline{1})^{(2)}{}_{(1)} \cap \sum_{i\in\mathbb{F}_p^*} \Psi_\alpha(Q_{i\alpha}) \subset \sum_{i\in\mathbb{F}_p^*} \Psi_\alpha\big((\operatorname{rad} Q(\alpha)\big)_{i\alpha} \subset \Psi_\alpha(L_{(1)}(\alpha)). \end{aligned}$$

It is well known (see the proof of [S-F, Theorem 4.7.2]) that the n-fold products of these elements generate the homogeneous part $H(2;\underline{1})^{(2)}{}_n$. As a consequence, $L(\alpha)_{(n)} \subset L_{(n)} + H + \operatorname{rad} L(\alpha)$ for $n \geq 0$.

Recall that $\pm\alpha \in \Gamma_-$. The roots sticking out of $Q(\alpha)$ are determined by the vectors $D_H(x_1), D_H(x_2)$. This shows that $\alpha(D_H(x_1x_2)) \in \{\pm 1\}$. As $H(2;\underline{1})_{(0)} = FD_H(x_1^2) + FD_H(x_1x_2) + FD_H(x_2^2) + H(2;\underline{1})_{(1)}$, we get

$$Q(\alpha) = Q_{-2\alpha} + Q_{0\cdot\alpha} + Q_{2\alpha} + \operatorname{rad} Q(\alpha) \subset Q_{-2\alpha} + Q_{0\cdot\alpha} + Q_{2\alpha} + H + L_{(1)}.$$

(2) Note that

$$L_\alpha + L_{2\alpha} = Q_\alpha + Q_{2\alpha} = (\operatorname{rad} Q(\alpha))_\alpha + (\operatorname{rad} Q(\alpha))_{2\alpha} \subset L_{(1)}$$

as α is Witt and $L_{-\alpha}$ sticks out of Q.

Now suppose $L_{2\alpha} \subset L_{(2)} + \operatorname{rad} L(\alpha)$. Let us normalize $\Psi_\alpha(T) = Fx\partial$ and let $u_2 \in L_\alpha \cap L_{(1)}$ for which $\Psi_\alpha(u_2) = x^2\partial$ and $u_3 \in L_{2\alpha} \cap L_{(2)}$ for which $\Psi_\alpha(u_3) = x^3\partial$. Then $u_n := (\operatorname{ad} u_2)^{n-2}(u_3) \in L_{(n)}$ $(n \geq 3)$ has the property that $\Psi_\alpha(u_n) = (\operatorname{ad} \Psi_\alpha(u_2))^{n-2}(\Psi_\alpha(u_3))$ spans $Fx^{n+1}\partial$. Claim (2) follows.

Next suppose $L_{2\alpha} \not\subset L_{(2)} + \operatorname{rad} L(\alpha)$. Choose arbitrary $-\beta \in \Gamma_{-1}$. Put $\mathfrak{g} = L[\alpha, \beta]$. By checking the possibilities in Theorem 20.1.1 we intend to show that

$$2\alpha - \beta \text{ is solvable and } 2\alpha - \beta \notin \Delta, \text{ or } 2\alpha - \beta \text{ is Witt and } 2\alpha - \beta \notin \Gamma_-, \\ \text{or } [L_{2\alpha}, L_{-\beta}] \subset L_{(1)}$$

holds. If $2\alpha - \beta$ is solvable and $2\alpha - \beta \notin \Delta$, then $L_{2\alpha-\beta} = Q_{2\alpha-\beta} \subset L_{(1)}$ by Lemma 20.5.4. If $2\alpha - \beta$ is Witt and $2\alpha - \beta \notin \Gamma_-$, then $L_{2\alpha-\beta} \subset Q_{2\alpha-\beta} \subset L_{(1)}$ by the same lemma. So, when we have accomplished this intermediate claim, we conclude $[L_{2\alpha}, L_{(-1)}] \subset L_{(1)}$, whence $L_{2\alpha} \subset L_{(2)}$, contradicting our present assumption.

(a) Cases (1) and (4) of Theorem 20.1.1 are impossible, because in these cases all roots are solvable or classical. In case (2) there is only one non-solvable root direction. Therefore $-\beta = -\alpha$ (as both stick out of Q and α is Witt), and $2\alpha - \beta = \alpha \notin \Gamma_-$ is Witt. In case (3) $\mathbb{F}_p\alpha \cup \mathbb{F}_p\beta$ contain all non-solvable roots. Then $2\alpha - \beta \notin \Delta$ is solvable.

(b) Consider case (5) of Theorem 20.1.1. Note that $(\mathbb{F}_p\alpha + \mathbb{F}_p\beta) \cap \Delta = \{0\}$.

If $\mathfrak{g} = W(2; \underline{1})$, then only $\mathbb{F}_p^*\alpha \cup \mathbb{F}_p^*\beta \cup \mathbb{F}_p^*(\alpha - \beta)$ are non-solvable (Proposition 20.2.1(1)). Thus $2\alpha - \beta$ is solvable.

If $\mathfrak{g} = S(3; \underline{1})^{(1)}$, then we may by symmetry take $\partial_2 \in \mathfrak{g}_{-\alpha}$, $\partial_1 \in \mathfrak{g}_{-\beta}$. Then $\alpha(x_1\partial_1 - x_3\partial_3) = 0$ and $\alpha(x_2\partial_2 - x_3\partial_3) = 1$. Theorem 17.2.4(1) applies and yields

$$\begin{aligned} \mathfrak{g}_{2\alpha} \subset & \sum_{\substack{1 \le i \le p-1,\ 0 \le j \le p-1 \\ j-i+1 \equiv 2,\ 2i+j \le 3}} F D_{1,3}(x_1^i x_2^j x_3^i) \\ & + \sum_{\substack{0 \le i \le p-2,\ 0 \le j \le p-1 \\ j-i-1 \equiv 2,\ 2i+j \le 2}} F D_{1,2}(x_1^{i+1} x_2^j x_3^i) \\ & + \sum_{\substack{0 \le i \le p-2,\ 0 \le j \le p-1 \\ j-i-1 \equiv 2,\ 2i+j \le 2}} F D_{2,3}(x_1^i x_2^j x_3^{i+1}) + \mathfrak{g}_{(2)} = \mathfrak{g}_{(2)}. \end{aligned}$$

This gives $\Psi_{\alpha,\beta}([L_{2\alpha}, L_{-\beta}]) \subset \mathfrak{g}_{(1)} \cap \mathfrak{g}_{2\alpha-\beta} \subset \Psi_{\alpha,\beta}(\operatorname{rad} Q(2\alpha - \beta))_{2\alpha-\beta}$. Lemma 20.5.4 yields

$$[L_{2\alpha}, L_{-\beta}] \subset \big(\operatorname{rad} Q(2\alpha - \beta)\big)_{2\alpha-\beta} + (\ker \Psi_{\alpha,\beta})_{2\alpha-\beta} \subset L_{(1)}.$$

If $\mathfrak{g} \subset H(4; \underline{1})$, then no root is Witt (Proposition 20.2.1(3)). So this case is impossible.

If $\mathfrak{g} = K(3; \underline{1})$, then $-\alpha, -\beta \in \Gamma_{-1}$ (Lemma 20.5.5). The only non-solvable roots are $\mathbb{F}_p^*\alpha \cup \mathbb{F}_p^*\beta \cup \mathbb{F}_p^*(\alpha + \beta) \cup \mathbb{F}_p^*(\alpha - \beta)$ (Proposition 20.2.1(4)). Then $2\alpha - \beta$ is solvable.

(c) Suppose Proposition 20.3.2(4) occurs. Since α is Witt, one has $(\mathbb{F}_p\alpha+\mathbb{F}_p\beta)\cap\Gamma_- = -\alpha + \mathbb{F}_p\eta$ for some $\eta \in \Delta$. In particular, $-\beta = -\alpha + i_0\eta$, and hence $2\alpha - \beta = \alpha - i_0\eta \notin \Gamma_-$ is Witt. This solves case (6) with $\mathfrak{g} = W(1;\underline{2})$ and cases (7) and (9) of Theorem 20.1.1.

If $\mathfrak{g} = H(2;\underline{1};\Phi(\tau))^{(1)}$, then all roots are solvable. So this is impossible.

If $\mathfrak{g} = H(2;\underline{1};\Phi(1))$ or if $\mathfrak{g}$ is as in (8), then Proposition 20.2.3 shows that the only non-solvable roots are $\mathbb{F}_p^*\alpha \cup \mathbb{F}_p^*\beta \cup \mathbb{F}_p^*(\alpha-\beta)$. Then $2\alpha-\beta$ is solvable. Computing traces on $\mathfrak{g}/\mathfrak{g}_{(0)}$ we obtain that $\alpha+\beta$ vanishes on $\mathfrak{t}\cap\mathfrak{g}$. As $\alpha \notin \Delta$ this gives $2\alpha-\beta \notin \Delta$.

This finishes case (6) and settles cases (8) of Theorem 20.1.1.

(d) There remains the case that $L[\alpha,\beta]$ is ruled by Proposition 20.2.5. If $d = 0$, an argument as before gives that $2\alpha-\beta$ is solvable.

So assume $d \neq 0$. We take the notations of $\mathcal{D}$, ζ, κ, m from that proposition. Note that $\mathfrak{g}_{(1)} := \mathfrak{g} \cap \big(H(2;\underline{1}) \otimes \mathcal{O}(r;\underline{1})_{(1)} + \mathrm{Id} \otimes \mathcal{D}_{(1)}\big)$ is a $\Psi_{\alpha,\beta}(T)$-invariant nilpotent ideal of $\bar{Q}(\alpha,\beta)$. It contains $D_H(x_1)\otimes\mathcal{O}(r;\underline{1})_{(1)}$. Then $[\mathfrak{g}_{(1)},\mathfrak{g}_{-\kappa}]$ contains $D_H(x_1)\otimes 1$. Taking preimages, we find $u \in L_{(1)}$, $v \in L_{-\kappa}$ so that $[u,v] \notin Q$. This can only be if $-\kappa \in \cup_{-i<-1}\Gamma_{-i}$, and this gives $\mathbb{F}_p^*\kappa \cap \Gamma_{-1} = \emptyset$ (Proposition 20.3.2(2)).

As a consequence, $\alpha,\beta \notin \mathbb{F}_p^*\kappa$. After an adjustment we may take $-\alpha = \zeta - m\kappa$ and $-\beta = -\zeta - m\kappa$. Then $2\alpha - \beta = -3\zeta + m\kappa$. This root is not contained in $\mathbb{F}_p(\pm\zeta - m\kappa) \cup \mathbb{F}_p\zeta \cup \mathbb{F}_p\kappa$. Hence $2\alpha - \beta$ is solvable.

(3) Now suppose $-\alpha \in \Gamma_{-2}$. There are roots $-\beta, -\alpha+\beta \in \Gamma_{-1}$ and nonzero x, y such that $x \in L_{-\beta} \cap (L_{(-1)} \setminus Q)$, $y \in L_{-\alpha+\beta} \cap (L_{(-1)} \setminus Q)$, and $[x,y] \notin L_{(-1)}$. Proposition 20.3.2(5) shows that only $L[\alpha,\beta] \cong K(3;\underline{1})$ or cases (10), (11) of Theorem 20.1.1 can occur. Proposition 20.2.5(5) gives that in the latter cases one has $[x,y] \in Q$. So we are in the case when $L[\alpha,\beta] \cong K(3;\underline{1})$. Proposition 20.2.1(4) shows that α is Witt.

The proof of [S-F, Theorem 4.7.5] gives that the n-fold products of the homogeneous space $K(3;\underline{1})_1$ generates the homogeneous space $K(3;\underline{1})_n$. Since $K(3;\underline{1})_1 \cap C_{K(3;\underline{1})}(FD_K(x_3)) = \{0\}$, Lemma 20.5.5 gives $K(3;\underline{1})_1 \subset {\Psi_{\alpha,\beta}}^{-1}(L_{(1)} \cap L(\alpha,\beta))$. We conclude for $n \geq 1$

$$K(3;\underline{1})_{(n)} = \sum_{i\geq n} K(3;\underline{1})_i \subset \sum_{i\geq n} \gamma_i\big(\Psi_{\alpha,\beta}(L_{(1)} \cap L(\alpha,\beta))\big)$$
$$\subset \Psi_{\alpha,\beta}(L_{(n)} \cap L(\alpha,\beta)).$$

Next we have to check that the filtrations of $\Psi_\alpha(L(\alpha)) \cong W(1;\underline{1})$ and $K(3;\underline{1})$ match. We may normalize (see the respective part (e) of the proof of Proposition 20.2.1) $\Psi_{\alpha,\beta}(T) = FD_K(x_1x_2) \oplus FD_K(x_3)$. Since $\Psi_{\alpha,\beta}([x,y]) \notin K(3;\underline{1})_{(-1)}$ by Lemma 20.5.5, we have $\alpha(D_K(x_1x_2)) = 0$. Theorem 17.2.8(3a) applies and shows for $n \geq 0$

$$\Psi_{\alpha,\beta}(L(\alpha)_{(n)}) \subset \sum_{i\geq n+1} FD_K({x_3}^i) + \mathrm{rad}\, K(3;\underline{1})(\alpha) \subset K(3;\underline{1})_{(2n)} + \mathrm{rad}\, K(3;\underline{1})(\alpha).$$

The claim follows.

(4) If $L_{(-2)}$ were different from L, there would be roots α and β and nonzero x, y such that $x \in L_\alpha \cap L_{(-1)}, x \notin Q, y \in L_\beta \cap L_{(-2)}, y \notin L_{(-1)}$ and $[x, y] \notin L_{(-2)}$. Proposition 20.3.2(5) shows that only $L[\alpha, \beta] \cong K(3; \underline{1})$ or cases (10) and (11) of Theorem 20.1.1 can occur. Proposition 20.2.5(5) gives that in the latter cases one has $[x, y] \in Q$. So we are in the case when Lemma 20.5.5 applies. Then $\Psi_{\alpha,\beta}(x) \in K(3; \underline{1})_{(-1)} \setminus K(3; \underline{1})_{(0)}, \Psi_{\alpha,\beta}(y) \notin K(3; \underline{1})_{(-1)}$. But in this case, $\alpha \in \Gamma_{-1}, \beta \in \Gamma_{-2}$, whence $\alpha + \beta \notin \Gamma_-$ (Proposition 20.2.1), which gives $[x, y] \in Q$, a contradiction. Hence only $r = 1, 2$ are possible.

Due to the previous results we have $s \geq p - 2 \geq 3$ if $r = 1$ and $s \geq 2(p - 2) \geq 6$ if $r = 2$. □

Lemma 20.5.7. *Let $i = 1, 2$. For every $\alpha \in \Gamma_{-i}$ and $u \in (L_\alpha \cap L_{(-i)}) \setminus L_{(-i+1)}$ there is $v \in L_{-2\alpha} \cap L_{(2i)}$ such that $\alpha([u, [u, v]]) \neq 0$.*

Proof. Suppose $i = 1$ and $\alpha \in \Gamma_{-1}$. Then α is Witt or Hamiltonian. We may normalize $\Psi_\alpha(T) = Fx\partial$ and $\Psi_\alpha(T) = FD_H(x_1x_2)$ in the respective cases. If α is Witt, then may take $\Psi_\alpha(u) = \partial$, and due to Lemma 20.5.6 there is an eigenvector $v \in L_{(2)}$, for which $\Psi_\alpha(v) = x^3\partial$. We obtain that $\Psi_\alpha([u, [u, v]]) = 6x\partial \neq 0$. One argues similarly if α is Hamiltonian.

Suppose that $i = 2$ and $\alpha \in \Gamma_{-2}$. Due to Lemma 20.5.6 α is Witt. We may normalize $\Psi_\alpha(T) = Fx\partial$ and take $\Psi_\alpha(u) = \partial$, and due to Lemma 20.5.6 there is an eigenvector $v \in L_{(4)}$, for which $\Psi_\alpha(v) = x^3\partial$. We obtain that $\Psi_\alpha([u, [u, v]]) = 6x\partial \neq 0$. □

20.6 Determining $\mathfrak{G}(L, T)$

We will next investigate the graded algebra $\mathrm{gr}\,(L + T)$ associated with the filtration under consideration. Due to Proposition 3.5.3(1), this graded algebra satisfies properties $(g1) - (g3)$ of Notation 3.5.2. Let $M(\mathrm{gr}\,(L + T))$ denote the ideal defined in Notation 3.5.2. Lemma 20.5.7 implies $M(\mathrm{gr}\,(L+T)) = \{0\}$. Proposition 3.5.3(3) now yields that $\mathrm{gr}\,(L + T)$ also satisfies Property $(g4)$ of Notation 3.5.2. Since $s \geq r + 2 \geq 3$, Theorem 3.5.6 applies with $A(\mathrm{gr}\,(L+T))_{r+1} \supset \mathrm{gr}_{r+1}([L_{(-1)}, L_{(r+2)}]) \neq \{0\}$. Observe that according to this theorem

$$\mathfrak{A} := A(\mathrm{gr}\,(L + T))$$

is $\mathrm{gr}_0\,(L + T)$-simple. Set

$$\mathfrak{T} := \mathrm{gr}_0\, T.$$

Proposition 20.6.1.

(1) $\mathfrak{A} = \oplus_{i=-r}^{s'} \mathfrak{A}_i$ *with $r \in \{1, 2\}$ and $s' > r$.*

(2) $\mathfrak{A}_{-r}$, $\mathfrak{A}_{s'}$ *are* $\mathrm{gr}_0(L+T)$*-irreducible and* $[\mathfrak{A}_0, \mathfrak{A}_{-r}] = \mathfrak{A}_{-r}$, $[\mathfrak{A}_0, \mathfrak{A}_{s'}] = \mathfrak{A}_{s'}$ *hold.*

(3) $\mathfrak{A}_i = \mathrm{gr}_i L$ *for* $i < 0$ *and* $\mathfrak{A}_{i-1} = [\mathfrak{A}_{-1}, \mathfrak{A}_i]$ *holds for all* $i \le s'$.

(4) *Let* $i > 0$. *For every* $x \in \mathfrak{A}_{-i,-\alpha} \setminus \{0\}$ *there is* $y \in \mathfrak{A}_{2i,2\alpha}$ *such that* $\alpha([x,[x,y]]) \neq 0$. *For every* $y \in \mathfrak{A}_i \setminus \{0\}$ *there is* $x \in \mathfrak{A}_{-1}$ *such that* $[x,y] \neq 0$.

Proof. Theorem 3.5.6 states that $\mathfrak{A}$ is graded with height $s' \geq s-1 > r$. Statements (2) and (3) follow from Theorem 3.5.6, while Statement (4) is a direct consequence of Lemma 20.5.7 and Property $(g1)$ for $\mathrm{gr}\, L$. □

Lemma 20.6.2.

(1) $\mathfrak{A}_{-1}$ *is* $(\mathfrak{A}_0 + \mathfrak{T})$*-irreducible.*

(2) $\mathfrak{A}$ *is* $\mathfrak{T}$*-simple.*

(3) *If* α *is a non-solvable root, then* $\sum_{i \in \mathbb{F}_p^*} (\mathrm{gr}\, L)_{i\alpha} \subset \mathfrak{A}$.

(4) $(\mathrm{gr}_0 L)_\mu \subset \mathfrak{A}_0$ *for all* $\mu \notin \Delta$.

Proof. (1) Let V be a minimal $(\mathfrak{A}_0 + \mathfrak{T})$-invariant subspace of $\mathfrak{A}_{-1}$ and $-\kappa \in \Gamma_{-1}$ with $V_{-\kappa} \neq \{0\}$. Note that $M := [V,V] + V + \sum_{i \geq 0} \mathfrak{A}_i$ is a graded subalgebra of $\mathfrak{A}$. Proposition 20.6.1 shows that for nonzero $x \in V_{-\kappa}$ there is $y \in \mathfrak{A}_{2,2\kappa}$ such that $\kappa([x,[x,y]]) \neq 0$. Since $[y, \mathfrak{A}] \subset \sum_{i \geq 0} \mathfrak{A}_i \subset M$, we have that y and hence $h := [x,[x,y]]$ annihilates $\mathfrak{A}/M$. As $\mathfrak{A}$ is an ideal of $\mathrm{gr}\, L$, we obtain that h acts nilpotently on $(\mathrm{gr}\, L)/M$. This implies that the Fitting-1-component $(\mathrm{gr}\, L)_1(\mathrm{ad}\, h)$ is contained in M. Then M contains the non-zero $(\mathrm{gr}\, L)$-ideal $(\mathrm{gr}\, L)_1(\mathrm{ad}\, h) + [(\mathrm{gr}\, L)_1(\mathrm{ad}\, h), (\mathrm{gr}\, L)_1(\mathrm{ad}\, h)]$. Since $[\mathfrak{T}, h] = \{0\}$, this ideal is $\mathfrak{T}$-invariant. We mentioned, however, that $\mathfrak{A}$ is $\mathrm{gr}(L+T)$-simple. This can only be when $\mathfrak{A} = M$, whence $\mathfrak{A}_{-1} = V$. This is the claim.

(2) Let J be a minimal $\mathfrak{T}$-invariant ideal of $\mathfrak{A}$. Taking in Lemma 3.5.4(1) $\mathfrak{A} + \mathfrak{T}$ for L we obtain that $J \subset \sum_{i \leq -2} \mathfrak{A}_i$ or $\mathfrak{A}_{-1} \subset J$. The first case is impossible by Proposition 20.6.1(4). Proposition 20.6.1(2),(3) then proves $J = \mathfrak{A}$.

(3) If α is non-solvable and non-classical, then there is $l \in \mathbb{F}_p^*$ with $l\alpha \in \Gamma_{-k}$ (suitable $k > 0$). Proposition 20.6.1(4) shows that there are elements $x \in \mathfrak{A}_{-k,l\alpha}$, $y \in \mathfrak{A}_{k,-l\alpha}$ such that $\alpha([x,y]) \neq 0$. As $\mathfrak{A}$ is an ideal in $\mathrm{gr}\, L$, we get $\sum_{i \in \mathbb{F}_p^*} (\mathrm{gr}\, L)_{i\alpha} \subset \mathfrak{A}$.

Suppose α is classical. Take (after an adjustment of α) $h' \in [L_\alpha, L_{-\alpha}]$ such that $\alpha(h') \neq 0$. Then $h' \notin L_{(1)}$, hence there is $\beta \in \Gamma_{-1}$ with $\beta(h') \neq 0$. Clearly, $\mathfrak{g} := L[\alpha, \beta]$ is not of the types (1)–(4) of Theorem 20.1.1 because there are two non-solvable root directions, they do not annihilate each other, and one of them is non-classical. In case (5) we have $\bar{Q}(\alpha, \beta) = \mathfrak{g}_{(0)}$ (Proposition 20.2.1). In this case one obtains $\mathfrak{g}_{(0)} = [\mathfrak{g}_{(-1)}, \mathfrak{g}_{(1)}] + \mathfrak{g}_{(1)}$. Then there is $h_0 \in \mathfrak{A}_0 \cap (\mathrm{gr}_0 H)$ for which $\alpha(h_0) \neq 0$. Argue as before to obtain the claim.

In cases (6) – (9) one has that $\Psi_{\alpha,\beta}(H) \subset \bar{Q}(\alpha,\beta)$ satisfies $TR(\Psi_{\alpha,\beta}(H),\mathfrak{g}) \leq 1$. This implies that $H/\big(H \cap (\ker\,\Psi_{\alpha,\beta}) + \mathrm{nil}\,H\big)$ is 1-dimensional and, as $\alpha(\mathrm{gr}_0\,H) \neq 0$, that there is $h_0 \in [\mathfrak{A}_{-1,\beta}, \mathfrak{A}_{1,-\beta}] \cap (\mathrm{gr}_0\,H)$ for which $\alpha(h_0) \neq 0$. Argue as before to obtain the claim.

We are left with cases (10) and (11). Since α is classical, one has $\Psi_{\alpha,\beta}(h') \in (\mathrm{Der}\,\mathfrak{s})^{(1)} \otimes \mathcal{O}(r;\underline{1})$ $(r = 1,2)$. Since $\beta(h') \neq 0$, one also has $\Psi_{\alpha,\beta}(L(\beta)) \subset (\mathrm{Der}\,\mathfrak{s})^{(1)} \otimes \mathcal{O}(r;\underline{1}) + \Psi_{\alpha,\beta}(H)$. Now one argues as in the previous case.

(4) Let $\mu \notin \Delta$. The case when μ is non-solvable has been dealt with in the preceding assertion. If μ is solvable, then Lemma 20.5.4 shows $L_\mu \subset L_{(1)}$, which means $(\mathrm{gr}_0\,L)_\mu = \{0\}$. □

Theorem 3.5.7 proves the existence of an isomorphism $\mathfrak{A} \cong \mathfrak{S}(L,T) \otimes \mathcal{O}(m';\underline{n})$, where $\mathfrak{S} := \mathfrak{S}(L,T)$ is simple. Note that $\mathcal{O}(m';\underline{n}) \cong \mathcal{O}(m;\underline{1})$ for a suitable m. Due to Theorem 3.6.1 one may take the isomorphism

$$\mathfrak{A}(L,T) \cong \mathfrak{S}(L,T) \otimes \mathcal{O}(m;\underline{1}), \quad \mathfrak{S} := \mathfrak{S}(L,T) \text{ simple}, \tag{20.6.1}$$

in such a way that

$$\mathfrak{T} = \mathrm{gr}_0\,T \subset (\mathrm{Der}\,\mathfrak{S}) \otimes F + \mathrm{Id} \otimes W(m;\underline{1}).$$

Since $\mathfrak{A}$ is $\mathfrak{T}$-simple, one has

$$\dim\,\pi_2(\mathfrak{T})/\pi_2(\mathfrak{T}) \cap W(m,\underline{1})_{(0)} = m.$$

Choose accordingly toral elements $t_1,\dots,t_m$ which span this space. Theorem 3.6.1 gives a realization

$$\mathfrak{T} = \big(\mathfrak{T}_0 \otimes F\big) \oplus \bigoplus_{i=1}^{m} F\mathrm{Id} \otimes (1+x_i)\partial_i, \quad \mathfrak{T}_0 \subset \mathrm{Der}\,\mathfrak{S} \text{ a torus.} \tag{20.6.2}$$

Set $\mathfrak{T}_1 := \oplus_{i=1}^{m} F\mathrm{Id} \otimes (1+x_i)\partial_i$ which is a subtorus of $\mathfrak{T}$. With this normalization, $\mathfrak{S} \cong \mathfrak{S} \otimes F$ is $\mathfrak{T}$-invariant and

$$\mathfrak{S}(L,T) \otimes F = C_{\mathfrak{A}(L,T)}(\mathfrak{T}_1). \tag{20.6.3}$$

The restriction $\mu \mapsto \mu|_{\mathfrak{T}_0 \otimes F}$ defines a bijection $\Gamma(\mathfrak{S}(L,T) \otimes F, \mathfrak{T}) \to \Gamma(\mathfrak{S}(L,T), \mathfrak{T}_0)$. Theorem 3.5.7(1) applies in the case under consideration and yields that $\mathfrak{S}$ is graded with a grading induced by that of $\mathfrak{A}$. In particular, the grading of $\mathfrak{S}$ is $\mathfrak{T}$-invariant. Now Proposition 20.6.1 and part of Lemma 20.6.2 have analogues for $\mathfrak{S}$.

Proposition 20.6.3.

(1) $\mathfrak{S} = \oplus_{i=-r}^{s'} \mathfrak{S}_i$.

(2) $\mathfrak{S}_{-r}$, $\mathfrak{S}_{s'}$ *are* $\mathfrak{S}_0$*-irreducible and* $[\mathfrak{S}_0, \mathfrak{S}_{-r}] = \mathfrak{S}_{-r}$, $[\mathfrak{S}_0, \mathfrak{S}_{s'}] = \mathfrak{S}_{s'}$ *hold.*

(3) $\mathfrak{S}_{i-1} = [\mathfrak{S}_{-1}, \mathfrak{S}_i]$ *holds for all* $i \le s'$.

(4) $\mathfrak{S}_{-1}$ *is* $(\mathfrak{S}_0 + \mathfrak{T}_0)$*-irreducible.*

(5) *Let* $i > 0$. *For every* $x \in \mathfrak{S}_{-i,-\alpha} \setminus \{0\}$ *there is* $y \in \mathfrak{S}_{2i,2\alpha}$ *such that* $\alpha([x,[x,y]]) \ne 0$. *For every* $y \in \mathfrak{S}_i \setminus \{0\}$ *there is* $x \in \mathfrak{S}_{-1}$ *such that* $[x,y] \ne 0$.

(6) $\mathrm{ann}_{\mathfrak{S}}(\mathfrak{S}_{-1}) = \mathfrak{S}_{-r}$.

(7) *If* $d \in \mathrm{Der}_0\, \mathfrak{S}$ *satisfies* $d(\mathfrak{S}_{-1}) = 0$, *then* $d = 0$.

Proof. Assertions (1) and part of (2) and (3)–(5) follow from the respective assertions for $\mathfrak{A}$. [S-F, Proposition 3.3.5] yields the remaining part of (2), namely that $\mathfrak{S}_{-r}, \mathfrak{S}_{s'}$ are $\mathfrak{S}_0$-irreducible. It also gives assertion (6) (observing that $\mathfrak{S}^- := \sum_{i<0} \mathfrak{S}_i$ is generated by $\mathfrak{S}_{-1}$).

Let $d \in \mathrm{Der}_0\, \mathfrak{S}$ satisfy $d(\mathfrak{S}_{-1}) = 0$. Then $d(\mathfrak{S}_i) = \{0\}$ for $i < 0$ by Property $(g2)$. For $i \ge 0$ we conclude inductively $0 = d(\mathfrak{S}_{i-1}) = d([\mathfrak{S}_{-1}, \mathfrak{S}_i]) = [\mathfrak{S}_{-1}, d(\mathfrak{S}_i)]$ and $d(\mathfrak{S}_i) = \{0\}$ by Property $(g1)$. □

Lemma 20.6.4. *Suppose* $\alpha \notin \Delta$, $\beta \in \Delta$. *Then*

(1) $Q(\alpha,\beta)^{(1)} \cap \mathrm{rad}_T\, Q(\alpha,\beta) \subset L_{(1)}$;

(2) $Q(\alpha,\beta)^{(1)} \cap \mathrm{rad}\, Q(\alpha,\beta)$ *acts nilpotently on* L/Q;

(3) $\mathfrak{S}_0(\alpha,\beta)^{(1)} \in \{\{0\}, \mathfrak{sl}(2), \mathfrak{sl}(2) \otimes \mathcal{O}(1;\underline{1})\}$.

Proof. (a) The algebra $Q[\alpha,\beta] = Q(\alpha,\beta)/\mathrm{rad}_T\, Q(\alpha,\beta)$ is ruled by Proposition 20.3.3. In case (A) all sections $Q(\mu)$ are solvable. Lemma 20.5.4 yields

$$Q(\alpha,\beta) \subset Q(\beta) + L_{(1)}.$$

Since $Q(\beta) = L(\beta)$ acts trigonalizably by Theorem 17.1.2, the lemma also shows $Q(\beta)^{(1)} \subset L_{(1)}$. Thus $Q(\alpha,\beta)^{(1)} \subset L_{(1)}$.

In case (B) one has (see Lemma 20.5.4) $Q(\alpha,\beta) \subset Fe_\mu + Fe_{-\mu} + Q(\beta) + L_{(1)}$ for some root $\mu \notin \Delta$ and $Fe_\mu + Fe_{-\mu} + F[e_\mu, e_{-\mu}] \cong \mathfrak{sl}(2)$ holds. Then (Theorem 17.1.2 and Lemma 20.5.4) $Q(\alpha,\beta)^{(1)} \subset Fe_\mu + Fe_{-\mu} + F[e_\mu, e_{-\mu}] + Q(\beta)^{(1)} + L_{(1)} = Fe_\mu + Fe_{-\mu} + F[e_\mu, e_{-\mu}] + L_{(1)}$. Therefore $Q(\alpha,\beta)^{(1)} \cap \mathrm{rad}\, Q(\alpha,\beta) \subset L_{(1)}$ and $Q(\alpha,\beta)^{(1)}/Q(\alpha,\beta)^{(1)} \cap L_{(1)} \cong \mathfrak{sl}(2)$ hold.

Since β does vanish on H, cases (C) and (F) do not occur. Since α does not vanish on H, case (E) is impossible. Assertions (1) and (2) of the lemma are now proved if $Q(\alpha,\beta)$ is as in cases (A), (B), (C), (E), and (F).

(b) Consider case (D), i.e., $Q[\alpha,\beta] \cong \mathfrak{sl}(2) \otimes \mathcal{O}(1;\underline{1})$. We prove that $Q(\alpha,\beta)^{(1)} \cap \mathrm{rad}_T\, Q(\alpha,\beta)$ is contained in $L_{(1)}$. Note that (Lemma 20.5.4)

$$Q(\beta)^{(1)} + \sum_{i \in \mathbb{F}_p^*} \sum_{j \in \mathbb{F}_p} Q_{i\alpha + j\beta} \cap \mathrm{rad}_T\, Q(\alpha,\beta) \subset L_{(1)}.$$

Recall that (after an adjustment of α) $Q_{i\alpha+j\beta} \not\subset L_{(1)}$ only if $i = 0, \pm 1$ and $\dim Q_{\pm\alpha+j\beta}/Q_{\pm\alpha+j\beta} \cap L_{(1)} = 1$ for all $j \in \mathbb{F}_p$. Thus, considering any $u \in Q_{l\beta} \cap Q(\alpha,\beta)^{(1)} \cap \operatorname{rad}_T Q(\alpha,\beta)$, write

$$u = \sum_j [u_{\alpha+j\beta}, u_{-\alpha+(l-j)\beta}]$$

with all $u_{\pm\alpha+s\beta} \in Q_{\pm\alpha+s\beta}$. In the case under consideration there are $y_j \in Q_{j\beta}$ with $\alpha(y_j) = 1$. So there are $u_\alpha \in Q_\alpha$ and $r_j \in F^*$ such that $u_{\alpha+j\beta} \equiv [u_\alpha, r_j y_j]$ (mod $L_{(1)}$) for all j. Then

$$\begin{aligned} u = \sum_j [u_{\alpha+j\beta}, u_{-\alpha+(l-j)\beta}] &\equiv \sum_j [[u_\alpha, r_j y_j], u_{-\alpha+(l-j)\beta}] \\ &\equiv -[u_\alpha, \sum_j [u_{-\alpha+(l-j)\beta}, r_j y_j]] \pmod{L_{(1)}}, \end{aligned}$$

whence

$$-[u_\alpha, \sum_j [u_{-\alpha+(l-j)\beta}, r_j y_j]] \in \operatorname{rad}_T Q(\alpha,\beta).$$

This can only be if $u_\alpha \in \operatorname{rad}_T Q(\alpha,\beta)$ or $\sum_j [u_{-\alpha+(k-j)\beta}, r_j y_j] \in \operatorname{rad}_T Q(\alpha,\beta)$. In both cases Lemma 20.5.4 shows that the respective element is contained in $L_{(1)}$. Then, however, $u \in L_{(1)}$ holds, which is the first claim.

(c) To prove the second claim of this lemma in case $Q[\alpha,\beta] \cong \mathfrak{sl}(2) \otimes \mathcal{O}(1;\underline{1})$ we choose $\kappa \in \Gamma_{-1}$ arbitrary and put

$$W_\kappa := \sum_{i,j\in\mathbb{F}_p} L_{\kappa+i\alpha+j\beta}/Q_{\kappa+i\alpha+j\beta}.$$

Note that H distinguishes the elements of $\kappa + \mathbb{F}_p(\alpha + j\beta)$ (arbitrary fixed j) as $\alpha(H) \neq 0$. Proposition 20.3.2(3),(4) gives

$$\begin{aligned} \dim W_\kappa = &\sum_{j\in\mathbb{F}_p} \sum_{i\in\mathbb{F}_p^*} \dim L_{\kappa+i(\alpha+j\beta)}/Q_{\kappa+i(\alpha+j\beta)} \\ &+ \sum_{j\in\mathbb{F}_p} \dim L_{\kappa+j\beta}/Q_{\kappa+j\beta} \le 2p + p < p^2 - 1. \end{aligned}$$

Let U denote a composition factor of the $Q(\alpha,\beta)^{(1)}$-module W_κ with representation ϱ_U and set $G_U := \varrho_U(Q(\alpha,\beta)^{(1)})$. Due to (b)

$$Q(\alpha,\beta)^{(1)} \cap \operatorname{rad}_T Q(\alpha,\beta) \subset L_{(1)} \cap Q(\alpha,\beta) \subset \ker \varrho_U$$

holds. Thus we have the natural surjective homomorphism

$$\mathfrak{sl}(2) \otimes \mathcal{O}(1;\underline{1}) \longrightarrow G_U.$$

In particular, ${G_U}^{(1)} = G_U$. Note that $\mathfrak{sl}(2) \otimes \mathcal{O}(1;\underline{1})$ has a descending sequence of ideals $\mathfrak{sl}(2) \otimes \mathcal{O}(1;\underline{1})_{(i)}$ and the quotient $\mathfrak{sl}(2) \otimes \mathcal{O}(1;\underline{1})_{(i)}/\mathfrak{sl}(2) \otimes \mathcal{O}(1;\underline{1})_{(i+1)}$

is a simple 3-dimensional $\mathfrak{sl}(2) \otimes \mathcal{O}(1;\underline{1})$-module. Consequently, all factors of every composition series of $\mathfrak{sl}(2) \otimes \mathcal{O}(1;\underline{1})$ as an $\mathfrak{sl}(2) \otimes \mathcal{O}(1;\underline{1})$-module are 3-dimensional. As this is also true in particular for any composition series through $\ker \varrho_U$, we get $C(G_U) = \{0\}$.

Note that $Q(\beta)$ is a CSA of $Q(\alpha,\beta)$ of toral rank 1, which acts trigonalizably on L. Then $G_U(\beta)$ is so as well. Theorem 11.3.1 gives a list of possible algebras. As ${G_U}^{(1)} = G_U$ and $C(G_U) = \{0\}$, in that theorem only the cases where G_U is simple or where G_U has a non-central abelian ideal can occur. In the first case one obtains $Q(\alpha,\beta)^{(1)} \cap \operatorname{rad} Q(\alpha,\beta) \subset \ker \varrho_U$. In the second case Theorem 11.3.4 applies and shows that G_U either is solvable or contains a non-trivial center. Both cases are impossible. As a result, $Q(\alpha,\beta)^{(1)} \cap \operatorname{rad} Q(\alpha,\beta)$ annihilates all composition factors, hence acts nilpotently on W_κ. As this is true for all $\kappa \in \Gamma_{-1}$, it acts nilpotently on $L_{(-1)}/Q$. Finally we recall that L is generated $L_{(-1)}$ and Q.

(d) To prove the final claim on $\mathfrak{S}_0(\alpha,\beta)$ we observe that there is an embedding

$$\mathfrak{S}_0(\alpha,\beta)^{(1)} \hookrightarrow Q(\alpha,\beta)^{(1)}/Q(\alpha,\beta)^{(1)} \cap L_{(1)}.$$

We obtained in (a) and (b) that $Q(\alpha,\beta)^{(1)}/Q(\alpha,\beta)^{(1)} \cap L_{(1)}$ is one of $\{0\}$, $\mathfrak{sl}(2)$, $\mathfrak{sl}(2) \otimes \mathcal{O}(1;\underline{1})$. In the first case the statement is obvious. In the remaining cases these algebras are generated by non-solvable roots and therefore Lemma 20.6.2 yields an isomorphism

$$Q(\alpha,\beta)^{(1)}/Q(\alpha,\beta)^{(1)} \cap L_{(1)} \cong \mathfrak{A}_0(\alpha,\beta).$$

Equation (20.6.3) gives

$$\mathfrak{S}_0(\alpha,\beta) = C_{\mathfrak{A}_0(\alpha,\beta)}(\mathfrak{T}_1).$$

Then $\mathfrak{S}_0(\alpha,\beta) = \mathfrak{A}_0(\alpha,\beta)$, or $\mathfrak{S}_0(\alpha,\beta) = \mathfrak{A}_0(i\alpha + j\beta)$ for some i, j, or $\mathfrak{S}_0(\alpha,\beta) \subset \operatorname{gr}_0 H$. The claim follows. □

Lemma 20.6.5. *Let $\alpha,\beta \in \Delta$, $-\kappa \in \Gamma_{-l}(\mathfrak{S},\mathfrak{T}_0)$ $(l > 0)$. Assume $\alpha(\mathfrak{S}_{0,\beta}) \neq 0$, $\kappa(\mathfrak{S}_{0,\alpha}) \neq 0$. The following holds.*

(1) *$(s\beta + \kappa)(\mathfrak{S}_{0,\alpha}) \neq 0$ for all $s \in \mathbb{F}_p$, $\kappa(\mathfrak{S}_{0,\beta}) \neq 0$.*

(2) *Set $I_{l,\kappa} := \sum_{i,j,r,s \in \mathbb{F}_p} [\mathfrak{S}_{-l,-\kappa+i\alpha+j\beta}, \mathfrak{S}_{l,\kappa+r\alpha+s\beta}]$. Then $\alpha(I_{l,\kappa} \cap \mathfrak{S}_{0,\beta}) = 0$.*

Proof. (a) Let V denote a composition factor of the $(\mathfrak{S}_0(\alpha+\beta) + \mathfrak{T}_0)$-module $\sum_{i,j \in \mathbb{F}_p} \mathfrak{S}_{-l,-\kappa+i\alpha+j\beta}$ which meets the root $-\kappa$.

Suppose there are $i_0, j_0 \in \mathbb{F}_p$ such that $-\kappa + i_0\alpha + j_0\beta \notin \Gamma(V,\mathfrak{T}_0)$. Since $\kappa(\mathfrak{S}_{0,\alpha}) \neq 0$, we have $-\kappa + \mathbb{F}_p\alpha \subset \Gamma(V,\mathfrak{T}_0)$. Proposition 20.3.2(4) with $\zeta = -\kappa + i_0\alpha$ and $\eta \in \mathbb{F}_p^*\beta$ cannot occur. Part (3) of this proposition then applies and shows that $\operatorname{gr} H$ distinguishes the roots, whence

$$(-\kappa + i_0\alpha + \mathbb{F}_p\beta) \cap \Gamma(V,\mathfrak{T}_0) = -\kappa + i_0\alpha.$$

Choose $r \neq i_0$. Then $-\kappa + r\alpha + (i_0 - r)(\alpha + \beta) = -\kappa + i_0\alpha + (i_0 - r)\beta \notin \Gamma(V, \mathfrak{T}_0)$. As $\alpha(\mathfrak{S}_{0,\beta}) \neq 0$, one has $\mathfrak{S}_{0,\alpha+\beta} \neq \{0\}$. Arguing as before (substituting i_0 by r and β by $\alpha + \beta$) we obtain

$$-\kappa + (2i_0 - r)\alpha + 2(i_0 - r)\beta = -\kappa + r\alpha + 2(i_0 - r)(\alpha + \beta) \notin \Gamma(V, \mathfrak{T}_0).$$

As $\alpha(\mathfrak{S}_{0,\beta}) \neq 0$, one has $(-\kappa + i\alpha)(\mathfrak{S}_{0,\beta}) \neq 0$ for all but one i at the most. For all these, one has $-\kappa + i\alpha + \mathbb{F}_p\beta \subset \Gamma(V, \mathfrak{T}_0)$. As a consequence, $(-\kappa + i_0\alpha)(\mathfrak{S}_{0,\beta}) = 0$, $(-\kappa + (2i_0 - r)\alpha)(\mathfrak{S}_{0,\beta}) = 0$. But this contradicts the assumption $\alpha(\mathfrak{S}_{0,\beta}) \neq 0$. As a result,

$$-\kappa + \mathbb{F}_p\alpha + \mathbb{F}_p\beta \subset \Gamma(V, \mathfrak{T}_0).$$

By Proposition 20.3.2 with $\zeta = -\kappa - s\beta$, $\eta = \alpha$ we conclude $(\kappa + s\beta)(\mathfrak{S}_{0,\alpha}) \neq 0$. Setting $\zeta = -\kappa$ and $\eta = \beta$ we obtain $\kappa(\mathfrak{S}_{0,\beta}) \neq 0$. This is the first claim. We also conclude, as all spaces $\mathfrak{S}_{-l,-\kappa+i\alpha+j\beta}$ are 1-dimensional, that $\sum_{i,j\in\mathbb{F}_p} \mathfrak{S}_{-l,-\kappa+i\alpha+j\beta} = V$ is irreducible of dimension p^2.

(b) Suppose there are i_0, j_0 with $-2\kappa + i_0\alpha + j_0\beta \in \Gamma_{-2l}$. As $(-2\kappa + i_0\alpha + j_0\beta)(\mathfrak{S}_{0,\alpha}) = (-2\kappa + j_0\beta)(\mathfrak{S}_{0,\alpha}) \neq 0$ and $\kappa(\mathfrak{S}_{0,\beta}) \neq 0$ by (1), we argue as before to show that $-2\kappa \in \Gamma_{-2l}$. This contradicts Proposition 20.3.2(2). Consequently, $(-2\kappa + \mathbb{F}_p\alpha + \mathbb{F}_p\beta) \cap \Gamma_{-2l} = \emptyset$.

As a result, $[V, V] = \{0\}$. Therefore there is a mapping

$$\sum_{r,s\in\mathbb{F}_p} \mathfrak{S}_{l,\kappa+r\alpha+s\beta} \to \mathcal{C}_+(V, \mathfrak{S}_0(\alpha, \beta) + \mathfrak{T}_0),$$

the space of Cartan prolongations.

(c) Consider the case that $\mathfrak{S}_0(\alpha, \beta) + \mathfrak{T}_0$ is solvable and let G_p denote the p-envelope of the image of $\mathfrak{S}_0(\alpha, \beta) + \mathfrak{T}_0$ in $\mathfrak{gl}(V)$. Then V is induced by a 1-dimensional subspace, which means (see Corollary 3.3.7)

$$V \cong F \otimes \mathcal{O}(2; \underline{1}), \quad G_p \hookrightarrow \big(\mathrm{gl}(F) \otimes \mathcal{O}(2; \underline{1})\big) \rtimes \big(\mathrm{Id} \otimes W(2; \underline{1})\big)$$

such that $\pi_2(G_p)$ is a transitive subalgebra of $W(2; \underline{1})$. Since G_p is solvable, $\pi_2(G_p)$ cannot be semisimple. Theorem 8.1.6(3) now implies that $\pi_2(I_{l,\kappa}/\operatorname{ann} V) = \{0\}$. Consequently, $I_{l,\kappa}/\operatorname{ann} V \subset \mathrm{gl}(F) \otimes \mathcal{O}(2; \underline{1})$ is abelian. If $\alpha(I_{l,\kappa} \cap \mathfrak{S}_{0,\beta}) \neq 0$, this would imply $\mathfrak{S}_{0,\alpha} \subset \operatorname{ann} V$. But $\kappa(\mathfrak{S}_{0,\alpha}) \neq 0$ holds, a contradiction.

(d) Consider the case that $\mathfrak{S}_0(\alpha, \beta) + \mathfrak{T}_0$ is not solvable and let G denote the image of $\mathfrak{S}_0(\alpha, \beta) + \mathfrak{T}_0$ in $\mathrm{gl}(V)$. Note that $\mathfrak{S}_0(\alpha, \beta)$ may be considered a T-invariant subalgebra of a T-invariant homomorphic image $\mathrm{gr}_0\, L(\alpha, \beta)$ of $Q(\alpha, \beta)$. Proposition 20.3.3(E) applies. Then

$$H(2; \underline{1}; \Phi(\tau))^{(1)} \subset G/\operatorname{rad} G \subset \operatorname{Der} H(2; \underline{1}; \Phi(\tau))^{(1)} = H(2; \underline{1}; \Phi(\tau))^{(1)}{}_{[p]}.$$

Let G_p denote the p-envelope of G in $\mathfrak{gl}(V)$. Theorem 16.3.15 applies. As in the former case, Theorem 8.1.6(3) shows that either $\pi_2(I_{l,\kappa}/\operatorname{ann} V) = \{0\}$ or

$\pi_2(I_{l,\kappa}/\operatorname{ann} V)$ is $\mathcal{O}(2;\underline{1})$-invariant and is a minimal ideal of $\pi_2(G_p)$. The latter implies that $\pi_2(I_{l,\kappa}/\operatorname{ann} V) \cong H(2;\underline{1};\Phi(\tau))^{(1)}$ which, however, is not $\mathcal{O}(2;\underline{1})$-invariant. Thus $\pi_2(I_{l,\kappa}/\operatorname{ann} V) = \{0\}$ holds, and we obtain the claim as in (c). □

Lemma 20.6.6. *Let $\alpha, \beta \in \Delta$, $\kappa \in \Gamma \setminus \Delta$. Assume $\alpha(\mathfrak{S}_{0,\beta}) \neq 0$, $\kappa(\mathfrak{S}_{0,\alpha}) \neq 0$. Then $\alpha([\mathfrak{S}_{0,\kappa}, \mathfrak{S}_{0,\beta-\kappa}]) = 0$.*

Proof. (a) Let us assume that the claim is not true. Then $\mathfrak{S}_0(\kappa,\beta)^{(1)}$ carries the solvable root β. Therefore it can only be that

$$\mathfrak{S}_0(\kappa,\beta)^{(1)} \cong \mathfrak{sl}(2) \otimes \mathcal{O}(1;\underline{1})$$

(Lemma 20.6.4). We conclude $\kappa([\mathfrak{S}_{0,\kappa}, \mathfrak{S}_{0,\beta-\kappa}]) \neq 0$.

(b) Set

$$G := \langle \mathfrak{S}_{0,\alpha} \cup \mathfrak{S}_{0,\kappa} \cup \mathfrak{S}_{0,\beta-\kappa} \rangle$$

the algebra generated by the displayed set. Since $\alpha([\mathfrak{S}_{0,\kappa}, \mathfrak{S}_{0,\beta-\kappa}]) \neq 0$, $\kappa([\mathfrak{S}_{0,\kappa}, \mathfrak{S}_{0,\beta-\kappa}]) \neq 0$, $(\beta-\kappa)([\mathfrak{S}_{0,\kappa}, \mathfrak{S}_{0,\beta-\kappa}]) \neq 0$, we have $G^{(1)} = G$.

Since G is contained in a 3-section, one has $TR(G) \leq 3$. On the other hand, $\kappa([\mathfrak{S}_{0,\kappa}, \mathfrak{S}_{0,-\kappa}]) \neq 0$ (because $\mathfrak{S}_0(\kappa)$ cannot be solvable), $\alpha(H) = 0$, $\alpha(\mathfrak{S}_{0,\beta} \cap G) \neq 0$. Therefore $TR(\mathfrak{S}_0(\beta) \cap G, G) \geq 2$. Let I be a maximal ideal. Since G is perfect, $\bar{G} := G/I$ is simple.

Suppose $TR(\bar{G}) = 2$. Then $\bar{G}(\beta)$ is a trigonalizable CSA of maximal toral rank. If $\bar{G}(\alpha,\beta)$ is solvable, then it is a solvable 1-section with respect to $\bar{G}(\beta)$ and $\alpha(\bar{G}(\beta)) \neq 0$ holds. Corollary 17.3.3(1) yields that $\bar{G}_\alpha$ consists of (ad)-nilpotent elements. But κ is a root on $\bar{G}$. So this case is impossible. If $\bar{G}(\alpha,\beta)$ is not solvable, then $G(\alpha,\beta)$ has core $H(2;\underline{1};\Phi(\tau))^{(1)}$ (as $\alpha,\beta \in \Delta$). There is an element $t \in G_\alpha$ satisfying $\beta(t) \neq 0$, $\alpha(t) = 0$. But then $TR(\bar{G}) = 3$ holds contrary to the present assumption.

We now have $TR(\bar{G}) = 3$. Then $TR(G) = 3$ and hence $\mathfrak{t} := \mathfrak{T}_0 \cap (G_{[p]} + C(G_{[p]} + \mathfrak{T}_0))$ is a torus of maximal toral rank in $G_{[p]} + C(G_{[p]} + \mathfrak{T}_0)$ (Theorem 1.2.9). Let $\bar{G}_{[p]}$ denote a minimal p-envelope of $\bar{G}$ and $I_{[p]}$ the p-envelope of I in $G_{[p]}$. Then

$$(G_{[p]} + C(G_{[p]} + \mathfrak{T}_0))/I_{[p]} \supset G_{[p]}/I_{[p]},$$

where $G_{[p]}/I_{[p]}$ is a p-envelope of $\bar{G}$ and $G_{[p]}/I_{[p]} \cong \bar{G}_{[p]} \oplus C$ with C central in $G_{[p]}/I_{[p]}$ (Theorem 1.1.6(4),(2)). Then we also have

$$(G_{[p]} + C(G_{[p]} + \mathfrak{T}_0))/I_{[p]} \cong \bar{G}_{[p]} \oplus C'$$

with C' centralizing the algebra. Since $\bar{G}_{[p]}$ is a minimal p-envelope of the simple algebra $\bar{G}$, the definition yields that $\bar{G}_{[p]}$ is centerless. Then C' is the full center and thus a restricted ideal, and hence there is a restricted homomorphism $G_{[p]} + C(G_{[p]} + \mathfrak{T}_0) \twoheadrightarrow \bar{G}_{[p]}$. Let $\bar{\mathfrak{t}}$ denote the image of $\mathfrak{t}$ in $\bar{G}_{[p]}$. By construction, $\bar{\mathfrak{t}}$ is a torus of maximal dimension. All $\bar{\mathfrak{t}}$-roots in $\bar{G}$ are classical or solvable and there is a non-solvable root. Theorem 20.4.5 yields that $\bar{G}$ is classical. But then $\bar{G}$ is restricted, $\bar{\mathfrak{t}} \subset \bar{G}$ is a CSA and no root vanishes on it. This contradicts the existence of the root β on $\bar{G}$. □

Set

$$\mathfrak{H} := C_{\mathfrak{S}_0}(\mathfrak{T}_0), \qquad \mathfrak{P} := \sum_{\mu\in\Delta} \mathfrak{S}_{0,\mu}.$$

Note that $\mathfrak{H} \otimes 1 = (\mathfrak{S}_0 \otimes 1) \cap (\mathrm{gr}_0\, H)$ and $\mathfrak{H} \subset \mathfrak{P}$.

Proposition 20.6.7. *$\mathfrak{P}$ is an abelian CSA of $\mathfrak{S}_0$.*

Proof. (a) Let $\mu \notin \Delta$ and $\mathfrak{S}_{0,\mu} \neq \{0\}$. Due to Lemma 20.6.4(3) (with $\beta = 0$) one has $\mu([\mathfrak{S}_{0,\mu}, \mathfrak{S}_{0,-\mu}]) \neq 0$. Therefore $C_{\mathfrak{S}_0}(\mathfrak{H}) \subset \mathfrak{P}$. Due to Theorem 17.1.2 and Lemma 20.5.4 one has $\mathfrak{P} \subset C_{\mathfrak{S}_0}(\mathfrak{H})$, whence $\mathfrak{P} = C_{\mathfrak{S}_0}(\mathfrak{H})$.

Suppose $\mathfrak{P}$ is nilpotent. Set $\mathfrak{t}_0$ the torus spanned by $\mathfrak{T}_0$ and the degree derivation on $\mathfrak{S}$. Theorem 17.1.2 applied to $\mathfrak{S}$ yields that $\mathfrak{P}^{(1)}$ acts nilpotently on $\mathfrak{S}$. Then $\mathfrak{P}^{(1)} \otimes F$ acts nilpotently on $\mathrm{gr}\, L$, and Lemma 20.5.4 now yields that $\mathfrak{P}^{(1)} \otimes F = \{0\}$.

(b) So assume that $\mathfrak{P}$ is not nilpotent. Then there are $\alpha, \beta \in \Delta$ with $\mathfrak{S}_{0,\alpha} \neq \{0\}$ and $\alpha(\mathfrak{S}_{0,\beta}) \neq 0$. Proposition 20.6.3(3) proves

$$\mathfrak{S}_0(\alpha,\beta) = \sum_{-\kappa\in\Gamma_{-1}(\mathfrak{S},\mathfrak{T}_0)} \sum_{r,s\in\mathbb{F}_p} [\mathfrak{S}_{-1,-\kappa}, \mathfrak{S}_{1,\kappa+r\alpha+s\beta}].$$

If $-\kappa \in \Gamma_{-1}(\mathfrak{S},\mathfrak{T}_0)$ satisfies $\kappa(\mathfrak{S}_{0,\alpha}) \neq 0$, then Lemma 20.6.5 gives $[\mathfrak{S}_{-1,-\kappa}, \mathfrak{S}_{1,\kappa+\beta}] \subset \ker \alpha$.

(c) Let $-\kappa \in \Gamma_{-1}(\mathfrak{S},\mathfrak{T}_0)$ satisfy $\kappa(\mathfrak{S}_{0,\alpha}) = 0$. Since $\mathfrak{S}_{0,\alpha} \neq \{0\}$, there is $\mu \in \Gamma_{-1}(\mathfrak{S},\mathfrak{T}_0)$ such that $\mu(\mathfrak{S}_{0,\alpha}) \neq 0$ (Lemma 20.5.2, Proposition 20.6.3(6)). Set

$$\Omega := \{\mu \in \Gamma(\mathfrak{S},\mathfrak{T}_0) \mid \mu(\mathfrak{S}_{0,\alpha}) \neq 0\} \neq \emptyset.$$

The simplicity of $\mathfrak{S}$ implies (recall that $r \leq 2$)

$$\mathfrak{S}_{-1,-\kappa} = \sum_{\mu\in\Omega} [\mathfrak{S}_{-2,\mu}, \mathfrak{S}_{1,-\kappa-\mu}] + [\mathfrak{S}_{-1,\mu}, \mathfrak{S}_{0,-\kappa-\mu}].$$

Then

$$\begin{aligned}
[\mathfrak{S}_{-1,-\kappa}, \mathfrak{S}_{1,\kappa+\beta}] &\subset \sum_{\mu\in\Omega} [[\mathfrak{S}_{-2,\mu}, \mathfrak{S}_{1,-\kappa-\mu}], \mathfrak{S}_{1,\kappa+\beta}] + [[\mathfrak{S}_{-1,\mu}, \mathfrak{S}_{0,-\kappa-\mu}], \mathfrak{S}_{1,\kappa+\beta}] \\
&\subset \sum_{\mu\in\Omega} [[\mathfrak{S}_{-2,\mu}, \mathfrak{S}_{1,\kappa+\beta}], \mathfrak{S}_{1,-\kappa-\mu}] + [\mathfrak{S}_{-2,\mu}, \mathfrak{S}_{2,\beta-\mu}] \\
&\quad + [[\mathfrak{S}_{-1,\mu}, \mathfrak{S}_{1,\kappa+\beta}], \mathfrak{S}_{0,-\kappa-\mu}] + [\mathfrak{S}_{-1,\mu}, \mathfrak{S}_{1,\beta-\mu}].
\end{aligned}$$

Let us consider the first summand. We may assume $\mu \in \Gamma_{-2}(\mathfrak{S},\mathfrak{T}_0)$ and $\mu+\kappa+\beta \in \Gamma_{-1}(\mathfrak{S},\mathfrak{T}_0)$. Lemma 20.6.5(1) shows that $(\mu+\beta)(\mathfrak{S}_{0,\alpha}) \neq 0$. Then $(\mu+\beta+\kappa)(\mathfrak{S}_{0,\alpha}) = (\mu+\beta)(\mathfrak{S}_{0,\alpha}) \neq 0$ holds. Lemma 20.6.5(2) shows that the first summand is contained in $\ker\alpha$. One argues similarly to show that the second and forth summands are contained in $\ker\alpha$. Since $\mathfrak{S}_0(\beta)$ is abelian, α is linear on this space. Therefore there is $\mu \in \Omega \cap \Gamma_{-1}(\mathfrak{S},\mathfrak{T}_0)$, such that $\alpha([[\mathfrak{S}_{-1,\mu},\mathfrak{S}_{1,\kappa+\beta}],\mathfrak{S}_{0,-\kappa-\mu}]) \neq 0$. Set $\gamma := -\kappa-\mu \in \Omega$. Lemma 20.6.6 gives $\gamma \in \Delta$.

(d) Our construction implies that $-\kappa, -\kappa-\gamma = \mu \in \Gamma_{-1}$. Proposition 20.3.2(3),(4) shows that $\kappa(\mathfrak{S}_{0,\gamma}) \neq 0$. Then Lemma 20.6.5(1) gives (with $\alpha \to \gamma$, $\beta \to \alpha$) that $\kappa(\mathfrak{S}_{0,\alpha}) \neq 0$, contradicting the present assumption. As a consequence, $\mathfrak{P}$ is abelian. □

Lemma 20.6.8. $TR(\mathfrak{S}) = TR(\mathfrak{P},\mathfrak{S})$.

Proof. (a) Corollary 1.4.7 applied to $L+T$ states that $\mathrm{ad}_{\mathrm{gr}\,(L+T)}\,T = \mathrm{ad}_{\mathrm{gr}\,(L+T)}\,\mathfrak{T}$ is a torus of maximal dimension in $(\mathrm{gr}\,(L+T))_{[p]}$, where we take $(\mathrm{gr}\,(L+T))_{[p]}$ the p-envelope of $\mathrm{ad}\,\mathrm{gr}\,(L+T)$ in $\mathrm{Der}\,\mathrm{gr}\,(L+T)$. Next Theorem 1.2.9 implies that

$$(\mathrm{ad}_{\mathrm{gr}\,(L+T)}\,\mathfrak{T}) \cap \Big((\mathfrak{S}\otimes F)_{[p]} + C\big((\mathrm{ad}_{\mathrm{gr}\,(L+T)}\,\mathfrak{T}) + (\mathfrak{S}\otimes F)_{[p]}\big)\Big)$$

is a torus of maximal toral rank in $(\mathfrak{S}\otimes F)_{[p]} + C((\mathrm{ad}_{\mathrm{gr}\,(L+T)}\,\mathfrak{T}) + (\mathfrak{S}\otimes F)_{[p]}) \subset \mathrm{Der}\,\mathrm{gr}\,(L+T)$. Let $\mathfrak{S}_{[p]}$ denote the p-envelope in $\mathrm{Der}\,\mathfrak{S}$. The restriction map $\mathrm{ad}_{\mathrm{gr}\,(L+T)} \to \mathrm{ad}_{\mathfrak{S}}$ yields a homomorphism

$$(\mathfrak{S}\otimes F)_{[p]} + C\big((\mathrm{ad}_{\mathrm{gr}\,(L+T)}\,\mathfrak{T}) + (\mathfrak{S}\otimes F)_{[p]}\big) \to \mathrm{Der}\,\mathfrak{S}$$

with central kernel and image $\mathfrak{S}_{[p]}$. Consequently (see also Theorem 1.2.8(3)),

$$(\mathrm{ad}_{\mathfrak{S}}\,\mathfrak{T}) \cap \mathfrak{S}_{[p]}$$

is a torus of maximal toral rank in $\mathfrak{S}_{[p]}$.

(b) Note that

$$(\mathrm{ad}_{\mathfrak{S}}\,\mathfrak{T}) \cap \mathfrak{S}_{[p]} \subset C_{\mathfrak{S}_{[p]}}(\mathfrak{T}) = \mathfrak{H} + \sum_{n\in\mathbb{Z}}\sum_{\mu\in\Gamma}\sum_{i>0} \mathfrak{S}_{n,\mu}{}^{[p]^i}.$$

If $n \neq 0$, then $\mathfrak{S}_{n,\mu}$ is $[p]$-nilpotent for all μ. If $\mu \notin \Delta$, then $\mathfrak{S}_{0,\mu}$ acts nilpotent on $\mathfrak{S}_{-1}$ (Proposition 20.3.2). Proposition 20.6.3(7) implies that it acts nilpotently on $\mathfrak{S}$. Since $\mathfrak{T}$ is a torus of maximal toral rank in $\mathrm{gr}\,(L+T)$, $\mathfrak{T}^{[p]} = \mathfrak{T}$ holds and $C_{\mathfrak{S}_{[p]}}(\mathfrak{T})$ is nilpotent. Lemma 1.1.1 gives $(h+\sum_{n\in\mathbb{Z}}\sum_{\mu\in\Gamma}\sum_{i>0} x_{n,\mu}{}^{[p]^i})^{[p]^e} \in \mathfrak{P}_{[p]}+\mathfrak{H}_{[p]} = \mathfrak{P}_{[p]}$ for $e \gg 0$. We conclude

$$(\mathrm{ad}_{\mathfrak{S}}\,\mathfrak{T}) \cap \mathfrak{S}_{[p]} = (\mathrm{ad}_{\mathfrak{S}}\,\mathfrak{T}) \cap \sum_{\mu\in\Delta}\sum_{i\geq 0} \mathfrak{S}_{0,\mu}{}^{[p]^i} = (\mathrm{ad}_{\mathfrak{S}}\,\mathfrak{T}) \cap \mathfrak{P}_{[p]}.$$

This is the claim. □

For any $\alpha \in \Gamma$ we set

$$\mathfrak{S}_{0,\bar{\alpha}} := \sum_{\mu \in \Delta} \mathfrak{S}_{0,\alpha+\mu}, \quad \mathfrak{S}_0(\bar{\alpha}_1, \ldots, \bar{\alpha}_r) := \sum_{i_1,\ldots,i_r \in \mathbb{F}_p} \sum_{\mu \in \Delta} \mathfrak{S}_{0,\sum i_k \alpha_k + \mu}.$$

Note that $\mathfrak{S}_{0,\bar{\alpha}}$ is the full root space with respect to $\mathfrak{P}$ (if it is nonzero) and then $\mathfrak{P}$ has toral rank 1 in $\mathfrak{S}_0(\bar{\alpha})$.

Lemma 20.6.9. *Let $\alpha \notin \Delta$, $\mathfrak{S}_{0,\alpha} \neq \{0\}$. The following holds.*

(1) $0, \pm\bar{\alpha}$ *are the only $\mathbb{F}_p$-multiples of $\bar{\alpha}$, which are $\mathfrak{P}$-roots on $\mathfrak{S}_0$.*

(2) $\mathfrak{S}_{0,\bar{\alpha}}$ *acts nilpotently on $\mathfrak{S}$.*

(3) $\mathfrak{S}_0(\bar{\alpha})/\operatorname{rad}\mathfrak{S}_0(\bar{\alpha}) \cong \mathfrak{sl}(2)$.

(4) $\mathfrak{S}_0(\bar{\alpha}, \bar{\beta})/\operatorname{rad}\mathfrak{S}_0(\bar{\alpha}, \bar{\beta}) \in \{\mathfrak{sl}(2), \mathfrak{sl}(2) \oplus \mathfrak{sl}(2), A_2, B_2, G_2\}$ *for every* $\beta \in \Gamma(\mathfrak{S}_0, \mathfrak{T}_0)$.

Proof. (1) Let $\bar{\alpha}$ and $k\bar{\alpha}$ with $k \neq 0, \pm 1$ be $\mathfrak{P}$-roots on $\mathfrak{S}_0$. Then there are roots α and $k\alpha + \mu$ on $\mathfrak{S}_0$ with $\mu \in \Delta$. Applying Proposition 20.3.3 to $Q[\alpha, \mu]$ shows that $Q(k\alpha + \mu)$ is solvable. Then $Q_{k\alpha+\mu} \subset L_{(1)}$ (Lemma 20.5.4) and hence $k\alpha + \mu$ is not a root on $Q/L_{(1)}$. But then $k\alpha + \mu$ is not a root on $\mathfrak{S}_0$.

(2) Take $x \in \mathfrak{S}_{0,\bar{\alpha}}$ and write $x = \sum_{\mu \in \Delta} x_{\alpha+\mu}$, where $x_{\alpha+\mu} \in \mathfrak{S}_{0,\alpha+\mu}$. We conclude from (1) that $[x_{\alpha+\mu}, x_{\alpha+\nu}] \in \mathfrak{S}_{0,2\bar{\alpha}} = \{0\}$. Therefore it is sufficient to prove the claim for any $x_{\alpha+\mu}$. Due to Proposition 20.3.2, if $\kappa \in \Gamma_{-1}$, no string $\kappa + \mathbb{F}_p(\alpha + \mu)$ is completely contained in Γ_{-}. Therefore $x_{\alpha+\mu}$ acts nilpotently on $\mathfrak{S}_{-1}$. Then it acts nilpotently on $\mathfrak{S}$ by Proposition 20.6.3(7).

(3) Since by assumption $\mathfrak{S}_{0,\alpha} \neq \{0\}$, then $\mathfrak{S}_0(\alpha)$ cannot be solvable (Lemma 20.5.4). Hence $\mathfrak{S}_0(\bar{\alpha})$ is also not solvable. We mentioned that $\mathfrak{P}$ is an abelian CSA of toral rank 1 in $\mathfrak{S}_0(\bar{\alpha})$. Theorem 11.1.1 shows that $\mathfrak{S}_0(\bar{\alpha})/\operatorname{rad}\mathfrak{S}_0(\bar{\alpha})$ has a unique minimal ideal of type $\mathfrak{sl}(2)$, $W(1; \underline{n})$ or $H(2; \underline{1}; \Phi)^{(2)}$. On the other hand, only the $\mathfrak{P}$-roots $0, \pm\bar{\alpha}$ do occur. Then the minimal ideal is isomorphic to $\mathfrak{sl}(2)$. Since this algebra has no outer derivations, the claim follows.

(4) As before, $\mathfrak{S}_0(\bar{\alpha})$ is not solvable. We may assume that $\bar{\alpha}, \bar{\beta}$ are $\mathbb{F}_p$-independent. Let $\bar{U}$ be the sum of all minimal ideals of $\mathfrak{S}_0(\bar{\alpha}, \bar{\beta})/\operatorname{rad}\mathfrak{S}_0(\bar{\alpha}, \bar{\beta})$, and $\bar{\mathfrak{P}}$ the image of $\mathfrak{P}$ in $\mathfrak{S}_0(\bar{\alpha}, \bar{\beta})/\operatorname{rad}\mathfrak{S}_0(\bar{\alpha}, \bar{\beta})$. Let $\mathfrak{t} \subset \bar{\mathfrak{P}}_{[p]}$ denote the torus of maximal dimension and put $\mathfrak{t}_{\bar{U}} := \mathfrak{t} \cap \bar{U}_{[p]}$. Due to Lemma 20.6.8 and Theorem 1.2.9 $\mathfrak{t}_{\bar{U}}$ has maximal toral rank in $\bar{U}_{[p]}$. Then $0 \neq TR(\mathfrak{t}_{\bar{U}}, \bar{U}_{[p]}) \leq TR(\mathfrak{P}, \mathfrak{S}_0(\bar{\alpha}, \bar{\beta})) = 2$ holds (Theorem 1.3.11(2)). The following cases can occur:

- $TR(\mathfrak{t}_{\bar{U}}, \bar{U}_{[p]}) = 2$ and $\bar{U}$ has more than one summand: then $\bar{U} = \bar{U}_1 \oplus \bar{U}_2$ with $TR(\bar{\mathfrak{P}} \cap \bar{U}_i, \bar{U}_i) = 1$ for $i = 1, 2$. The maximal torus $\mathfrak{t}$ of the p-envelope of $\bar{\mathfrak{P}} \cap \bar{U}$ is 2-dimensional. Then we have $C_{\mathfrak{S}_0(\bar{\alpha}, \bar{\beta})/\operatorname{rad}\mathfrak{S}_0(\bar{\alpha}, \bar{\beta})}(\mathfrak{t}) = \bar{\mathfrak{P}}$. Theorem 10.6.4(3) gives that every $\bar{U}_i$ is simple, whence $\bar{U}_i \cong \mathfrak{sl}(2)$.

- $TR(\mathfrak{t}_{\bar U}, \bar U_{[p]}) = 2$ and $\bar U$ is the unique minimal ideal: as before Theorem 10.6.4(3) gives that $\bar U$ is simple. As every 1-section is classical, the classification of these algebras yields that $\bar U$ is classical simple of absolute toral rank $TR(\bar U) = 2$.
- $TR(\mathfrak{t}_{\bar U}, \bar U_{[p]}) = 1$: As a minimal ideal, $\bar U$ is of the form $\bar U \cong \bar U_0 \otimes \mathcal{O}(r;\underline{1})$, where $\bar U_0$ is simple. One may normalize $\mathfrak{t} = (Fh \otimes 1) \oplus F(d \otimes 1 + \mathrm{Id} \otimes t_0)$, where Fh is a maximal torus in $\bar U_0$, d is a toral derivation in $\mathrm{Der}\,\bar U_0$ and t_0 is toral derivation in $W(r;\underline{1})$. Moreover, $\bar U_0$ has absolute toral rank 1, and every 1-section with respect to $\mathfrak{t}$ is solvable or classical. Then it can only be that $\bar U_0 \cong \mathfrak{sl}(2)$, whence $\mathfrak{S}_0(\bar\alpha, \bar\beta)/\mathrm{rad}\,\mathfrak{S}_0(\bar\alpha, \bar\beta)$ maps into $\mathfrak{sl}(2) \otimes \mathcal{O}(r;\underline{1}) + \mathrm{Id} \otimes W(r;\underline{1})$ and $d = 0$. If $r \neq 0$, then $Fh \otimes \mathcal{O}(r;\underline{1})$ is contained in a 1-section and has p distinct $\mathfrak{t}$-roots. This is not true by (1). Therefore $r = 0$ and $\mathfrak{S}_0(\bar\alpha, \bar\beta)/\mathrm{rad}\,\mathfrak{S}_0(\bar\alpha, \bar\beta) \cong \mathfrak{sl}(2)$ hold.

In all cases $\bar U$ has no outer derivations. □

Set

$$\mathfrak{P}_0 := \{x \in \mathfrak{P} \cap \mathfrak{S}_0{}^{(1)} \mid x \text{ acts nilpotently on } \mathfrak{S}\}.$$

Lemma 20.6.10.

(1) *There exists a linear mapping* $\Theta : \mathfrak{P} \cap \mathfrak{S}_0{}^{(1)} \to \mathfrak{H} \cap \mathfrak{S}_0{}^{(1)}$ *such that* $u - \Theta(u) \in \mathfrak{P}_0$ *for all* $u \in \mathfrak{P} \cap \mathfrak{S}_0{}^{(1)}$.
(2) $[[\mathfrak{P}_0, \mathfrak{S}_{0,\alpha}], \mathfrak{S}_{0,-\alpha}] \subset \mathfrak{P}_0$ *for all* $\alpha \notin \Delta$.
(3) $[\mathfrak{P} \cap \mathfrak{S}_0{}^{(1)}, \mathfrak{S}_{0,\alpha}] = \mathfrak{S}_{0,\tilde\alpha}$ *for all* $\alpha \notin \Delta$ *with* $\mathfrak{S}_{0,\alpha} \neq \{0\}$.
(4) $\big(\mathrm{rad}\,\mathfrak{S}_0(\bar\alpha)\big)_{\tilde\alpha} = [\mathfrak{P}_0, \mathfrak{S}_{0,\alpha}]$ *for all* $\alpha \notin \Delta$ *with* $\mathfrak{S}_{0,\alpha} \neq \{0\}$.

Proof. Let $\beta \in \Delta$ be such that $\mathfrak{S}_\beta \cap \mathfrak{S}_0{}^{(1)} \neq \{0\}$. Since $\mathfrak{P}$ is abelian, one has

$$\mathfrak{S}_\beta \cap \mathfrak{S}_0{}^{(1)} = \sum_{\alpha \notin \Delta} [\mathfrak{S}_{0,\alpha}, \mathfrak{S}_{0,\beta-\alpha}].$$

Let $\alpha \notin \Delta$ be any root satisfying $[\mathfrak{S}_{0,\alpha}, \mathfrak{S}_{0,\beta-\alpha}] \neq \{0\}$. Due to Lemma 20.6.4(3) there is an isomorphism

$$\varphi : \mathfrak{S}_0(\alpha, \beta)^{(1)} \cong \mathfrak{sl}(2) \otimes \mathcal{O}(1;\underline{1}),$$

It is not hard to see that for a suitable $\mathfrak{sl}(2)$-triple (e, h, f) one may take

$$\varphi\big(\mathfrak{S}_{0,\alpha}\big) = Fe \otimes 1, \quad \varphi\big(\mathfrak{H} \cap \mathfrak{S}_0(\alpha,\beta)^{(1)}\big) = Fh \otimes 1, \quad \varphi\big(\mathfrak{S}_{0,-\alpha}\big) = Ff \otimes 1,$$
$$\varphi\big(\mathfrak{P} \cap \mathfrak{S}_0(\alpha,\beta)^{(1)} \cap \mathfrak{S}_0(\beta)\big) = Fh \otimes \mathcal{O}(1;\underline{1}).$$

Note that $\alpha(\mathfrak{T}_1) = \beta(\mathfrak{T}_1) = 0$. Therefore (cf. Equation (20.6.3))

$$\mathfrak{S}_0(\alpha,\beta) + \mathrm{gr}_0 H = C_{\mathrm{gr}_0 Q(\alpha,\beta)}(\mathfrak{T}_1) = \mathrm{gr}_0\, Q(\alpha,\beta) \cong Q(\alpha,\beta)/L_{(1)}(\alpha,\beta),$$

and in Proposition 20.3.3 only the case

$$Q[\alpha,\beta] \cong \mathfrak{sl}(2) \otimes \mathcal{O}(1;\underline{1})$$

can occur. This gives

$$\begin{aligned} &Q(\alpha,\beta)^{(1)}/Q(\alpha,\beta)^{(1)} \cap \operatorname{rad}_T Q(\alpha,\beta) \\ &\qquad \cong Q[\alpha,\beta]^{(1)} = Q[\alpha,\beta] \cong \mathfrak{sl}(2) \otimes \mathcal{O}(1;\underline{1}) \cong \mathfrak{S}_0(\alpha,\beta)^{(1)}. \end{aligned}$$

(1) Let $\alpha \notin \Delta$ and $0 \neq u \in [\mathfrak{S}_{0,\alpha}, \mathfrak{S}_{0,\beta-\alpha}]$. There is $r \in F$ with $\varphi(u) - r(h \otimes 1) \in \mathfrak{sl}(2) \otimes \mathcal{O}(1;\underline{1})_{(1)} = \operatorname{rad}(\mathfrak{sl}(2) \otimes \mathcal{O}(1;\underline{1}))$. Set $h_\alpha := \varphi^{-1}(rh \otimes 1) \in \mathfrak{H} \cap \mathfrak{S}_0(\alpha,\beta)^{(1)}$. Observe that the inverse image of $\operatorname{rad}(\mathfrak{sl}(2) \otimes \mathcal{O}(1;\underline{1}))$ in $Q(\alpha,\beta)^{(1)}$ coincides with $Q(\alpha,\beta)^{(1)} \cap \operatorname{rad} Q(\alpha,\beta)$. Lemma 20.6.4 yields that this space acts nilpotently on L/Q. As a consequence, $u - h_\alpha$ acts nilpotently on $\mathfrak{S}_{-1}$. Proposition 20.6.3(7) gives $u - h_\alpha \in \mathfrak{P}_0$.

Next assume that $h_0, h_0' \in \mathfrak{H} \cap {\mathfrak{S}_0}^{(1)}$ are such that $u - h_0, u - h_0' \in \mathfrak{P}_0$. Since $\mathfrak{P}$ is abelian, $h_0 - h_0' \in \mathfrak{H}$ acts nilpotently on $\mathfrak{S}_{-1}$, which is possible only if $h_0 - h_0' = 0$. Thus the element $\Theta(u)$ is well defined by the requirement $u - \Theta(u) \in \mathfrak{P}_0$. As $\mathfrak{P}$ is an abelian subalgebra, one can extend Θ to a linear mapping.

(2) Since $\mathfrak{P}_0 = \sum_{\beta \in \Delta} \mathfrak{P}_0 \cap \mathfrak{S}_0(\beta)$ by the deliberations in (1), it is sufficient to prove $[[x, \mathfrak{S}_{0,\alpha}], \mathfrak{S}_{0,-\alpha}] \subset \mathfrak{P}_0$ for all $\alpha \notin \Delta$ and for any $x \in \mathfrak{P}_0 \cap \mathfrak{S}_0(\beta)$. By the introductory remark it is sufficient to prove this for those α for which $\mathfrak{S}_0(\alpha,\beta)^{(1)} \cong \mathfrak{sl}(2) \otimes \mathcal{O}(1;\underline{1})$ holds. Then

$$\begin{aligned} \varphi([[x, \mathfrak{S}_{0,\alpha}], \mathfrak{S}_{0,-\alpha}]) \subset [[h \otimes \mathcal{O}(1;\underline{1})_{(1)}, e \otimes \mathcal{O}(1;\underline{1})], f \otimes \mathcal{O}(1;\underline{1})] = \\ Fh \otimes \mathcal{O}(1;\underline{1})_{(1)}. \end{aligned}$$

This is the claim.

(3) One inclusion is obvious. To prove "$\supset$" we consider any 2-section $\mathfrak{S}_0(\alpha,\beta)$ with $\beta \in \Delta$ having the property that $\mathfrak{S}_{0,\alpha}$ and $\mathfrak{S}_{0,\alpha+\beta}$ are nonzero. In Lemma 20.6.4(3) the case $\mathfrak{S}_0(\alpha,\beta)^{(1)} \cong \mathfrak{sl}(2) \otimes \mathcal{O}(1;\underline{1})$ occurs. In this type of section there is $x_\beta \in \mathfrak{S}_{0,\beta} \cap \mathfrak{S}_0(\alpha,\beta)^{(1)} \subset \mathfrak{P} \cap {\mathfrak{S}_0}^{(1)}$ such that $\alpha(x_\beta) \neq 0$. Therefore $\mathfrak{S}_{0,\alpha+\beta} = [x_\beta, \mathfrak{S}_{0,\alpha}]$ is true. As a result,

$$\mathfrak{S}_{0,\bar{\alpha}} = \sum_{\beta \in \Delta} \mathfrak{S}_{0,\alpha+\beta} = \sum_{\beta \in \Delta} [x_\beta, \mathfrak{S}_{0,\alpha}] \subset [\mathfrak{P} \cap \mathfrak{S}_0^{(1)}, \mathfrak{S}_{0,\alpha}].$$

(4) Our assumption $\mathfrak{S}_{0,\alpha} \neq \{0\}$ implies that $\mathfrak{S}_0(\alpha)$ is not solvable. Recall that $\mathfrak{P}_0 \subset \mathfrak{S}_0(\bar{\alpha})$ acts nilpotently on $\mathfrak{S}_0(\bar{\alpha})$. From Lemma 20.6.9(3) we conclude that $\mathfrak{S}_{0,\bar{\alpha}}/\operatorname{rad} \mathfrak{S}_{0,\bar{\alpha}}$ is 1-dimensional. This enforces

$$[\mathfrak{P}_0, \mathfrak{S}_{0,\alpha}] \subset \big(\operatorname{rad} \mathfrak{S}_0(\bar{\alpha})\big)_{\bar{\alpha}}.$$

To prove equality we observe that

$$\mathfrak{S}_{0,\bar\alpha} = [\mathfrak{P}\cap{\mathfrak{S}_0}^{(1)}, \mathfrak{S}_{0,\alpha}] = [\mathfrak{P}_0, \mathfrak{S}_{0,\alpha}] + [\mathfrak{H}\cap{\mathfrak{S}_0}^{(1)}, \mathfrak{S}_{0,\alpha}]$$
$$\subset [\mathfrak{P}_0, \mathfrak{S}_{0,\alpha}] + \mathfrak{S}_{0,\alpha}.$$

As $\mathfrak{S}_{0,\alpha}\not\subset \operatorname{rad}\mathfrak{S}_0(\bar\alpha)$ is 1-dimensional, this gives the required inclusion. □

Lemma 20.6.11. *One has* $\mathfrak{S}_{0,\beta} = \{0\}$ *for all* $\beta\in\Delta\setminus\{0\}$ *and* $(\operatorname{rad}\mathfrak{S}_0(\bar\alpha))_{\bar\alpha} = \{0\}$ *for all* $\bar\alpha\neq 0$.

Proof. (a) Set

$$\mathfrak{R} := \sum_{\bar\alpha\neq 0}(\operatorname{rad}\mathfrak{S}_0(\bar\alpha))_{\bar\alpha} + \mathfrak{P}_0.$$

In a first step we show that $\mathfrak{R}$ is an ideal of $\mathfrak{S}_0$ which acts nilpotently on $\mathfrak{S}$. Lemma 20.6.9(4) shows that

$$\big(\operatorname{rad}\mathfrak{S}_0(\bar\alpha)\big)_{\bar\alpha} = \big(\operatorname{rad}\mathfrak{S}_0(\bar\alpha,\bar\beta)\big)\cap\mathfrak{S}_{0,\bar\alpha}$$

for all $\bar\alpha\neq 0$ with $\mathfrak{S}_{0,\bar\alpha}\neq\{0\}$ and all β. Therefore

$$[\big(\operatorname{rad}\mathfrak{S}_0(\bar\alpha)\big)_{\bar\alpha}, \mathfrak{S}_{0,\bar\beta}] \subset \big(\operatorname{rad}\mathfrak{S}_0(\bar\alpha+\bar\beta)\big)_{\bar\alpha+\bar\beta} \ \text{ if } \bar\alpha+\bar\beta\neq 0.$$

Next consider roots $\bar\alpha\neq 0$, $\bar\beta = -\bar\alpha$. For any $\mu\in\Delta$ with $\mathfrak{S}_{0,\alpha-\mu}\neq\{0\}$ one has $\big(\operatorname{rad}\mathfrak{S}_0(\bar\alpha)\big)_{\bar\alpha} = [\mathfrak{P}_0, \mathfrak{S}_{0,\alpha-\mu}]$ (Lemma 20.6.10(4)). We also observe that $\mathfrak{S}_{0,\alpha-\mu}\neq\{0\}$ if and only if $\mathfrak{S}_{0,-\alpha+\mu}\neq\{0\}$. Then, by Lemma 20.6.10(2),

$$[\big(\operatorname{rad}\mathfrak{S}_0(\bar\alpha)\big)_{\bar\alpha}, \mathfrak{S}_{0,-\bar\alpha}] = \sum_{\mu\in\Delta}[\big(\operatorname{rad}\mathfrak{S}_0(\bar\alpha)\big)_{\bar\alpha}, \mathfrak{S}_{0,-\alpha+\mu}]$$
$$= \sum_{\mu\in\Delta \text{ with } \mathfrak{S}_{0,\alpha-\mu}\neq\{0\}}[[\mathfrak{P}_0, \mathfrak{S}_{0,\alpha-\mu}], \mathfrak{S}_{0,-\alpha+\mu}] \subset \mathfrak{P}_0.$$

So $\mathfrak{R}$ is an ideal of $\mathfrak{S}_0$.

The set $\cup_{\bar\alpha\neq 0}(\operatorname{rad}\mathfrak{S}_0(\bar\alpha))_{\bar\alpha}\cup\mathfrak{P}_0$ is closed under Lie multiplication and consists of elements only which act nilpotently on $\mathfrak{S}$. Thus $\mathfrak{R}$ acts nilpotently on $\mathfrak{S}$.

(b) Suppose $\mathfrak{S}_{-2} = \{0\}$. Then $\mathfrak{S}_{-1}$ is $\mathfrak{S}_0$-irreducible (Proposition 20.6.3(2)). Note that $[\mathfrak{R}, \mathfrak{S}_{-1}]\neq\mathfrak{S}_{-1}$ is $\mathfrak{S}_0$-invariant. Hence $[\mathfrak{R}, \mathfrak{S}_{-1}] = \{0\}$. This gives $\mathfrak{R} = \{0\}$ (Proposition 20.6.3(6)). As a consequence, $\big(\operatorname{rad}\mathfrak{S}_{0,\bar\alpha}\big)_{\bar\alpha} = \{0\}$ for all $\bar\alpha\neq 0$ and $\mathfrak{P}_0 = \{0\}$. The latter means $\mathfrak{P}\cap{\mathfrak{S}_0}^{(1)}\subset\mathfrak{H}$, whence

$$\mathfrak{S}_{0,\bar\alpha} = [\mathfrak{P}\cap{\mathfrak{S}_0}^{(1)}, \mathfrak{S}_{0,\alpha}] = [\mathfrak{H}, \mathfrak{S}_{0,\alpha}] = \mathfrak{S}_{0,\alpha}$$

whenever $\alpha\notin\Delta$ and $\mathfrak{S}_{0,\alpha}\neq\{0\}$. In particular, $\mathfrak{S}_{0,\alpha+\gamma} = \{0\}$ for all $\gamma\in\Delta\setminus\{0\}$.

Now let $\beta\in\Delta$ and $x\in\mathfrak{S}_{0,\beta}$. One has $[x, \mathfrak{S}_{0,\alpha}] = \{0\}$ for all $\alpha\in\Delta$ as $\mathfrak{P}$ is abelian. If $\alpha\notin\Delta$ and $\mathfrak{S}_{0,\alpha}\neq\{0\}$, then $[x, \mathfrak{S}_{0,\alpha}]\subset\mathfrak{S}_{0,\alpha+\beta} = \{0\}$. Therefore

x centralizes $\mathfrak{S}_0$. Take $x_n \in \sum_{i\geq 0} F x^{[p]^i}$ the $[p]$-nilpotent part of x and observe that this is a nilpotent element in $\mathrm{Der}_0\,\mathfrak{S}$ which centralizes $\mathfrak{S}_0$. As before we get $[x_n, \mathfrak{S}_{-1}] = \{0\}$ and hence $x_n = 0$ (Proposition 20.6.3(7)). This means that x is semisimple, whence is contained in $\sum_{i>0} F x^{[p]^i}$ ([S-F, §2.3, Definition]). But then $[\mathfrak{T}, x] = \{0\}$ and hence $\beta = 0$ holds. This is the claim.

(c) Next assume that $\mathfrak{S}_{-2} \neq \{0\}$. As $\mathfrak{S}_{-3} = \{0\}$, we obtain that $\mathfrak{S}_{-2}$ is $\mathfrak{S}_0$-irreducible (Proposition 20.6.3(2)) and hence $[\mathfrak{R}, \mathfrak{S}_{-2}] = \{0\}$. Set $V := [[\mathfrak{R}, \mathfrak{S}_{-1}], [\mathfrak{R}, \mathfrak{S}_{-1}]] \subset \mathfrak{S}_{-2}$. Note that V is $\mathfrak{S}_0$-invariant. If $V \neq \{0\}$, then we obtain $V = \mathfrak{S}_{-2}$. This gives $\{0\} \neq [\mathfrak{S}_{-2}, \mathfrak{S}_1] \subset [\mathfrak{R}, \mathfrak{S}_{-1}] \subsetneq \mathfrak{S}_{-1}$. But $[\mathfrak{S}_{-2}, \mathfrak{S}_1]$ is $(\mathfrak{S}_0 + \mathfrak{T}_0)$-invariant and $\mathfrak{S}_{-1}$ is $(\mathfrak{S}_0 + \mathfrak{T}_0)$-irreducible. So this case is impossible.

As a result, $V = \{0\}$. We obtain

$$[[\mathfrak{R}, [\mathfrak{R}, \mathfrak{S}_{-1}]], \mathfrak{S}_{-1}] \subset [\mathfrak{R}, \mathfrak{S}_{-2}] + [[\mathfrak{R}, \mathfrak{S}_{-1}], [\mathfrak{R}, \mathfrak{S}_{-1}]] = \{0\}.$$

Proposition 20.6.3(6) yields $[\mathfrak{R}, [\mathfrak{R}, \mathfrak{S}_{-1}]] = \{0\}$. In particular, for $\mu \in \Delta$, $x \in \mathfrak{S}_{0,\mu} \cap {\mathfrak{S}_0}^{(1)}$ and all $\kappa \in \Gamma_{-1}$ one has $(\mathrm{ad}\,(x - \Theta(x)))^2(\mathfrak{S}_{-1,\kappa}) = \{0\}$.

Suppose $\mu \neq 0$. Considering root space components one obtains $(\mathrm{ad}\,x)^2(\mathfrak{S}_{-1,\kappa}) = \{0\}$. By Proposition 20.3.2, this is only possible if $(\mathrm{ad}\,x)(\mathfrak{S}_{-1,\kappa}) = \{0\}$ for all κ, whence $[x, \mathfrak{S}_{-1}] = \{0\}$. Proposition 20.6.3(6) gives $x = 0$. If $\mu = 0$, then $x - \Theta(x) \in \mathfrak{H}$ acts nilpotently on $\mathfrak{S}_{-1,\kappa}$. As this space is 1-dimensional, $x - \Theta(x)$ annihilates $\mathfrak{S}_{-1,\kappa}$. Then $x - \Theta(x) = 0$ holds as well.

As a result, $\mathfrak{P}_0 = \{0\}$. This gives $\mathrm{rad}\,\mathfrak{S}_0(\bar{\alpha}) = \{0\}$ by Lemma 20.6.10(4). As in (b) we now conclude that $\mathfrak{S}_{0,\beta} = \{0\}$ for all $\beta \in \Delta \setminus \{0\}$. □

Lemma 20.6.12.

(1) *$\mathfrak{H}$ is a torus of maximal dimension in $\mathfrak{S}_{[p]}$.*

(2) *$\mathfrak{H}$ distinguishes the $\mathfrak{T}_0$-roots on $\mathfrak{S}_{-1}$.*

(3) *$\mathfrak{S}_{-1}$ is $\mathfrak{S}_0$-irreducible.*

Proof. (1) Lemma 20.6.11 shows that $\mathfrak{P} = \mathfrak{H}$, while Lemma 20.6.8 then gives $TR(\mathfrak{S}) = TR(\mathfrak{H}, \mathfrak{S})$. It remains to show that $\mathfrak{H}$ acts semisimply on $\mathfrak{S}$. To do so we take arbitrary $h \in \mathfrak{H}$ and let $h_n \in \mathfrak{H}_{[p]}$ denote the $[p]$-nilpotent part. Then h_n acts nilpotently on every $\mathfrak{S}_{-1,\kappa}$. As these spaces are 1-dimensional (if nonzero), they are annihilated by h_n. But then h_n annihilates $\mathfrak{S}_{-1}$ and Proposition 20.6.3(7) yields $h_n = 0$.

(2)(a) Let $-\alpha, -\beta \in \Gamma(\mathfrak{S}_{-1}, \mathfrak{T}_0)$ be $\mathbb{F}_p$-independent and suppose that $\eta := \alpha - \beta$ vanishes on H. We are not in case (3) of Proposition 20.3.2, so we are in case (4). Then $\alpha(Q_\eta) \neq 0$. This gives $Q_\eta \not\subset L_{(1)}$. Since $\eta(\mathfrak{T}_1) = 0$, we conclude $\mathfrak{S}_{0,\eta} \neq \{0\}$, this contradicting Lemma 20.6.11. Consequently, H distinguishes α and β.

(b) Next we intend to show that $\alpha - \beta$ does not vanish on $(\mathrm{gr}_0\,H) \cap \mathfrak{A}$. We first check the possibilities of $Q[\alpha, \beta]$ in Proposition 20.3.3. In case (C) $[Q_{\mu_1}, Q_{-\mu_1}] + [Q_{\mu_2}, Q_{-\mu_2}]$

distinguishes the roots. Then Lemma 20.6.2(4) proves that $\mathfrak{A} \cap (\mathrm{gr}_0\, H)$ distinguishes the roots on $\sum_{\mu \in \mathbb{F}_p\alpha + \mathbb{F}_p\beta} \mathfrak{A}_{-1,\mu}$. Case (D) does not occur, since in this case $\alpha - \beta$ vanishes on H, in case (E) all roots are solvable. In case (F) we argue as in case (C). As a result, only cases (A) and (B) remain under consideration.

Let $-\mu \in (\mathbb{F}_p\alpha + \mathbb{F}_p\beta) \cap \Gamma(\mathfrak{A}_{-i}, \mathfrak{T})$ $(i > 0)$ be arbitrary. Due to Proposition 20.6.1(4) there are $x \in \mathfrak{A}_{-i,-\mu}$, $y \in \mathfrak{A}_{i,\mu}$ such that $\mu([x, y]) \neq 0$. Choose inverse images $\tilde{x} \in L_{(-i),-\mu} \setminus L_{(-i+1)}$, $\tilde{y} \in L_{(i),\mu} \setminus L_{(i+1)}$ so that $\mathrm{gr}_{-i}\, \tilde{x} = x$, $\mathrm{gr}_i\, \tilde{y} = y$. Then $\mathrm{gr}_0\, [\tilde{x}, \tilde{y}] = [x, y] =: h_\mu$ and $\mu(h_\mu) \neq 0$.

We now check the possibilities in Theorem 20.1.1. Cases (1) and (2) cannot occur, since α, β are non-solvable $\mathbb{F}_p$-independent roots. In case (3) we have $\alpha(h_\alpha) \neq 0$, $\beta(h_\beta) \neq 0$, $\beta(h_\alpha) = 0$, so the claim follows. Case (4) does not occur, since α is non-solvable and non-classical. In case (5) with $L[\alpha, \beta] \not\cong K(3; \underline{1})$ the algebra $Q[\alpha, \beta]$ has not the required form. If $L[\alpha, \beta] \cong K(3; \underline{1})$, then $(\mathrm{gr}_0\, H) \cap \mathfrak{A}$ distinguishes the roots on the (-1)-space. In case (6) with $L[\alpha, \beta]$ being $W(1; \underline{2})$ or $H(2; \underline{1}; \Phi(\tau))^{(1)}$ the roots $-\alpha, -\beta$ are not distinguished by H, so this is impossible. If $L[\alpha, \beta] \cong H(2; \underline{1}; \Phi(1))$, then $[Q_{\alpha-\beta}, Q_{-\alpha+\beta}]$ distinguishes the roots $-\alpha, -\beta$ (see also Proposition 20.2.3 and its proof). Argue as before. In case (7) the roots $-\alpha, -\beta$ are not distinguished by H. In case (8) we argue as for $H(2; \underline{1}; \Phi(1))$. In case (9) $\mathfrak{s}$ must be non-classical because α is non-classical. If $\mathfrak{s} \cong W(1; \underline{1})$, then the roots $-\alpha, -\beta$ are not distinguished by H. If $\mathfrak{s} \cong H(2; \underline{1})^{(2)}$, then $\{-\alpha, -\beta\} \subset \pm\alpha + \mathbb{F}_p\eta$ with $\eta \in \Delta$. So the roots $-\alpha, -\beta$ are not distinguished by H or are distinguished by $(\mathrm{gr}_0\, H) \cap \mathfrak{A}$. In cases (10) and (11) we, by symmetry, may take $\alpha \notin \mathbb{F}_p\kappa$. Then $\Psi_{\alpha,\beta}(h_\alpha) \in \mathfrak{s} \otimes \mathcal{O}(r; \underline{1})$. In particular, $\kappa(h_\alpha) = 0$. Note that $\alpha(h_\alpha) \neq 0$, $\kappa(h_\kappa) \neq 0$.

(c) As a consequence, $\alpha - \beta$ does not vanish on $(\mathrm{gr}_0\, H) \cap \mathfrak{A}$. All involved roots vanish on $\mathfrak{T}_1$, whence the statements hold for $C_{\mathfrak{A}}(\mathfrak{T}_1) = \mathfrak{S}$. So $\mathfrak{H}$ distinguishes α, β.

(3) As a consequence of (2), the $\mathfrak{H}$-module $\mathfrak{S}_{-1} = \oplus\, \mathfrak{S}_{-1,\mu}$ decomposes into the direct sum of its $\mathfrak{T}_0$-root spaces. Then $\mathfrak{S}_{-1}$ is not only $(\mathfrak{S}_0 + \mathfrak{T}_0)$-irreducible but $\mathfrak{S}_0$-irreducible. □

Lemma 20.6.13. *Let $u \in \mathfrak{S}_{-1,-\lambda}$ be a nonzero $\mathfrak{T}_0$-root vector. Then $\mathfrak{S}_{-1} = [\mathfrak{S}_0, u]$ holds.*

Proof. Using induction one has to prove the following:

$$\text{if } e_\alpha \in \mathfrak{S}_{0,\alpha},\ e_\beta \in \mathfrak{S}_{0,\beta}, \text{ then } [e_\beta, [e_\alpha, u]] \in [\mathfrak{S}_{0,\alpha+\beta}, u].$$

Choose according to Proposition 20.6.3(5)

$$w_2 \in \mathfrak{S}_{2,2\lambda} \text{ satisfying } \lambda([u, [u, w_2]]) \neq 0$$

Put $w_1 := [u, w_2] \in \mathfrak{S}_{1,\lambda}$, $h := [u, w_1] \in \mathfrak{S}_{0,0}$. One has $\lambda(h) \neq 0$.

(a) Suppose $(\mathrm{ad}\, u)^2(e_\alpha) = (\mathrm{ad}\, u)^2(e_\beta) = 0$.

If $[u,[e_\beta,w_1]]\neq 0$, then, as $\dim\mathfrak{S}_{0,\beta}=1$, one has $[u,[e_\beta,w_1]]=qe_\beta$ with $q\in F^*$. In this case the following holds

$$[e_\beta,[e_\alpha,u]]=q^{-1}[[u,[e_\beta,w_1]],[e_\alpha,u]]=q^{-1}[u,[[e_\beta,w_1],[e_\alpha,u]]]\in[u,\mathfrak{S}_{0,\alpha+\beta}].$$

So we may assume $[u,[e_\beta,w_1]]=0$.

If $[[u,e_\beta],w_1]]\neq 0$, then $[u,[[u,e_\beta],w_2]]=[[u,e_\beta],w_1]]\neq 0$. Argue as before so that we may assume $[[u,e_\beta],w_1]]=0$. Consequently,

$$0=[[u,e_\beta],w_1]=[[u,w_1],e_\beta]=\beta(h)e_\beta.$$

This gives $\beta(h)=0$. Next we compute

$$\begin{aligned}0&=[u,[[u,e_\beta],w_1]]=[[u,e_\beta],[u,w_1]]=-[h,[u,e_\beta]]\\&=-(\beta-\lambda)(h)[u,e_\beta]=\lambda(h)[u,e_\beta].\end{aligned}$$

As $\lambda(h)\neq 0$, this gives $[e_\beta,[e_\alpha,u]]=[[e_\beta,e_\alpha],u]\in[\mathfrak{S}_{0,\alpha+\beta},u]$.

(b) By symmetry we now may assume $(\operatorname{ad}u)^2(e_\beta)\neq 0$. At first we recall that $\mathfrak{S}_{-3}=\{0\}$. Hence $(\operatorname{ad}u)^3(e_\alpha)=0$. Let $w_{2,\beta}\in\mathfrak{S}_{2,2\lambda+\beta}$ be arbitrary (at the moment we allow the case $\mathfrak{S}_{2,2\lambda+\beta}=\{0\}$). We obtain a congruence modulo $[\mathfrak{S}_0,u]$

$$\begin{aligned}0&\equiv(\operatorname{ad}u)^2([[e_\alpha,u],w_{2,\beta}])=2[[u,[e_\alpha,u]],[u,w_{2,\beta}]]+[[e_\alpha,u],(\operatorname{ad}u)^2(w_{2,\beta})]\\&=2[u,[[e_\alpha,u],[u,w_{2,\beta}]]]-[[e_\alpha,u],(\operatorname{ad}u)^2(w_{2,\beta})]\equiv-[[e_\alpha,u],(\operatorname{ad}u)^2(w_{2,\beta})].\end{aligned}$$

Note that $(\operatorname{ad}u)^2(w_{2,\beta})=re_\beta$ with $r\in F$. If $r\neq 0$, then we obtain $[e_\beta,[e_\alpha,u]]\in[\mathfrak{S}_0,u]$. Thus we may assume $(\operatorname{ad}u)^2(w_{2,\beta})=0$.

Choose $\tilde u\in L_{(-1),-\lambda}\setminus L_{(0)}$, so that $\operatorname{gr}_{-1}\tilde u=u$. Our assumption $(\operatorname{ad}u)^2(e_\beta)\neq 0$ implies that $L[\lambda,\beta]\cong K(3;\underline{1})$ (Propositions 20.3.2(5) and 20.2.5(5)). One may normalize the torus so that $\Psi_{\lambda,\beta}(T)=FD_K(x_1x_2)+FD_K(x_3)$ (Theorem 7.5.13). We also may assume after an application of an automorphism of $K(3;\underline{1})$, that $\Psi_{\lambda,\beta}(\tilde u)\in F^*D_K(x_1)+K(3;\underline{1})_{(0)}$ (Lemma 20.5.5). The root spaces $K(3;\underline{1})_{-1,-\lambda}$ and $K(3;\underline{1})_{-1,-\lambda+\beta}$ are then spanned by $D_K(x_1)$ and $D_K(x_2)$, respectively. We get

$$\begin{aligned}-\lambda(D_K(x_1x_2))D_K(x_1)&=[D_K(x_1x_2),D_K(x_1)]=-D_K(x_1),\\(-\lambda+\beta)(D_K(x_1x_2))D_K(x_2)&=[D_K(x_1x_2),D_K(x_2)]=D_K(x_2),\end{aligned}$$

whence

$$-\lambda(D_K(x_1x_2))=-1,\quad(-\lambda+\beta)(D_K(x_1x_2))=1.$$

Then $(2\lambda+\beta)(D_K(x_1x_2))=4$. Since

$$K(3;\underline{1})_2=\sum_{i=0}^{4}FD_K(x_1^ix_2^{4-i})+\sum_{i=0}^{2}FD_K(x_1^ix_2^{2-i}x_3)+FD_K(x_3^2),$$

the root space $K(3;\underline{1})_{2,2\lambda+\beta}$ is spanned by $D_K(x_2^4) = (1/3)[D_K(x_1x_2^2), D_K(x_2^3)]$. Note that $\lambda(D_K(x_3)) = 1$, $\beta(D_K(x_3)) = 0$. Then

$$D_K(x_2^4) \in [K(3;\underline{1})_{1,\lambda}, K(3;\underline{1})_{1,\lambda+\beta}].$$

By Lemma 20.5.5, there is $w \in L_{(2),2\lambda+\beta}$ such that $\Psi_{\lambda,\beta}(w) = D_K(x_2^4)$. Then

$$\begin{aligned}\Psi_{\lambda,\beta}\big((\operatorname{ad}\tilde{u})^2(w)\big) &\equiv [D_K(x_1), [D_K(x_1), D_K(x_2^4)]] \\ &= 12 D_K(x_2^2) \not\equiv 0 \qquad (\operatorname{mod} K(3;\underline{1})_{(1)}).\end{aligned}$$

In particular, $(\operatorname{ad}\tilde{u})^2(w) \notin L_{(1)}$ holds, whence $w \notin L_{(3)}$. Take $w_{2,\beta} := \operatorname{gr}_2 w$. Since $\beta(\mathcal{T}_1) = 0$, we have $\operatorname{gr}_2 w \in \mathfrak{S}_{2,2\lambda+\beta}$. We obtain

$$(\operatorname{ad} u)^2(w_{2,\beta}) = \operatorname{gr}_0\big((\operatorname{ad}\tilde{u})^2(w)\big) \neq 0,$$

which contradicts the result of (b). So this case is impossible. □

We conclude

Theorem 20.6.14. *Let L be a finite dimensional simple Lie algebra over an algebraically closed field F of characteristic $p > 3$ and assume that all tori of maximal dimension in the minimal p-envelope of L are standard. Suppose T is such a torus, which in addition is optimal. Assume further that $Q(L,T) \neq L$. Then $\mathfrak{S}(L,T)$ is, as a graded algebra, isomorphic to a restricted simple Lie algebra of Cartan type.*

Proof. (a) Lemmas 20.6.11, 20.6.9, and 20.6.12 show that $\mathfrak{S}_0(\bar{\alpha}) \cong \mathfrak{sl}(2) \oplus (\mathfrak{H} \cap (\ker\bar{\alpha}))$ for all nonzero $\bar{\alpha}$, $\mathfrak{S}_0(\bar{\alpha},\bar{\beta})$ is classical reductive, and $\mathfrak{H}$ is a maximal torus. Then

$$\mathfrak{S}_0 = \mathfrak{H} + \mathfrak{S}_0{}^{(1)},\ \mathfrak{S}_0{}^{(1)} = \mathfrak{S}_0{}^{(2)},\ \operatorname{rad}\mathfrak{S}_0 = C(\mathfrak{S}_0)$$

hold. Moreover, if $\alpha, \beta \in \Gamma(\mathfrak{S}_0, \mathfrak{T}_0)$ coincide on $\mathfrak{H}$, then due to Lemma 20.6.9(4) they are equal. This means that the decompositions of $\mathfrak{S}_0$ with respect to $\mathfrak{T}_0$ and $\mathfrak{H}$ coincide. Theorem 4.1.2 applies and shows that $\mathfrak{S}_0{}^{(1)}/C(\mathfrak{S}_0{}^{(1)}) = \oplus I_j$ is the direct sum of classical simple algebras (if it is nonzero).

Choose a set of simple roots for $\oplus I_j$, and put

$$\mathfrak{b}^+ := \mathfrak{H} + \sum_{\mu>0} \mathfrak{S}_{0,\mu}, \quad \mathfrak{b}^- := \mathfrak{H} + \sum_{\mu<0} \mathfrak{S}_{0,\mu}.$$

We know that every root vector $e_\alpha \in \mathfrak{S}_0$, $\alpha \neq 0$, satisfies $(\operatorname{ad} e_\alpha)^2(\mathfrak{S}_{-1}) = \{0\}$. Proposition 20.6.3(7) then yields that $(\operatorname{ad} e_\alpha)^p = 0$. Hence there is a root vector

$$u \in \mathfrak{S}_{-1,-\lambda},\ [\mathfrak{b}^+, u] \subset Fu.$$

With this choice of u, Lemma 20.6.13 gives $\mathfrak{S}_{-1} = [\mathfrak{b}^-, Fu]$. Therefore a highest weight vector of $\mathfrak{b}^+$ induces a rank 1 transformation on $\mathfrak{S}_{-1}$. Theorem 5.7.3 proves that $\mathfrak{S}_0$ is one of $\mathfrak{gl}(q)$, $\mathfrak{sl}(q)$, $\mathfrak{sp}(q)$, $\mathfrak{sp}(q) \oplus F\mathrm{Id}$, where $q = \dim\mathfrak{S}_{-1}$.

We check the assumptions of KAC' Recognition Theorem 5.6.1. We just determined $\mathfrak{S}_0$. In Lemma 20.6.12 we proved that $\mathfrak{S}_{-1}$ is an irreducible $\mathfrak{S}_0$-module. Proposition 20.6.3(5) shows that for all $j > 0$, if $x \in \mathfrak{S}_j$ and $[x, \mathfrak{S}_{-1}] = \{0\}$, then $x = 0$. It also shows that, if $x \in \mathfrak{S}_{-j}$ and $[x, \mathfrak{S}_1] = \{0\}$, then $x = 0$ $(j = 1, 2)$.

Consequently, Theorem 5.6.1 applies and shows that $\mathfrak{S}$ is one of the following.

- simple classical with standard grading; but then $r = s$ which is not true in our case;
- $\mathfrak{pgl}(kp)$ with standard grading; this algebra is not simple;
- a graded Cartan type algebra with its natural grading or the reverse; since $r < s$, the reverse grading does not occur;
- Melikian with natural grading or the reverse; then the grading has depth 3 or bigger, while we have $r \leq 2$.

Consequently, $\mathfrak{S} \cong X(q; \underline{n})^{(2)}$ with $X = W, S, H, K$.

It remains to show that $\mathfrak{S}$ is restricted. Recall (Lemma 20.6.12) that $\mathfrak{H}$ is a torus of maximal dimension in the semisimple p-envelope $\mathfrak{S}_{[p]}$. Its dimension can be computed as the dimension of a maximal torus in the simple classical algebra ${\mathfrak{S}_0}^{(1)}/C({\mathfrak{S}_0}^{(1)})$ plus the dimension of the center of $\mathfrak{S}_0$. In all cases this dimension coincides with the dimension of the torus $(\sum_{i=1}^{q} F x_i \partial_1) \cap X(q; \underline{n})^{(2)}$. As a consequence, this torus is a torus of maximal dimension in $X(q; \underline{n})^{(2)}{}_{[p]}$, and therefore every 1-section with respect to this torus has absolute toral rank at most 1. It is not hard to see that this can only happen if $n_i = 1$ for all $i = 1, \dots, q$. Then $\mathfrak{S} \cong X(q; \underline{1})^{(2)}$ is restricted. □

20.7 Completing the classification

In a final step we detect the structure of L from that of $\mathfrak{S}(L, T)$.

Theorem 20.7.1. *If $\mathfrak{S}(L, T) \cong K(2q+1; \underline{1})^{(1)}$ is of Contact type, then $L \cong K(2q+1; \underline{n})^{(1)}$ is of Contact type as well with the same number q and suitable $\underline{n}$.*

Proof. In the present case one has $\mathfrak{S} \cong K(2q + 1; \underline{1})^{(1)}$ with the natural grading. Hence $\mathfrak{S}_0$ contains a central element $z \neq 0$ related to $D_K(x_{2q+1})$. Let $t \in Q$ be an element which is mapped onto $z \otimes 1$. Then t acts on $\mathrm{gr}_{-1} L \cong \mathfrak{S}_{-1} \otimes \mathcal{O}(m; \underline{1})$ with the single eigenvalue -1. But then it acts on every $\mathrm{gr}_k L$ with the single eigenvalue k. Choose subspaces V_i $(i = -2, \dots, 2)$ in the weight spaces $L_i(t)$, so that

$$L_{(i)} = L_{(i+1)} \oplus V_i.$$

Choose subspaces $R_i \subset V_i$ $(i = -2, -1, 0)$ such that

$$(R_i + L_{(i+1)})/L_{(i+1)} = \mathfrak{S}_i \otimes \mathcal{O}(m; \underline{1})_{(1)}.$$

Since L is simple, we have

$$L = \sum_{i \not\equiv 0\,(p)} L_i(t) + [L_i(t), L_{-i}(t)] = \sum_{i \not\equiv 0\,(p)} L_i(t) + [V_2, V_{-2}] + [V_1, V_{-1}] + L_{(1)}.$$

Consequently, $L_{(0)} = [V_2, V_{-2}] + [V_1, V_{-1}] + L_{(1)}$. Turn this information into information on $\operatorname{gr} L$. Since $\operatorname{gr}_{-i} L \subset \mathfrak{A}$ $(i = 1, 2)$, we get

$$\operatorname{gr}_0 L = [\operatorname{gr}_2 L, \operatorname{gr}_{-2} L] + [\operatorname{gr}_1 L, \operatorname{gr}_{-1} L] \subset \mathfrak{A}_0 = \mathfrak{S}_0 \otimes \mathcal{O}(m; \underline{1}).$$

As a consequence, $\mathfrak{S} \otimes \mathcal{O}(m; \underline{1})_{(1)}$ is an ideal of $\operatorname{gr} L$. Therefore

$$\begin{aligned}
&[R_{-2}, V_0] \subset L_{-2}(t) \cap (R_{-2} + L_{(-1)}) = R_{-2} + L_{-2}(t) \cap L_{(-1)} \subset R_{-2} + Q,\\
&[R_{-2}, V_1] \subset L_{-1}(t) \cap (R_{-1} + Q) \subset R_{-1} + Q,\\
&[R_{-1}, Q] \subset R_{-1} + Q,\\
&[R_{-1}, V_{-1}] \subset L_{-2}(t) \cap (R_{-2} + L_{(-1)}) = R_{-2} + L_{-2}(t) \cap L_{(-1)} \subset R_{-2} + Q,\\
&[R_0, V_{-2}] \subset L_{-2}(t) \cap (R_{-2} + L_{(-1)}) = R_{-2} + L_{-2}(t) \cap L_{(-1)} \subset R_{-2} + Q,\\
&[R_0, V_{-1}] \subset R_{-1} + Q,\\
&[R_0, Q] \subset R_0 + L_{(1)}.
\end{aligned}$$

Considering weights with respect to $\operatorname{ad} t$, we obtain

$$[R_{-2}, V_{-2}] \subset Q, \qquad [R_{-2}, V_{-1}] \subset Q.$$

By definition one has

$$[Q, V_{-1}] \subset L_{(-1)} = V_{-1} + Q.$$

We conclude that

$$L'_{(0)} := R_{-2} + R_{-1} + Q$$

is a subalgebra containing Q and

$$L'_{(-1)} := V_{-1} + L'_{(0)}$$

is $L'_{(0)}$-invariant. Next observe that

$$[R_{-2} + R_{-1} + R_0, V_{-1} + L'_{(0)}] \subset [R_{-2} + R_{-1} + R_0, V_{-1}] + [L'_{(0)}, L'_{(0)}] \subset L'_{(0)}.$$

Therefore $L'_{(1)} := \{u \in L'_{(0)} \mid [u, L'_{(-1)}] \subset L'_{(0)}\}$ contains $R_{-2} + R_{-1} + R_0$. By Lemma 20.6.12(3), $L'_{(-1)}/L'_{(0)} \cong V_{-1}/R_{-1} \cong \mathfrak{S}_{-1}$ is $L'_{(0)}$-irreducible. Moreover, one obtains for the graded algebra associated with the standard filtration (Definition 3.5.1) given by the pair $(L'_{(0)}, L'_{-1})$

$$L/L'_{(-1)} \oplus L'_{(-1)}/L'_{(0)} \oplus L'_{(0)}/L'_{(1)} \cong V_{-2}/R_{-2} \oplus V_{-1}/R_{-1} \oplus V_0/R_0.$$

The above-listed inclusions show that the non-positive part of this graded algebra is isomorphic as an algebra to $\mathfrak{S}_{-2} \oplus \mathfrak{S}_{-1} \oplus \mathfrak{S}_0$. Since $\sum_{n \geq 0} \mathfrak{S}_n$ is a maximal subalgebra of $\mathfrak{S}$, it is easy to conclude that $L'_{(0)}$ is a maximal subalgebra of L. The Weak Recognition Theorem 5.5.3 applies. Then $L \cong K(2q + 1; \underline{n})^{(1)}$. □

We are left with the case that $\mathfrak{S}$ is of Witt, Special, or Hamiltonian type. **In particular $r = 1$ holds**, which we will assume from now on.

Lemma 20.7.2. *There are T-invariant subspaces V_{-1}, V_0 with the properties*

$$L = V_{-1} + Q, \quad Q \cap V_{-1} = \{0\},$$

$$V_0 \subset Q, \quad L_{(1)} \cap V_0 = \{0\}, \quad (V_0 + L_{(1)})/L_{(1)} = \mathfrak{A}(L,T)_0.$$

For $i = -1, 0$, let R_i denote the subspace of V_i, which is mapped bijectively onto $\mathfrak{S}_i \otimes \mathcal{O}(m;\underline{1})_{(1)}$, i.e.,

$$(R_i + L_{(i+1)})/L_{(i+1)} = \mathfrak{S}_i \otimes \mathcal{O}(m;\underline{1})_{(1)}.$$

The spaces V_{-1}, V_0 can be chosen such that the following statements hold.

(1) $V_{-1} \subset \sum_{\mu\in\Gamma_{-1}(L,T)} L_\mu$,
(2) $\sum_{\mu\notin\Delta} Q_\mu \subset V_0 + L_{(1)}$,
(3) $[V_{-1}, V_{-1}] \subset Q$, *and* $[V_{-1}, V_{-1}] \subset L_{(1)}$ *if* $\mathfrak{S}(L,T)$ *is Witt or Special,*
(4) $[Q, V_{-1}] \subset V_{-1} + V_0 + L_{(1)}$,
(5) $[L_{(1)}, V_{-1}] \subset V_0 + L_{(1)}$,
(6) $[Q, V_0] \subset V_0 + L_{(1)}$,
(7) $R_{-1} + V_0 + L_{(1)} = [V_0, R_{-1}] + V_0 + L_{(1)} = [R_0, V_{-1}] + V_0 + L_{(1)}$,
(8) $R_0 + L_{(1)} = [V_0, R_0] + L_{(1)}$.

Proof. (a) For any root $\alpha \in \Gamma_{-1}(L,T)$ one has $\dim L_\alpha/Q_\alpha = 1$ (Proposition 20.3.2(1)). Choose $x_\alpha \in L_\alpha$ such that $L_\alpha = Fx_\alpha + Q_\alpha$ and set $V_{-1} := \sum_{\alpha\in\Gamma_{-1}(L,T)} Fx_\alpha$. Choose a T-invariant subspace $V_0 \subset Q$ such that $V_0 \cap L_{(1)} = \{0\}$ and $V_0 + L_{(1)}/L_{(1)} \cong \mathfrak{A}_0$.

Statement (1) is true by construction, and (2) follows from Lemma 20.6.2(4) and the definition of V_0.

(b) For any $\mu \in \Gamma_i(\operatorname{gr} L, \mathfrak{T})$ put $\bar\mu := \mu|\mathfrak{H}$ (where we identity $\mathfrak{H} \cong \mathfrak{H}\otimes F \subset \mathfrak{T} \cong T$). Let $u = \sum u_j \otimes f_j + \mathrm{Id}\otimes d \in (\operatorname{gr}_i (L+T))_\mu$ be a nonzero root vector. If there is some $u_j \neq 0$, then $0 \neq u_j \in \mathfrak{S}_{i,\bar\mu}$. Otherwise, $i = 0$, and $\mu(\mathfrak{H}) = 0$ and hence $\bar\mu = 0$. In this case one has $\mathfrak{H} \subset \mathfrak{S}_{0,\bar\mu}$. Consequently, $\mathfrak{S}_{i,\bar\mu} \neq \{0\}$ whenever $\mu \in \Gamma_i(\operatorname{gr} L, T)$. This means $\bar\mu \in \Gamma_i(\mathfrak{S}, \mathfrak{H})$.

To prove (3), take $u \in V_{-1,\alpha}$, $v \in V_{-1,\beta}$. Decompose $[u,v] = y + z$ with $y \in V_{-1} \cap L_{\alpha+\beta}$, $z \in Q_{\alpha+\beta}$. Suppose $y \neq 0$, whence $\alpha + \beta \in \Gamma_{-1}(\operatorname{gr} L, T)$. By our introductory remark this gives $\bar\alpha + \bar\beta = \overline{\alpha+\beta} \in \Gamma_{-1}(\mathfrak{S}, \mathfrak{H})$. Recall that $\mathfrak{S} \cong X(q;\underline{1})^{(2)}$ with $X = W, S, H$, and $\mathfrak{H} \subset \mathfrak{S}_0$ is a maximal torus (Theorem 20.6.14, Lemma 20.6.12). Then $\mathfrak{H}$ is conjugate under $\operatorname{Aut}\mathfrak{S}$ to $X(q;\underline{1})^{(2)} \cap \big(\sum_{i=1}^q Fx_i\partial_i\big)$. Lemma 5.5.1(1) shows that this is impossible.

Next suppose that $\mathfrak{S}$ is Witt or Special. If $[u,v] \notin L_{(1)}$, then $\alpha + \beta \in \Gamma_0(\operatorname{gr} L, T)$. By our introductory remark, $\bar\alpha + \bar\beta = \overline{\alpha+\beta} \in \Gamma_0(\mathfrak{S}, \mathfrak{H})$. As before, Lemma 5.5.1(2) shows that this is impossible.

(c) In order to prove (4), take $u \in L_{(i),\beta} \setminus L_{(i+1)}$, $v \in V_{-1,\alpha}$. If $i > 1$, then $[u, v] \in L_{(1)}$. If $i = 1$, then properties of $\mathfrak{A}$ show that $[\mathrm{gr}_1\, u, \mathrm{gr}_{-1}\, v] \in \mathfrak{A}_0$. In this case $[u, v] \in V_0 + L_{(1)}$.

Now suppose $i = 0$. Then $\beta \in \Gamma_0(\mathrm{gr}\, L, T)$. Decompose $[u, v] = y + z$ with $y \in V_{-1} \cap L_{\alpha+\beta}$, $z \in Q_{\alpha+\beta}$. Suppose $z \notin L_{(1)}$. If $\mathfrak{S}$ is of Witt or Special type, we argue as in (b), apply Lemma 5.5.1(2), and derive the contradiction $Q_{\alpha+\beta} \subset L_{(1)}$. If $\mathfrak{S}$ is Hamiltonian, we again argue as in (b) but face a slight problem when applying Lemma 5.5.1. In this case Lemma 5.5.1(4) applies and shows that $p = 5$ and $\bar{\beta} = 2\bar{\alpha}$. We conclude that $\overline{\alpha + \beta} = 3\bar{\alpha} \neq 0$. In particular, $\alpha + \beta \notin \Delta$. Part (2) of this lemma gives $z \in V_0 + L_{(1)}$.

(d) Assertions (5) and (6) are direct consequences from the fact that $\mathfrak{A}$ is an ideal of $\mathrm{gr}\, L$.

(e) Since $R_{-1} + Q$ is the full preimage of $\mathfrak{S}_{-1} \otimes \mathcal{O}(m; \underline{1})_{(1)}$ under the mapping $L \to L/Q \cong \mathfrak{A}_{-1}$, and V_0 is mapped onto $\mathfrak{A}_0$ by the mapping $Q \to Q/L_{(1)}$, we obtain

$$R_{-1} + Q = [V_0, R_{-1}] + Q.$$

By use of (4) we obtain

$$[V_0, R_{-1}] \subset (R_{-1} + Q) \cap (V_{-1} + V_0 + L_{(1)}) = R_{-1} + V_0 + L_{(1)}.$$

Since now $R_{-1}, [V_0, R_{-1}] \subset V_{-1} + V_0 + L_{(1)}$, the first equation gives

$$R_{-1} + Q \cap (V_{-1} + V_0 + L_{(1)}) = [V_0, R_{-1}] + Q \cap (V_{-1} + V_0 + L_{(1)}).$$

Finally observe that $Q \cap (V_{-1} + V_0 + L_{(1)}) = V_0 + L_{(1)}$.

The second equation of (7) is proved analogously.

(g) Assertion (8) is a direct consequence from the fact that $\mathfrak{A}$ is an ideal of $\mathrm{gr}\, L$. □

Lemma 20.7.3. $L = V_{-1} + V_0 + [V_{-1}, V_{-1}] + L_{(1)}$.

Proof. Applying Lemma 20.7.2 we obtain

$$\begin{aligned}
[L, V_{-1}] &= [V_{-1} + Q, V_{-1}] \subset [V_{-1}, V_{-1}] + V_{-1} + V_0 + L_{(1)}, \\
[L, [V_{-1}, V_{-1}]] &\subset [L, V_{-1}] \subset [V_{-1}, V_{-1}] + V_{-1} + V_0 + L_{(1)}, \\
[L, V_0] &= [V_{-1} + Q, V_0] \subset V_{-1} + V_0 + L_{(1)}, \\
[L, L_{(1)}] &= [V_{-1} + Q, L_{(1)}] \subset V_0 + L_{(1)},
\end{aligned}$$

whence

$$[L, V_{-1} + V_0 + [V_{-1}, V_{-1}] + L_{(1)}] \subset V_{-1} + V_0 + [V_{-1}, V_{-1}] + L_{(1)}.$$

Therefore $V_{-1} + V_0 + [V_{-1}, V_{-1}] + L_{(1)}$ is a non-zero ideal of L. The claim follows. □

Theorem 20.7.4. *If $\mathfrak{S}(L,T) \cong X(q;\underline{1})^{(1)}$ with $X = W$, S is of Witt or Special type, then $L \cong X(q;\underline{n};\Psi)^{(1)}$ for suitable $\underline{n}$.*

Proof. Due to Lemma 20.7.2(3) one has $[V_{-1},V_{-1}] \subset L_{(1)}$. Then $L = V_{-1}+V_0+L_{(1)}$, $Q = V_0 + L_{(1)}$ and

$$[R_{-1},L] + [R_0,L] \subset R_{-1} + V_0 + L_{(1)}.$$

As a consequence, $L'_{(0)} := R_{-1} + Q$ is a subalgebra of L and $R_{-1} + R_0 \subset L'_{(1)} := \{u \in L'_{(0)} \mid [u,L] \subset L'_{(0)}\}$. Then $L/L'_{(0)} \cong \mathfrak{S}_{-1}$ and $L'_{(0)}/L'_{(1)} \cong \mathfrak{S}_0$. The Weak Recognition Theorem 5.5.3 proves the claim. □

In the only remaining case of **Hamiltonian algebras** we have to work a little harder to construct the maximal subalgebra $L'_{(0)}$. Asume now that $\mathfrak{S} = \mathfrak{S}(L,T) \cong H(2q;\underline{1})^{(2)}$. Note that

$$\mathfrak{S} \otimes \mathcal{O}(m;\underline{1}) \subset \operatorname{gr} L \subset \big((\operatorname{Der}\mathfrak{S}) \otimes \mathcal{O}(m;\underline{1})\big) \rtimes \big(\operatorname{Id} \otimes W(m;\underline{1})\big).$$

Looking at the homogenous component of degree 0 we get

$$H(2q;\underline{1})_0 \otimes \mathcal{O}(m;\underline{1}) \subset \operatorname{gr}_0 L \subset \big((CH(2q;\underline{1}))_0 \otimes \mathcal{O}(m;\underline{1})\big) \rtimes \big(\operatorname{Id} \otimes W(m;\underline{1})\big).$$

Let as before $\pi_2 : ((\operatorname{Der}\mathfrak{S}) \otimes \mathcal{O}(m;\underline{1})) \rtimes (\operatorname{Id} \otimes W(m;\underline{1})) \to W(m;\underline{1})$ denote the projection into the second summand, set

$$\mathcal{D} := \pi_2(\operatorname{gr}_0 L) \subset W(m;\underline{1}).$$

Lemma 20.6.2(4) gives $\mathcal{D} = \sum_{\mu\in\Delta} \mathcal{D}_\mu$.

We may normalize $\mathfrak{H} = \sum_{i=1}^{q} FD_H(x_i x_{i+q})$. The multiplication on L defines a bilinear mapping $V_{-1} \times V_{-1} \to Q$ (Lemma 20.7.2(3)). For convenience we will write the elements of $\operatorname{gr} L$ as $\bar{x} := x + L_{(i)}$ for $x \in L_{(i-1)} \setminus L_{(i)}$. Lemma 20.7.2 yields that this defines a linear isomorphism $V_{-1} \to \mathfrak{A}_{-1}$. We define by use of this isomorphism a $\mathfrak{T}$-invariant skew-symmetric bilinear mapping

$$\Lambda : \mathfrak{A}_{-1} \times \mathfrak{A}_{-1} \to \operatorname{gr}_0 L, \quad \Lambda(\bar{u},\bar{v}) := \overline{[u,v]} = [u,v] + L_{(1)} \quad u,v \in V_{-1}.$$

Put $\Lambda_2 := \pi_2 \circ \Lambda$.

Theorem 20.7.5. *If $\Lambda_2 = 0$, then $L \cong H(2q;\underline{n};\Psi)^{(2)}$ for some $\underline{n}$. This occurs in particular, if $\mathfrak{H} \subsetneq \mathfrak{T}_0$.*

Proof. Our assumption in combination with Lemma 20.7.3 implies $L = V_{-1} + V_0 + L_{(1)}$. The short proof of Theorem 20.7.4 applies and gives the claim.

Now suppose $\mathfrak{H} \subsetneq \mathfrak{T}_0$. This can only be if $\sum_{i=1}^{2q} D_H(x_i x_{i+q}) =: t \in \mathfrak{T}_0$. Note that $t \otimes 1$ acts on every $\operatorname{gr}_k L$ as $k\operatorname{Id}$. Then we have $[V_{-1},V_{-1}] \subset \sum_{k>1} L_{(k)}$. This implies $\Lambda = 0$. □

Lemma 20.7.6. *If* $\Lambda_2 \neq 0$, *then*

$$\operatorname{gr}_0 L = \big(H(2q;\underline{1})_0 \otimes \mathcal{O}(m;\underline{1})\big) \rtimes \big(\mathrm{Id} \otimes \mathfrak{D}\big).$$

Proof. Let z denote the degree derivation on $\mathfrak{S}$. Suppose

$$\big(H(2q;\underline{1})_0 \otimes \mathcal{O}(m;\underline{1})\big) \rtimes \big(\mathrm{Id} \otimes \mathfrak{D}\big) \neq \operatorname{gr}_0 L.$$

Then there is a root vector $w = z \otimes \prod (1 + x_i)^{a_i} + \mathrm{Id} \otimes d_\alpha \in (\operatorname{gr}_0 L)_\alpha$, where $\alpha((1 + x_i)\partial_i) = a_i$ (see Equation (20.6.2)) and $d_\alpha \in \mathfrak{D}_\alpha$. If $d_\alpha \neq 0$, then $\alpha \in \Delta$.

Note that $\operatorname{gr}_0 L$ contains $\mathfrak{H} \otimes \prod (1+x_i)^{a_i}$. Since $L(\alpha)$ acts trigonalizably if $d_\alpha \neq 0$, one has that $(\operatorname{ad} w)^j(\mathfrak{H} \otimes \prod (1 + x_i)^{a_i}) = \mathfrak{H} \otimes {d_\alpha}^j(\prod (1 + x_i)^{a_i})$ acts nilpotently for every $j \geq 1$. This can only be if ${d_\alpha}^j(\prod (1 + x_i)^{a_i}) \in \mathcal{O}(m;\underline{1})_{(1)}$. We now have $w^{[p]} \equiv z \otimes 1 + \mathrm{Id} \otimes {d_\alpha}^p \pmod{z \otimes \mathcal{O}(m;\underline{1})_{(1)}}$ in both cases $d_\alpha \neq 0$ and $d_\alpha = 0$. Note that $w^{[p]^e} \in \mathfrak{T}$, but $w^{[p]^e} \notin \mathfrak{H} \otimes 1 + \mathfrak{T}_1$ (for $e \gg 0$). Choose a root vector $\tilde{w} \in L_{(0)}$ which is mapped onto w. Then $\tilde{w}^{[p]^e} \in T$ has the property that $\operatorname{gr}_0(\tilde{w}^{[p]^e}) = (\operatorname{gr}_0 \tilde{w})^{[p]^e} \notin \mathfrak{H} \otimes 1 + \mathfrak{T}_1$. Lemma 20.7.5 yields $\Lambda_2 = 0$. □

Assume $\boldsymbol{\Lambda_2 \neq 0}$ from now on. Lemma 20.7.6 proves $\mathfrak{T} = \mathfrak{H} \otimes F \oplus \mathfrak{T}_1$.

Lemma 20.7.7. *The following holds.*

(1) $[\Lambda(\bar{u},\bar{v}),\bar{w}] = [\Lambda(\bar{u},\bar{w}),\bar{v}] + [\bar{u},\Lambda(\bar{v},\bar{w})]$ *holds for all* $u, v, w \in V_{-1}$.

(2) *For all* $f \in Q \setminus L_{(1)}$ *and* $u, v \in V_{-1}$ *one has*

$$[\bar{f},\Lambda(\bar{u},\bar{v})] - \Lambda([\bar{f},\bar{u}],\bar{v}) - \Lambda(\bar{u},[\bar{f},\bar{v}]) \in \mathfrak{A}_0.$$

Proof. (1) One computes

$$\begin{aligned}
[\Lambda(\bar{u},\bar{v}),\bar{w}] &= [[u,v] + L_{(1)}, w + Q] = [[u,v],w] + Q \\
&= [[u,w],v] + [u,[v,w]] + Q \\
&= [[u,w] + L_{(1)}, v + Q] + [u + Q,[v,w] + L_{(1)}] \\
&= [\Lambda(\bar{u},\bar{w}),\bar{v}] + [\bar{u},\Lambda(\bar{v},\bar{w})].
\end{aligned}$$

(2) Take u, v, f root vectors with respective roots α, β, γ. For any $\mu \in \Gamma_i(\operatorname{gr} L, \mathfrak{T})$ we put $\bar{\mu} := \mu|\mathfrak{H}$. If $\mu \notin \Delta$, then $\bar{\mu} \in \Gamma_i(\mathfrak{S},\mathfrak{H})$ by Lemma 20.6.2(4). If $\mu \in \Delta$, then $\bar{\mu} = 0$.

Suppose $\bar{\alpha} + \bar{\beta} + \bar{\gamma} \neq 0$, which implies $\alpha + \beta + \gamma \notin \Delta$. Then

$$\Lambda_2([\overline{f},\overline{u}],\overline{v}), \Lambda_2(\overline{u},[\overline{f},\overline{v}]) \in \mathfrak{D}_{\alpha+\beta+\gamma} = \{0\}.$$

This gives $\Lambda([\overline{f},\overline{u}],\overline{v}), \Lambda(\overline{u},[\overline{f},\overline{v}]) \in \mathfrak{A}_0$. Moreover, $\bar{\gamma} \neq 0$ or $\bar{\alpha} + \bar{\beta} \neq 0$ then holds, which gives $\bar{f} \in \mathfrak{A}_0$ or $\Lambda(\overline{u},\overline{v}) \in \mathfrak{A}_0$. In both cases we have $[\bar{f},\Lambda(\overline{u},\overline{v})] \in \mathfrak{A}_0$. The required inclusion is true in this case.

So we may consider the case $\bar{\alpha} + \bar{\beta} + \bar{\gamma} = 0$. If $\bar{\gamma} = 2\bar{\alpha}$, then $-3\bar{\alpha} = \bar{\beta}$ would be a root on $\mathfrak{S}_{-1}$. However, since $\bar{\alpha}$ is a root on $\mathfrak{S}_{-1}$ and $\mathfrak{S}$ is Hamiltonian, the only multiples of $\bar{\alpha}$ which are roots on $\mathfrak{S}_{-1}$ are $\pm\bar{\alpha}$. Consequently, $\bar{\gamma} \neq 2\bar{\alpha}$, and symmetrically, $\bar{\gamma} \neq 2\bar{\beta}$ hold. Lemma 5.5.1(4) now shows that $\bar{\alpha} + \bar{\gamma}, \bar{\beta} + \bar{\gamma} \notin \Gamma_0(\mathfrak{S}, \mathfrak{H})$. This gives $\alpha + \gamma, \beta + \gamma \notin \Gamma_0(\operatorname{gr} L, \mathfrak{T})$.

Choose $u' \in V_{-1,\alpha+\gamma}$, $v' \in V_{-1,\beta+\gamma}$ satisfying $\overline{u'} = [\bar{f}, \bar{u}]$, $\overline{v'} = [\bar{f}, \bar{v}]$. Observe that

$$0 = \overline{u'} - [\bar{f}, \bar{u}] = (u' + Q) - [f + L_{(1)}, u + Q] = u' - [f, u] + Q,$$

which means $u' - [f, u] \in Q$. The preceding deliberations show that $u' - [f, u] \in L_{(1)}$. Similarly we have $v' - [f, v] \in L_{(1)}$. Using this and Lemma 20.7.2(5) we obtain

$$\begin{aligned}
&[\bar{f}, \Lambda(\bar{u}, \bar{v})] - \Lambda([\bar{f}, \bar{u}], \bar{v}) - \Lambda(\bar{u}, [\bar{f}, \bar{v}]) \\
&\quad = [f + L_{(1)}, [u, v] + L_{(1)}] - ([u', v] + L_{(1)}) - ([u, v'] + L_{(1)}) \\
&\quad = [f, [u, v]] - [u', v] - [u, v'] + L_{(1)} \\
&\quad = [[f, u] - u', v] + [u, [f, v] - v'] + L_{(1)} \\
&\quad \in \big([L_{(1)}, V_{-1}] + L_{(1)}\big)/L_{(1)} \subset (V_0 + L_{(1)})/L_{(1)} = \mathfrak{A}_0.
\end{aligned}$$

□

Lemma 20.7.8. *Let $\alpha, \beta \in \Delta$ be such that $Q(\alpha, \beta)$ is solvable. Then $\operatorname{gr}_0 Q(\alpha, \beta)$ is abelian.*

Proof. We may assume that α, β are $\mathbb{F}_p$-independent (see Lemma 20.5.4). Recall that $\mathfrak{T}_1 \cap (\operatorname{Id} \otimes W(m; \underline{1})_{(0)}) = \{0\}$. As a consequence one may adjust the toral elements so that $\alpha(\operatorname{Id} \otimes (1 + x_i)\partial_i) = \delta_{i,1}$ and $\beta(\operatorname{Id} \otimes (1 + x_i)\partial_i) = \delta_{i,2}$. Put in Equation (8.2.1) from Volume 1, p. 452

$$\begin{aligned}
V &:= \mathfrak{S}_{-1} \otimes F[x_1, x_2], \\
G &:= \mathfrak{S}_0 \otimes F[x_1, x_2] + \operatorname{Id} \otimes \overline{\mathfrak{D}} + F\operatorname{Id} \otimes (1 + x_1)\partial_1 + F\operatorname{Id} \otimes (1 + x_2)\partial_2, \\
\overline{\mathfrak{D}} &:= \Lambda_2(V, V)/\{d \in \Lambda_2(V, V) \mid d\big(F[x_1, x_2]\big) = \{0\}\}.
\end{aligned}$$

By assumption $\overline{\mathfrak{D}}$ is a homomorphic image of a subalgebra of $\operatorname{gr}_0 Q(\alpha, \beta)$, whence is solvable. Lemma 20.7.7 ensures that Equation (8.2.1) is satisfied when one puts $t = \sum_{i=1}^{q} D_H(x_i x_{i+q})$. Theorem 8.2.5 applies and yields $\overline{\mathfrak{D}} = \{0\}$. As a result, $\Lambda_2(V, V)$ annihilates $F[x_1, x_2]$.

Observe that $Q(\alpha, \beta) \subset L_{(0)}$ and $\operatorname{gr}_0 Q(\alpha, \beta) = \mathfrak{H} \otimes F[x_1, x_2] + \operatorname{Id} \otimes \mathfrak{D}(\alpha, \beta)$. Due to Lemma 20.7.7 one has

$$\begin{aligned}
\mathfrak{D}(\alpha, \beta) &= \Lambda_2(\mathfrak{A}_{-1}, \mathfrak{A}_{-1})(\alpha, \beta) = \Lambda_2\Big(\sum_{i,j \in \mathbb{F}_p} \mathfrak{A}_{-1, i\alpha + j\beta}, \mathfrak{S}_{-1} \otimes F\Big) \\
&= \Lambda_2(\mathfrak{S}_{-1} \otimes F[x_1, x_2], \mathfrak{S}_{-1} \otimes F).
\end{aligned}$$

We have proved that this space annihilates $F[x_1, x_2]$. One may recall that $\mathfrak{D}(\alpha,\beta) \subset \sum_{i=1}^{m} F[x_1, x_2](1 + x_i)\partial_i$. Therefore $\mathfrak{D}(\alpha,\beta)^{(1)} = \{0\}$ is true. Since $\mathfrak{H}$ is abelian, one obtains that $\operatorname{gr}_0 Q(\alpha,\beta)$ is abelian. □

Lemma 20.7.9. *$\mathfrak{D}$ is $\pi_2(\mathfrak{T}_1)$-simple.*

Proof. (a) Put $Q(\Delta) := \sum_{\mu\in\Delta} Q_\mu \subset L_{(0)}$. Let $\alpha,\beta \in \Delta$. If $Q(\alpha,\beta)$ is solvable, then $Q(\alpha,\beta)^{(1)} \subset L_{(1)}$ by Lemma 20.7.8. If $Q(\alpha,\beta)$ is not solvable, then Proposition 20.3.3 gives $Q[\alpha,\beta] \cong H(2;\underline{1};\Phi(\tau))^{(1)}$. In this case there is $u \in Q_\beta$ with $\alpha(u) \neq 0$. Then $Q_\alpha \subset Q(\Delta)^{(1)}$ holds.

Now take $\alpha \in \Delta$ and assume $Q_\alpha \not\subset Q(\Delta)^{(1)}$. The above tells us that $[Q_\alpha, Q_\beta] \subset L_{(1)}$ for all $\beta \in \Delta$. This means $Q(\Delta) = Q(\Delta)^{(1)} + C$, where $C \subset Q(\Delta)$ satisfies $[C, Q(\Delta)] \subset L_{(1)}$. Since $Q(\Delta)/Q(\Delta) \cap L_{(1)} \cong \big(\mathfrak{H} \otimes \mathcal{O}(m;\underline{1})\big) \rtimes \big(\mathrm{Id} \otimes \mathfrak{D}\big)$, we conclude

$$\mathfrak{D} = \mathfrak{D}^{(1)} + C(\mathfrak{D}).$$

(b) Put in Equation (8.2.1) in Volume 1, p. 452

$$V := \mathfrak{S}_{-1} \otimes \mathcal{O}(m;\underline{1}),$$
$$G := \mathfrak{S}_0 \otimes \mathcal{O}(m;\underline{1}) + \mathrm{Id} \otimes \mathfrak{D} + \mathfrak{T}_1.$$

Lemma 20.7.7 ensures that Equation (8.2.1) is satisfied with $t = \sum_{i=1}^{q} D_H(x_i x_{i+q})$. Theorem 8.2.5 applies and yields (as $\mathfrak{D} \neq \{0\}$ by the assumption $\Lambda_2 \neq 0$) that $\mathfrak{D} + \pi_2(\mathfrak{T}_1)$ is semisimple and $\gamma_\infty(\mathfrak{D})$ is the unique minimal ideal. Due to (a) this gives that $C(\mathfrak{D}) = \{0\}$ and hence $\mathfrak{D} = \gamma_\infty(\mathfrak{D})$ is $\pi_2(\mathfrak{T}_1)$-simple. □

Recall that Λ_2 is invariant under $\operatorname{gr}_0 L$. There is a root α for which $\pm\alpha \in \Gamma_{-1}(\mathfrak{A},\mathfrak{T})$, because otherwise $\Lambda_2 = 0$. Fix such a root and fix root vectors $u_{-1,\alpha} \in V_{-1,\alpha}$, $u_{-1,-\alpha} \in V_{-1,-\alpha}$. Since $\mathfrak{A}_{-1}$ is irreducible as a module for $\operatorname{gr}_0 L + \mathfrak{T}$ and Λ_2 is invariant under this algebra, we have $\mathfrak{D} = \Lambda_2(u_{-1,\alpha} \otimes \mathcal{O}(m;\underline{1}), u_{-1,-\alpha} \otimes F)$. Define a surjective $\mathfrak{D}$-module homomorphism

$$\Lambda_2' : \mathcal{O}(m;\underline{1}) \to \mathfrak{D}, \quad f \mapsto \Lambda_2(u_{-1,\alpha} \otimes f, u_{-1,-\alpha} \otimes 1).$$

Lemma 20.7.10. $\ker \Lambda_2' = \{f \in \mathcal{O}(m;\underline{1}) \mid \mathfrak{D}(f) = \{0\}\}$.

Proof. For $f \in \mathcal{O}(m;\underline{1})$ satisfying $\mathfrak{D}(f) = \{0\}$ one has

$$\{0\} = \Lambda_2'(\mathfrak{D}(f)) = [\mathfrak{D}, \Lambda_2'(f)],$$

whence $\Lambda_2'(f) \in C(\mathfrak{D}) = \{0\}$.

Let conversely $f \in \ker \Lambda_2'$. Since $\ker \Lambda_2'$ is $\pi_2(\mathfrak{T}_1)$-invariant, we may take $f \in \mathcal{O}(m;\underline{1})_\alpha$ a root vector. Note that all $\pi_2(\mathfrak{T}_1)$-root spaces of $\mathcal{O}(m;\underline{1})$ are 1-dimensional. Take any root β. Then $\Lambda_2'(\mathcal{O}(m;\underline{1})(\alpha,\beta))$ does not carry the root α.

Since $\mathfrak{H} \otimes \mathcal{O}(m;\underline{1})(\alpha,\beta)$ is an abelian $\mathfrak{T}$-invariant ideal of $(\mathrm{gr}_0\, L)(\alpha,\beta)$, there is a surjective $\mathfrak{T}$-invariant homomorphism

$$\Lambda_2'(\mathcal{O}(m;\underline{1})(\alpha,\beta)) = \mathcal{D}(\alpha,\beta) \to Q[\alpha,\beta].$$

Therefore the root α is missing on $Q[\alpha,\beta]$. Consequently, one has $Q[\alpha,\beta] \not\cong H(2;\underline{1};\Phi(\tau))^{(1)}$. Proposition 20.3.3 now yields that $Q(\alpha,\beta)$ is solvable, Lemma 20.7.8 then gives that $(\mathrm{gr}_0\, L)(\alpha,\beta)$ is abelian. In particular,

$$\{0\} = [\mathfrak{H} \otimes f, \mathrm{Id} \otimes \mathcal{D}_\beta] = \mathfrak{H} \otimes \mathcal{D}_\beta(f).$$

This is the claim. □

Let us determine $\mathcal{D}$. According to Lemma 20.7.9 there is a presentation

$$\mathcal{D} = \mathfrak{d} \otimes F[y_1,\dots,y_s], \quad \mathfrak{d} \text{ is simple}, \quad \mathcal{D} \text{ is } \pi_2(\mathfrak{T}_1)-\text{simple},$$

and this gives a decomposition (Theorem 3.6.1)

$$\pi_2(\mathfrak{T}_1) = (\mathfrak{t}_0 \otimes 1) \oplus (\mathrm{Id} \otimes \mathfrak{t}_1), \quad \mathfrak{t}_1 = \sum_{i=1}^{s} F(1+y_i)\partial_{y_i}.$$

Decompose $\mathcal{O}(m;\underline{1}) = \mathcal{O}(m-s;\underline{1}) \otimes \mathcal{O}(s;\underline{1})$ likewise, where the first factor is annihlated by $\mathfrak{t}_1$ and the second by $\mathfrak{t}_0$. Since $\pi_2(\mathfrak{T}_1) \cap W(m;\underline{1})_{(0)} = \{0\}$, this gives

$$\mathfrak{t}_0 = \sum_{i=s+1}^{m} F(1+x_i)\partial_i, \quad \mathfrak{t}_1 = \sum_{i=1}^{s} F(1+x_i)\partial_i,$$

$$\mathfrak{d} = C_{\mathcal{D}}(\mathfrak{t}_1) \subset \sum_{i=1}^{m} F[x_{s+1},\dots,x_m](1+x_i)\partial_i,$$

and looking at the root lattice that result gives us

$$\mathcal{D} = F[x_1,\dots,x_s]\mathfrak{d}.$$

The multiplication in $\mathcal{D} \cong \mathfrak{d} \otimes F[y_1,\dots,y_s]$ in combination with the identification $y_i = x_i$ for $i = 1,\dots,s$ gives even more

$$\mathfrak{d} \subset \sum_{i=s+1}^{m} F[x_{s+1},\dots,x_m]\partial_i.$$

Lemma 20.7.10 then yields $F[x_1,\dots,x_s] \subset \ker \Lambda_2'$.

Note that $\mathfrak{d} = C_{\mathcal{D}}(\mathfrak{t}_1) = \Lambda_2'(F[x_{s+1},\dots,x_m])$. Observe that $\mathfrak{t} := \mathfrak{d}_{[p]} \cap \pi_2(\mathfrak{T}_1)$ is a torus of maximal dimension. Let $\mathfrak{d}(\lambda)$ be a $\mathfrak{t}$-1-section. This is a full $\pi_2(\mathcal{T}_1)$-section of $\mathfrak{d}$. If it were not solvable, then it would contain a $\pi_2(\mathcal{T}_1)$-2-section of core

$H(2;\underline{1};\Phi(\tau))^{(1)}$. But then $TR(\mathfrak{d}(\lambda)) \geq 2$ which cannot be true (Theorem 1.3.11(3)). Hence all $\mathfrak{t}$-1-sections are solvable. Then Theorems 19.4.5 or 19.4.8 apply. Suppose Theorem 19.4.5 holds whence $\mathfrak{d} \cong S(m;\underline{n};\Phi(\tau))^{(1)}$ with $m \geq 3$. Theorem 7.2.2(3) shows that $\operatorname{Der}\mathfrak{d} = \mathfrak{d}_{[p]}$. In particular, the restriction of $\pi_2(\mathfrak{T}_1)$ to $\mathfrak{d}$ is contained in $\mathfrak{t}$. But then the former description $\mathfrak{d} = \Lambda_2'(F[x_{s+1},\dots,x_m])$ shows that all root spaces are 1-dimensional. We know, however, that the root spaces in $S(m;\underline{n};\Phi(\tau))^{(1)}$ with respect to a torus of maximal dimension have dimension $m-1 \geq 2$. So this cannot happen. Then Theorem 19.4.8 applies and shows that $\mathfrak{d} \cong L(G,0,f)$ is a BLOCK algebra. In particular, $\dim\mathfrak{d} = |G|-1 = p^r-1$ for some r. Next observe that $F[x_{s+1},\dots,x_m] \cap (\ker \Lambda_2') = \operatorname{ann}_{\mathcal{O}(m;\underline{1})}(\mathfrak{t}_1) \cap (\ker \Lambda_2')$ is a $\pi_2(\mathcal{T}_1)$-invariant subalgebra of $\mathcal{O}(m;\underline{1})$ (Lemma 20.7.10). Since $\pi_2(\mathcal{T}_1)$ acts transitively, $F[x_{s+1},\dots,x_m] \cap (\ker \Lambda_2')$ is a full truncated polynomial ring. Then the former description of $\mathfrak{d}$ gives

$$p^r - 1 = \dim \mathfrak{d} = p^{m-s} - p^q$$

where $p^q = \dim\big(F[x_{s+1},\dots,x_m] \cap \ker \Lambda_2'\big)$. As a result, $r = m-s$ and $q = 1$. We conclude

$$\mathfrak{d} \cong L(G,0,f),\ \dim \mathfrak{d} = p^{m-s} - 1.$$

Then $TR(\mathfrak{d}) = m - s$, whence $\dim \mathfrak{t} = m-s$, i.e.,

$$\mathfrak{t} = \mathfrak{t}_0 = \sum_{i=s+1}^{m} F(1+x_i)\partial_i,\ \ker \Lambda_2' = F[x_1,\dots,x_s].$$

Next we recall the realization of $L(G,0,f)$ given via Θ in Equation 16.1.1. Namely, choose a basis $(\alpha_{s+1},\dots,\alpha_m)$ of G, and write $\alpha = \sum_{k=s+1}^m a_k\alpha_k$. We set $z^\alpha := \prod_{k=s+1}^m (1+x_i)^{a_i}$ and put

$$\Theta(u_\alpha) := \sum_{k=s+1}^{m} z^\alpha f(\alpha,\alpha_k) z_k \partial_k, \quad \mathfrak{g} := \Theta(L(G,0,f)).$$

This gives rise to mappings

$$\rho : F[x_{s+1},\dots,x_m] \xrightarrow{\Lambda_2'} \mathfrak{d} \cong L(G,0,f) \xrightarrow{\Theta} \mathfrak{g}.$$

This becomes a $\mathfrak{g}$-module homomorphism, if one puts $\rho(f)(g) := \Lambda_2'(f)(g)$. Define another mapping

$$\phi : F[x_{s+1},\dots,x_m] \to \mathfrak{g},\quad \phi(z^\alpha) := \sum_{k=s+1}^{m} z^\alpha f(\alpha,\alpha_k) z_k \partial_k.$$

We compute (with $\alpha = \sum a_k\alpha_k$, $\beta = \sum b_k\alpha_k$, $z^\alpha := \prod(1+x_k)^{a_k}$, $z^\beta := \prod(1+x_k)^{b_k}$)

$$\phi((\phi(z^\alpha)(z^\beta)) = \phi(z^{\alpha+\beta} f(\alpha,\beta)) = z^{\alpha+\beta} f(\alpha,\beta)\sum_k f(\alpha+\beta,\alpha_k) z_k\partial_k,$$

$$\begin{aligned}[\phi(z^\alpha),\phi(z^\beta)] &= [z^\alpha \sum_k f(\alpha,\alpha_k)z_k\partial_k, z^\beta\sum_l f(\beta,\alpha_l)z_l\partial_l] \\ &= z^{\alpha+\beta} f(\alpha,\beta)\sum_l f(\beta,\alpha_l)z_l\partial_l - z^{\alpha+\beta} f(\beta,\alpha)\sum_k f(\alpha,\alpha_k)z_k\partial_k \\ &= z^{\alpha+\beta} f(\alpha,\beta)\sum_k f(\alpha+\beta,\alpha_k)z_k\partial_k.\end{aligned}$$

This shows that ϕ is a module homomorphism as well. Moreover, since $\ker\phi = F = F[x_{s+1},\dots,x_m]\cap(\ker\Lambda_2') = \ker\rho$ and $F[x_{s+1},\dots,x_m]/F \cong \mathfrak{d}$ is an irreducible $\mathfrak{d}$-module, it is an irreducible $\mathfrak{g}$-module. Hence there is $r\in F^*$ such that $\rho - r\phi$ vanishes on this module. We obtain

$$\begin{aligned}\rho\big(\Lambda_2'(z^\alpha)(z^\beta)\big) &= [\rho(z^\alpha),\rho(z^\beta)] = r^2[\phi(z^\alpha),\phi(z^\beta)] \\ &= r^2\phi\big(\phi(z^\alpha)(z^\beta)\big) = r^2 f(\alpha,\beta)\phi(z^{\alpha+\beta}) = rf(\alpha,\beta)\rho(z^{\alpha+\beta}).\end{aligned}$$

This gives $\Lambda_2'(z^\alpha) = rz^\alpha\sum_{k=s+1}^m f(\alpha,\alpha_k)z_k\partial_k$. We summarize.

Lemma 20.7.11. *The mapping $\Lambda_2 : \mathfrak{A}_{-1}\times\mathfrak{A}_{-1}\to W(m;\underline{1})$ can be described by the following.*

(a) *Using the notation $y_1,\dots,y_{2q}$ for the canonical generators of $\mathfrak{S}\cong H(2q;\underline{1})^{(2)}$ define a skew-symmetric bilinear form $\mu : \mathfrak{S}_{-1}\times\mathfrak{S}_{-1}\to F$, $\mu(y_i,y_j) := \sigma(i)\delta_{j,i'}$.*

(b) *There is s with $0 < s \le m$ and a skew-symmetric $(m-s)\times(m-s)$-matrix $(q_{i,j})$ with entries in F, which has $\mathbb{F}_p$-rank $(m-s)$.*

(c) *Make the right choice of generators $x_1,\dots,x_m$ for $\mathcal{O}(m;\underline{1})$ and set $z^a := \prod_{i=1}^m(1+x_i)^{a_i}$.*

Then there is $r\in\mathbb{F}^*$ so that

$$\Lambda_2(u\otimes z^a, v\otimes z^b) = r\mu(u,v)z^{a+b}\sum_{i,j=s+1}^m (a_i+b_i)q_{i,j}z_j\partial_j.$$

Proof. (a) One mainly has to pursue the preceding deliberations. There is a root α for which $\pm\alpha\in\Gamma_{-1}(\mathfrak{A},\mathfrak{T})$, because otherwise $\Lambda_2 = 0$. Fix such a root and fix root vectors $u_{-1,\alpha}\in V_{-1,\alpha}$, $u_{-1,-\alpha}\in V_{-1,-\alpha}$. Since $\mathfrak{A}_{-1}$ is irreducible as a module for

$\mathrm{gr}_0 L + \mathfrak{T}$ and Λ_2 is invariant under this algebra and root spaces are 1-dimensional, one has

$$\Lambda_2(u \otimes z^a, v \otimes z^b) = \Lambda_2(u_{-1,-\alpha} \otimes z^{a+b}, s(u,v)u_{-1,-\alpha} \otimes 1) = s(u,v)\Lambda_2'(z^{a+b}),$$

where $s(\cdot,\cdot)$ is a skew-symmetric non-degenerate $\mathfrak{S}_0$-invariant bilinear form. Then there is $r' \in \mathbb{F}^*$ so that $s(u,v) = r'\mu(u,v)$.

(b) Next choose the generators $x_1, \dots, x_m$ as above. In particular, $\ker \Lambda_2' = F[x_1, \dots, x_s]$ and $\mathcal{D} = F[x_1, \dots, x_s]\mathfrak{d}$ for some subalgebra

$$\mathfrak{d} \subset \sum_{i=s+1}^{m} F[x_{s+1}, \dots, x_m]\partial_i.$$

This implies $\Lambda_2'(g)(f) = 0$ for $f \in F[x_1, \dots, x_s]$ and $g \in \mathcal{O}(m;\underline{1})$. Then one has for $d \in \mathcal{D}$ that $fd, f\Lambda_2'(g) \in \mathcal{D}$, $d(f) = 0$, and

$$\begin{aligned}[d, \Lambda_2'(fg) - f\Lambda_2'(g)] &= \Lambda_2'(d(fg)) - [fd, \Lambda_2'(g)] \\ &= \Lambda_2'((fd)(g)) - [fd, \Lambda_2'(g)] \\ &= 0.\end{aligned}$$

As a consequence, $\Lambda_2'(fg) - f\Lambda_2'(g) \in C(\mathcal{D}) = \{0\}$, whence Λ_2' is $F[x_1, \dots, x_s]$-invariant.

(c) The algebra $\mathfrak{d}$ is isomorphic to some $L(G, 0, f)$, where $|G| = p^{m-s}$ and f is a non-degenerate biadditive form. Choose a basis $(\alpha_{s+1}, \dots, \alpha_m)$ and put $q_{i,j} := f(\alpha_i, \alpha_j) \in F$. Since f is non-degenerate, this matrix has $\mathbb{F}_p$-rank $(m-s)$. With $\alpha = \sum_{i=s+1}^m a_i\alpha_i$ $(a_i \in \mathbb{F}_p)$ one gets $f(\alpha, \alpha_j) = \sum_{i=s+1}^m a_i q_{i,j}$. The above deliberations give $\Lambda_2'(z^a) = r'' z^a \sum_{i,j=s+1}^m a_i q_{i,j} z_j \partial_j$ for all $z^a \in F[x_{s+1}, \dots, x_m]$. Set $r := r'r''$. □

Lemma 20.7.12.

(1) $\Lambda_2(\mathfrak{S}_{-1} \otimes F[x_1, \dots, x_s]_{(1)} F[x_{s+1}, \dots, x_m], \mathfrak{S}_{-1} \otimes F) \subset W(m;\underline{1})_{(0)}$.

(2) $\Lambda_2(\mathfrak{S}_{-1} \otimes F[x_{s+1}, \dots, x_m]_{(2)}, \mathfrak{S}_{-1} \otimes F) \subset W(m;\underline{1})_{(0)}$.

Proof. The first assertion follows immediately from Lemma 20.7.11.

To prove the second claim we stick to elements in $F[x_{s+1}, \dots, x_m]$. Note that $F[x_{s+1}, \dots, x_m]_{(2)}$ is spanned by the set $\{\prod_{i=s+1}^m z_i^{a_i} - \sum_{i=s+1}^m a_i z_i + (-1 + \sum_{i=s+1}^m a_i)\}$. The lemma then gives (as $z^a \equiv 1 \pmod{\mathcal{O}(m;\underline{1})_{(1)}}$)

$$\Lambda_2\Big(u \otimes (z^a - \sum_{i=s+1}^m a_i z_i + (-1 + \sum_{i=s+1}^m a_i)), v \otimes 1\Big)$$

$$= r\mu(u,v)\Big(z^a \sum_{j,k=s+1}^m a_j q_{j,k} z_k \partial_k - \sum_{i=s+1}^m a_i z_i \sum_{k=s+1}^m q_{i,k} z_k \partial_k\Big) \in W(m;\underline{1})_{(0)}.$$

□

Lemma 20.7.13. *Put* $L'_{(0)} := R_{-1} + \mathrm{Nor}_Q\,(R_{-1} + V_0 + L_{(1)})$. *Then* $L'_{(0)}$ *is a maximal subalgebra of* L. *It contains* $V_0 + L_{(1)}$.

Proof. (a) Lemma 20.7.2(7),(5) gives $V_0 + L_{(1)} \subset \mathrm{Nor}_Q\,(R_{-1} + V_0 + L_{(1)})$.

(b) Let us prove that $L'_{(0)}$ is a subalgebra. In order to show that $[R_{-1}, R_{-1}] \subset L'_{(0)}$ we argue in the associated graded algebra. Since $R_{-1} \subset V_{-1}$ and $(R_{-1} + Q)/Q = \mathfrak{S}_{-1} \otimes \mathcal{O}(m;\underline{1})_{(1)}$, $[R_{-1}, R_{-1}]$ is contained in the inverse image of $\Lambda_2\big((\mathfrak{S}_{-1} \otimes \mathcal{O}(m;\underline{1})_{(1)}), (\mathfrak{S}_{-1} \otimes \mathcal{O}(m;\underline{1})_{(1)})\big)$ under the canonical mapping

$$Q \to Q/L_{(1)} \cong \mathrm{gr}_0\, L \to \mathcal{D}.$$

According to Lemma 20.7.12, $\Lambda_2((\mathfrak{S}_{-1} \otimes \mathcal{O}(m;\underline{1})_{(1)}), (\mathfrak{S}_{-1} \otimes \mathcal{O}(m;\underline{1})_{(1)}))$ is contained in $\mathcal{D} \cap W(m;\underline{1})_{(0)}$. This means

$$\mathrm{gr}_0\,\big([R_{-1}, R_{-1}]\big) \subset \mathfrak{A}_0 \rtimes \big(\mathrm{Id} \otimes W(m;\underline{1})_{(0)}\big).$$

Therefore $\mathrm{gr}_0\,([R_{-1}, R_{-1}])$ stabilizes $\mathfrak{S}_{-1} \otimes \mathcal{O}(m;\underline{1})_{(1)}$, and this gives (Lemma 20.7.2(4))

$$[[R_{-1}, R_{-1}], R_{-1}] \subset (R_{-1} + Q) \cap (V_{-1} + V_0 + L_{(1)}) = R_{-1} + V_0 + L_{(1)}.$$

As, by Lemma 20.7.2(6), $[[R_{-1}, R_{-1}], V_0] \subset [Q, V_0] \subset V_0 + L_{(1)}$, we conclude $[R_{-1}, R_{-1}] \subset \mathrm{Nor}_Q\,(R_{-1} + V_0 + L_{(1)})$.

As $V_0 + L_{(1)} \subset \mathrm{Nor}_Q\,(R_{-1} + V_0 + L_{(1)})$, it is now easy to conclude that $L'_{(0)}$ is a subalgebra.

(c) By definition one has $L'_{(0)} \cap V_{-1} = R_{-1}$. So we have $L'_{(0)} \neq L$. In order to prove that $L'_{(0)}$ is a maximal subalgebra we consider any subalgebra K containing $L'_{(0)}$. Note that

$$L/(L'_{(0)} + Q) \cong \mathfrak{A}_{-1}/\big(\mathfrak{S}_{-1} \otimes \mathcal{O}(m;\underline{1})_{(1)}\big) \cong \mathfrak{S}_{-1}$$

is an irreducible non-trivial $\mathfrak{S}_0$-module. Consequently, $(V_0 + L_{(1)})$ acts irreducibly and non-trivially on $L/(L'_{(0)} + Q)$.

Observe that $U_1 := (K + Q)/(L'_{(0)} + Q)$ is a $(V_0 + L_{(1)})$-module. If it is $\{0\}$, then we have $K \subset L'_{(0)} + Q$, and this gives $K = L'_{(0)} + K \cap Q = R_{-1} + K \cap Q$. Lemma 20.7.2 yields

$$[K \cap Q, R_{-1}] \subset K \cap [Q, V_{-1}] \subset (R_{-1} + K \cap Q) \cap (V_{-1} + V_0 + L_{(1)}) = R_{-1} + V_0 + L_{(1)}.$$

From this it is not hard to conclude that $K \cap Q \subset \mathrm{Nor}_Q\,(R_{-1} + V_0 + L_{(1)})$. Hence we have $K = L'_{(0)}$ in this case.

Then suppose $U_1 \neq \{0\}$, whence $L = K + Q$. Note that $[V_0, L] \subset V_{-1} + V_0 + L_{(1)}$ and

$$\begin{aligned}[V_0 + L_{(1)}, K + Q] &\subset [V_0, K] + Q \subset (V_{-1} + V_0 + L_{(1)}) \cap K + Q \\ &= (V_{-1} \cap K) + Q.\end{aligned}$$

As $(V_0 + L_{(1)}) \cdot U_1 = U_1 \neq \{0\}$, this gives

$$L = (V_{-1} \cap K) + L'_{(0)} + Q = (V_{-1} \cap K) + R_{-1} + Q = (V_{-1} \cap K) \oplus Q,$$

and hence $V_{-1} \subset K$. Lemma 20.7.3 implies $K = L$. □

Theorem 20.7.14. *Suppose $\mathfrak{S}(L,T) \cong H(2q;\underline{1})^{(2)}$ is of Hamiltonian type. Then L is of Hamiltonian type.*

Proof. (a) In view of Theorem 20.7.5 we may assume $\mathrm{gr}_0 L = \mathfrak{A}_0 + \mathrm{Id} \otimes \mathcal{D}$, where $\mathcal{D} = F[x_1,\dots,x_s]\mathfrak{d}$ and $\mathfrak{d} \cong L(G,0,f)$ as in the preceding reasonings. Note that $\mathfrak{d} \cong H(2r;\underline{t};\Psi)^{(2)}$ is of Hamiltonian type. It has been stated in Remark 19.4.10 that $\mathfrak{d} \cap W(m-s;\underline{1})_{(0)}$ is the unique subalgebra of maximal dimension, whence $\mathfrak{d}_{(0)} := \mathfrak{d} \cap W(m-s;\underline{1})_{(0)}$ is mapped onto $H(2r;\underline{t};\Psi)^{(2)}{}_{(0)}$ under this isomorphism. Let $\mathfrak{d}_{(1)}$ denote the inverse image of $H(2r;\underline{t};\Psi)^{(2)}{}_{(1)}$.

Let $\mathcal{O}' := \{f \in \mathcal{O}(m;\underline{1})_{(1)} \mid \Lambda_2(\mathfrak{S}_{-1} \otimes f, \mathfrak{S}_{-1} \otimes 1) \subset W(m;\underline{1})_{(0)}\}$ and $R'_{-1} \subset R_{-1}$ a vector space which maps isomorphically onto $\mathfrak{S}_{-1} \otimes \mathcal{O}'$. Note that

$$F[x_1,\dots,x_s]_{(1)} F[x_{s+1},\dots,x_m] + F[x_{s+1},\dots,x_m]_{(2)} \subset \mathcal{O}'.$$

Choose $V' \subset \sum_{\mu\in\Delta} Q_\mu$ to be a vector space which is mapped isomorphically onto $\mathrm{Id} \otimes (\mathcal{D} \cap W(m;\underline{1})_{(0)})$ and $V'' \subset V'$ a vector space which is mapped isomorphically onto $\mathrm{Id} \otimes \big(F[x_1,\dots,x_s]_{(1)}\mathcal{D} + \mathfrak{d}_{(1)}\big)$.

By our construction, $L'_{(0)} = R_{-1} + V_0 + V' + L_{(1)}$ and

$$L/L'_{(0)} \cong \mathfrak{S}_{-1} \oplus (\mathfrak{d}/\mathfrak{d}_{(0)}) \cong \mathfrak{S}_{-1} \oplus \big(H(2r;\underline{t};\Psi)^{(2)}/H(2r;\underline{t};\Psi)^{(2)}{}_{(0)}\big).$$

Next one proves, using Lemma 20.7.2,

$$[R'_{-1} + R_0 + L_{(1)}, L] + [V'', Q] \subset L'_{(0)}.$$

Consequently, $R'_{-1} + R_0 + L_{(1)} \subset L'_{(1)}$. Since

$$[(F[x_1,\dots,x_s]_{(1)}\mathcal{D} + \mathfrak{d}_{(1)}), \mathcal{D}] \subset F[x_1,\dots,x_s]_{(1)}\mathcal{D} + \mathfrak{d}_{(0)} \subset W(m;\underline{1})_{(0)},$$

one obtains $[V'', V_{-1}] \subset R'_{-1}$, and hence $[V'', L] \subset L'_{(0)}$. Then $V'' \subset L_{(1)}$. Thus $L'_{(0)}/L'_{(1)}$ is an image of the space

$$\begin{aligned}(R_{-1}/R'_{-1}) &\oplus (V_0/R_0) \oplus (\mathfrak{d}_{(0)}/\mathfrak{d}_{(1)})\\ &\cong (\mathfrak{S}_{-1} \otimes U) \oplus \mathfrak{S}_0 \oplus \big(H(2r;\underline{t};\Psi)^{(2)}{}_{(0)}/H(2r;\underline{t};\Psi)^{(2)}{}_{(1)}\big)\\ &\cong (\mathfrak{S}_{-1} \otimes U) \oplus \mathfrak{sp}(2q) \oplus \mathfrak{sp}(2r),\end{aligned}$$

where $U \subset \sum_{i=s+1}^{m} Fx_i$ is a subspace.

The action of $\mathfrak{S}_{-1} \otimes U$ on $\mathfrak{S}_{-1}$ is given by Λ_2 and maps this space into $\mathfrak{d}/\mathfrak{d}_{(0)}$. Its action on $\mathfrak{d}/\mathfrak{d}_{(0)}$ is given by the action of $\mathfrak{d}$ on $\mathcal{O}(m;\underline{1})$ and maps $\mathfrak{d}/\mathfrak{d}_{(0)}$ into $\mathfrak{S}_{-1}$. The subalgebra $\mathfrak{sp}(2q)$ acts only non-trivially on the first summand, while $\mathfrak{sp}(2r)$ acts only non-trivially on the second summand.

Let us determine the dimension of U. Define

$$\zeta : \mathfrak{d} \times \mathcal{O}(m;\underline{1}) \to F, \quad d(f) - \zeta(d, f) \in \mathcal{O}(m;\underline{1})_{(1)}.$$

One has $\mathfrak{d}_{(0)} = \mathfrak{d} \cap W(m;\underline{1})_{(0)}$, hence $\zeta(\mathfrak{d}_{(0)}, \mathcal{O}(m;\underline{1})) = 0$. Note that

$$F[x_1, \dots, x_s]_{(1)} F[x_{s+1}, \dots, x_m] + F[x_{s+1}, \dots, x_m]_{(2)} \subset \ker \zeta.$$

Next let $f = \sum_{l=s+1}^{m} r_l x_l \in \ker \zeta$. By Lemma 20.7.11 this assumption gives $\sum_{j=s+1}^{m} q_{i,j} z_j \partial_j(f) \in \mathcal{O}(m;\underline{1})_{(1)}$ for all i, and then

$$0 = \sum_{j=s+1}^{m} q_{i,j} r_j = -\sum_{j=s+1}^{m} r_j q_{j,i},$$

$$\Lambda_2(u \otimes f, v \otimes 1) = r\mu(u,v) \sum_{l=s+1}^{m} r_l z_l \sum_{j=s+1}^{m} q_{i,j} z_j \partial_j \in W(m;\underline{1})_{(0)}.$$

That means $f \in \mathcal{O}'$. Hence ζ defines a bilinear pairing $(\mathfrak{d}/\mathfrak{d}_{(0)}) \times U$, which is non-degenerate in the second argument. We conclude $\dim U \leq \dim \mathfrak{d}/\mathfrak{d}_{(0)} = 2r$.

(b) We intend to apply WILSON's Theorem 5.7.3. Since $\mathfrak{S}_0$ is the classical simple symplectic algebra $\mathfrak{sp}(2q)$ and $\mathfrak{S}_{-1}$ is the canonical $2q$-dimensional representation space, there is a root vector $u \in \mathfrak{S}_0$ satisfying $(\operatorname{ad} u)^2(\mathfrak{S}_0) \neq \{0\}$ and $\dim [u, \mathfrak{S}_{-1}] = 1$. Take f to be a root vector in V_0 with $\overline{f} = u \otimes 1$. The assumption $[(\operatorname{ad} f)^2(L'_{(0)}), L] \subset L'_{(0)}$ would in particular imply the following inclusion in $\operatorname{gr} L$

$$[[u \otimes 1, [u \otimes 1, \mathfrak{S}_0 \otimes F]], \mathfrak{S}_{-1} \otimes F] \subset \big(\mathfrak{S}_{-1} \otimes \mathcal{O}(m;\underline{1})_{(1)}\big) \cap (\mathfrak{S}_{-1} \otimes F) = \{0\},$$

which implies (using Theorem 20.6.3(6)) $(\operatorname{ad} u)^2(\mathfrak{S}_0) = \{0\}$, a contradiction.

Note that

$$[f, Q] \subset [V_0, Q] \subset V_0 + L_{(1)} \subset L'_{(0)}.$$

This gives rise to linear isomorphisms

$$\big([f, L] + L'_{(0)}\big)/L'_{(0)} \cong \big([f, V_{-1}] + L'_{(0)}\big)/L'_{(0)} \cong [f, V_{-1}]/\big([f, V_{-1}] \cap L'_{(0)}\big).$$

The linear mapping $L \to L/(R_{-1} + Q) \cong \mathfrak{S}_{-1}$ induces a surjective linear mapping $[f, V_{-1}] \twoheadrightarrow [u, \mathfrak{S}_{-1}]$, whose kernel is given by $[f, V_{-1}] \cap (R_{-1} + Q)$. Next we observe that

$$\begin{aligned}[f, V_{-1}] \cap (R_{-1} + Q) &\subset [V_0, V_{-1}] \cap (R_{-1} + Q) \subset (V_{-1} + V_0 + L_{(1)}) \cap (R_{-1} + Q) \\ &= R_{-1} + V_0 + L_{(1)} \subset L'_{(0)}.\end{aligned}$$

Therefore we have

$$[f, V_{-1}] \cap (R_{-1} + Q) = [f, V_{-1}] \cap L'_{(0)}$$

and hence

$$\big([f, L] + L'_{(0)}\big)/L'_{(0)} \cong [f, V_{-1}]/\big([f, V_{-1}] \cap L'_{(0)}\big) \cong [u, \mathfrak{S}_{-1}].$$

Consequently, $\dim\,([f, L] + L'_{(0)})/L'_{(0)} = \dim\,[u, \mathfrak{S}_{-1}] = 1$.

The left multiplication in L induces a representation Γ of $L'_{(0)}$ on $L/L'_{(0)}$

$$\Gamma(x)\colon y + L'_{(0)} \mapsto [x, y] + L'_{(0)} \quad \text{for all} \quad x \in L'_{(0)}, y \in L.$$

The rank of $\Gamma(x)$ is given by $\dim\,\big([x, L]+L'_{(0)}\big)/L'_{(0)}$. The preceding results mean that there is $f \in L'_{(0)}$ such that $(\mathrm{ad}_{L'_{(0)}/L'_{(1)}} f)^2 \neq 0$ and $\operatorname{rank}\Gamma(f) = 1$. Thus we have shown that the assumptions of Theorem 5.7.3 are valid. As a consequence $L'_{(0)}/L'_{(1)}$ is one of $\mathfrak{gl}(L/L'_{(0)})$, $\mathfrak{sl}(L/L'_{(0)})$, $\mathfrak{sp}(L/L'_{(0)})$, $\mathfrak{sp}(L/L'_{(0)}) \oplus \mathrm{Id}$. The dimension of $L'_{(0)}/L'_{(1)}$ does not exceed

$$(2q)(2r) + q(2q - 1) + r(2r - 1) = (q + r)\big(2(q + r) - 1\big).$$

As $\dim L/L'_{(0)} = 2(q + r)$, only $L'_{(0)}/L'_{(1)} \cong \mathfrak{sp}(L/L'_{(0)})$ is possible. The Weak Recognition Theorem proves that L is of Hamiltonian type. □

We conclude

Theorem 20.7.15. *Every simple finite dimensional Lie algebra L over an algebraically closed field of characteristic $p > 3$ is of classical, Cartan, or Melikian type.*

Proof. Theorem 18.3.8 shows that L is Melikian if there is a torus of maximal dimension in the minimal p-envelope of L which is non-standard.

So we are left with the case that all tori of maximal dimension in the minimal p-envelope of L are standard. If there is such a torus for which all roots are solvable, then Theorems 19.4.5 and 19.4.9 show that L is of Special or Hamiltonian type.

By Theorem 20.4.5, if there is a torus T of maximal dimension in the minimal p-envelope of L such that all T-roots are solvable or classical and there is a classical root, then L is of classical type.

Theorem 20.1.9 tells us, that there is a torus T of maximal dimension such that all T-roots are proper. We are now in the case that the assumptions on T needed in Sections 20.5–20.7 are fulfilled. Theorems 20.7.1, 20.7.4, 20.7.5, and 20.7.14 prove that L is of the Cartan type. □

20.8 Epilogue

In the last section, more is proved than necessary for the classification. This additional insight gives information on the absolute toral rank of some algebras.

Theorem 20.8.1. $TR(K(2q+1;\underline{n})^{(1)}) = |\underline{n}| - q$.

Proof. Let T be a torus of maximal dimension. Since every such torus is standard, one may assume that T is optimal. Pursuing the constructions one is lead to the case of Theorem 20.7.1. Then

$$K(2q+1;\underline{1})^{(1)} \otimes \mathcal{O}(m;\underline{1}) \subset \operatorname{gr}(L+T) \subset \big(K(2q+1;\underline{1}) \otimes \mathcal{O}(m;\underline{1})\big) \rtimes W(m;\underline{1})$$

and $\mathfrak{T} = \mathfrak{T}_0 \otimes F + \mathrm{Id} \otimes (\sum_{i=1}^{m} F(1+x_i)\partial_i)$ (Equation (20.6.2)). In the present case one has $\mathrm{Der}_0\, K(2q+1;\underline{1}) = K(2q+1;\underline{1})_0$, and therefore $\mathfrak{T}_0$ is a torus of maximal dimension in $K(2q+1;\underline{1})_0$. By Corollary 7.5.16 and Theorem 20.7.1 we have $TR(L) = \dim T = \dim \mathfrak{T}_0 + m = q+1+m$.

In the proof of Theorem 20.7.1 we mentioned that $\operatorname{gr}_0 L \subset \mathfrak{A}_0$. Then we have

$$K(2q+1;\underline{1})^{(1)} \otimes \mathcal{O}(m;\underline{1}) \subset \operatorname{gr} L \subset K(2q+1;\underline{1}) \otimes \mathcal{O}(m;\underline{1}).$$

Let $\delta = 0$ if $2q+4 \not\equiv 0\ (p)$ and $\delta = 1$ if $2q+4 \equiv 0\ (p)$. Counting dimensions gives $(p^{2q+1}-\delta)p^m \le \dim L = p^{|\underline{n}|} - \delta \le p^{2q+1+m}$. Then only $(1/2)p^{2q+1+m} < p^{|\underline{n}|} \le p^{2q+1+m}$ is possible, and this gives $|\underline{n}| = 2q+1+m$. This is the claim. □

Next consider algebras of Hamiltonian type.

Theorem 20.8.2. $TR(H(2q;\underline{n};\Phi(l))) = |\underline{n}| - q + 1$.

Proof. Let T be a torus of maximal dimension in $H(2q;\underline{n};\Phi(l))_{[p]}$. Since every such torus is standard, one may assume that T is optimal. Let us pursue the constructions of Chapter 20. One has $\mathfrak{T} = \mathfrak{T}_0 \otimes F + \mathrm{Id} \otimes (\sum_{i=1}^{m} F(1+x_i)\partial_i)$ (Equation (20.6.2)).

(a) Suppose $\mathfrak{H} \subsetneq \mathfrak{T}_0$. Then $TR(L) = \dim T = q+1+m$. Theorem 20.7.5 shows that $\Lambda_2 = 0$. Then

$$H(2q;\underline{1})^{(2)} \otimes \mathcal{O}(m;\underline{1}) \subset \operatorname{gr} L \subset CH(2q;\underline{1}) \otimes \mathcal{O}(m;\underline{1}).$$

Estimating dimensions we get (with $\delta \in \{0,1,2\}$)

$$(p^{2q}-2)p^m \le \dim L = p^{|\underline{n}|} - \delta \le (p^{2q}+2q)p^m,$$

and this gives $|\underline{n}| = 2q+m$. The claim follows.

(b) Suppose $\mathfrak{H} = \mathfrak{T}_0$. Observe that $\tilde{H}$ acts on L by derivations. Consider $\operatorname{gr}\tilde{H}$ a subalgebra in the graded algebra $\mathrm{Der}(\operatorname{gr} L)$. Clearly one has $[\operatorname{gr}\tilde{H}, \mathfrak{T}] = \{0\}$.

Let I be any minimal (gr $\tilde{H}$)-invariant ideal of gr L. Clearly it contains the unique minimal $\mathfrak{T}$-invariant ideal $H(2q;\underline{1})^{(2)} \otimes \mathcal{O}(m;\underline{1})$ and must be of the form $H(2q;\underline{1})^{(2)} \otimes \mathcal{O}(m+t;\underline{1})$. But that is only possible if $t = 0$. Hence one may regard gr $\tilde{H}$ as derivations of $H(2q;\underline{1})^{(2)} \otimes \mathcal{O}(m;\underline{1})$. As they are annihilated by $\mathfrak{T}$, they are contained in $CH(2q;\underline{1})_{(0)} \otimes F + \mathfrak{T}$. The present assumption $\mathfrak{H} = \mathfrak{T}_0$ gives gr $\tilde{H} \subset H(2q;\underline{1})_{(0)} \otimes F + \mathfrak{T}$. As a consequence, $L_{[p]}$ is contained in $L'_{(0)} + T + (\mathrm{Der}\, L)_{(1)}$, where we now consider the natural filtration of $H(2q;\underline{n};\Phi(l))$. This contradicts Theorem 7.1.3(4). □

The determination of the absolute toral rank for a Hamiltonian algebra of the first type still remains open. The following example shows that there seems to be no relation between the absolute toral rank and the number of generators or the dimension. Define on $\mathcal{O}(2r+2s;\underline{1})$ a Poisson algebra structure in the following way: let $(\alpha_{i,j})_{2r+1 \le i,j \le 2r+2s}$ be a skew-symmetric non-singular matrix with entries in F; put

$$\{x_i, x_j\} = \begin{cases} \sigma(i)\delta_{j,i'}, & i,j \le 2r, \\ \alpha_{i,j}(1+x_i)(1+x_j), & 2r+1 \le i,j \le 2r+2s, \\ 0, & \text{otherwise.} \end{cases}$$

As we mentioned in Volume 1, pp. 339/340, the iterated derived algebra modulo center is a Hamiltonian algebra of the first type. Since the matrix $(\alpha_{i,j})$ is non-degenerate, one even has

$$\Big(\mathcal{O}(2r+2s;\underline{1})/F, \{\cdot,\cdot\}\Big)^{(\infty)} \cong H(2r+2s;\underline{1};\Psi)^{(2)}.$$

This algebra contains the direct sum of the algebra $\mathfrak{g}_1$ generated by $F[x_1,\dots,x_{2r}]$ and $\mathfrak{g}_2$ generated by $F[x_{2r+1},\dots,x_{2r+2s}]$. It is not hard to see that $\mathfrak{g}_1^{(2)} \cong H(2r;\underline{1})^{(2)}$ and $\mathfrak{g}_2^{(2)} \cong L(G,0,f)$ is a BLOCK algebra, where $|G| = p^{2s}$ and f is given by the matrix $(\alpha_{i,j})$. Then one has $TR(H(2r+2s;\underline{1};\Psi)^{(2)}) \ge r+2s$. Clearly, the choice of r, s is completely arbitrary.

The isomorphism problem for Hamiltonian algebras is quite involved. In Volume 1, pp. 324/325 we mentioned SKRYABIN's solution. The deliberations of Chapter 20 might give a different access. Namely, there we described in detail the structure of the graded algebra connected with a T-invariant filtration. It is well known that filtered deformations are ruled by cohomology groups. I wonder if one can compute the relevant cohomology groups and describe the isomorphism types of Hamiltonian algebras by them.

Notation

Introduction

$L_{[p]}$, $H = C_L(T)$, $\tilde{H} = C_{L_{[p]}}(T)$, p. x

Chapter 16

$\gamma_k(\mathfrak{r})$, p. 30

Chapter 17

$Q_{[p]}$ denotes the p-envelope of a subalgebra Q in $L_{[p]}$, p. 46
$\mathrm{rad}_p\, Q_{[p]}$ is the maximal ideal of $Q_{[p]}$ consisting of $[p]$-nilpotent elements, p. 46
$\mathrm{nil}\, Q := Q \cap \mathrm{rad}_p\, Q_{[p]}$, p. 46

$\Psi_\alpha : L_{[p]}(\alpha) \longrightarrow L_{[p]}[\alpha] = L_{[p]}(\alpha)/\mathrm{rad}\, L_{[p]}(\alpha)$, $\Psi_{L_{[p]},\alpha}$, $\Psi_{L,\alpha}$, p. 58
$Q(\alpha)$, $Q(L_{[p]}(\alpha))$, $Q(L(\alpha))$, p. 58
$Q(\alpha)_{(1)}$, $Q(L_{[p]}(\alpha))_{(1)}$, $Q(L(\alpha))_{(1)}$, p. 58
$\mathrm{rad}_T\, L(\alpha_1, \dots, \alpha_s)$ is the maximal T-invariant solvable ideal of $L(\alpha_1, \dots, \alpha_s)$, p. 81
$L[\alpha_1, \dots, \alpha_s] := L(\alpha_1, \dots, \alpha_s)/\mathrm{rad}_T\, L(\alpha_1, \dots, \alpha_s)$, p. 81
$\Psi_{\alpha_1,\dots,\alpha_s} : L(\alpha_1, \dots, \alpha_s)_{[p]} + T \longrightarrow \mathrm{Der}\, \tilde{S}$, p. 81

Chapter 20

$\Gamma_{prop}(\cdot,\cdot)$ is the set of proper roots, p. 167
$\Gamma_{imp}(\cdot,\cdot) := \Gamma(\cdot,\cdot) \setminus \Gamma_{prop}(\cdot,\cdot)$ is the set of improper roots, p. 167
$\bar{Q}_\alpha := \Psi_{\alpha,\beta}(Q_\alpha)$, $\bar{Q}(\alpha) = \Psi_{\alpha,\beta}(Q(\alpha))$, $\bar{Q}(\alpha,\beta) := \sum_{\mu \in (\mathbb{F}_p\alpha + \mathbb{F}_p\beta)\setminus\{0\}} \bar{Q}(\mu)$, p. 176
$\Delta := \Delta(L,T) := \{\mu \in \Gamma \mid \mu(H) = 0\}$, p. 176
$Q := Q(L,T) := \sum_{\alpha \in \Gamma\setminus\{0\}} Q(\alpha)$, p. 183
$\Gamma_- := \{\mu \in \Gamma \mid L_\mu \not\subset Q_\mu\}$, $\Gamma_-(L(\alpha,\beta),T) := \Gamma(L(\alpha,\beta),T) \cap \Gamma_-$, p. 184
$Q[\alpha,\beta] := Q(\alpha,\beta)/\mathrm{rad}_T\, Q(\alpha,\beta)$, p. 185
$\Gamma_n := \{\mu \in \Gamma \mid L_\mu \cap L_{(n)} \not\subset L_{(n+1)}\}$, p. 195
$\Gamma_+ := \cup_{n>0} \Gamma_n$, p. 195
$\Delta_n := \Delta \cap \Gamma_n$, p. 195

$\mathfrak{A} := A(\mathrm{gr}\,(L+T))$, p. 201
$\mathfrak{T} := \mathrm{gr}_0\, T$, p. 201
$\mathfrak{H} := C_{\mathfrak{S}_0}(\mathfrak{T}_0)$, p. 209
$\mathfrak{P} := \sum_{\mu \in \Delta} \mathfrak{S}_{0,\mu}$, p. 209
$\mathfrak{S}_{0,\bar{\alpha}} := \sum_{\mu \in \Delta} \mathfrak{S}_{0,\alpha+\mu}$, p. 211
$\mathfrak{S}_0(\bar{\alpha}_1, \dots, \bar{\alpha}_r) := \sum_{i_1,\dots,i_r \in \mathbb{F}_p} \sum_{\mu \in \Delta} \mathfrak{S}_{0,\sum i_k\alpha_k+\mu}$, p. 211
$\mathfrak{P}_0 := \{x \in \mathfrak{P} \cap {\mathfrak{S}_0}^{(1)} \mid x$ acts nilpotently on $\mathfrak{S}_{-1}\}$, p. 212

Bibliography

[B-W 88] R. E. Block and R. L. Wilson, Classification of the Restricted Simple Lie Algebras, *J. Algebra,* **114**: 115–259, 1988.

[Blo 58] R. E. Block, New simple Lie algebras of prime characteristic, *Trans. Amer. Math. Soc.,* **89**: 421–449, 1958.

[Pre 94] A. A. Premet, A Generalization of WILSON's Theorem on Cartan Subalgebras of Simple Lie Algebras, *J. Algebra,* **167**: 641–703, 1994.

[P-Sk 99] A. A. Premet and S. Skryabin, Representations of restricted Lie algebras and families of associative $\mathcal{L}$-algebras, *J. reine angew. Math.*, **507**: 189–218, 1999.

[P-S 04] A. A. Premet and H. Strade, Simple Lie Algebras of Small Characteristic: IV. Solvable and classical roots, *J. Algebra,* **278**: 766–833, 2004.

[P-S 08] A. A. Premet and H. Strade, Simple Lie Algebras of Small Characteristic: VI. Completion of the classification, *J. Algebra,* **320**: 3559–3604, 2008.

[Sch 60] R. D. Schafer, Nodal noncommutative Jordan algebras and simple Lie algebras of characteristic p, *Trans. Amer. Math. Soc.,* **94**: 310–326, 1960.

[Str 89/2] H. Strade, The classification of the simple modular Lie algebras: I. Determination of the two-sections, *Ann. of Math.,* **130**: 643–677, 1989.

[S-F] H. Strade and R. Farnsteiner, *Modular Lie Algebras And Their Representations*, Monographs and Textbooks in Pure and Applied Mathematics 116, New York/Basel, Marcel Dekker, 1988

[Wil 77] R. L. Wilson, Cartan subalgebras of simple Lie algebras, *Trans. Amer. Math. Soc.*, **234**: 435–446, 1977; Correction, *Trans. Amer. Math. Soc.,* **305**: 851–855, 1988.

Bei Fragen zur Produktsicherheit wenden Sie sich bitte an:
If you have any questions regarding product safety,
please contact:

Walter de Gruyter GmbH
Genthiner Straße 13
10785 Berlin
productsafety@degruyterbrill.com